AF700367

BIOFUELS FOR AVIATION

BIOFUELS FOR AVIATION

Feedstocks, Technology and Implementation

Edited by

CHRISTOPHER J. CHUCK

Department of Chemical Engineering, University of Bath, Bath, United Kingdom

ELSEVIER

AMSTERDAM • BOSTON • HEIDELBERG • LONDON
NEW YORK • OXFORD • PARIS • SAN DIEGO
SAN FRANCISCO • SINGAPORE • SYDNEY • TOKYO

Academic Press is an imprint of Elsevier

Academic Press is an imprint of Elsevier
125 London Wall, London EC2Y 5AS, UK
525 B Street, Suite 1800, San Diego, CA 92101-4495, USA
50 Hampshire Street, 5th Floor, Cambridge, MA 02139, USA
The Boulevard, Langford Lane, Kidlington, Oxford OX5 1GB, UK

British Library Cataloguing-in-Publication Data
A catalogue record for this book is available from the British Library

Library of Congress Cataloging-in-Publication Data
A catalog record for this book is available from the Library of Congress

ISBN: 978-0-12-804568-8

For Information on all Academic Press publications visit our website at http://www.elsevier.com/

Publisher: Joe Hayton
Acquisition Editor: Raquel Zanol
Editorial Project Manager: Ana Claudia Abad Garcia
Production Project Manager: Kiruthika Govindaraju
Designer: Matthew Limbert

Contents

II

THE SCIENCE AND TECHNOLOGY OF DEVELOPING BIOFUELS FOR AVIATION

8. Pyrolysis of Biomass for Aviation Fuel

R.W. JENKINS, A.D. SUTTON AND D.J. ROBICHAUD

9. Towards an Aviation Fuel Through the Hydrothermal Liquefaction of Algae

S. RAIKOVA, C.D. LE, J.L. WAGNER, V.P. TING AND C.J. CHUCK

10. Aviation Turbine Fuels Through the Fischer—Tropsch Process

A. DE KLERK

11. Current and Potential Aviation Additives for Higher Biofuel Blends in Jet A-1

S.L. REPETTO, J.F. COSTELLO AND D. PARMENTER

III

TESTING, ASSESSMENT AND THE FUTURE

List of Contributors

M.J. Allen Plymouth Marine Laboratory, Prospect Place, The Hoe, Plymouth, United Kingdom

M. Anand CSIR – Indian Institute of Petroleum, Dehradun, Uttarakhand, India

D.M. Anderson Pacific Northwest National Laboratory (PNNL), Richland, WA, United States

R. Baldassin, Jr. Interdisciplinary Center for Energy Planning (NIPE), State University of Campinas (UNICAMP), Campinas, SP, Brazil

A. Bauen E4tech Management Consultancy, London, United Kingdom

A. Bergman Department of Biology and Biological Engineering, Chalmers University of Technology, Gothenburg, Sweden; Novo Nordisk Foundation Center for Biosustainability, Chalmers University of Technology, Gothenburg, Sweden

K.P. Brooks Pacific Northwest National Laboratory (PNNL), Richland, WA, United States

F. Burton LanzaTech, Skokie, IL, United States

M.G. Butcher Pacific Northwest National Laboratory (PNNL), Richland, WA, United States

L. Canoira Department of Energy & Fuels, ETS Ingenieros de Minas y Energía, Universidad Politécnica de Madrid, Madrid, Spain

H. Cantarella Agronomic Institute of Campinas (IAC), Campinas, SP, Brazil

R.S. Capaz Universidade Estadual de Campinas (UNICAMP), Campinas, SP, Brazil

C.J. Chuck Department of Chemical Engineering, University of Bath, United Kingdom

L.A.B. Cortez School of Agricultural Engineering (FEAGRI), State University of Campinas (UNICAMP), Campinas, SP, Brazil; Interdisciplinary Center for Energy Planning (NIPE), State University of Campinas (UNICAMP), Campinas, SP, Brazil

J.F. Costello Faculty of Health and Applied Sciences, University of the West of England, Bristol, United Kingdom

A. de Klerk Department of Chemical and Materials Engineering, University of Alberta, Edmonton, AB, Canada

G.E. Dorrington School of Engineering, RMIT University, Melbourne, VIC, Australia

S.A. Farooqui CSIR – Indian Institute of Petroleum, Dehradun, Uttarakhand, India

T.T. Franco Interdisciplinary Center for Energy Planning (NIPE), State University of Campinas (UNICAMP), Campinas, SP, Brazil; School of Chemical Engineering (FEQ), State University of Campinas (UNICAMP), Campinas, SP, Brazil

J.G. Frye Pacific Northwest National Laboratory (PNNL), Richland, WA, United States

R. Handler Michigan Technological Institute, Houghton, MI, United States

L. Harmon LanzaTech, Skokie, IL, United States

J.E. Holladay Pacific Northwest National Laboratory (PNNL), Richland, WA, United States

L. Hudson Sustainable Aviation Group, London, United Kingdom

A. Jefferson Sustainable Aviation Group, London, United Kingdom

R.W. Jenkins Chemistry Division, Los Alamos National Laboratory, Los Alamos, NM, United States

S.B. Jones Pacific Northwest National Laboratory (PNNL), Richland, WA, United States

M. Lapuerta Grupo de Combustibles y Motores, ETS Ingenieros Industriales, Universidad de Castilla La Mancha, Ciudad Real, Spain

C.D. Le Department of Oil Refining and Petrochemistry, Hanoi University of Mining and Geology, Hanoi, Vietnam

R.L.V. Leal Interdisciplinary Center for Energy Planning (NIPE), State University of Campinas (UNICAMP), Campinas, SP, Brazil

G.-S.J. Lee Pacific Northwest National Laboratory (PNNL), Richland, WA, United States

M. McManus Department of Mechanical Engineering, University of Bath, United Kingdom

M.A.F.D. Moraes "Luiz de Queiroz" College of Agriculture (ESALQ), University of São Paulo (USP), Piracicaba, SP, Brazil

A.M. Nassar AGROICONE, São Paulo, SP, Brazil

L. Nattrass E4tech Management Consultancy, London, United Kingdom

F.E.B. Nigro Polytechnic School (EPUSP), University of São Paulo (USP), São Paulo, SP, Brazil

L.A.H. Nogueira Interdisciplinary Center for Energy Planning (NIPE), State University of Campinas (UNICAMP), Campinas, SP, Brazil

J. Owen LanzaTech, Skokie, IL, United States

I. Palou-Rivera LanzaTech, Skokie, IL, United States

D. Parmenter Airbus Operations Ltd, Bristol, United Kingdom

J. Plaza Kopius Energy Solutions LLC, Seattle, WA, United States

J. Ponitka Deutsches Biomasseforschungszentrum gGmbH – DBFZ, Leipzig, Germany

S. Raikova Centre for Doctoral Training, Centre for Sustainable Chemical Technologies, Department of Chemical Engineering, University of Bath, Bath, United Kingdom

S.L. Repetto Department of Chemical Engineering, University of Bath, Bath, United Kingdom

D.J. Robichaud National Bioenergy Center, National Renewable Energy Laboratory, Golden, CO, United States

A.W. Schäfer University College London, London, United Kingdom

U. Schuchardt Institute of Chemistry (IQ), State University of Campinas (UNICAMP), Campinas, SP, Brazil

J.E.A. Seabra Universidade Estadual de Campinas (UNICAMP), Campinas, SP Brazil

D. Shonnard Michigan Technological Institute, Houghton, MI, United States

V. Siewers Department of Biology and Biological Engineering, Chalmers University of Technology, Gothenburg, Sweden; Novo Nordisk Foundation Center for Biosustainability, Chalmers University of Technology, Gothenburg, Sweden

S. Singh Joint Bioenergy Institute, Emeryville, CA, United States; Biomass Science and Conversion Technology Department, Sandia National Laboratories, Livermore, CA, United States

A.K. Sinha CSIR – Indian Institute of Petroleum, Dehradun, Uttarakhand, India

L.J. Snowden-Swan Pacific Northwest National Laboratory (PNNL), Richland, WA, United States

A.D. Sutton Chemistry Division, Los Alamos National Laboratory, Los Alamos, NM, United States

D. Thrän Department Bioenergy, Helmholtz Center for Environmental Research – UFZ, Leipzig, Germany; Deutsches Biomasseforschungszentrum gGmbH – DBFZ, Leipzig, Germany; University of Leizpig, IIRM, Leipzig, Germany

V.P. Ting Department of Chemical Engineering, University of Bath, Bath, United Kingdom

J.L. Wagner Centre for Doctoral Training, Centre for Sustainable Chemical Technologies, Department of Chemical Engineering, University of Bath, Bath, United Kingdom

Preface

The aviation industry uses approximately 10.8 exajoules (EJ) of fossil fuel energy every year, produces 700 m tonnes of CO_{2eq}, and accounts for nearly 12% of global transportation emissions. In addition to its detrimental environmental impact, fossil-derived jet fuel is increasingly expensive and subject to market volatility. These factors are driving the development of renewable fuels for aviation. Recent decades have seen accelerated improvements in aircraft design that have produced huge efficiency savings. However, due to current infrastructure and safety requirements and the need for extended range versus payload, an energy-dense liquid fuel will still be necessary for the foreseeable future. Thus, to reduce the environmental impact of aviation in the short to medium term the production of an advanced biofuel is essential.

The aviation sector is relatively small in comparison with road transport; consequently less research and development has been invested in producing suitable alternatives in this area. Yet the challenges of developing an effective fuel are arguably greater: critical aviation fuel requirements are not met by traditional biofuels such as bioethanol and biodiesel, and a new approach is needed.

The aim of this book is to guide readers through the latest research and the status of industrial development in this complex and important subject, and to consider the future prospects of aviation biofuels. Leading international scientists, engineers, policy makers, and economists have contributed chapters alongside industrial stakeholders across the supply chain to survey the current state-of-the-art in this area.

In the first section an overview of the aviation sector is provided. In chapter 1, the necessity of creating suitable biofuels for aviation and the remaining challenges in delivering a sustainable solution are discussed. Chapter 2 considers the available feedstocks for the production of biofuels, their geographical distribution, and the potential of future biomass sources. Finally, chapter 3 discusses the certification and performance of a future aviation biofuel, including what an aviation biofuel must look like and where deviation away from the current international standards would potentially be acceptable.

The major technological routes to an aviation biofuel are presented in the second section. Aviation biofuels are not one homogeneous mixture, and there are numerous chemical and biological routes to a range of fuel-like molecules that could potentially be used. The diagram overleaf summarizes the major routes presented in this section. In chapter 4, the evidence for using biodiesel as a blending agent in Jet A-1 is presented, including further data suggesting certain FAME molecules can fit the Jet A-1 fuel performance criteria if fractionated. In chapter 5, an alternative lipid conversion technology is presented, where the catalysts and production routes to producing hydroprocessed esters and fatty acids (HEFA) are examined. Chapters 6 and 7 discuss the chemical and biological conversion of sugars and small oxygenates. The current status of the promising alcohol-to-jet (ATJ) route is presented in chapter 6, including the organisms currently under development to produce the alcohols from sustainable biological feedstocks. The latest research into the production of hydrocarbons suitable for aviation through solely biological routes, via advanced metabolic engineering, is presented in chapter 7.

A further route to aviation fuels is through the thermal conversion of biomass to produce either crude bio-oils that can be upgraded similarly to crude oils in a refinery, or the production of novel hydrocarbon ranges that can be blended with existing jet fuels. In chapter 8, we examine the pyrolysis of lignocellulose and developing strategies to direct the upgrading steps to aviation range hydrocarbons. While pyrolysis is suitable for dry lignocellulosic biomass, the direct conversion of wet feedstocks—specifically algae—also has merit. The hydrothermal liquefaction of both microalgae and macroalgae is explored in chapter 9. Arguably the most established route to an aviation biofuel is through the gasification and subsequent Fischer–Tropsch (F-T) processing of the syngas. The production of syngas and its commercial challenges are presented in chapter 10. In the final chapter in the section, the suitability of using additives to counteract some of the negative effects of using aviation biofuels is considered.

In the final section we assess the impact of aviation biofuels on the wider world. In chapter 12, we examine the potential environmental impact of some of the more established routes, with an overview of Life Cycle Analysis (LCA) studies in the area. Some of the key policy challenges and opportunities to bring aviation biofuels to market are considered in chapter 13. Finally, chapters 14 and 15 describe case studies of bringing an aviation biofuel to market in Brazil and in the United Kingdom, examining the specific challenges, the infrastructure currently in place, and considering what a future aviation biofuel sector could look like.

We hope that by giving an overview of the whole aviation biofuel area, presenting the myriad of challenges and technological solutions to delivering a sustainable fuel, this will be a useful reference point for the engineers, scientists, industrialists, and researchers committed to delivering a sustainable future for the aviation sector.

Dr C.J. Chuck

February 2016.

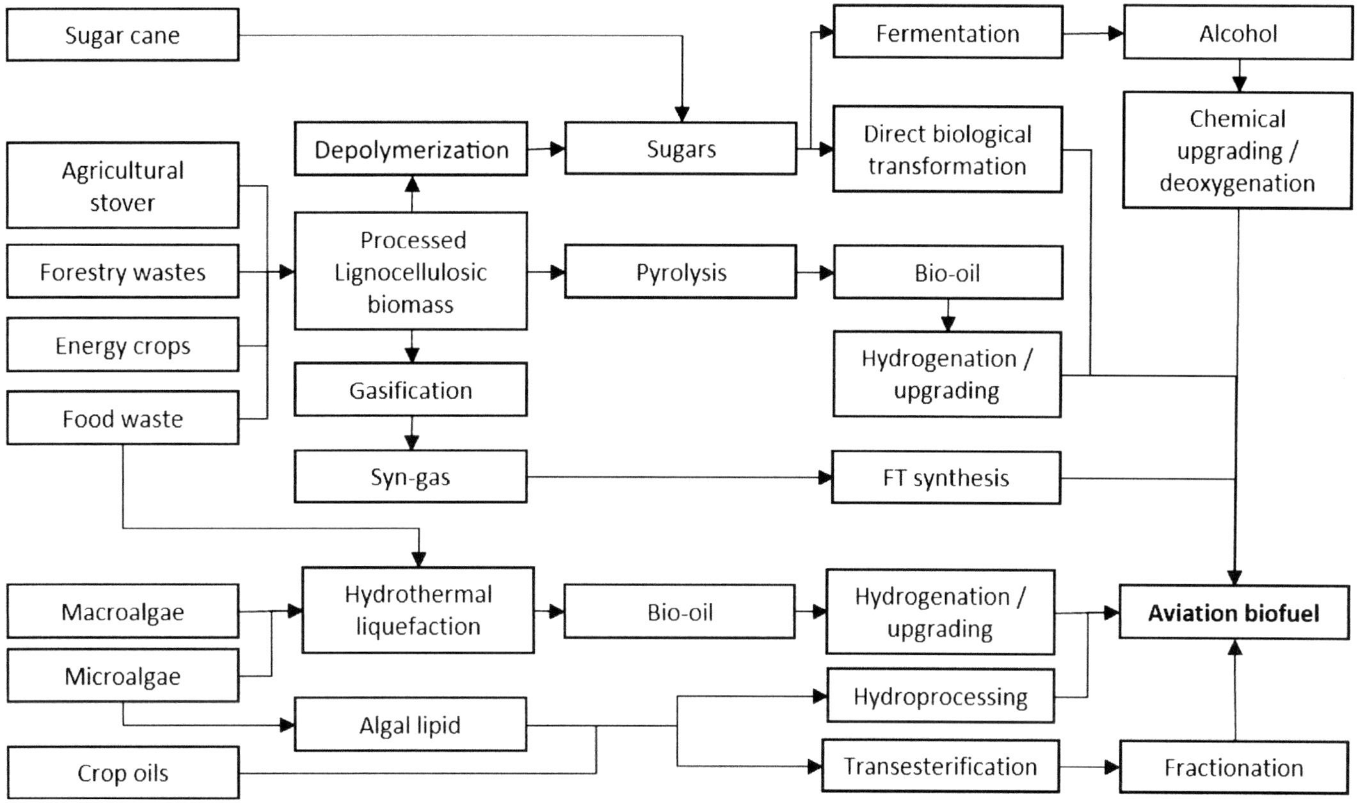

The major routes to aviation biofuels presented in the book.

SECTION I

AN OVERVIEW OF THE SECTOR

CHAPTER

1

The Prospects for Biofuels in Aviation

A.W. Schäfer

University College London, London, United Kingdom

1.1 INTRODUCTION

Since the introduction of jet engine aircraft in the early 1950s, world air transportation revenue traffic volume has experienced unprecedented growth. The shift to vastly more productive aircraft, in combination with infrastructure investments, has led to an increase in revenue tonne-kilometres from 5 billion in 1952 to 650 billion in 2011, an average rate of 8.9% per year [1,2]. Today, air transportation accounts for about 10% of the passenger kilometres travelled by all major motorized modes, and for around 40% of the interregional transport of goods by value [3,4]. In the future, air transportation is expected to continue to grow in both absolute and relative terms.

The historical growth in air transportation was entirely fuelled with petroleum-derived jet fuel. Unlike any other sector, air transportation heavily depends on this high-energy-density fuel. Petroleum-derived jet fuels contain the largest amount of chemical energy per unit volume of all fuels, and the largest amount of chemical energy per unit weight of all liquid fuels. These desirable characteristics imply a minimal sacrifice in terms of space and weight for on-board fuel storage, variables that directly translate into payload capacity and range. Yet, the search for alternatives has a long history.

Several factors motivated the quest for petroleum-derived jet fuel substitutes. One stimulus was the requirement for higher-performance aircraft. In the 1950s, liquid hydrogen was studied for use in subsonic and supersonic military aircraft. Hydrogen's nearly threefold energy content per unit weight compared to jet fuel was projected to translate into longer range and higher flight altitudes [5]. Thereafter, oil supply shortages and associated price increases during the Oct. 1973 to Mar. 1974 Organization of the Petroleum Exporting Countries (OPEC) oil embargo were the impetus for large-scale alternative fuel research. In the aftermath of the first oil crisis, liquid hydrogen, and liquefied natural gas (LNG) were examined as alternative aviation fuels first at Lockheed in the US [6], then in the former Soviet Union (eg, Sosounov [7]), and later in Germany (eg, Deutsche Airbus [8]). The detailed Lockheed studies estimated that the fuel efficiency benefits of liquid hydrogen

Biofuels for Aviation.
DOI: http://dx.doi.org/10.1016/B978-0-12-804568-8.00001-9

would increase with aircraft size and mission length. While cryogenic fuel-based narrow body, short-distance aircraft would experience energy intensities similar to their jet fuel-based counterparts, the largest energy intensity reductions (up to one-third) would emerge for large, long-distance, hydrogen-fuelled aircraft. During this entire period, and for nearly 100 years, the perennial fear of peak oil – the point in time when half of the world's oil resources will have been depleted and during which some analysts expect oil prices to skyrocket – has also contributed to the search for alternatives, especially during periods of high oil prices.

In response to concerns about oil import dependence, governments have established a number of measures that affect the demand for, and especially the supply of, transportation fuels. Key supply side measures have included funding research with the goal of developing alternative fuels, supporting the development of unconventional oil resources, and forming public-private partnerships to commercialize the production of synthetic fuels. However, as oil prices declined after the second oil crisis, so did government support for, and industry interest in, alternative fuels. The United States serves as an illustrative example. In 1982, Exxon abandoned a $5 billion investment in the Colony Shale Oil Project, of which it held 60% of the shares [9]. In 1985, the United States Congress abolished the Synthetic Fuels Corporation, a public-private partnership with the aim of developing 2 million barrels of synthetic fuels per day from coal and shale oil [10]. Also in 1985, funding for the Aquatic Species Programme – established during the Carter Administration with the goal of producing fuels from algae – was drastically reduced in preparation for discontinuation in 1996 [11].

Oil-importing countries have also learned to deal with the traditional concern of oil price volatility. Mainly due to the replacement of oil in non-transport sectors and efficiency improvements in general, the oil dependence of many economies has continuously declined.

However, a new challenge has emerged that is fundamentally different in nature. Burning 1 kg of jet fuel generates around 3.2 kg of CO_2, the most abundant greenhouse gas (GHG).[1] Since the beginning of industrialization, the atmospheric concentration of CO_2 has increased from 280 ppm in 1850 to around 400 ppm in 2015. Today, aviation generates around 2.5% of energy use-related CO_2 emissions [12].

In light of the projected continuous increase in the consumption of fossil fuels to satisfy world energy needs, the atmospheric concentration of GHG emissions (primarily CO_2) would continue to rise, further driving the anthropogenic greenhouse effect and thus increasing the mean Earth temperature. The projected implications would be unparalleled in human civilization. The thermal expansion of oceans and the melting of glaciers and ice sheets are expected to result in a sea level rise of several metres through 2100, depending on the amount of fossil fuels consumed and CO_2 emissions generated [13]. In addition, the frequency of extreme weather events, such as heavy rains and tropical storms, heat and cold waves, floods and droughts, is projected to continue to increase. These direct impacts are likely to cause a number of secondary effects such as increasing income inequality, mass migration, and terrorism. Fig. 1.1 depicts the historical growth trend in atmospheric concentration of CO_2 emissions.

[1]In addition to CO_2, air transportation contributes to climate change in other ways. Nitrogen oxide emissions reduce the atmospheric methane concentration but produce tropospheric ozone, another GHG. Their overall effect is net warming. The impact of line-shaped contrails on the greenhouse effect is greatly increased if they transition into cirrus clouds.

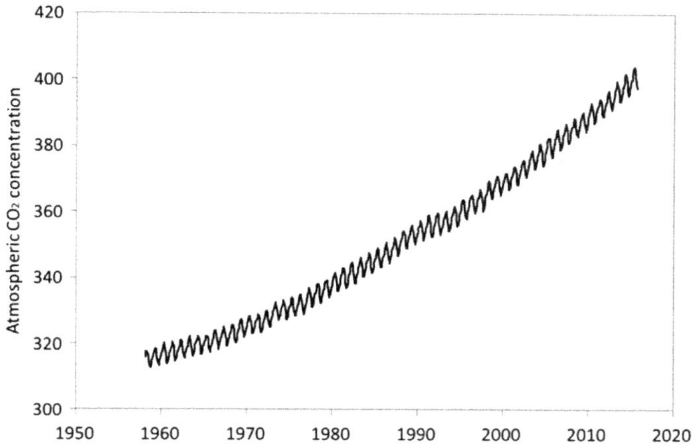

FIGURE 1.1 Atmospheric concentration of CO_2 as measured at Mauna Loa, Hawaii. *Source: National Oceanic and Atmospheric Administration (NOAA), ESRL data. Trends in atmospheric carbon dioxide. http://www.esrl.noaa.gov/gmd/ccgg/trends/data.html, 2015 [14].*

In response to these concerns, 197 countries committed in the Paris Agreement of the United Nations Framework Convention on Climate Change in 2015 to limit 'the increase in the global average temperature to well below 2°C above pre-industrial levels and to pursue efforts to limit the temperature increase to 1.5°C above pre-industrial levels' [15]. This target would require the atmospheric concentration of CO_2 to be below 450 ppm in Fig. 1.1. For this to happen, global CO_2 emissions would have to peak around 2020 and subsequently decline. Clearly, such a dramatic diversion of historical emission trends requires all sectors of the economy to participate, including aviation. This tight constraint on using the Earth's fossil fuel endowments also implies that a third of global oil reserves, half of global gas reserves, and over 80% of global coal reserves, would have to remain unused to meet the target of 2°C [16] – the concerns have shifted from peak oil to excessive use of fossil fuels.

Meanwhile, since the early 2000s, there has been a growing consensus that, though technologically feasible, a transition toward hydrogen aircraft and the supporting fuel infrastructure would be extremely capital-intensive. The commercialization of biodiesel fuels for road transportation has exemplified a more practical and economic approach for reducing aviation CO_2 emissions through a liquid fuel with similar specifications to petroleum products. The resulting interest in biomass-based 'drop-in' fuels, which would not require any infrastructure modifications, was initially in the form of vegetable oils and biodiesel as a jet fuel extender. However, first-generation biofuels have demonstrated three key disadvantages. Although they provide a large oil replacement potential, they show less promise in significantly reducing GHG emissions. Also, because the feedstocks of these fuels can be used for producing both food and fuel, they intensify the food versus fuel conflict. Finally, due to their poor low-temperature properties, only low blend levels can be used in the aviation sector.

These limitations have drawn attention to second-generation biofuels, which offer a significant GHG emission reduction potential, largely avoid the food versus fuel trade-off, and can be designed to exhibit petroleum-derived jet fuel specifications. To enhance their deployment, governments have sponsored ambitious biofuel programmes and imposed consumption targets. Recently, several commercial-scale cellulosic ethanol plants have begun operating in the US, Europe, and Brazil [17]. In contrast, progress on biofuels for air transportation has been slower. Although three synthetic fuel production routes have been certified for use in aviation since 2009, only one process relying on hydrogenation of vegetable oils has started commercial-scale production. The largest potential for drop-in biofuels is rooted in a process that makes use of the more abundant low-cost cellulosic feedstocks. However, due to the capital-intensive synthetic fuel plants and the long time scales involved in scaling up production, the associated economic risk is high. The economic risk of all biofuels is further increased by a highly variable oil price and their comparatively high production costs. As a result of these factors, several biofuel companies went bankrupt, despite having developed promising novel gasification technologies. To overcome the commercialization challenge of producing synthetic jet fuels from biomass, government support will be required.

This chapter provides an overview of the demand for air transportation, the supply technologies that could help reduce CO_2 emissions, and the role biomass-based synthetic fuels could play in significantly reducing aviation's carbon footprint. It concludes by addressing the challenges outlined above in greater detail.

1.2 GROWTH IN AIR TRANSPORTATION, FUEL USE, AND CO_2 EMISSIONS

Growth in air transportation and the economy undergo predictable patterns. At early stages of air transportation development, growth in air transportation exceeds that of the economy. With rising economic development and a growing base, growth rates in air transportation decline in order to converge toward those of the population in a mature market. This development – an initial strong growth and a subsequent decline in growth rates – is typical for all transportation systems and consumer goods at large, and can be observed both longitudinally and cross-sectionally. On a longitudinal dimension, United States air traffic volume grew at a rate of 10.5% per year between 1952 and 1980, whereas this rate has declined to 3.5% per year since 1980 [18]. On a cross-sectional dimension, the highest growth rates in air traffic currently occur in the emerging economies of Latin America and Asia, whereas the lowest rates are being observed in the domestic markets of the United States and the European Union.

Compared to the observed growth in global air traffic of 5.4% per year since 1980, the growth in fuel demand has been slower. Reliable, global-scale fuel consumption data is difficult to obtain for the period prior to 1980, but since then, jet fuel consumption has increased at a rate of 2.1% per year, effectively doubling every 33 years [19,20]. The difference in the growth rates of air transportation demand and fuel consumption of 3.3% per year can be attributed to fuel efficiency improvements. The latter have consisted of technological improvements, increases in passenger and freight load factors, and operational changes such as increases in stage length. Because petroleum-derived jet fuel has been used nearly exclusively since the introduction of jet aircraft, direct CO_2 emissions have grown in proportion to jet fuel consumption.

Econometric models predicting future levels of air transportation demand typically take per person income, population growth, airfares, and other factors such as viability of competing modes, exchange rates, etc. as variables. Mainly because of an anticipated continuous increase in per person income and population growth, most projections suggest the global demand growth to remain at recent rates of around 5% per year, at least over the next 20 years (as shown in Fig. 1.2). Most future

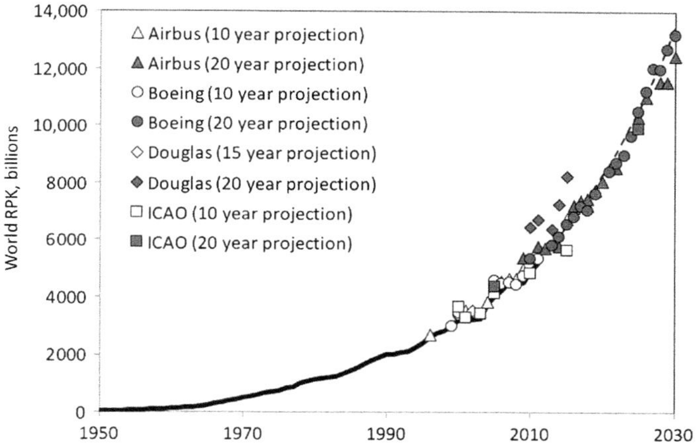

FIGURE 1.2 Historical growth in passenger air transportation demand (1950–2011) and projections for the future. *Source: Schäfer and Waitz (2014) [21]. The dashed line represents an annual growth by 5%.*

growth is expected to occur in the emerging markets of Africa, Asia, and Latin America, where income and population are projected to grow most rapidly. Airbus and Boeing project annual average growth rates in excess of 6% per year. In contrast, growth is projected to be at a more moderate 3% per year or less in the more mature markets within the industrialized world.

Future advances in fleet fuel efficiency will then determine the amount of jet fuel consumed and, in combination with its fuel composition, the amount of CO_2 emissions generated. Assuming the 2.1% annual growth in fuel use continues, along with a roughly 4% annual growth in air traffic, the 2011 global air transportation sector energy use of 10.8 EJ would more than double to 24.3 EJ in 2050. If continuously relying on petroleum-derived jet fuel, direct CO_2 emissions in 2050 would result in 1.8 billion tonnes. At the same time, a 450 ppm atmospheric stabilization target, which would likely limit the temperature increase to below 2°C, would allow total anthropogenic CO_2 equivalent emissions to be below 20 billion tonnes in 2050. This would mean direct air transportation CO_2 emissions would account for as much as 10% of total energy sector CO_2 emissions. Yet, continuing to grow at a rate of 2.1% per year in light of an annual demand growth of 4% already requires significant annual improvements in fuel efficiency; otherwise, the air transportation share of CO_2 emissions would be even larger.

1.3 GOVERNMENT AND INDUSTRY INITIATIVES TO CONTROL AIR TRANSPORTATION CO_2 EMISSIONS

Whereas regulations of aircraft emissions of local air pollutants and noise date back to the early 1970s, efforts to control CO_2 emissions have emerged only recently. Since the early 2000s, governments and industry have explored, specified, or introduced policies or targets to control air transportation CO_2 emissions. These targets, directly or indirectly, also affect the development and use of aviation biofuels.

1.3.1 Airline Industry and Government Targets

In 2009, the International Air Transport Association (IATA) announced a carbon neutral growth target beginning in 2020, with a subsequent CO_2 emissions reduction to 50% of the 2005 level by 2050. These targets are ambitious in light of the anticipated demand growth described above, and rely on a substantial introduction of low-carbon fuels in addition to improved technology, operational measures, and emissions trading with other sectors, at least in the interim [22]. US Government agencies, such as the US Federal Aviation Administration (FAA), also adopted the vision of carbon neutral growth starting in 2020. In addition, the FAA aims to reduce the climate impact of all aviation emissions relative to the 2005 levels over the same period [23]. Most recently, the US Environmental Protection Agency (EPA) has started a process to regulate aircraft CO_2 emissions [24]. Moreover, the International Civil Aviation Organization (ICAO), a UN body, has begun to develop aircraft CO_2 emissions standards for new aircraft types and to explore a global, market-based mechanism on international routes [25,26]. It is expected that ICAO will introduce a global market-based measure in 2016 that aims to maintain international aviation-related CO_2 emissions at the 2020 level through a global CO_2 tax.

1.3.2 European Emissions Trading Scheme

The most concrete step towards controlling CO_2 emissions was implemented by the European Union, which has included commercial aviation in its CO_2 Emissions Trading System (ETS) since 2012. However, following massive protests from a number of non-European nations, the EU suspended the inclusion until 2016, with the expectation that ICAO would develop a market-based mechanism for aviation CO_2 emissions reduction that would be put in place by 2020 [27]. Under the ETS, biofuels are credited with zero CO_2 emissions if they satisfy the lifecycle GHG emission reductions relative to petroleum-derived jet fuels as specified by the European Commission (EC) Renewable Energy Directive (RED) discussed below.

1.3.3 ACARE Targets

Driven largely by the aspiration to expand the competitiveness of the European air transportation industry, and the European economy at large, the Advisory Council for Aeronautics Research in Europe (ACARE) – established in 2001 and composed of key stakeholders from industry, regulators, and research/academia – generated two visions, one in 2001 for 2020 [28], and another in 2011 for 2050 [29]. Both visions are ambitious and address several objectives, including travel quality and affordability, safety, industrial leadership, and environmental impact. They also include a set of recommendations on how best to realize them. On the environmental domain, the Flightpath 2050 document specifies a goal of 2050 technologies to reduce CO_2 emissions per revenue passenger kilometre (RPK) by 75% to support the above-specified IATA target, NO_X emissions by 90%, and perceived noise by 65% compared to the average new aircraft in 2000. The two visions illustrate well how the way of thinking about alternative fuels has changed. Whereas the 2020 vision still expects that 'low-polluting cryogenic fuels are becoming affordable' in 2020, Flightpath 2050 declares that the 'dependence on crude oil is reduced by drop-in liquid fuels from other sources at a competitive cost'.

1.3.4 Biofuel Targets

Biofuel programme development dates back to the 1970s, including the sugarcane-to-ethanol programme in Brazil or corn-to-ethanol production in the US. In the meantime, biofuel programmes have been introduced in many other parts of the world, including the European Union. Common to nearly all of these early programmes is their focus on first-generation biofuels, with the implicit focus on road transportation. Only recently have these programmes shifted their focus to second-generation biofuels.

In the US, the Renewable Fuels Standard (RFS) requires surface transportation fuels to contain a minimum volume of biomass-based fuels. At its inception in 2005, the total amount of renewable fuels was projected to increase from 11.1 billion gallons in 2009 to 36 billion gallons in 2022 [30]. Whereas the amount of corn-based ethanol was mandated to saturate at 15 billion gallons in 2015, other, more advanced biofuels with a lifecycle GHG emission reduction potential of at least 50% or 60%, depending on the type of fuel, are required to fill the gap. Despite the comparatively modest requirements for advanced biofuel volumes through 2015, the demands could not be met. In the European Union, the Renewable Energy Directive (RED) requires renewables to account for at least 20% of total energy supply by 2020 [31]. While the country targets range from 10% to 40%, each country must also ensure that at least 10% of its transport fuels are from renewable sources. Similar to the RFS, biofuels counted toward the 10% target must satisfy certain lifecycle GHG emission reductions compared to petroleum fuels, increasing from 35% in 2010 to 60% in 2018. Although the RED is directed to the entire transportation sector, biofuels have been consumed nearly exclusively in road transportation. In contrast, the European Advanced Biofuels Flight Path – launched in 2011 by the EC in close collaboration with Airbus, leading European airlines, and European biofuel producers – specifically targets aviation. Its objective is to facilitate funding of biofuel production plants to produce 2 million tonnes of sustainable biofuels (corresponding to about 4% of current jet fuel consumption) for the European aviation sector in 2020 [32].

1.4 TRANSPORTATION SYSTEM CO_2 INTENSITY

CO_2 intensity is the product of CO_2 emissions per unit energy, that is, the CO_2 emissions factor of the fuel, and the energy intensity of the transportation mode. Fig. 1.3 depicts the relationship between CO_2 intensity, here lifecycle CO_2 emissions per revenue tonne-km (RTK), over the average amount of cargo carried for four freight transportation

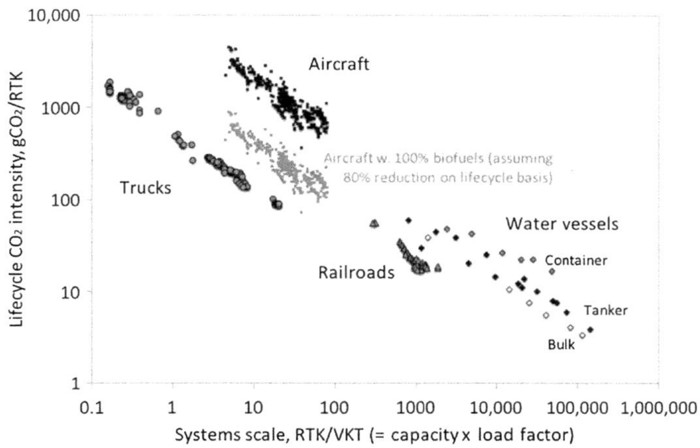

FIGURE 1.3 CO_2 intensity of freight transportation modes. *Source: Adapted from: M. Gucwa, A. Schäfer, The impact of scale on energy intensity in freight transportation, Transportation Research Part D: Transport and Environment 23 (2013) 41–49 [33].*

modes (RTK over vehicle-km traveled (VKT)). Because all modes considered here are fuelled with oil products with an only slightly different CO_2 emissions factor, the differences across modes can nearly exclusively be attributed to those in energy intensity. As can be seen, energy intensity declines with an increasing amount of load carried, in other words, increasing vehicle capacity ($RTK_{available}/VKT$) and increasing freight load factor ($RTK/RTK_{available}$). The inverse relationship between energy intensity and vehicle capacity can be attributed to the square-cube law, which applies to all modes but aircraft [33]. Energy intensity also declines with a rising freight load factor ($RTK/RTK_{available}$). Although the higher payload mass increases E/VKT, it can be shown that it reduces VKT/RTK more strongly, thus resulting in an overall decline in E/RTK. Multiplying energy intensity by the lifecycle CO_2 emissions factor of the respective petroleum-based fuel leads to the lifecycle CO_2 intensity shown in Fig. 1.3.

The CO_2 intensity of each mode develops within one respective envelope. All other factors being equal, those transportation systems carrying the smallest amount of goods (light trucks) account for the highest level of energy and CO_2 intensity, whereas those exploiting scale to the largest extent (large water vessels) account for the lowest amount of energy used and CO_2 emitted per tonne-km. However, other important factors affect energy and CO_2 intensity too. Generally, the lowest energy and CO_2 intensity of ships results from their large scale, as manifested by their higher energy intensity at a given payload (eg, 1000 tonnes) compared to railways. This does not come as a surprise in light of the low (wheel-on-rail-based) rolling resistance of trains compared to a ship's hull resistance. Similarly, the compared-to-trucks higher energy intensity of dedicated freight aircraft mainly results from their one order of magnitude higher speed of operation. In addition, the payload weight accounts for less than 30% of the maximum take-off weight of dedicated freight aircraft, which is significantly less than the payload weight-to-gross vehicle weight ratio for heavy trucks of about 67%.

1.5 AIRCRAFT CO_2 INTENSITY

Aircraft CO_2 intensity is the product of CO_2 emissions per unit energy (the CO_2 emissions factor) and the energy intensity of the aircraft. Fuels with higher amounts of carbon per unit energy generate a larger amount of CO_2 emissions during combustion. To ensure a holistic perspective, the CO_2 emissions factor has to also account for those emissions released during the production and transportation of the fuel to its final destination. The resulting lifecycle-based emissions factor of petroleum-based jet fuels then increases the combustion-related CO_2 emissions factor by around 21% [34]. In contrast, aircraft energy intensity is affected by the technological characteristics of the vehicle components and can be derived from the Breguet range equation. Although this relationship only applies to cruise flight, it offers a perspective for understanding the determinants of aircraft energy use. The right-hand-side of Eq. (1.1) describes the two factors affecting aircraft CO_2 intensity per RPK (CO_2/RPK) for passenger aircraft.

$$\frac{CO_2}{RPK} = CO_2EF \frac{Q \times SFC}{PAX \times V \times L/D} \frac{W_F}{\ln\left(W_F/(W_0 - W_F)\right)} \tag{1.1}$$

According to Eq. (1.1), aircraft CO_2 intensity is affected (all other factors equal) by the carbon content of the fuel (CO_2EF) and several factors determining energy intensity. The latter consist of engine efficiency (or engine specific fuel consumption, ie, *SFC*), the number of passengers (*PAX*), the aerodynamic efficiency (or lift-to-drag ratio, ie, L/D), and the structural weight (ie, the difference between take-off weight (W_0) and fuel weight before take-off (W_F)). Q represents the lower heating value of jet fuel (42.8 MJ/kg).

1.6 REDUCING AIR TRANSPORTATION CO_2 INTENSITY

The opportunities for reducing aircraft CO_2 intensity can be seen from Eq. (1.1). All other factors being equal, aircraft CO_2 intensity can be reduced through less carbon-containing fuels on a lifecycle basis (lower CO_2EF), such as second-generation fuels from biomass, and through a lower aircraft CO_2 intensity. Aircraft energy intensity can be reduced in multiple ways, including a higher engine efficiency. In fact, improvements in jet engine efficiency have been the single largest contributor to the historical reduction of aircraft CO_2 intensity since around 1970 [35]. Other measures for reducing aircraft energy intensity include a larger number of passengers, higher aerodynamic efficiency, and a lower structural weight. Note that the variables in Eq. (1.1) are interrelated, which typically results in compensating effects. For example, while a larger wingspan increases L/D, it also increases structural weight.

Whereas an increase in aircraft size cannot yield a decline in the energy and CO_2 intensity of the aircraft fleet, natural fleet turnover, a combination of fuel-saving technologies, airline operational changes, and enhanced air traffic management (which reduce the right-hand-side variables in Eq. (1.1)) can. A recent analysis of the US domestic narrow body passenger aircraft fleet concluded that fleet energy and CO_2 intensity could be reduced by 2% per year through 2050 at oil prices between \$50 and \$100 per barrel at zero marginal costs [12]. Fig. 1.4 depicts various CO_2 intensity trajectories with different levels of technology implementation; the cost-effective trajectory is represented by a dotted line.

This mitigation potential could translate into carbon neutral growth or absolute reductions of CO_2 emissions within the mature domestic markets of the industrialized world, where demand growth is projected to be 2% per year or less. However, mainly because of the strong demand growth in the developing world, which translates into global growth rates on the order of 5% per year, technology improvements would continue to be outpaced

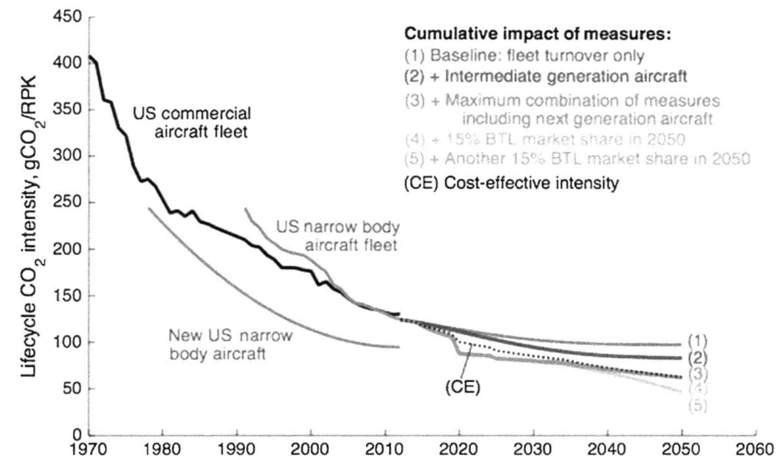

FIGURE 1.4 CO_2 intensity of the US narrow body aircraft fleet, historical development (1970–2012), and future projections (2013–2050). *Source: Schäfer et al. (2016) [12].*

on a global scale. Yet, stronger CO_2 intensity reductions could be achieved with biomass-based fuels. Trajectories 4 and 5 in Fig. 1.4 represent a biofuel uptake of 15 and 30% of all jet fuels in 2050, which translates into a CO_2 intensity reduction of 2.2%–2.6% per year. In theory, even further reductions could be achieved with higher penetration levels of cellulosic biofuels.

In addition to the continuous improvement in mainstream technologies, various conceptual designs of electric and hybrid electric aircraft have recently emerged. These radically different propulsion systems promise significant fuel burn and emission reductions, especially if combined with novel concepts such as distributed propulsion. However, in order for (hybrid) electric propulsion systems to be integrated into commercial aircraft, technological breakthroughs are required. For example, battery energy density needs to be increased by at least one order of magnitude, as would the specific power for (LNG-based) fuel cell systems [36]. The introduction of such propulsion systems would be further delayed by the risk adversity of the aircraft industry. Given the high development costs of commercial aircraft, on the order of \$10–20 billion, the aircraft industry manages risk by incremental changes to previous designs [4]. This evolutionary progression means that radical design changes can only occur over long-term periods. Hence, over the medium-term (2020–50), advanced biofuels seem to be the most promising option for drastically reducing aircraft lifecycle CO_2 intensity.

1.7 BIOFUELS FOR AVIATION: OPPORTUNITIES AND CHALLENGES

While hundreds of millions of years were required to form fossil fuels from biomass by nature, industrial plants dramatically accelerate nature's geological clock to convert biomass into synthetic fuels. Multiple conversion pathways exist. Until today, three process routes have been certified by ASTM International [37]. Biomass-based Fischer–Tropsch (F–T) fuels were certified in 2009. This process transforms lignocellulosic feedstocks into a synthesis gas, which, after purification, is converted into a liquid fuel using Fischer–Tropsch (F–T) synthesis. Hydrogenated Vegetable Oils, or Hydroprocessed Esters and Fatty Acids (HEFA), were certified in 2011. This process, which can be performed with existing refinery infrastructure, hydrotreats the triglycerides that compose waste fat and fatty acids (such as vegetable oil) to remove oxygen and double carbon bonds to paraffinic alkanes. In contrast to the two thermochemical processes, Direct-Sugar-to-Hydrocarbon, which was certified in 2014, follows a biochemical process. It consists of an enzymatic hydrolysis of cellulosic feedstocks with subsequent fermentation into hydrocarbon molecules and upgrading to jet fuel. Because the resulting paraffin only consists of one carbon chain length, the maximum certified blending ratio with petroleum-derived jet fuel is only 10%, as compared to 50% for the other two certified processes. As this book goes to press, the certification process of five additional process routes is ongoing [37], and many more are under development. Among the three approved processes, only HEFA fuels have been produced at commercial scale.

Careful assessments for HEFA and F–T fuels show a significant GHG emission reduction potential compared to petroleum-derived jet fuel under certain conditions. Provided no land-use changes are required for cultivating feedstocks, lifecycle GHG emission reductions of 70–90% could be achieved. On the other extreme, if tropical rainforests were converted to farmland, lifecycle GHG emissions could multiply by a factor of about eight compared to petroleum-derived jet fuel. In between the two extremes, the conversion of grassland and logged-over forests to cropland can result in modest (if any) lifecycle GHG emission reductions [38]. The binding constraint of avoiding land-use changes requires the use of waste biomass and algae feedstocks to the largest extent possible. Following such a strategy could result in greatly reduced CO_2 intensities, as exemplified in Fig. 1.3, for an assumed 80% reduction of lifecycle CO_2 emissions. If, in addition, assuming the energy intensity of the aircraft fleet could eventually be cut in half as a result of natural fleet turnover, fuel efficiency improvements, airline operational changes, and enhanced air traffic management as discussed above for narrow body aircraft, the lifecycle CO_2 intensity trajectories in Fig. 1.3 would be comparable to those from trucks operating under today's conditions (not shown in Fig. 1.3).

The requirement for avoiding land-use changes may limit the feedstock supply to, and thus the scalability of, the HEFA process. In contrast, biofuels based on more abundant, low-cost cellulosic feedstocks (municipal solid waste, agricultural residues, and industry waste) yield a significantly larger potential. For example, in the US, cellulosic biomass residues from agriculture, forests, and urban wood amounted to nearly 230 million tonnes in 2012, and is expected to grow to some 300 million tonnes in 2030 [39]. These amounts, which correspond to 3.4–4.4 EJ of biomass – or, using a 50% conversion efficiency of biomass to F–T fuels, to 13–33 billion gallons of synthetic fuel – greatly exceed the nearly 11 billion gallons of jet fuel consumed by the US commercial airline fleet in domestic passenger operations in 2012.

However, while the biomass-based F–T process is proven at a demonstration plant size, scaling up the capital-intensive process has been challenging. These challenges range from obtaining high-quality feedstocks at the larger quantities required (thus coming from different geographic areas), to overcoming potential technical problems associated with the scale-up, to accessing and maintaining financing for the potentially long transition period from demonstration plant size to full commercial scale. In addition, during the transition period, government policies and oil prices may change, as discussed in the introduction of this chapter.

Fig. 1.5 depicts the variability of the jet fuel spot price and the projected cost range of biofuels from low-cost agricultural and industrial wastes via thermochemical and biochemical processes [20,40,41]. Given a spot market price

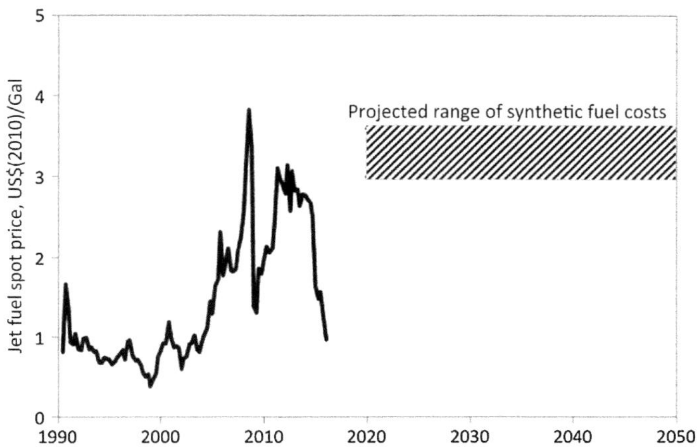

FIGURE 1.5 Historical jet fuel spot price and projected range of production costs of biofuels via biochemical and thermochemical processes. The cost range of biofuels is based on [40] and [41].

of around $1 per gallon at the end of 2015, a carbon tax of around $250 per tonne of CO_2 (around $920 per tonne of carbon) would need to be imposed in order to achieve cost-effectiveness of synthetic biomass-based fuels. This tax is extremely high, and it is difficult to imagine its implementation. Yet, the low spot market price of jet fuel mirrors low oil prices, which are likely to increase again once overcapacities in the oil market cease to exist.

1.8 CONCLUSIONS

Biofuels can play a critical role in reducing air transportation CO_2 emissions. The anticipated continuous strong growth in air transportation demand of around 5% per year over the next several decades cannot be compensated by fuel efficiency improvements alone. Alternative fuel options exist, but there are challenges. Liquid hydrogen could strongly reduce the CO_2 intensity of aircraft and reverse the CO_2 growth trend. However, although technologically feasible, a transition toward a hydrogen infrastructure would be extremely capital-intensive. In addition, potential disruptive technologies such as hybrid/electric propulsion systems would require technological breakthroughs to be introduced into commercial aircraft and thus represent only long-term opportunities for reducing energy use and CO_2 emissions. In contrast, second-generation biofuels from waste and renewable feedstocks could help bridge the gap and enable the airline industry to reach carbon-neutral growth in the medium-term, and even reduce CO_2 emissions over the longer term.

Up until 2015, three second-generation production processes have been certified for blends with petroleum-derived jet fuels, and many other processes are either expected to receive certification or are still being explored. Among the certified processes, only HEFA fuels are already being produced at industrial scale. However, the underlying process may not be sufficiently scalable due to its reliance on waste fats and vegetable oil feedstocks that may compete for food production or induce CO_2-intensive land-use changes. In contrast, the other two licensed processes build upon abundant cellulosic fuels. Among those, F–T fuels are closest to commercialization, but the capital-intensive process, the potentially long time periods required for scaling up the process, and variable oil prices require policy measures.

Multiple policy targets, measures, and Research and development (R&D) programmes that affect aircraft manufacturers, airlines, and fuel producers directly or indirectly are already in place or about to be introduced. However, in many countries, energy policies supporting the deployment of biofuels have focused on road transportation via cellulosic ethanol and renewable diesel through mandatory production quotas and fiscal incentives. Removing these distortions by opening up the biofuel market to other modes is critical, as is coordination across countries to ensure consistency of sustainability standards. This is especially important for airlines, which purchase fuel in multiple countries with potentially conflicting standards and could be credited against market-based measures.

The history books are full of energy policies that were guided exclusively by oil price developments. However, oil price may not be a reliable ally in the fight against climate change. The large risk associated with scaling up fuel production from demonstration to commercial scale of capital-intensive synthetic fuel plants requires coordinated, predictable, and long-term government policies to ensure investor confidence. These policies could consist of subsidies, mandatory production quotas, or other measures.

Although comparatively abundant, competition for cellulosic feedstocks may ultimately arise between road and air transportation. However, once a level playing field exists, that is, subsidies and regulations are applied equally to both sectors, the market will decide about the quantities consumed in each sector. Given the reduced degrees of freedom for fuel shifting in air transportation, this sector's willingness to pay for biofuels may be larger and thus attract a larger share.

References

[1] International Civil Aviation Organization (ICAO), Civil Aviation Statistics of the World, ICAO, Montreal, QC, 1975–1999.

[2] United Nations (UN), United Nations statistical yearbook, New York, 2004, 2006, 2008, 2010, 2013.

[3] Air Transport Action Group (ATAG), The Economic and Social Benefits of Air Transport, ATAG, Geneva, 2005.

[4] A. Schäfer, J.B. Heywood, H.D. Jacoby, I.A. Waitz, Transportation in a Climate-Constrained World, MIT Press, Cambridge MA, 2009.

[5] A. Silverstein, E.W. Hall, Liquid Hydrogen as Jet Fuel for High-Altitude Aircraft, NACA Research Memorandum, National Advisory Committee for Aeronautics, Washington, DC, 1955.

[6] G.D. Brewer, Liquid Hydrogen Aircraft, CRC Press, Boca Raton FL, 1990.

[7] V. Sosounov, Some aspects of hydrogen and other alternative fuels for application in air-breathing engines, Central Institute of Aviation Motors (CIAM), Moscow, USSR. IX International Symposium on Air Breathing Engines, 3–8 September 1989, Athens, Greece, 1989.

[8] Deutsche Airbus, Euro-Quebec Hydro-Hydrogen pilot project: aerospace applications. Final Report. Doc No: EQHP MBB 200-1, 2000.

[9] New York Times (NYT), Exxon abandons shale oil project. 3 May 1982. http://www.nytimes.com/1982/05/03/business/exxon-abandons-shale-oil-project.html>, 1982.

[10] C.Q. Almanac, Congress Dismantles Synthetic Fuels Program, fortyfirst ed., Congressional Quarterly, Washington, DC, 1985, pp. 212–2141986. http://library.cqpress.com/cqalmanac/cqal85-1147715

[11] J. Sheehan, T. Dunahay, J. Benemann, P. Roessler, A Look Back at the US Department of Energy's Aquatic Species Program—Biodiesel from Algae, National Renewable Energy Laboratory, Goldon, CO, 1998.

[12] A. Schäfer, A.D. Evans, T.G. Reynolds, L. Dray, Costs of mitigating CO_2 emissions from passenger aircraft, Nature Climate Change 6 (2016) 412–417, <http://www.nature.com/nclimate/journal/vaop/ncurrent/full/nclimate2865.html>.

[13] Intergovernmental Panel on Climate Change (IPCC), Climate Change 2013: The Physical Science Basis, WHO & UNEP, 2013<http://www.climatechange2013.org>

[14] National Oceanic and Atmospheric Administration (NOAA), ESRL data. Trends in atmospheric carbon dioxide. <http://www.esrl.noaa.gov/gmd/ccgg/trends/data.html>, 2015.

[15] United Nations (UN), Framework convention on climate change, Conference of the Parties, Twenty-first session, Adoption of the Paris Agreement. <https://unfccc.int/resource/docs/2015/cop21/eng/l09.pdf>, 2015.

[16] C. McGlade, P. Ekins, The geographical distribution of fossil fuels unused when limiting global warming to 2°C, Nature 517 (2015) 187–190.

[17] European Biofuels Technology Platform (EBTP), Cellulosic ethanol. <http://biofuelstp.eu/cellulosic-ethanol.html>, no date.

[18] Airlines for America (A4A), Economic report. Washington, DC, 1950–2011.

[19] International Energy Agency (IEA), Energy balances from OECD and Non-OECD Countries, IEA/OECD, Paris, 2015.

[20] Energy Information Administration (EIA), International energy statistics. <http://www.eia.gov/cfapps/ipdbproject/iedindex3.cfm?tid=5&pid=54&aid=2>, 2015.

[21] A. Schäfer, I.A. Waitz, Air Transportation and the Environment, Transp Policy 34 (2014) 1–4.

[22] International Air Transport Association (IATA), Reducing emissions from aviation through carbon-neutral growth from 2020, A Position Paper presented by the Global Aviation Industry. 2013.

[23] Federal Aviation Administration (FAA), Aviation Environmental and Energy Policy Statement, Department of Transportation, Washington, DC, 2012.

[24] Environmental Protection Agency (EPA), EPA takes first steps to address greenhouse gas emissions from aircraft. <http://yosemite.epa.gov/opa/admpress.nsf/0/4A0CC9026F4CBCC285257E60005C15F8>, 2015.

[25] International Civil Aviation Organization (ICAO), Assembly resolutions in force (as of 4 October 2013). ICAO Doc 10022, Montréal, Canada, 2013.

[26] International Civil Aviation Organization (ICAO), Annual report of the ICAO Council, 2014.

[27] European Commission (EC), Reducing emissions from aviation (2015). <http://ec.europa.eu/clima/policies/transport/aviation/index_en.htm>, 2015.

[28] European Commission (EC), European aeronautics: a vision for 2020. Meeting society's needs and

winning global leadership. Report of the Group of Personalities. 2001.

[29] European Commission (EC), Flightpath 2050: Europe's vision for aviation. Report of the high level of group on aviation research. 2011.

[30] Environmental Protection Agency (EPA), Program overview for renewable fuel standard program. <http://www.epa.gov/renewable-fuel-standard-program/program-overview-renewable-fuel-standard-program>, no date.

[31] European Commission (EC), Renewable energy directive. <https://ec.europa.eu/energy/en/topics/renewable-energy/renewable-energy-directive>, no date.

[32] European Commission (EC), 2 million tons per year: a performing biofuels supply chain for EU aviation. August 2013 Update. <https://ec.europa.eu/energy/sites/ener/files/20130911_a_performing_biofuels_supply_chain.pdf>, 2013.

[33] M. Gucwa, A. Schäfer, The impact of scale on energy intensity in freight transportation, Transport. Res. D Transport Environ. 23 (2013) 41–49.

[34] R.W. Stratton, H. Min Wong, J. Hileman, Quantifying variability in life cycle greenhouse gas inventories of alternative middle distillate transportation fuels, Environ. Sci. Technol. 45 (2011) 4637–4644.

[35] J.J. Lee, S.P. Lukachko, I.A. Waitz, A. Schäfer, Historical and future trends in aircraft performance, cost, and emissions, Annu. Rev. Energy Environ. 26 (2001) 167–200.

[36] M.K. Bradley, C.K. Droney, Subsonic ultra green aircraft research phase II: N + 4 Advanced Concept Development, NASA/CR-2012-217556. 2012.

[37] International Air Transport Association (IATA), IATA 2014 Report on Alternative Fuels, 9th Edition, Montreal-Geneva, 2014.

[38] J. Hileman, et al., Near-term feasibility of alternative jet fuels, RAND and MIT; <http://web.mit.edu/aeroastro/partner/reports/proj17/altfuelfeasrpt.pdf>, 2009.

[39] Department of Energy (DOE), US billion-ton update: biomass supply for a bioenergy and bioproducts industry. R.D. Perlack, B.J. Stokes (Leads), ORNL/TM-2011/224. Oak Ridge National Laboratory. 2011.

[40] National Academies, Liquid Transportation Fuels from Coal and Biomass: Technological Status, Costs, and Environmental Impacts, National Academies Press, Washington, DC, 2009.

[41] International Energy Agency, Technology roadmap—biofuels for transport, 2011.

CHAPTER

2

Feedstocks for Aviation Biofuels

C.J. Chuck[1], M. McManus[2], M.J. Allen[3] and S. Singh[4,5]

[1]Department of Chemical Engineering, University of Bath, United Kingdom, [2]Department of Mechanical Engineering, University of Bath, United Kingdom, [3]Plymouth Marine Laboratory, Prospect Place, The Hoe, Plymouth, United Kingdom, [4]Joint Bioenergy Institute, Emeryville, CA, United States, [5]Biomass Science and Conversion Technology Department, Sandia National Laboratories, Livermore, CA, United States

Presented throughout this book are a number of chemical and biological processes to produce effective aviation biofuels, however, a cost effective solution is highly dependent on the feedstock suitability and availability. In this chapter the major biomass resources available for fuel production are presented, including the global production, the regional availability, key challenges in delivering suitable quantities of biomass as well as key forecasts on the availability for bioenergy ahead to 2050.

2.1 FEEDSTOCKS FOR BIO-DERIVED AVIATION FUELS

2.1.1 Fermentation Sugars

The dominant global feedstock for biofuel production is simple sugars derived from agricultural starches, mainly in the United States, and from sugarcane predominantly grown in Brazil. Sugarcane is one of the most photosynthetically efficient crops, cultured mainly in warm to tropical regions. The stalks of the cane can constitute over 70% of the mature plant and contain up to 20% sucrose by weight. Sugarcane is the world's largest grown crop, with approximately 1.8 billion tonnes cultivated in 2013 (Table 2.1) [1]. While the majority of this crop is used directly for sugar production, increasingly large amounts of the cane are being used for biofuels such as ethanol (Table 2.2).

Sucrose is mainly derived from sugarcane, however, a number of other crops are also cultivated for simple sugars, including sugar beet, sorghum, and whey. These types of crops offer an inexpensive source of fermentable sugar. These sugar crops only need to be milled prior to fermentation, and the extraction of sugars is relatively simple and inexpensive [3]. The fuels produced can be derived directly from the sugar hydrolysate, or more commonly, from

Biofuels for Aviation.
DOI: http://dx.doi.org/10.1016/B978-0-12-804568-8.00002-0

TABLE 2.1 Worldwide Production of Sugarcane

Country or Region	Hectares Under Cultivation	Average Yield (Tonnes/ Hectare)	Total Production (Tonnes)
World	26,522,734	70.77	1,877,105,112
Africa	1,533,641	63.36	97,168,645
Americas	13,571,208	74.13	1,005,988,943
Asia	11,036,446	67.49	744,857,430
Europe	64	85.16	5450
Oceania	381,375	76.26	29,084,644
BY COUNTRY			
Brazil	9,835,169	75.17	739,267,042
India	5,060,000	67.43	341,200,000
China	1,827,300	69.03	126,136,000
Thailand	1,321,600	75.74	100,096,000
Pakistan	1,128,800	56.48	63,749,900
Mexico	782,801	78.16	61,182,077
Colombia	405,737	85.96	34,876,332
Indonesia	450,000	74.89	33,700,000
Philippines	435,405	73.21	31,874,000
United States	368,588	75.71	27,905,943
Australia	329,303	82.40	27,136,082
Guatemala	261,520	100.70	26,334,667
Argentina	370,000	64.05	23,700,000
Vietnam	309,300	64.72	20,018,400
South Africa	325,000	55.38	18,000,000
Egypt	139,600	115.33	16,100,000
Cuba	361,300	39.86	14,400,000
Peru	82,205	133.72	10,992,240
Myanmar	156,500	63.26	9,900,000
Bolivia	167,662	48.11	8,065,889
El Salvador	78,200	91.80	7,179,000
Ecuador	101,066	70.83	7,158,265
Nicaragua	71,567	98.18	7,026,599
Sudan	69,804	97.39	6,797,900
Venezuela	130,000	51.54	6,700,000
Iran	71,000	87.32	6,200,000
Honduras	72,500	84.08	6,096,000
Kenya	85,000	69.41	5,900,000
Paraguay	116,000	47.80	5,544,797

Data adapted from the Food and Agricultural Organisation of the United Nations Food and Agricultural Organisation of the United Nations. <http://faostat.fao.org> in United Nations, 2015.

TABLE 2.2 Bioethanol from Sugarcane Producers Globally

Country	Ethanol Production (Tonnes)	Estimated Land Area Used for Bioethanol (Hectares)	Approximate Percentage of Total Sugarcane Land Area Used for Biofuels
Brazil	18,487,610	3,124,227	32%
China	1,896,548	320,498	18%
Thailand	925,874	156,464	12%
Argentina	477,870	80,755	22%
India	462,937	78,232	2%

Data modified from [1] Food and Agricultural Organisation of the United Nations. <http://faostat.fao.org> in United Nations, 2015; [2] Going Global 2015 Ethanol Industry Outlook. <http://www.ethanolrfa.org> in Renewable Fuels Association, 2015, assuming all bioethanol production from that country is produced from sugarcane with a yield of 7500 L/ha per hectare.

the molasses produced as a by-product of food-grade sugar production [4].

Sucrose can be produced efficiently in large volumes, especially in Brazil. As such, the production of ethanol via fermentation has a positive energy balance and low environmental impact relative to gasoline production, with greenhouse gas (GHG) savings ranging from 46–90% [5]. As such, Brazilian-produced sucrose is the main feedstock for fermentation technologies in the aviation field, such as farnesene production by Amyris [6]. However, sugarcane production is dependent on the specific climate, and despite its high photosynthetic efficiency, the land area needed for significant cultivation is still substantial and in direct competition with other food crops. In addition, any aviation biofuel produced from sugarcane would be in direct competition with bioethanol as a gasoline substitute.

An alternative to the production of fuel from sucrose is to ferment simple food starches. While the technology of fermentation is similar to the sugar process, *Saccharomyces cerevisiae* cannot directly use starch, and therefore the polysaccharide must be hydrolyzed to release the fermentable glucose. To enable this, the starch is extracted through a milling process, gelatinized, and then hydrolyzed enzymatically by two enzymes, α-amylase and amyloglucosidase, prior to fermentation [3].

The predominant biofuel crop, mainly used in the United States, is corn grain, though other starches such as wheat starch or cassava starch are used in a limited capacity across the globe. The main producer of biofuel from starch is the United States, where approximately 60% of the global ethanol market is produced from corn grain [7].

While the technology of fermentation is well established, the use of starch-based technology for ethanol production seems to be reaching a maximum. Approximately 36% of US grain production is used for livestock feed, with 40% diverted for biofuel use. However, even if all US corn was diverted to ethanol production, this would only account for approximately 15% of domestic fuel use [7]. In addition, while corn ethanol has a positive energy balance, it has poor GHG savings. Much of the uncertainty associated with this is due to calculation methodology choices or regional variability [5,8]. It has also been predicted that the impact of increasing starch-based biofuel production could have a large impact on food prices. Agricultural commodities could increase in price by up to 34% and livestock by 6% by 2020 at current trends [9].

2.1.2 Lignocellulosic Feedstocks

Due to limited yields and region-specific growth of sugarcane as well, second-generation lignocellulosic biofuels are being intensely studied. Lignocellulose can be derived from one of four main sources: agricultural residues, wood residues, food waste, and specific energy crops. While assessments vary substantially, it is estimated that approximately 2.5–6 billion tonnes of lignocellulose biomass could be available for fuel production without significantly impacting food supplies.

2.1.2.1 Agricultural Residue

Agricultural residue, or agro-residue, is the lignocellulosic material including the stalks and leaves left over once the edible portion of the plant, such as the grain, has been extracted. The production of agricultural residue is huge, with an estimated total of between 1.5 and 3.9 billion tonnes of residue produced globally each year [10,11]. A number of factors are important in assessing the suitability of agricultural residues as a feedstock for biofuels. These include the total crop yield, the amount of residue produced, the composition of the crop, and the inherent energy content (Table 2.3).

The view that all agricultural residue is a waste product, and therefore is free for the production of biofuels, is overly simplistic. A number of competing uses of residue exist that could restrict the availability of this feedstock to the fuel market. These range from ploughing the residue back into the land, the use of

TABLE 2.3 Agro-Residue Production and Composition

Crop	Global Production of Edible Portion ('000,000 Tonnes)	Residue to Crop Ratio	Estimate of Annual Residue Production 2010 ('000,000 Tonnes)						Moisture Content (wt%)	Calorific Value (GJ/Tonne)	Cellulose	Hemicellulose	Lignin	Ash
			Africa	America	Asia	Europe	Oceanic	World						
Barley	144.8	2.25	10	24	30	111	11	186		16.9				
Cassava	276.7	0.088								18.4				
Coconut	62.0	0.419							9–10	18.1–18.6	44		33	
Cotton	44.5	1.5							12	12.4–16.3				
Maize	0.0	2.0	0–64	149–445	34–246	29–85	0.2–0.5	203–840	11–15	12.4–16.8	31–43	21–43	8–13	4.3
Millet	1016.7		23	0.4	24	0.5	0.06	48						
Oats	29.9	2.5	0.2	5	1	12	1	20		17.4				
Oil palm	55.8	2.6								9.8–17.9				
Potatoes	368.1	0.25	5	10	38	27	0.4	8	12	17.4				
Rapeseed	72.5								8	16.5				
Rice	745.7	1.5	21–35	37–56	668–949	3.9–6.5	1.7–0.3	731–1,045	10–14	12.6–16.3	32–47	19–27	5–24	12
Rye	16.7	2.25	0.1	4	3	35	0.8	39		17.4				
Sesame	4.8	1.5							9	15.9				
Sorghum	61.4	1.07	32	34	15	1	2	84		13.9				
Soybean	276.4	0.85								16.3				
Sugar beet	250.2		3	8	9	38	0	58						
Sugarcane bagasse	1877.1	0.3	11–22	88–241	75–156	0	6–8	181–428	50	15.8–18.1	65		18	2.4
Sunflower	44.8	5.0								18.6				
Wheat	713.2	1.5	5.3–33	63–169	145–439	133–306	8–34	354–981	9	15.9	35–45	20–30	8–15	10.1
TOTAL	**6,602**							**1,470–3,836**						

Data compiled from several sources; where there is a large discrepancy between sources in the estimated values of production, a range is given in: [10] N. Sarkar, S.K. Ghosh, S. Bannerjee, K. Aikat, Bioethanol production from agricultural wastes: an overview, Renew. Energy, 37 (2012) 19–27; [11] A. Gupta, J.P. Verma, Sustainable bio-ethanol production from agro-residues: a review, Renew. Sustain. Energy Rev. 41 (2015) 550–567; [12] M. Ahiduzzaman, A.A. Islam, Z. Yaakob, J.A. Ghani, N. Anuar, Agricultural residues from crop harvesting and processing: a renewable source of bio-energy, in: Biomass and Bioenergy, Springer, 2014, pp. 323–337; [13] L. Wang, Sustainable Bioenergy Production, CRC Press, Boca Raton, FL, 2014.

straw as a low-grade fuel for domestic use, and, increasingly, the use of residue for electricity generation on a larger scale [14]. Residue is also seen as a key future source of bulk chemicals and materials, where there is little alternative in the displacement of fossil resources [15].

2.1.2.2 *Food Waste*

Nearly 4 billion tonnes of food is produced globally each year. However, almost 1.2–2 billion tonnes of this is wasted before human consumption [16]. Food wastage falls into two rough categories. The first includes poor transportation, harvesting, and storage practices. This is the predominant mechanism of wastage observed in the developing world. The second category applies to the developed world, where food disposed by supermarkets due to arbitrary sell-by dates, exacting aesthetic standards, and general waste by the consumer, are the main routes of wastage [16]. Food disposed of by the latter mechanisms is commonly termed urban food waste (UFW), is collectable, and could be directed towards biofuel production [17]. There are currently an estimated 600 million tonnes of UFW produced annually (Fig. 2.1) [18].

Food waste is an extremely heterogeneous feedstock, and the variability depends on the time of year, the locality, and the source. However, food waste has already been heavily processed, does not contain lignin, and contains high levels of fermentable material. As such, food waste is relatively easier to convert than most other lignocellulosic materials. Numerous reports have demonstrated the suitability of food waste as a feedstock for fermentations and further chemical transformation [19,20]. A number of key issues remain, however. While food waste is extremely prevalent, there are huge infrastructure issues with collecting and sorting the waste for use. This carries a large energetic penalty. Currently, the major use of food waste is for electricity production through anaerobic digestion. If food waste served as a feedstock for aviation biofuels, then it would compete with this market as well as alternative liquid fuels.

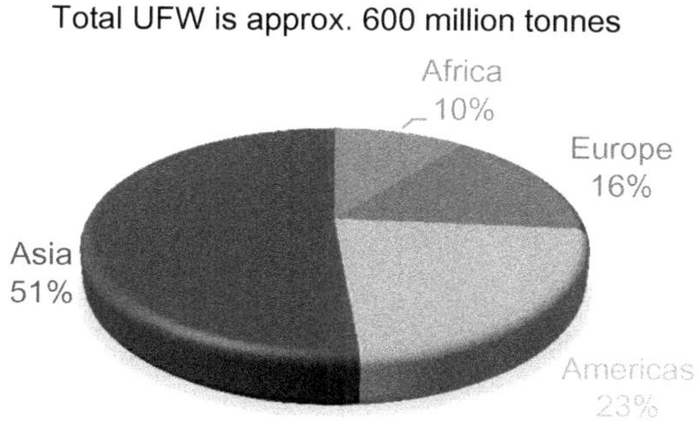

FIGURE 2.1 Urban food waste produced as percentage of continent. *Source: Data adapted from C.S.K. Lin, L.A. Pfaltzgraff, L. Herrero-Davila, E.B. Mubofu, S. Abderrahim, J.H. Clark, et al., Food waste as a valuable resource for the production of chemicals, materials and fuels, Curr. Situat. Global Perspect. Energy Environ. Sci. 6 (2013) 426–464; B.K. Adhikari, S. Barrington, J. Martinez, Predicted growth of world urban food waste and methane production, Waste Manag. Res. 24 (2006) 421–433.*

2.1.2.3 *Dedicated Energy Crops*

Beyond wastes and residues, a number of plants can be cultivated specifically to use their lignocellulose for fuel production. A high lignin and cellulose content is desirable due to a higher heating value, though lignin can be more difficult to process depending on the method used. Plants with high lignin contents can stand upright at low water contents, therefore, their biomass has a lower water content and dries more efficiently on the stem [21].

Common energy crops include perennial grasses such as Miscanthus or switchgrass, and short rotation forestry crops such as willow or poplar [9]. Key characteristics of these crops are their low fertilizer-input requirements, low energy cultivation, and ability to grow on marginal land (Table 2.4).

TABLE 2.4 Energy Crops and Their Vital Statistics [9,21–24]

Energy Crop	Productivity (Tonne/ha per Year)	Suitable Geographic Location
Miscanthus (*Miscanthus* sp.)	5–43	Temperate
Switchgrass (*Panicum virgatum*)	5–35	Temperate
Elephant grass (*Pennisetum purpureum*)	22–31	Temperate, subtropical, tropical
Canary grass (*Phalaris arundinacea*)	2–10	Temperate
Giant reed (*Arundo donax*)	10–30	Temperate, subtropical, tropical
Energy cane (*saccharum* sp.)	33–400	Similar to sugarcane
Eucalyptus	10–21	Temperate, subtropical, tropical
Willow	5–11	Temperate
Poplar	2–34	Temperate
Alfafa	1–17	Temperate
Sorghum fibre	16–43	Warm climates worldwide

2.1.3 Triglyceride Oils

An alternative to lignocellulose are triglyceride oils, commonly used in the production of biodiesel, or more recently, hydrocarbons through a hydrodeoxygenation reaction. Global production of edible oils was roughly 160 million tonnes worldwide, with approximately 26 million tonnes of this feedstock being used for biodiesel and hydrotreated fuel production [1,25] (Fig. 2.2).

The most common oil crops for biodiesel are rapeseed in the EU, soybean in the United States and Latin America, and palm and coconut in tropical Asian countries (such as Malaysia and Indonesia). Each oil type has a different blend of fatty acids, which gives the oils different properties. Typically, palm oil contains more saturated esters, rapeseed oil is mainly monounsaturated, and soybean oils are predominantly polyunsaturated. Almost all terrestrial oils have carbon chain lengths of 16 and 18 carbons, though coconut oil is mainly made up of smaller chain esters (C_{8-16}) [26].

Similar to biofuel production from first-generation sugar feedstocks, the use of vegetable oil is widely seen as unsustainable, competes with food, and has questionable environmental benefits [27]. In addition, it has been demonstrated that biofuel expansion could raise the price of oil crops from 9–27% by 2020 [9]. Other impacting factors include the intensive use of fertilizers and other chemicals, vast deforestation, and a reduction in water and soil quality, especially with reference to palm oil plantations in Southeast Asia [28]. Due to these effects and the overall demand necessary to provide fuel and food, a range of alternative second-generation feedstocks have been investigated (Table 2.5).

One of the most promising sources of alternative oils are second-generation sources that can be grown on degraded or marginal land, and therefore have the potential to be non competitive with food. The leading plant currently under cultivation is Jatropha (typically *Jatropha curcas*). Jatropha is a shrub that contains seeds with high oil content and is drought resistant. It is predominantly grown in India, though cultivation in Africa and other similar climates has proven successful [33]. Approximately 13 million hectares of Jatropha have now been planted almost exclusively for biodiesel production. Similar plants are also being grown on a trial basis, including Karanja, castor, camelina, and linseed [9].

Because a large proportion of cooking oils are not directly consumed, they can be collected after use and repurposed for fuel

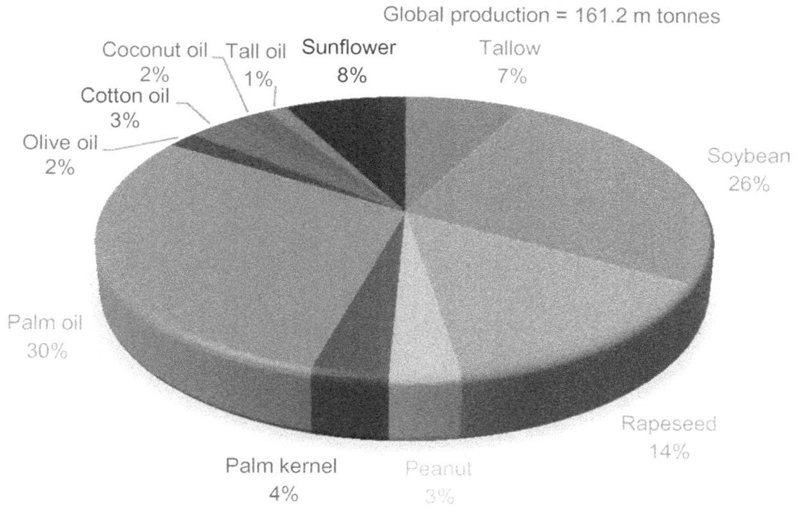

FIGURE 2.2 Global production of edible oils in 2013. *Source: Reproduced from C.S.K. Lin, L.A. Pfaltzgraff, L. Herrero-Davila, E.B. Mubofu, S. Abderrahim, J.H. Clark, et al., Food waste as a valuable resource for the production of chemicals, materials and fuels, Curr. Situat. Global Perspect. Energy Environ. Sci. 6 (2013) 426–464.*

TABLE 2.5 Predominant Biofuel- Producing Crops

Crop	Biomass Yield (Tonne/ha)	Seed Oil Content	Oil Productivity (L/ha per a)	Source
FIRST GENERATION				
Corn			168–187	United States, China, Brazil India, Europe
Cotton		14–20%	327–421	Dry tropics and subtropics
Soybean	2.8	21%	374–514	United States, China, South America
Sunflower		32–49%	702–982	Temperate regions
Rapeseed	3.1	35%	1029–1216	Worldwide
Coconut			2338–2806	Tropical
Palm oil	18.4	40%	3742–6080	Southeast Asia
SECOND GENERATION				
Jatropha	Up to 20 (seed yield)	50–60%	1310–1871	Tropical and subtropical
Camelina	Up to 2.3 (seed yield)	20%	561–608	Colder temperate climates

(*Continued*)

TABLE 2.5 (Continued)

Crop	Biomass Yield (Tonne/ha)	Seed Oil Content	Oil Productivity (L/ha per a)	Source
Karanja (Pongamia)	6 (seed yield)	30–40%	1800–3600	South east Asia including India
Waste coffee	–	Up to 20% of waste coffee grounds	8 m tonnes of waste coffee grounds worldwide	Worldwide
Waste cooking oil	–	–	25 m tonnes worldwide	Worldwide
Animal fats	–	–	Estimated 5 m tonnes collected	Worldwide

Aadapted from [9] D.P. Ho, H.H. Ngo, W. Guo, A mini review on renewable sources for biofuel, Bioresour. Technol. 169 (2014) 742–749; [26] S.K. Hoekman, A. Broch, C. Robbins, E. Ceniceros, M. Natarajan, Review of biodiesel composition, properties, and specifications, Renew. Sustain. Energy Rev. 16 (2012) 143–169; [29] R.W. Jenkins, N.E. Stageman, C.M. Fortune, C.J. Chuck, Effect of the type of bean, processing, and geographical location on the biodiesel produced from waste coffee grounds, Energy Fuels, 28 (2014) 1166–1174; [30] M.Y. Koh, T.I.M. Ghazi, A review of biodiesel production from Jatropha curcas *L. oil, Renew. Sustain. Energy Rev. 15 (2011) 2240–2251; [31] A. Kumar, S. Sharma, An evaluation of multipurpose oil seed crop for industrial uses (*Jatropha curcas *L.): a review, Ind. Crops Prod., 28 (2008) 1–10; [32] M. Berti, R. Wilckens, S. Fischer, A. Solis, B. Johnson, Seeding date influence on camelina seed yield, yield components, and oil content in chile, Ind. Crops Prod. 34 (2011) 1358–1365.*

production. It has been estimated that approximately 25 million tonnes of waste cooking oil (WCO) could be collected worldwide [34]. However, due to the frying process, WCOs are not as stable as virgin oils. Also, they contain significantly elevated levels of free fatty acids, water, and other solid impurities. This can have significant effects on the conversion technology, and as such, WCOs tend to need significant pretreatment [35]. Other waste oil products are available. For example, the animal processing industries produce large amounts of animal fats or tallow, and some food wastes, such as coffee grounds, contain high levels of triglyceride (Table 2.5).

2.1.3.1 *Oleaginous Yeasts*

While second-generation crops can be grown on marginal land, to produce oil economically they tend to need fertile, arable land, and require similar fertilizer inputs to the first-generation crops. Waste oils, including from coffee, cooking oil, and animal fats, only make a small fraction of the demand for biodiesel and advanced biofuels. One possible solution has been to cultivate oleaginous yeasts on lignocellulosic residues. A number of studies have demonstrated the suitability of these yeast oils for hydrocarbon production [36,37]. Yeast oils offer many advantages over alternatives, including a shorter life cycle, a far higher oil yield, and a production process that is less affected by location, season, and climate change [38]. Unlike algae, yeast do not require light, have shorter doubling times, and reach much higher cell densities (10–100 g/L in 3–7 days). The oil accumulation can reach 80% of the cell by weight, and they also produce a wide range of fermentation products in addition to oils which could be sold as co-products. Over 1600 yeast species are known, but only 40 are known to be oleaginous. Predominantly, these yeasts belong to ascomycete genera *Candida*, *Cyberlindnera*, *Geotrichum*, *Kodamaea*, *Lipomyces*, *Magnusiomyces*, *Metschnikowia*, *Trigonopsis*, *Wickerhamomyces*, and *Yarrowia*, and to basidiomycete genera *Cryptococcus*, *Guehomyces*, *Leucosporidiella*, *Pseudozyma*, *Rhodosporidium*, *Rhodotorula*, and *Trichosporon* [39].

Most yeast oils have a fatty acid composition equivalent to rapeseed oil, with elevated levels of oleic acid, linoleic acid, and palmitic acid. While the percentage of lipid accumulation within the single cell is of importance, the lipid coefficient (defined as the gram of lipid produced per gram of sugar consumed) is the key parameter for biofuel production. The theoretical maximums for the lipid yields are 0.32 and 0.34 g/g from glucose and xylose, respectively. However, due to other cellular processes requiring glucose, even under ideal conditions for lipid production, the lipid yield on glucose is rarely more than 0.22 g/g [40].

While there has been little economic analysis undertaken on the production of biofuels from single cell oils (SCOs), the production costs will derive largely from the cost of the initial feedstock, though the extraction, conversion, and fermentation will also carry significant costs. Recent assessments have placed the cost of yeast oils between \$1200–3000 $tonne^{-1}$ [40,41].

2.1.4 Algae

Algae represent a sustainable source of biomass that can be cultivated in aquatic environments without exerting undue pressure on agricultural land usage. Indeed, marine, freshwater, brackish, and industrial waste streams have all been debated as suitable media for the mass cultivation of algae for fuel production. Initially, the majority of research into algal-derived biofuels focused on the production, and subsequent extraction, of lipids directly from the algae (as drop-in biodiesel replacements), followed by fermentation of the remaining sugars to ethanol or butanol. Such fuels are suited primarily to microalgae, which offer rapid growth rates, metabolic and physiological versatility, and high lipid content. More recently, attention has been given to algal-derived biofuels, which rely on the direct thermal processing and/or fermentation of algal biomass (microalgae or macroalgae) into aviation biofuels.

2.1.4.1 Oleaginous Microalgae and Third-Generation Biofuels

Land-based crops have a much lower lipid yield per unit area than could potentially be achieved by algae. For example, some species of microalgae can accumulate over 60% lipid by dry weight. This equates to nearly 90,000 L/ha a in comparison to rapeseed (1200 L/ha a) and soybean (450 L/ha a), if a standard freshwater alga containing 30 wt% lipid was to be cultivated.

The microalga *Botryococcus braunii* has received particular attention as a biofuel production platform as it is one of the few organisms known to have contributed directly to existing oil shale deposits [42]. It exists in three races, with each exhibiting distinct hydrocarbon profiles: Race A, with C_{21-33} odd-numbered alkadienes and alkatrienes (around 60% of dry mass); Race B, with C_{30-37} triterpenoid and C_{31-34} methylated squalenes (up to 65% dry cell mass); and Race L, with C_{40} lycopadiene (up to 8% dry cell mass). *Botryococcus braunii* accumulates lipids internally, and crucially, externally in a extracellular matrix, which raises the potential of continual harvesting by milking. Unfortunately, despite its obvious impact over geological timescales, growth rates (it typically doubles over three days) are too poor to warrant a serious industrial deployment. For this reason, microalgae exhibiting high growth rates with acceptable membrane lipid contents have been favoured for biofuel production. The lipid profile from microalgae is generally far more variable than terrestrial or yeast-derived oils, and is dependent not only on the species of microalgae, but also on the culture conditions. For example, some strains of *Chlorella vulgaris* can produce C_6 chains, while long chains of up to C_{24} are observed in the halophyte *Isochrysis galbana*. Highly unsaturated esters are also common among microalgae, especially marine species. For example, strains of *Chlorella salina* produce C_{22} esters with up to six double bonds [43].

While attaining biomass can be achieved with some strains in a few days, the lipid accumulation generally takes up to 14–21 days, and is often associated with nutrient (nitrate, in particular) starvation.

Currently there are no major commercial algal lipid fuel producers, however, there are numerous estimations on the cost of production which have been published over the last few decades. Varying estimates on size, variable cost of nutrients, the algal species used, and technologies used for the growth, extraction, and conversion of the lipids have led to wildly different estimates on the predicted cost of the fuels. Riberio and de Silva surveyed the techno-economic indicators of microalgae biofuel technologies, comparing and contrasting the differing production methods [44]. These different estimates have given a large range of lipid costs, varying between $0.43 per litre, assuming an oil content of 50 wt%, oil yield of 30 g/m^2 d, and culturing of the algae in an open raceway pond bioreactor. The cost of generating the algal oil was subsidized by the use of co-products from the same pond, akin to the refinery approach taken by the existing oil industry. The highest cost given was $24.60 per litre of algal oil in which the economics of a range of oil by weight (15, 25, and 50%), and different production methods (open bioreactor, photobioreactor, and fermenter), were examined.

A recent techno-economic analysis of autotrophic microalgae for hydrotreated fuel hydroprocessed esters and fatty acids (HEFA) production estimated the cost of the algal lipid to be $1.87 L^{-1} for microalgae grown in open ponds, and $3.98 L^{-1} for that grown in a photobioreactor. The total cost of the HEFA was then estimated to be $2.16 L^{-1} for open-pond production and $4.52 L^{-1} for algae grown in a photobioreactor [45]. Though this is considerably more than deemed competitive, it demonstrates that the major cost for algal lipid is not in the chemical processing, but in the culturing and harvesting of the oleaginous algae.

In a meta-analysis of life cycle assessment (LCA) data from 30 investigations, Sills et al. demonstrated that the drying, dewatering, extraction, and conversion could account for up to double the energy produced from the lipid and biomass [46]. While a large research effort is underway to produce a sustainable algal lipid, it is unclear what breakthrough technology could reduce the costs sufficiently to produce a suitable feedstock for aviation fuel production.

2.1.4.2 Non-oleaginous Species for Thermochemical Upgrading

In order to reduce upstream and downstream processing costs, the concept of directly converting algal biomass through thermochemical methods has recently been suggested as a highly credible route for generating transport fuels. Under these conditions, oil is formed not only from the lipid fraction, but from the entire biomass, including the protein and carbohydrate fractions. This would allow the use of non lipid-producing algae with much faster growth rates, and would reduce the high costs of drying and lipid extraction. In addition, thermochemical conversion has many advantages: the nature of the algal biomass can be either micro- or macroalgal (eg, seaweed, see below), the impact of seasonal and global variations in productivity can be masked by the growth of multiple strains, and of course, there is the obvious reduction in processing expenditure. Large-scale microalgae growth for low-value products is typically undertaken in raceway ponds that are often prone to contamination by other microalgae and microbes. The impact on yield of such contamination is also reduced when the entire biomass is processed en bloc. It has been suggested that land-based open ponds located in desert areas harbour the potential for 90 exajoules (EJ) of production annually.

2.1.5 Seaweed

An alternative to microalgae is seaweed (macroalgae), which could provide a suitable source of biomass for further conversion to bioaviation fuels. While currently only approximately 16 million tonnes of seaweed biomass is cultivated each year globally, the potential is huge. Seaweed has a number of advantages over terrestrially cultured crops or algae. For example, macroalgae do not need a source of freshwater or fertilizers to grow, and they can form the basis of a stable ecosystem if grown responsibly [47]. Macroalgae have a higher solar efficiency (up to 8%), at least four times higher than any terrestrial crops [48]. Accordingly, seaweed can grow densely with between 1.5 and 131 tonnes dry weight per hectare observed. Macroalgae do not require land that can compete with food, and if cultivated correctly, can provide a diverse marine ecosystem. Macroalgae also inhabit every marine ecosystem on the planet, and could therefore be a significant part of any country's fuel production process. Estimates vary on the potential of macroalgae as a source of biomass due to technical restrictions associated with mass cultivation. Conservative estimates suggest offshore farms could produce 110 EJ, coastal farms 35 EJ, and open sea colonies harbour the potential for a staggering 6000 EJ of production, though the logistics of producing this much biomass remain unclear.

Challenges remain however. For example, the carbohydrate content is reasonably low, and the ash content, in comparison with microalgae, tends to be significant. This means that conversion through thermal processing [49] or fermentation is not particularly efficient [50]. The carbohydrate component also varies depending on the species of macroalgae. The composition varies substantially across species, however, seaweed can be loosely grouped into the three main classes: red, green, and brown (Table 2.6).

TABLE 2.6 Composition of a Variety of Species of Macroalgae

Algal Class Type	Green	Red	Brown
Example species	*Ulva lactuca*	*Gelidium amansii*	*Laminaria japonica*
	Ulva pertusa		*Sargassum fulvellum*
Carbohydrate content	25–65%	30–85%	40–60%
Carbohydrate types	Starch	Floredian starch	Laminarin
	Cellulose	Galactans	Mannitol
		Agar	Alginate
		Carrageenan	Fucoidan
		Cellulose	Cellulose
Lipid (%)	<6	<1.1	<2
Protein (%)	7–20	8–14	12–19
Ash (%)	18–25	3–9	Up to 46%
Sugars released on hydrolysis	Up to 60%	Up to 57%	Up to 37%
Main sugar composition	Glucose	Glucose	Glucose
		Galactose	Mannitol

Data adapted from [48] K.A. Jung, S.-R. Lim, Y. Kim, J.M. Park, Potentials of macroalgae as feedstocks for biorefinery, Bioresour. Technol. 135 (2013) 182–190; [50] N. Wei, J. Quarterman, Y.-S. Jin, Marine macroalgae: an untapped resource for producing fuels and chemicals, Trends. Biotechnol. 31 (2013) 70–77.

Aquatic biomass derived from both microalgae and macroalgae, has the potential to deliver appropriate levels of biomass as an energy resource, equivalent to, if not exceeding, the amount produced by terrestrial bioresources. Clearly, the potential exists for them to make a significant contribution to future fuel production, though the logistical aspects require heavy investment to produce economic fuels.

2.2 CHALLENGES AND OPPORTUNITIES

Predicting the future feedstock landscape is exceptionally difficult, partly due to a range of challenges, unpredictable changes in climate, and future technologies that are still being developed and researched. In this section, factors that could have a large impact on the development of aviation biofuels will be briefly discussed.

2.2.1 Genetic Engineering

Genetic engineering, especially of plant or algal feedstock, increases the potential production of the feedstock while offering reduced costs of harvesting and processing into fuel. The engineering of most food crop species is well established, and many of the feedstocks described previously, including cereals, grasses, and woody energy crops, have been transformed [7]. The two most common approaches are that of gene-gun-mediated transfer and use of *Agrobacterium tumefaciens* as a vector [51].

A key challenge is that among the thousands of species and cultivars, only a handful of strains are suitable for transformation [7]. Coupled with this, the development of suitable strains is handicapped by the lack of a ready market for the products while the conversion technology lags behind.

Some of the key advancements and developments have been in increasing the yield of feedstock per hectare [52] and genetically engineering plant cell walls to increase production of polysaccharides [53]. Production costs could be reduced by engineering the plants to produce certain cellulases and hemicellulases [7], while pretreatment costs could be reduced by reduction, or modification of, the lignin content [54]. There are also an increasing number of examples demonstrating increased oil production in first- and second-generation oilseed crops [55]. Microbial transformations have demonstrated increased yields of fuels from organisms capable of converting all sugar platforms (not just glucose) [56], while microalgae have been transformed to increase growth rates, cell density, and target molecules such as lipids [57]. Finally, key research has also been undertaken in combating climate change by improving abiotic stress tolerance, such as engineering drought resistance into plants [58].

Standard breeding strategies are also important, and are readily used for the improvement of feedstock species. Food crops have been improved in this manner for centuries, and novel energy crops, such as the woody energy crops and grasses, are essentially wild strains that are yet-to-be improved. While this process will take many years, selective plant breeding has already produced energy crops, such as energy cane, that are far higher yielding than the original cultivars [59].

At this point it is impossible to predict how selective breeding and genetic engineering will change the yields of the feedstocks. However, it seems likely that predictions based on current yields, even at their most optimistic, will prove to be achievable with the advancement of this area.

2.2.2 Biodiversity

A major criticism of a global biofuel industry is the disruption to pristine ecosystems and a corresponding reduction in biodiversity. There are significant differences in the impact of biofuel crops on the diversity of wildlife, some of which is due to a different emphasis on land use and land management [60]. There are many levels where the diversity of fauna can be inversely impacted. Genetically, for example, modified or aggressive genotypes used in non-native ecosystems can contaminate existing gene pools. At the species level, displacement of native species, habitat fragmentation, and even extinction, can occur when cultivating one species over all others. This is as much the case for large wildlife – destruction of Orangutan rainforest habitats, for example – as for the trees in the forest itself. On a larger scale, a landscape completely dominated by one type of biofuel plantation could lead to a simplified community and food chain system that reduces the resilience of an area to natural disaster and variation.

However, not all biofuel crops are the same in this regard, and negative impacts often assume the use of a pristine environment for bioenergy production rather than the use of co-products, existing managed forests, or farmland. While palm oil plantations have been demonstrated to lead to these negative effects on the local area, perennial biofuel feedstocks can improve soil quality, increase biodiversity, and form a suitable habitat for a range of species [61].

It seems clear, however, that while some biofuels can lead to an increase in biodiversity, in order to maintain an actual wealth of fauna and habitats, careful planning and regulation of biofuels, as well as any other agricultural crop, are required. At the least, this will impact estimates on available biomass, as well as add to the definition of what a truly 'sustainable' biofuel is. Such measures are beginning to be more closely monitored and regulated through mechanisms like the Renewable Energy Directive, for example [62].

2.2.3 Land Use Change Leading to Elevated Carbon Emissions

A further challenge with the use of biofuels is the conversion of land for their growth, as well as the perceived conversion of land for their growth. These are commonly described as either direct or indirect land use changes (LUC or ILUC). Estimates of the GHG impact of ILUC in particular can vary widely [63–65], and this has a significant impact on assignable carbon emissions. In extreme cases, the estimated or reported carbon footprint may span from below that of fossil fuel to over a hundred times worse. For example, Fargione et al. demonstrated that converting land such as rainforest or peatlands can incur a significant carbon debt, releasing up to 420 times more CO_2-equivalent emissions than the annual reduction achieved by the biomass cultured on that land. Estimates are particularly sensitive to the starting state of the land converted (the base line), leading to the conclusion that these debts are not incurred by biomass grown in aqueous environments, waste biomass (such as agricultural residue), or biofuel feedstocks grown on abandoned lands planted with perennial crops [66,67].

The ranges calculated for ILUC reflect the complexity of the system as well as the maturity of the method [63]. As the models have improved, the calculated GHG associated with biofuel has decreased. For example, US corn ethanol value in 2008 was reported to be 104 g/MJ CO_2eq, but was 15 g/MJ CO_2eq in 2012 [68].

2.2.4 Logistical Considerations

Key factors in the development of an economic biofuel are the logistical considerations of where the biomass is grown and how it is collected and processed. On processing, the fuel must be delivered either to a refinery for blending, or directly to the point-of-use. The expense of moving wet biomass long distances can add up to a significant amount of the final cost of the fuel, and this must be taken into account when selecting a suitable feedstock for biofuel production. The logistical considerations are heavily reliant on the locality; further examples are given in the final two chapters of this book.

2.2.5 Effect of Climate Change and Yield Improvement on Feedstock Availability

One key uncertainty in the future production of biofuels is the effect that the changing climate will have on land areas required for biofuel feedstock production. While overall increases in temperature, and increases in the extremes, will have a detrimental effect on the cultivation of biofeedstocks, a positive climate change effect may be expected in regions currently constrained by low temperatures. Haberl et al. ran several scenarios of climate change impact on yield; few gave decrease in crop yields if the assumption was made that CO_2 fertilization would occur as a result of higher atmospheric CO_2 concentrations [69].

Another impact on feedstock availability will be the potential for global yield improvement. If the United States is taken as an example, corn yield has increased from under 2 t/ha to almost 10 t/ha in 2006 [68,70]. Similar yield increases on a global scale could influence the availability of agricultural commodities.

Currently, transportation fuels equate to approximately 100 EJ of energy per year, of which aviation kerosene is responsible for 10–12 EJ. In almost all low-carbon energy scenarios, a significant contribution from

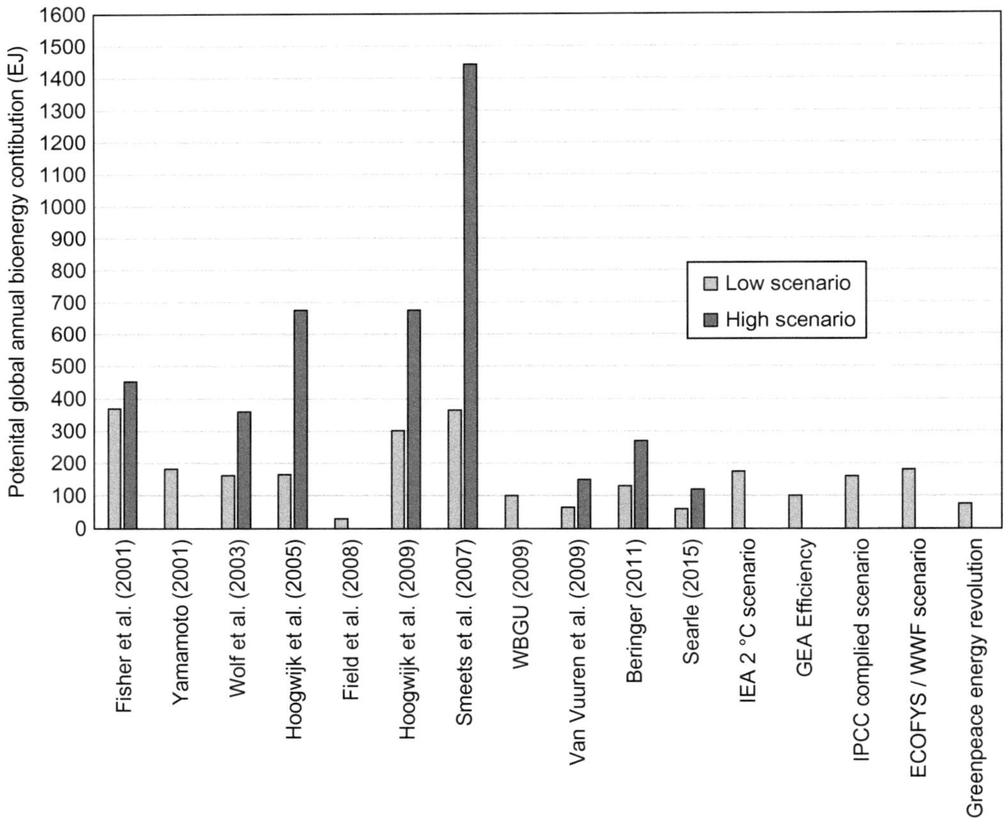

FIGURE 2.3 Low and high estimates for the level of biomass available for bioenergy in 2050 taken from references [71–78]; the final five scenarios are taken directly from reference [61].

bioenergy sourced from sustainable feedstocks is required [61]. A number of predications for bioenergy use as far forward as 2050 have been published, though understandably, due to differences in source data, assumptions on climate change, technology development, and policy interventions, the predictions vary substantially (Fig. 2.3).

If some of the more optimistic scenarios are accurate, then there is plenty of opportunity for biomass to be used as a primary energy source, for electricity as well as for liquid transport fuels. However, one of the lower estimates is given by Searle et al., who examined all the scenarios presented and found that most predictions had not properly considered the economics of feedstock production, actual land availability, and relevant biomass yields, and few studies used standard sustainability criteria to develop their models [71]. In this study the authors estimated that the maximum bioenergy available would be on the order of 60–120 EJ, with only 10–20 EJ being available for biofuels.

This study can obviously not predict whether a paradigmatic shift due to genetic engineering or the development of third-generation, non land-based biomass will alter the biofuel market. However, taking this value as an accurate depiction of 2050, there is simply not enough bioenergy feedstock to displace all transportation fuels, though solely displacing aviation kerosene, even with growth in the aviation sector, is theoretically possible. The extent to which feedstocks will be available for aviation will be heavily dependent on the development of alternative technologies for road transport, such as electric vehicles, and the regulatory framework in place at the time.

2.3 SUMMARY

In the following chapters, conversion technologies will be presented that demonstrate the range of technologies available to produce drop-in biofuels for aviation. While the properties of the fuel itself are key to the eventual replacement of fossil resources in this area, it is more likely that the availability and logistical considerations around the feedstock will be the key factors in what fuels are produced and where they are situated. While a large investment must still be made in research development and infrastructure to make aviation biofuels a reality, there is probably sufficient feedstock to produce aviation biofuels if the demand for road transport biofuels is reduced by significant technological advancements in electric and battery technology. It is also clear that no one feedstock will supply the entire aviation sector, and a range of technologies to develop aviation biofuels from lipids, lignocellulose, and marine resources will need to be developed co-currently to create aviation biofuels.

References

[1] Food and Agricultural Organisation of the United Nations. <http://faostat.fao.org> in United Nations, 2015.

[2] Going Global 2015 Ethanol Industry Outlook. <http://www.ethanolrfa.org> in Renewable Fuels Association, 2015.

[3] S.I. Mussatto, G. Dragone, P.M. Guimarães, J.P.A. Silva, L.M. Carneiro, I.C. Roberto, et al., Technological trends, global market, and challenges of bio-ethanol production, Biotechnol. Adv. 28 (2010) 817–830.

[4] W.M. Griffin, M.I.G. Scandiffio, Can Brazil replace 5% of the 2025 gasoline world demand with ethanol? Energy 34 (2009) 655–661.

[5] A.L. Borrion, M.C. McManus, G.P. Hammond, Environmental life cycle assessment of lignocellulosic conversion to ethanol: a review, Renew. Sustain. Energy Rev. 16 (2012) 4638–4650.

[6] M. Moreira, A.C. Gurgel, J.E. Seabra, Life cycle greenhouse gas emissions of sugar cane renewable jet fuel, Environ. Sci. Technol. 48 (2014) 14756–14763.

[7] M.B. Sticklen, Plant genetic engineering for biofuel production: towards affordable cellulosic ethanol, Nat. Rev. Genet. 9 (2008) 433–443.

[8] H. Von Blottnitz, M.A. Curran, A review of assessments conducted on bio-ethanol as a transportation fuel from a net energy, greenhouse gas, and environmental life cycle perspective, J. Clean. Prod. 15 (2007) 607–619.

[9] D.P. Ho, H.H. Ngo, W. Guo, A mini review on renewable sources for biofuel, Bioresour. Technol. 169 (2014) 742–749.
[10] N. Sarkar, S.K. Ghosh, S. Bannerjee, K. Aikat, Bioethanol production from agricultural wastes: an overview, Renew. Energy 37 (2012) 19–27.
[11] A. Gupta, J.P. Verma, Sustainable bio-ethanol production from agro-residues: a review, Renew. Sustain. Energy Rev. 41 (2015) 550–567.
[12] M. Ahiduzzaman, A.A. Islam, Z. Yaakob, J.A. Ghani, N. Anuar, Agricultural residues from crop harvesting and processing: a renewable source of bio-energy, Biomass and Bioenergy, Springer, Switzerland, 2014, pp. 323–337.
[13] L. Wang, Sustainable Bioenergy Production, CRC Press, Boca Raton, FL, 2014.
[14] T.L.T. Nguyen, J.E. Hermansen, L. Mogensen, Environmental performance of crop residues as an energy source for electricity production: the case of wheat straw in Denmark, Appl. Energy 104 (2013) 633–641.
[15] J.V. Haveren, E.L. Scott, J. Sanders, Bulk chemicals from biomass, Biofuels Bioprod. Bioref. 2 (2008) 41–57.
[16] T. Fox, C. Fimeche, Global Food: Waste Not, Want Not, Institute of Mechanical Engineers, London, 2013.
[17] C.S.K. Lin, L.A. Pfaltzgraff, L. Herrero-Davila, E.B. Mubofu, S. Abderrahim, J.H. Clark, et al., Food waste as a valuable resource for the production of chemicals, materials and fuels, Curr. Situat. Global Perspect. Energy Environ. Sci. 6 (2013) 426–464.
[18] B.K. Adhikari, S. Barrington, J. Martinez, Predicted growth of world urban food waste and methane production, Waste Manag. Res. 24 (2006) 421–433.
[19] S.R. Couto, M.A. Sanromán, Application of solid-state fermentation to food industry—a review, J. Food Eng. 76 (2006) 291–302.
[20] L.A. Sargeant, M. Mardell, K.M. Saad-Allah, A.H. Hussein, F. Whiffin, F. Santomauro, et al., Production of lipid from depolymerised lignocellulose using the biocontrol yeast, *Rhodotorula minuta*: the fatty acid profile remains stable irrespective of environmental conditions, Eur. J. Lipid Sci. Technol. (2015).
[21] I. Lewandowski, J.M.O. Scurlock, E. Lindvall, M. Christou, The development and current status of perennial rhizomatous grasses as energy crops in the US and Europe, Biomass Bioenergy 25 (2003) 335–361.
[22] L. López-Bellido, J. Wery, R.J. López-Bellido, Energy crops: prospects in the context of sustainable agriculture, Eur. J. Agron. 60 (2014) 1–12.
[23] S. Fazio, A. Monti, Life cycle assessment of different bioenergy production systems including perennial and annual crops, Biomass Bioenergy 35 (2011) 4868–4878.
[24] C. Pleguezuelo, V. Zuazo, C. Bielders, J. Bocanegra, F. PereaTorres, J. Martínez, Bioenergy farming using woody crops. A review, Agron. Sustain. Dev. 35 (2015) 95–119.
[25] REN21, Renewable 2013 Global Status Report. <http://www.ren21.net/status-of-renewables/global-status-report/>, 2015.
[26] S.K. Hoekman, A. Broch, C. Robbins, E. Ceniceros, M. Natarajan, Review of biodiesel composition, properties, and specifications, Renew. Sustain. Energy Rev. 16 (2012) 143–169.
[27] A. Gasparatos, P. Stromberg, K. Takeuchi, Sustainability impacts of first-generation biofuels, Anim. Front. 3 (2013) 12–26.
[28] K. Tan, K. Lee, A. Mohamed, S. Bhatia, Palm oil: addressing issues and towards sustainable development, Renew. Sustain. Energy Rev. 13 (2009) 420–427.
[29] R.W. Jenkins, N.E. Stageman, C.M. Fortune, C.J. Chuck, Effect of the type of bean, processing, and geographical location on the biodiesel produced from waste coffee grounds, Energy Fuels 28 (2014) 1166–1174.
[30] M.Y. Koh, T.I.M. Ghazi, A review of biodiesel production from *Jatropha curcas* L. oil, Renew. Sustain. Energy Rev. 15 (2011) 2240–2251.
[31] A. Kumar, S. Sharma, An evaluation of multipurpose oil seed crop for industrial uses (*Jatropha curcas* L.): a review, Ind. Crops Prod. 28 (2008) 1–10.
[32] M. Berti, R. Wilckens, S. Fischer, A. Solis, B. Johnson, Seeding date influence on camelina seed yield, yield components, and oil content in chile, Ind. Crops Prod. 34 (2011) 1358–1365.
[33] M.Y. Koh, T.I.M. Ghazi, A review of biodiesel production from *Jatropha curcas* L. oil, Renew. Sustain. Energy Rev. 15 (2011) 2240–2251.
[34] Z. Yaakob, M. Mohammad, M. Alherbawi, Z. Alam, K. Sopian, Overview of the production of biodiesel from waste cooking oil, Renew. Sustain. Energy Rev. 18 (2013) 184–193.
[35] M.G. Kulkarni, A.K. Dalai, Waste cooking oil an economical source for biodiesel: a review, Ind. Eng. Chem. Res. 45 (2006) 2901–2913.
[36] J.L. Wagner, V.P. Ting, C.J. Chuck, Catalytic cracking of sterol-rich yeast lipid, Fuel (2014), in press.
[37] R.W. Jenkins, L.A. Sargeant, F.M. Whiffin, F. Santomauro, D. Kaloudis, P. Mozzanega, et al., Cross-metathesis of microbial oils for the production of advanced biofuels and chemicals, ACS Sustain. Chem. Eng. 3 (2015) 1526–1535.
[38] Q. Li, W. Du, D. Liu, Perspectives of microbial oils for biodiesel production, Appl. Microbiol. Biotechnol. 80 (2008) 749–756.

[39] C.J. Chuck, F. Santomauro, L.A. Sargeant, F. Whiffin, T. Chantasuban, N.R.A. Ghaffar, et al., Liquid transport fuels from microbial yeasts–current and future perspectives, Biofuels 5 (2014) 293–311.

[40] C. Ratledge, Z. Cohen, Microbial and algal oils: do they have a future for biodiesel or as commodity oils, Lipid Technol. 20 (2008) 155–160.

[41] C. Huang, X.-f Chen, L. Xiong, L.-l Ma, Y. Chen, Single cell oil production from low-cost substrates: the possibility and potential of its industrialization, Biotechnol. Adv. 31 (2013) 129–139.

[42] S. Derenne, C. Largeau, M. Hetényi, A. Brukner-Wein, J. Connan, B. Lugardon, Chemical structure of the organic matter in a Pliocene maar-type shale: implicated *Botryococcus* race strains and formation pathways, Geochim. Cosmochim. Acta 61 (1997) 1879–1889.

[43] C. Chuck, J. Wagner, R. Jenkins, Biofuels from microalgae, Chem. Process. Sustain. Future (2014) 425.

[44] L.A. Ribeiro, P.P. da Silva, Surveying techno-economic indicators of microalgae biofuel technologies, Renew. Sustain. Energy Rev. 25 (2013) 89–96.

[45] R. Davis, A. Aden, P.T. Pienkos, Techno-economic analysis of autotrophic microalgae for fuel production, Appl. Energy 88 (2011) 3524–3531.

[46] D.L. Sills, V. Paramita, M.J. Franke, M.C. Johnson, T.M. Akabas, C.H. Greene, et al., Quantitative uncertainty analysis of life cycle assessment for algal biofuel production, Environ. Sci. Technol. 47 (2012) 687–694.

[47] S. Kraan, Mass-cultivation of carbohydrate rich macroalgae, a possible solution for sustainable biofuel production, Mitig. Adapt. Strategies Glob. Change 18 (2013) 27–46.

[48] K.A. Jung, S.-R. Lim, Y. Kim, J.M. Park, Potentials of macroalgae as feedstocks for biorefinery, Bioresour. Technol. 135 (2013) 182–190.

[49] P. Biller, A.B. Ross, Hydrothermal processing of algal biomass for the production of biofuels and chemicals, Biofuels 3 (2012) 603–623.

[50] N. Wei, J. Quarterman, Y.-S. Jin, Marine macroalgae: an untapped resource for producing fuels and chemicals, Trends. Biotechnol. 31 (2013) 70–77.

[51] E.W. Nester, *Agrobacterium*: nature's genetic engineer, Front. Plant Sci. 5 (2014) 730.

[52] S.P. Long, A. Marshall-Colon, X.-G. Zhu, Meeting the global food demand of the future by engineering crop photosynthesis and yield potential, Cell 161 (2015) 56–66.

[53] R.A. Burton, G.B. Fincher, Plant cell wall engineering: applications in biofuel production and improved human health, Curr. Opin. Biotechnol. 26 (2014) 79–84.

[54] G. Pilate, E. Guiney, K. Holt, M. Petit-Conil, C. Lapierre, J.-C. Leplé, et al., Field and pulping performances of transgenic trees with altered lignification, Nat. Biotechnol. 20 (2002) 607–612.

[55] A.S. Carlsson, L.-H. Zhu, M. Andersson, P. Hofvander, Platform crops amenable to genetic engineering – a requirement for successful production of bio-industrial oils through genetic engineering, Biocatal. Agric. Biotechnol. 3 (2014) 58–64.

[56] N. Wei, E.J. Oh, G. Million, J.H.D. Cate, Y.-S. Jin, Simultaneous utilization of cellobiose, xylose, and acetic acid from lignocellulosic biomass for biofuel production by an engineered yeast platform, ACS Synth. Biol. 4 (2015) 707–713.

[57] S.-H. Ho, X. Ye, T. Hasunuma, J.-S. Chang, A. Kondo, Perspectives on engineering strategies for improving biofuel production from microalgae — a critical review, Biotechnol. Adv. 32 (2014) 1448–1459.

[58] F. Marco, M. Bitrián, P. Carrasco, M.V. Rajam, R. Alcázar, A.F. Tiburcio, Genetic engineering strategies for abiotic stress tolerance in plants, Plant Biology and Biotechnology, Springer, 2015, pp. 579–609.

[59] S. Matsuoka, A.J. Kennedy, E.G.D. dos Santos, A.L. Tomazela, L.C.S. Rubio, Energy cane: its concept, development, characteristics, and prospects, Adv. Bot. 2014 (2014) 13.

[60] Y. Liu, Y. Xu, F. Zhang, J. Yun, Z. Shen, The impact of biofuel plantation on biodiversity: a review, Chin. Sci. Bull. 59 (2014) 4639–4651.

[61] B.E. Dale, J.E. Anderson, R.C. Brown, S. Csonka, V.H. Dale, G. Herwick, et al., Take a closer look: biofuels can support environmental, economic and social goals, Environ. Sci. Technol. 48 (2014) 7200–7203.

[62] B. Eickhout, G.J. van den Born, J. Notenboom, Mv Oorschot, J. Ros, D. Van Vuuren, et al., Local and Global Consequences of the EU Renewable Directive for Biofuels: Testing the Sustainability Criteria, Rapport nr. MNP 500143001, Netherlands Environmental Assessment Agency (MNP), Bilthoven, the Netherlands, 2008.

[63] M.C. McManus, C.M. Taylor, The changing nature of life cycle assessment, Biomass Bioenergy (2015).

[64] A. Broch, S.K. Hoekman, S. Unnasch, A review of variability in indirect land use change assessment and modeling in biofuel policy, Environ. Sci. Policy 29 (2013) 147–157.

[65] S. Ahlgren, L. Di Lucia, Indirect land use changes of biofuel production–a review of modelling efforts and policy developments in the European Union, Biotechnol. Biofuels. 7 (2014) 35.

[66] J. Fargione, J. Hill, D. Tilman, S. Polasky, P. Hawthorne, Land clearing and the biofuel carbon debt, Science 319 (2008) 1235–1238.

[67] V.H. Dale, K.L. Kline, J. Wiens, J. Fargione, Biofuels: Implications for Land use and Biodiversity, Ecological Society of America, Washington, DC, 2010.
[68] M. Finkbeiner, Indirect land use change–Help beyond the hype? Biomass Bioenergy 62 (2014) 218–221.
[69] H. Haberl, K.-H. Erb, F. Krausmann, A. Bondeau, C. Lauk, C. Müller, et al., Global bioenergy potentials from agricultural land in 2050: sensitivity to climate change, diets and yields, Biomass Bioenergy 35 (2011) 4753–4769.
[70] M.C. McManus, C.M. Taylor, A. Mohr, C. Whittaker, C.D. Scown, A.L. Borrion, et al., Challenge clusters facing LCA in environmental decision-making—what we can learn from biofuels, Int. J. Life Cycle Assess. 20 (2015) 1399–1414.
[71] S. Searle, C. Malins, A reassessment of global bioenergy potential in 2050, GCB Bioenergy 7 (2015) 328–336.
[72] C.B. Field, J.E. Campbell, D.B. Lobell, Biomass energy: the scale of the potential resource, Trends Ecol. Evol. 23 (2008) 65–72.
[73] E.M.W. Smeets, A.P.C. Faaij, I.M. Lewandowski, W.C. Turkenburg, A bottom-up assessment and review of global bio-energy potentials to 2050, Prog. Energy Combust. Sci. 33 (2007) 56–106.
[74] H. Yamamoto, J. Fujino, K. Yamaji, Evaluation of bioenergy potential with a multi-regional global-land-use-and-energy model, Biomass Bioenergy 21 (2001) 185–203.
[75] J. Wolf, P.S. Bindraban, J.C. Luijten, L.M. Vleeshouwers, Exploratory study on the land area required for global food supply and the potential global production of bioenergy, Agric. Syst. 76 (2003) 841–861.
[76] M. Hoogwijk, A. Faaij, B. Eickhout, B. de Vries, W. Turkenburg, Potential of biomass energy out to 2100, for four IPCC SRES land-use scenarios, Biomass Bioenergy 29 (2005) 225–257.
[77] M. Hoogwijk, A. Faaij, B. de Vries, W. Turkenburg, Exploration of regional and global cost–supply curves of biomass energy from short-rotation crops at abandoned cropland and rest land under four IPCC SRES land-use scenarios, Biomass Bioenergy 33 (2009) 26–43.
[78] D.P. van Vuuren, J. van Vliet, E. Stehfest, Future bioenergy potential under various natural constraints, Energy Policy 37 (2009) 4220–4230.

CHAPTER

3

Certification and Performance: What Is Needed from an Aviation Fuel?

G.E. Dorrington

School of Engineering, RMIT University, Melbourne, VIC, Australia

3.1 INTRODUCTION TO RELEVANT STANDARDS

All currently certified passenger and cargo-carrying aircraft fitted with gas turbine power plants, including turboprops and turbofans, are fuelled by Aviation Turbine Fuel (ATF) that is supplied through an internationally agreed-upon process complying with exacting standards [1]. In the civil sector, common grades of ATF include:

1. Jet A-1 and Jet A (in the US) complying with ASTM D1655 [2].
2. Jet A-1 complying with a UK Defence Standard (DEF STAN 91-91).
3. TS-1 (in Russia) complying with GOST 10227.
4. Jet Fuel No. 3 (in China) complying with GB6537.

For international civil airline operations, Jet A-1 is the most commonly available ATF across the world. Therefore, the following discussion focuses on this grade.

At the time of writing, the specification of Jet A-1 is defined by ASTM D1655 version 15d [2], which calls up 64 other ASTM standards, mainly test methods designed to demonstrate compliance. At its simplest, D1655 specifies a hydrocarbon mixture refined from petroleum, oil sand, or shale-derived feedstock. As might be expected, the requirements invoked by D1655 reflect the historic legacy of petroleum refinery capability as well as the needs of the aviation sector. Some of the main property requirements of a fuel complying with D1655 are presented in Table 3.1. The justification of these will be discussed in the next section.

ASTM D7566 [3] was first issued in 2009 in order to provide supply control of Synthesized Paraffinic Kerosene (SPK) derived from coal/natural gas through the Fischer–Tropsch (F–T) process [4]. This was followed in 2011 with approval of an annex covering SPK from Hydroprocessed Esters and Fatty Acids (HEFA) produced from bio-feedstock. At the time of writing, D7566-15c [3] has four

Biofuels for Aviation.
DOI: http://dx.doi.org/10.1016/B978-0-12-804568-8.00003-2

annexes covering four alternative production routes:

1. Annex A1: F–T Hydroprocessed SPK.
2. Annex A2: SPK from HEFA.
3. Annex A3: Synthesized Iso-Paraffins (SIP) from Hydroprocessed Fermented Sugars[1].
4. Annex A4: Synthesized Kerosene with Aromatics (SPK/A) Derived by Alkylation of Light Aromatics from Non-Petroleum Sources.

Other new annexes are currently in preparation, reflecting the gradual emergence of a diverse range of production routes from different feedstocks. In particular, an annex covering Alcohol-To-Jet (ATJ) in which the feedstock would be bio-ethanol or bio-butanol, is expected to be approved in 2016.

Each of the above production route specifications defines a product 'batch' that currently must be blended with fuel compliant with D1655. The maximum permissible blend fraction of F–T SPK (Annex A1), HEFA-SPK (Annex A2), and SPK/A (Annex A4) is currently restricted to 50% by volume. SIP (Annex A3) is currently restricted to just 10% maximum blend fraction.

The resulting Jet A-1 equivalent blends must adhere to the same property requirements listed in Table 3.1 and the additional requirements listed in Table 3.2. Furthermore, each annex has separate detailed batch requirements, Tables 3.3–3.5. Blends complying with these collective requirements are regarded within the aviation sector as 'drop-in' ATFs that may be employed 'seamlessly' without any

TABLE 3.1 Some Jet A-1 Requirements Specified by ASTM D1655-15d

Fuel Property Requirement	Value	ASTM Test Methods
Maximum aromatic content by volume	26.5%	D6379
Maximum total sulphur by mass	0.3%	D1266, D2622, D4294, D5453
Minimum flash point	38°C	D56, D93, D3828
Density minimum–maximum range at 15°C	775–840 kg/m^3	D1298, D4052
Maximum freezing point	−47°C	D5972, D7153, D7154, D2386
Maximum viscosity at −20°C	8.0 mm^2/s	D445, D7042
Minimum net heat of combustion	42.8 MJ/kg	D4529, D3338, D4809
Minimum smoke point	18 mm	D1322
Maximum naphthalene content by volume	3.0%	D1840
Thermal stability maximum filter pressure drop	25 mm Hg at 260°C	D3421
Maximum antioxidant additive	24 mg/L	
Icing inhibitor additive range by volume	0.07–0.15%	

[1]In ASTM D7566 version 15c, Annex A3 presumes the SIP is farnasane (2,6-trimethyldodecane, $C_{15}H_{32}$), and invokes a requirement that the actual farnasane component exceed 97% by mass and that a by-product of the production process, hexahydrofarnesol ($C_{15}H_{32}O$), be restricted to less than 1.5% by mass. Future versions of this annex will presumably be extended to any other biosynthesis products produced via this general route.

TABLE 3.2 Some Additional Post-Blending Requirements Specified by ASTM D7566-15c

Blend Property Requirement	Value	ASTM Test Methods
Minimum aromatic composition by volume	8.4%	D6379
Maximum viscosity at −40°C	12.0 mm^2/s	D445
Maximum wear in fuel lubricity test	0.85 mm	D5001

TABLE 3.3 Some Batch Requirements of ASTM D7566 Annexes A1 and A2 for F−T and HEFA-SPK

Batch Property Requirement	Value	ASTM Test Methods
Minimum flash point	38°C	D56, D3828
Density minimum−maximum range at 15°C	730−770 kg/m^3	D1298, D4052
Maximum freezing point	−40°C	D5972, D7153, D7154, D2386
Thermal stability maximum filter pressure drop	25 mm Hg at 325°C	D3421
Maximum cycloparaffin mass content	15%	D2425
Maximum aromatic mass content	0.5%	D2425

TABLE 3.4 Some Batch Requirements of ASTM D7566 Annex A3, SIP

Batch Property Requirement	Value	ASTM Test Methods
Minimum flash point	100°C	D93, D3828
Density minimum−maximum range at 15°C	765−780 kg/m^3	D1298, D4052
Maximum freezing point	−60°C	D5972, D7153, D7154, D2386
Thermal stability maximum filter pressure drop	25 mm Hg at 355°C	D3421
Maximum olefin mass content	300 mg (Br)/100 g	D2425
Maximum aromatic mass content	0.5%	D2425

TABLE 3.5 Some Batch Requirements of ASTM D7566 Annex A4, SPK/A

Batch Property Requirement	Value	ASTM Test Methods
Minimum flash point	38°C	D56, D3828
Density minimum−maximum range at 15°C	755−800 kg/m^3	D1298, D4052
Maximum freezing point	−40°C	D5972, D7153, D7154, D2386
Thermal stability maximum filter pressure drop	25 mm Hg at 325°C	D3421
Maximum cycloparaffin content	15%	D2425
Maximum aromatic mass content	20%	D2425

necessary changes to aircraft systems, airport infrastructure, or supply methods.

In principle, it would be possible for an organization to develop an all-new ATF (or ATF additive) that is not considered to be drop-in, but is still considered 'fit-for-purpose' by complying with ASTM D4054 [5]. However, this would require comprehensive engine and aircraft certification trials as well as the need to introduce separate fuel supply and bunkering facilities at airports, all of which would be relatively costly. Consequently, the main industry focus in recent years has been restricted to drop-in solutions [4].

Despite considerable production advances since the introduction of D7566 in 2009, the principle difficulty to date has been the ability to identify and develop a sustainable, drop-in, ATF that is commercially competitive with petroleum-derived Jet A-1. Aside from the challenges of biosynthesis and sustainable production scale-up, the major issue is the need to produce a product that complies with the exacting requirements of D7566. It is therefore useful to first review the historical reasons for these requirements, and then after that to identify possible concessions and/or likely future amendments.

3.2 DISCUSSION OF SOME DROP-IN FUEL PROPERTY REQUIREMENTS

3.2.1 Heating Value and Density

Perhaps the most important requirement of Jet A-1 and any equivalent D7566 ATF blend is the need to offer a minimum net heat of combustion (or so-called 'lower heating value'), H_f. The Breguet range equation [6] for a turbofan-powered aircraft demonstrates that for a given aircraft gross mass, m_0, fuel mass burn, m_f, propulsive efficiency, η_p, and lift-to-drag ratio, L/D, the range, R, is directly proportional to H_f:

$$R = \eta_p H_f (L/D) \ln\left(\frac{m_0}{m_0 - m_f}\right) \quad (3.1)$$

D7566 currently invokes the D1655 requirement that $H_f > 42.8$ MJ/kg (Table 3.1), which effectively ensures that current aircraft designs can achieve a prescribed range with specified minimum fuel reserves. This non-negotiable requirement also restricts the feasible fuel candidates. The Mendeleyev formula [7] for net heat release (quoted in kcal/kg) is $H_f = 81C + 300H\text{-}26(O\text{-}S)$, where C, H, O, and S are the percentages of carbon, hydrogen, oxygen, and sulphur, respectively. Therefore, to achieve the required D1655 heating value, the H/C mass ratio of a zero-oxygen-sulphur fuel must exceed ~0.14. Furthermore, no alcohol with less than 15 carbon atoms remains a feasible candidate, even when blended 50% with a fuel with $H_f = 45$ MJ/kg.

Both D1655 and D7566 require that the blended fuel density lie between 775 and 840 kg/m^3 (Table 3.1). However, the batch specification for the pre-blended (non-petroleum) fuel components deviates from this slightly: Annex A1 and Annex A2 permit 730–770 kg/m^3 for F–T SPK and HEFA-SPK; Annex A3 permits 765–780 kg/m^3 for SIP; and Annex A4 permits 755–800 kg/m^3. Using Ragozin's approximate density formula [7], this implies the hydrogen/carbon ratio, H/C, must be somewhere in the range ~0.16–0.18.

The requirement on maximum density is arguably not as stringent as that of the minimum density because aircraft fuel tanks do not have to be completely filled. Indeed, it is hard to find engineering reasons as to why an ATF with an excessively high density of (say) 850 kg/m^3 should be rejected. Conversely, it is obviously necessary to ensure the aircraft's maximum available fuel tank volume is not exceeded for a given fuel load. For some aircraft designs, fuel tank capacity actually limits range instead of fuel mass upload capability. While there is no explicitly stated requirement on the fuel heating value per unit volume, clearly, combining the aforementioned requirements on H_f and density, D7566 requires that it must exceed ~33 GJ/m^3.

3.2.2 Freezing Point and Fluidity at Low Temperatures

The requirement for the maximum fuel freezing point of −47°C for Jet A-1, is a consequence of the fact that currently nearly all long-range passenger aircraft flights occur in the lower stratosphere. Aircraft tank cooling occurs progressively throughout such flights during ascent and cruise-climb because the ambient air temperature falls with altitude in the troposphere approximately according to the standard lapse rate of 6.5 K/km. Since the duration of domestic flights within the US is typically less than ~8 h, Jet A (a separate fuel grade stored in US airports in different bunkers than Jet A-1) is therefore permitted to have a relaxed freezing point requirement of −40°C. Like D1566, D7566 allows blended fuel to have either a nominal minimum freezing point of −47°C or −40°C, depending on whether the product is considered equivalent to Jet A-1 or Jet A, respectively. The pre-blend fuel batches of F−T SPK, HEFA-SPK, and SPK/A are only required to have a minimum freezing point of −40°C, but Annex A3 requires SIP to have a freezing point of −60°C. According to D7566, it is also worth noting that the freezing point specification of Jet A-1 and Jet A is actually open to agreement between supplier and purchaser.

To further ensure fluidity in cold conditions, D1655 and D7655 require the fuel kinematic viscosity to be less than 8 mm^2/s at −20°C (Table 3.1). Since the Hagen−Poiseuille equation implies that the pressure drop in any pipe is proportional to dynamic viscosity (and fuel density is constrained), this requirement effectively ensures the aircraft fuel pumping power requirements cannot be breached by a fuel that is too viscous.

3.2.3 Combustion Cleanliness

ATF is burned at near-stoichiometric conditions in the primary zone of a gas turbine combustor. This hot primary flow is then diluted with excess air in the secondary zone to reduce its temperature prior to entry into the turbine. This staged flow arrangement also reduces convective heat transfer to the combustor outer walls. In general, paraffins (alkanes) offer the most desirable combustion characteristics and are relatively clean burning. Curiously, cycloparaffins also offer relatively clean combustion, but for historical reasons their content is limited to less than 15% in all four Annexes[2].

Olefins (alkenes) also have good combustion characteristics, but the tendency of olefins (especially dienes) to form gum deposits/residues [7−9] when fuel evaporates is of concern. According to D7566 the olefin content is usually restricted to 1% or less [3], although no actual upper limit is specified for Annexes A1, A2, and A4. The batch specification for SIP in Annex A3, however, limits olefin content through test D2710 that measures bromine reactivity.

Both D1655 and D7566 stipulate that the aromatic content of ATF must not exceed 26.5% by volume (Table 3.1). The driver for this requirement is that high aromatic-content fuels generally result in smoke formation and

[2]The justification for this is as quoted in ASTM D7566-15c as follows: 'Maximum cycloparaffin composition is based on current experience with the approved synthetic fuels and is within the range of what is typical for refined jet fuel'.

soot deposition. Naphthalenes or bicyclic aromatics are especially conducive to producing more soot, which often results in high levels of undesired radiation heat transfer. Hence, the naphthalene content is restricted to less than 3% [2,3]. Smoke formation is also controlled by a 'Smoke Point' test (ASTM D1322), which involves measuring the maximum height that a candle-like flame (supplied by fuel though a wick) remains smokeless.

Combustion efficiency and soot formation are also influenced by fuel injection, which involves spray atomization of the ATF in the combustor. Again, the aforementioned constraint on fuel viscosity partly ensures consistency of spray formation, although a lower limit on viscosity is not imposed. Curiously, there is no mention in D7566 of any constraint on liquid surface tension, which is the principle property governing spray droplet size [9].

3.2.4 Fuel System Compatibility

Unlike D1655, minimum aromatic content is constrained by D7566 to 8% after blending (Table 3.2). This minimum level is currently deemed necessary in order to ensure that the sealing properties of elastomers (in the form of numerous O-rings in a typical aircraft fuel system) are not compromised by different swelling properties [4]. To retain desired nitrile rubber volumetric swelling, aromatics such as Tetralin (1,2,3,4-tetrahydronaphthalene) may be added to aromatic depleted fuels (eg, [10,11]), although the deliberate use of such additives must also have the adverse effect of reducing the smoke point of the fuel. In this regard, it is also worth noting that Annexes A1 and A3 of D7655 restrict the batch aromatic content of F–T SPK and SIP, respectively, to less than 0.5% prior to blending. In effect, therefore, the 50% blend limit implies that the actual minimum aromatic content of D1655-compliant fuel needs to be in excess of 16% to permit F–T SPK blending.

Aircraft fuel systems rely on the fuel to lubricate their parts. Adequate ATF lubricity is assessed by a Ball-On-Cylinder Lubricity Evaluator (BOCLE) test, which involves measurement of wear caused by a sphere on a cylinder immersed in the fuel. ASTM D5001 requires that the resulting wear scar be limited to a diameter of 0.85 mm (Table 3.2). Severe hydroprocessing tends to eliminate fuel components that improve lubricity. Additives that also inhibit corrosion (such as HiTec 580 or Nalco 5403) are employed to achieve or regain acceptable lubricity, although their polar nature can have adverse effects on fuel filtration and, consequently, they are only used in small quantities (eg, less than 15 g/m^3) [3].

3.2.5 Thermal Stability, Flash Point, and Vapour Pressure

From a combustion standpoint, high fuel volatility is desirable to ensure good fuel ignitability. However, from a safety viewpoint, the opposite qualities of low volatility and storage stability are required. It is also necessary to avoid fuel decomposition before it reaches the combustor in fuel feed lines that are heated by the power plant. Both D1655 and D7566 require ATF to remain thermally stable up to a specified control temperature of 260°C [2,3]. However, for the pre-blending batches, Annexes A1, A2, and A4 require thermal stability at a control temperature of 325°C, and Annex A3 calls for 355°C. Compliance is tested (in ASTM D3241) by passing the fuel through a heated tube and filter and measuring the filter pressure drop caused by deposit formation. The maximum pressure drop imposed is 25 mm Hg in all cases, after a 2.5 h test duration.

Nearly all current aircraft fuel tanks allow air into the ullage, hence, any compliant ATF must have adequate oxidation stability. Antioxidant additives are employed to reduce the risk of oxidation, but added quantities are constrained to lie between 17 mg/L (minimum) and 24 mg/L (maximum) for each Annex of D7566.

Ignition (or flashing) of fuel vapour occurs at certain pressure, temperature, and oxygen concentration conditions. The flash point, as determined by the ASTM D56 test (Table 3.1), is the lowest temperature at which a fuel vapour air mixture ignites at standard sea level pressure. Both D1655 and D7566 specify a minimum flash point of 38°C (Table 3.1), but both standards state that a higher flash point may be agreed upon between purchaser and supplier. In the case of SIP, D7566 Annex A3 specifies a higher minimum flash point of 100°C. On the other hand, D7566 also states that HEFA-SPK with a lower flash point may be blended with D1655-compliant fuel.

Flash point is closely linked with vapour pressure. Fuels with vapour pressures less than that of decane (in the appropriate temperature range) are likely to comply with this requirement.

3.2.6 Sulphur Content and Other Contaminants

While sulphur compounds might improve fuel lubricity and increase heating value, D1655 and D7566 currently restrict total sulphur content to 0.3% by mass (Table 3.1). Past surveys of Jet A-1 composition indicate the actual sulphur content is 321–800 ppm [12]. Further reduction of sulphur content is desired by ICAO to prevent emission of sulphur oxides and sulphur-containing particulates [12]. Aside from this desired reduction, low sulphur content is required to prevent corrosion of turbine blades by sulphur oxides. Mercaptan (methanethiol, CH_4S) content is severely restricted to 0.003% [2,3] because it is known to be reactive with certain elastomers.

Several other fuel contaminants are restricted by D7566, including the following: existent gum to less than 7 mg per 100 mL; metals including phosphorus and halogens to 1 mg/kg; nitrogen to 2 mg/kg; water to 75 mg/kg; and electrical conductivity improvers to 5 mg/L. It should also be noted that D1566 [2] calls up a stringent requirement for Fatty Acid Methyl Ester (FAME) content to be less than 50 mg/kg, following reports of unapproved Jet A-1 supply contamination by FAME in 2009.

3.3 SPECULATION ON POSSIBLE FUTURE CHANGES IN ASTM 7566 REQUIREMENTS

3.3.1 Historical Trend

When comparing D7655-15c issued in 2015 [3] with D7655-13 issued in 2013 [13], aside from the introduction of Annexes A3 and A4, several requirement embellishments or overlays become apparent. For example, D7655-15c calls up an extended requirement on blend viscosity to be less than 12 mm^2/s at -40°C, but this requirement did not appear in D7655-13. On the other hand, the 2015 version offers the latitude of testing against Institute of Petroleum (IP) standards as an alternative compliance demonstration route. In summary, while new production routes have been approved and alternative test methods permitted, fuel blend specification has become progressively tighter over recent years. Concessions (or relaxation of requirements) are only likely to appear in D7655 when new/recent annexes are added or modified.

3.3.2 Reduced Aromatic and Cycloparaffin Content

The advantage of reducing aromatic content (to minimize formation of exhaust soot particulates) appears to be contradicted by the current need for a minimum aromatic content to maintain existing elastomer compatibility. This conflict might potentially be resolved by the introduction of new elastomer seals though ASTM D4054. Recent work [14] has shown that the properties of fluorocarbon and fluorosilicon rubbers are far less sensitive to fuel composition changes than nitrile rubber. The problem with such an introduction, however, is that all existing aircraft fuel systems would have to be suitably upgraded prior to introduction of low-aromatic fuel blends. Perhaps the only way such a major change could be enforced is by the introduction of stricter particulate emission constraints.

Removal of the constraint on minimum cycloparaffin content appears be a more likely concession, since it appears to be an ad hoc requirement that has not yet been challenged. In other words, it currently persists as a legacy of petroleum refinery experience.

3.3.3 New Generalized Annex for the Fermented Sugar Production Route

Currently, D7566 Annex A3 prohibits any significant olefin or aromatic content and effectively restricts production to farnasane. If it is supposed that the fermented sugar production route is generalized to include, for example, D-limonene or *p*-menthene (instead of farnasane), then either the Annex 3 requirements would need to be revised, or a new annex would have to be created. Arguably, the current olefin or aromatic content limits are not necessary, bearing in mind the current 10% blend limit. Instead, equitable constraints on thermal stability, propensity to form gum, and antioxidant additions should be sufficient to define a compliant product. Similarly, the preblend batch requirement for a minimum freezing point of −60°C (Table 3.5) appears to be unnecessarily excessive.

3.3.4 Flash Point Reduction

It is worth pointing out that in recent years On-Board Inerting Gas Generation Systems (OBIGGS) have been fitted in aircraft in order to reduce the oxygen/nitrogen ratios in fuselage tank ullages, and have thereby prevented the possibility of in-tank explosions. Despite this introduction, the historical trend towards improving safety measures in a parallel manner suggests it would be unlikely that reduced flash point concessions will be offered in the future, although such a reduction is feasible in principle and may deserve review.

3.3.5 Removal of Blend Fraction Constraint

Perhaps the most likely and significant future change to D7566 requirements is the current need for blending with D1655-compliant fuel produced by petroleum refineries. This requirement arguably involves a potential conflict of interest insomuch as it may not be in the commercial interests of a petroleum supplier to blend-in SPK produced by an independent bio-refiner, unless compelled to do so by some regulatory requirement. As mentioned previously, D7566-15c currently imposes the seemingly arbitrary constraint that F–T SPK and HEFA-SPK can only be blended up to 50%, and SIP blending is limited to 10%. It is noted that in practise, density, or aromatic content, or some other property such as viscosity, often limits the amount of SPK that can be blended below these stipulated limits [3]. However, these lower practical constraints are difficult to establish since the margins on the actual properties of Jet A-1 with respect to D1655 limits are not defined. For example, a lower acceptable limit in aromatic content is not explicitly defined in D1655. There must be cases, however, when blending beyond the stipulated 50/10% limits is viable. For example, if the aromatic content of a batch of Jet A-1 was 25% (say), then it would be possible to dilute it by a factor of three with a batch of F–T SPK before the stipulated 8.4% lower aromatic blend limit is reached.

Eventually, agreement on a 100% bio-refined ATF will probably be sought by some organizations in the bio-refinery sector in order to achieve commercial independence from the petroleum refinery sector, although the main challenge will still likely be production cost reduction [4].

3.3.6 Possible Introduction of Small Mass Fraction Blend Annexes

It is possible to conceive of new annexes to ASTM D7566 designed to permit more flexible production ATFs involving small blend fractions (up to 5%, say) of fuel batches from non-petroleum sources with Jet A-1. Such annexes could involve relaxed batch requirements, provided the final blend product does not deviate significantly from D1655-compliant ATF. For example, a bio-derived fuel with heating values as low as ~30 MJ/kg could be deemed permissible, provided it is blended with Jet A-1 with at least 43.5 MJ/kg at less than 5% by mass. Again the key issue in this regard is the exact margins between actual Jet A-1 properties and the requirements specified in D1566. If a new, tighter specification for petroleum-derived ATF is formulated, then this 'tight fuel' could be blended with batches of bio-derived ATF that comply with more relaxed requirements.

3.4 PROSPECTS FOR ALL-NEW, NON-DROP-IN FUELS

From an ideal performance standpoint, a low-cost fuel with an ultra-high heating value and sufficiently high density is needed by the aviation sector. Unfortunately, among the hydrocarbons there are few viable options to consider [15]. Liquid hydrogen has been proposed as the ultimate zero-carbon emission ATF solution (eg, [16]) because it has high energy content (~120 MJ/kg). Its density is relatively low (~70 kg/m^3), however, leading to designs with overly large, thermally insulated fuel tanks. The cryogenic delivery and storage of liquid hydrogen also poses significant operational problems. A more likely (intermediate) solution is the possible introduction of liquid methane-fuelled (including liquefied natural gas and liquefied bio-gas) aircraft (eg, [17]). Liquid methane has a slightly better heat content compared to Jet A-1 (~50 MJ/kg), moderately low density (~420 kg/m^3), and less problematic cryogenic storage issues than hydrogen [9]. Future aircraft designs with so-called 'blended-wing bodies' with improved lift-to-drag ratios may also offer increased wing tank volumes, thus offsetting the disadvantages associated with relatively low fuel density [18]. However, the capital investment for necessary aircraft development and airport infrastructure changes to permit liquid methane usage would probably be a formidable barrier to such a radical transformation. In summary, the prospects for an all-new, non-drop, alternative ATF appear to be poor in the foreseeable future, unless the price of petroleum-refined Jet A-1 rises rapidly, and/or unless strong environmental restrictions on net carbon emissions are imposed.

References

[1] Anonymous, Manual on Civil Aviation Jet Supply, ICAO, Doc. 9977 AN/489, International Civil Aviation Organization, Montreal, QC, 2012, ISBN 978-92-9249-105-5.

[2] Anonymous, Standard Specification for Aviation Turbine Fuel's, ASTM D1655-15d, ASTM International, West Conshohocken, PA, 2015.

[3] Anonymous, Standard Specification for Aviation Turbine Fuel Containing Synthesized Hydrocarbons, ASTM D7566-15c, ASTM International, West Conshohocken, PA, 2015.

[4] Anonymous, IATA 2013 Report on Alternative Fuels, eighth ed., International Air Transport Association, Montreal, QC, 2013, ISBN 978-92-9252-269-9.

[5] Anonymous, Standard Practice for Qualification of New Aviation Turbine Fuels and Fuel Additives, ASTM D4054-14, ASTM International, West Conshohocken, PA, 2014.

[6] D. Küchemann, The Aerodynamic Design of Aircraft, first ed., Pergamon, Oxford, 1978.

[7] N.A. Ragozin, Jet Propulsion Fuels, (W.E. Jones, Trans.), Pergamon, Oxford, 1961.

[8] W.R. Hawthorne, W.T. Olson, Design and Performance of Gas Turbine Power Plants, Oxford University Press, Oxford, 1960.

[9] A.H. Lefebvre, D.R. Ballal, Gas Turbine Combustion, Alternative Fuels and Emissions, third ed., CRC Press, Boca Raton, FL, 2010.

[10] J.R. Coleman, L.D. Gallop, Exposure of Nitrile Rubber to Fuels of Varying Aromatics Level, DREO Technical Note 82–86, Defence Research Establishment of Ottawa, Department of National Defence, Canada, May 1982.

[11] J.L. Graham, R.C. Striebich, K.J. Myers, D.K. Minus, W.E. Harrison, Swelling of nitrile rubber by selected aromatics blended in a synthetic jet fuel, Energy Fuels 20 (2006) 759–765.

[12] Anonymous, Additional Local Air Quality Benefits from Sustainable Alternative Fuels for Aircraft, ICAO Working Paper, CAAF/09-WP/05, Conference on Aviation and Alternative Fuels, Rio de Janerio, Brazil. 16–18 November 2009.

[13] Anonymous, Standard Specification for Aviation Turbine Fuel Containing Synthesized Hydrocarbons, ASTM D7566-13, ASTM International, West Conshohocken, PA, 2013.

[14] J. Otner, Material Compatibility presented in High Biofuel Blends in Aviation (HBBA) study and BioJetMap Workshop, Brussels, Belgium, 11 February 2015. See also: D. Ewing, Alternative Seals for Alternative Fuels, SAE Aerospace Engineering magazine, 8 June 2011.

[15] D. Daggett, O. Hadaller, R. Walther, Alternative Fuels and Their Potential Impact on Aviation, NASA/TM—2006-214365, October 2006.

[16] D. Verstraete, On the energy efficiency of hydrogen fuelled transport aircraft, Int. J. Hydrogen Energy 40 (23) (2015) 7388–7394.

[17] M.K. Bradley, C.K. Droney, Subsonic Ultra Green Aircraft Research Phase II: N + 4 Advanced Concept Development, NASA/CR-2012-217556, May 2012.

[18] R.T. Kawai, Benefit Potential for a Cost Efficient Dual Fuel BWB, AIAA paper 2013-0937 in 51st AIAA Aerospace Sciences Meeting including the New Horizons Forum and Aerospace Exposition, Texas, USA, 07–10 January 2013.

SECTION II

THE SCIENCE AND TECHNOLOGY OF DEVELOPING BIOFUELS FOR AVIATION

CHAPTER

4

The Suitability of Fatty Acid Methyl Esters (FAME) as Blending Agents in Jet A-1

M. Lapuerta[1] *and L. Canoira*[2]

[1]Grupo de Combustibles y Motores, ETS Ingenieros Industriales, Universidad de Castilla La Mancha, Ciudad Real, Spain, [2]Department of Energy & Fuels, ETS Ingenieros de Minas y Energía, Universidad Politécnica de Madrid, Madrid, Spain

4.1 AN INTRODUCTION TO FAME IN JET A-1

4.1.1 Current Restrictions of FAME Blends in Jet Fuel

The American Society for Testing Materials (ASTM) Subcommittee J on Aviation Fuels is comprised of members of the aviation fuels community and has developed the standard ASTM D1655 [1]. Aviation turbine fuels meeting ASTM D1655 are limited to hydrocarbon distillate products derived from a limited number of fossil sources. The listed sources of feedstocks for producing aviation turbine fuel include crude oil, natural gas liquid condensates, heavy oil, shale oil, and oil sands. ASTM has been working on approval of alternative fuels as blend components for ASTM D1655 through D02.0J.06 Section J.06 on Emerging Turbine Fuel, and has recently published the approved methods for the production of alternative jet fuels that are basically the Fischer–Tropsch hydroprocessed synthesized paraffinic kerosene (FT-SPK) and the hydroprocessed esters and fatty acids (HEFA).

The group has developed the ASTM D7566 Standard Specification for Aviation Turbine Fuel Containing Synthesized Hydrocarbons, which defines specifications for alternative fuels that can be used to blend semi-synthetic jet fuel meeting ASTM D1655 requirements [2]. As aviation turbine fuels must meet a myriad of requirements, both specified and implied, this group had a very daunting task. Annex A1 to ASTM D1655 defines how fuels from non-conventional sources are to be handled. It is assumed that jet fuel from conventional sources has properties that make it fit-for-purpose, and that fuels not from these sources must be thoroughly

Biofuels for Aviation.
DOI: http://dx.doi.org/10.1016/B978-0-12-804568-8.00004-4

evaluated before they can be used to blend ASTM D1655 jet fuel. These two alternative jet fuel blend components, when they meet all of the requirements of ASTM D7566, are allowed for blending ASTM D1655 jet fuel at up to 50% volumetric content. Other alternative jet fuel blend components are under evaluation by Section J.06 for inclusion in ASTM D7566. These include Alcohol-To-Jet (ATJ) fuels, alternative hydrocarbon fuels derived from pyrolysis processes, and other hydrocarbon fuels that come from other transformation technologies [3].

All of the alternative jet fuel blend components are hydrocarbons as these are the class of chemical compounds that are within the experience base of the airframe and turbine engine producers that bear the highest responsibility for aircraft design and manufacture. Other classes of chemical compounds that are not defined as fuel additives are expressly not allowed to be blend components for producing jet fuel as they may affect aircraft performance in ways that are not anticipated. ASTM D4054 [4] was recently amended to provide guidance on how new alternative jet fuel blend components need to be evaluated so that approval by regulatory bodies, airframe manufacturers, turbine engine manufacturers, and users (airlines, military, etc.) is timely. Approval of new jet fuel blend components is not swift, as SPK required years of effort to achieve an ASTM specification. But HEFA was approved much more quickly, and other alternative fuels are currently under evaluation and will be approved if they meet all fit-for-purpose requirements. The Commercial Aviation Alternative Fuel Initiative (CAAFI) is an independent US non-profit organization created by the United States Federal Aviation Administration, Airlines for America, the Aerospace Industries Association, and Airports Council International for the purpose of educating governments, industries, and the public about alternative aviation fuels. The CAAFI website (www.caafi.org) contains a link to the CAAFI ASTM D4054 Users' Guide. [5]. The CAAFI website also contains details as to the process of certifying alternative aviation fuels.

The aviation fuels community has considered the issue of FAME in jet fuel for some time. In Europe, jet fuel is transported in common carrier pipelines that also transport diesel fuel, which is required to contain biomass-derived components. FAME from diesel fuel trailed back in the pipeline and contaminated jet fuel that followed the diesel fuel. This resulted in concerns over the properties of jet fuel delivered to London Heathrow airport and nearly shut down the airport. In 2008, jet fuel contaminated by biodiesel led to an emergency fuel quarantine at Birmingham, UK, disrupting supplies and threatening closure of this UK airport.

In Europe, the Joint Inspection Group approved (on 2 May 2015) FAME as an Identified Incidental Material in jet fuel up to 50 ppm [2]. Efforts are underway to approve up to 100 ppm incidental material of jet with FAME, but final approval by the turbine engine manufacturers has not yet been completed and may not be approved due to concerns about performance of contaminated fuel in their equipment. As flight safety is the major concern, and there is no operational experience with fuel contaminated with FAME at the 100 ppm level, regulatory bodies are justifiably cautious.

4.1.2 Potential Benefits of FAME Blends in Jet Fuel

The actual situation of the production of renewable jet fuels, including the feedstock, the jet fuel input, and the technology used, was summarized recently [6], with a predominance of the HEFA technology over the ATJ or pyrolysis-to-jet and F–T technologies. Moreover, another research team evaluated some blends for use as jet biofuels in the tropics [7], and other authors reviewed the actual situation in aviation biofuels in the frame of the European ITAKA group [8]. More recently, a work was published on blends of sesquiterpanes and 5-methylundecane, generated from biomass sugars by a combination of fermentation and chemical catalysis, which may allow for their production at industrially relevant scales and that show good properties as jet fuel [9].

However, the literature on the production and use of biodiesel for the aviation sector is still scarce, and in some cases, contradictory. Dunn [10] studied the properties of a fuel obtained by blending 10−30% vol. of soybean FAME with JP-8 and JP-8 + 100. Dagaut and Gail [11] examined the oxidation behaviour of a blend of 20% vol. rapeseed FAME with Jet-A1. Korres et al. [12] compared the behaviour of the jet fuel JP-5 against fossil diesel and animal fat biodiesel in a diesel engine. Six FAME from jatropha, croton, calodendrum, coconut, sunflower, and soybean were prepared by Wagutu et al. [13], and they concluded among other findings that the coconut FAME most closely resemble fossil jet fuel properties.

All the accepted alternative jet fuels have a common drawback: they do not have any oxygen in their molecular structures since fatty acid methyl esters (FAME) are not approved additives. However, the presence of oxygen in a fuel has two main advantages: (1) it reduces the carbon content of the fuel, which in turn reduces its carbon footprint, and (2) it reduces the soot formation (emission) of the fuel. Numerous studies, both experimental and numerical, confirm that oxygenated fuel additives reduce soot formation in engines [14]. Llamas et al. [15] make the point that based on smoke point testing, jet fuel containing oxygenates should give lower total particulate matter (PM) emissions than conventional jet fuel. Moreover, a study has found that PM emissions from a plane's engine can fall by almost 40% when researchers blend jet fuel with oxygenated fuels; emissions from planes at large airports are significant sources of local air pollution, including fine PM that can increase people's risk of heart disease and asthma [16].

We believe that it is feasible to blend FAME components with conventional jet fuel at up to 10% and still meet ASTM D1655 specifications. References cited in our papers also indicate that FAME may meet some requirements of jet fuel. Our group has recently published research [15,17,18] suggesting that some specific FAME blends, such as low carbon number saturated fatty acid esters, could be reconsidered as a possible aviation fuel blend component.

We decided to focus our research on distillates of palm-like fruit oil FAME: babassu oil, coconut oil, and palm kernel oil blended with aviation fossil fuel up to 20% vol. Since Jet A-1 was not always available to us, we tested blends with both Jet A-1 and a straight-run atmospheric distillation cut kerosene. We followed the lead of the Virgin Airlines 2008 commercial flight from London to Amsterdam using babassu and coconut FAME. However we performed a vacuum distillation on babassu, palm kernel, and coconut FAME in order to lower, even to eliminate, the content of higher carbon number FAME in the blend, trying to limit the range to methyl myristate or lower, since low carbon number methyl esters would better fit aviation fuel requirements.

FAME is a well-known fuel that can be produced under more benign reaction conditions that do not require as much energy as the FT-SPK and HEFA. Only three low carbon number saturated methyl esters were blended with conventional jet fuel. Similarly to HEFA, the source of the energy crops must be plentiful, economical, and sustainable; however, the recent possibility of obtaining fatty acids from single cell organisms and using genetic engineering could be of great value in the future (see chapter: Feedstocks for Aviation Biofuels). Further research on FAME blends for aviation fuel is needed, and we suggest the stake holders and decision makers involved in the process for new aviation fuels admission – that is, ASTM, regulatory bodies, and fuel providers – not discard FAME as a possible aviation fuel blend component.

With FAME incorporation clearly an issue, there have been concerns in the aviation fuels community about FAME from palm kernel, babassu, coconut, and camelina being blended into jet fuel at up to 10% while still meeting jet fuel performance specifications. Wilson et al. [19] outlined how alternative fuels are approved and why participation in technical society activities such as the ASTM section on Emerging Aviation Turbine Fuels is important to researchers in this area. Wilson pointed out [19] that only hydrocarbons are allowed components in aviation fuels and that oxygenates such as FAME are not allowed as blend components. In their opinion, stating that FAME can meet jet fuel 'specifications' is misleading and ignores both the technical content of ASTM D1655 and ASTM D7566 (as well as Def Stan 91-91) and the implied fit-for-purpose requirements for aviation turbine fuels. The specifications listed in ASTM D1655 and ASTM D7566 are not all-encompassing. Meeting requirements of jet fuel specifications does not make a product Jet A or Jet A-1 fuel. Raw material sourcing and processing must also be considered, and in the case of alternative jet fuel blend components, additional, more stringent testing is included, such as the higher temperature testing requirement for ASTM D3241 Standard Test Method for Thermal Oxidation Stability of Aviation Turbine Fuels (JFTOT) [20].

Llamas et al. [17,18] argued that other tests not recognized by the aviation fuels community could be used to determine if FAME are appropriate as blend components for jet fuel. They proposed that based on a few parameters, FAME blends with jet fuel could be blended at up to 10% and still meet ASTM D1699 standards. Wilson [19] stated that jet fuel is a hydrocarbon fuel and oxygenated components are not allowed. Although this statement does not completely discard biomass-derived components as part of jet fuel now or in the future, it implies that biomass-derived jet fuel blend components must be converted into a hydrocarbon product with properties similar to what has been learned by experience over decades of operation of aircraft on hydrocarbon fuels of fossil origin. Wilson et al. [19] also criticized Llamas et al. [17,18] for not using appropriate tests. They contended that these potential jet fuel blend components must be tested and approved using established protocols as is being done for other potential biomass-based jet fuel blend components such as ATJ products.

ASTM in particular, and the aviation fuels community in general, agree that there is a need to develop alternatives to fossil-derived jet fuel. However, they state that these fuels must be hydrocarbon fuels, and that any other heteroatoms in alternative fuels should not be allowed. It is argued that, as it is simple to hydroprocess fatty oils into hydrocarbons, and even gasification/FT synthesis can be applied to biomass, there is essentially no restriction to the production of blend components for commercial and military aviation turbine engine fuels from a variety of biomass feedstocks.

Additionally ASTM D7566 includes very strict limits on the metals content of alternative fuels to be used for blending ASTM D1655-compliant jet fuel. Individual metals cannot exceed 0.1 ppm (100 ppb), and total metals cannot exceed 1 ppm (1000 ppb). Although there is a paucity of information available in the literature on metals content of FAME biofuels, the data presented by Blakey et al. [21] indicate that FAME fuels may not meet ASTM D7566 requirements. The Blakey et al. papers [21,22] also present information on the impact of FAME on the range of aircraft flight. These data indicate a 20% reduction in range of a Boeing 747 aircraft if fuelled with FAME blends. This impact will naturally depend on the specific properties of the FAME used for comparison, as lighter FAME products have proportionally higher oxygen content and thus lower energy content. Chuck et al. have investigated different jet biofuels, and interestingly, they have calculated their payload versus range [23]. This issue is further developed in Section 3.10.

While some authors have studied the properties of alternative aviation fuels related to their performance in a turbine (heating value, freezing point, oxidation, etc.) [10,12,24], there are no studies related to the effect of this fuel when in contact with airplane materials. Both the engine and the fuel system include many materials of different composition. It is crucial that they be compatible with the fuel they are in contact with in order to guarantee safety at all times.

4.1.3 Objectives

In conclusion, at the present moment FAME is not approved as a jet fuel additive. The maximum allowable level is 50 ppm, which is the officially accepted functional definition of 'Identified Incidental Material' [2]. However, based on the assumption that fatty acid esters could still serve as aviation fuel blend components, two different processes were tested, and they are outlined in this chapter.

In the first approach, FAME from oils rich in shorter-chain fatty acids – such as coconut, palm kernel, and babassu (a palm popular in Brazil) – were synthesized [25]. The fraction that approached the distillation range of fossil kerosene (from 175–185°C to 240–275°C at atmospheric pressure) was then separated by vacuum distillation. Because these pure distilled FAME (DFAME) did not meet aviation jet fuel standards, blends containing 5, 10, and 20% vol. of DFAME with fossil kerosene with and without additives were tested together with pure DFAME.

In the second method a FAME derived from camelina oil (commonly known as gold of pleasure oil) very rich in unsaturated methyl esters (84.7%wt.) was blended without previous distillation at 5, 10, and 20% vol. with fossil kerosene without additives. The excellent cold flow behaviour of these unsaturated methyl esters suggested direct blending without previous distillation. Linseed oil was also selected as a possible feedstock for FAME production because of its average low-chain acids and its high degree of unsaturation.

This chapter also offers a quantification of the potential of FAME blends for the reduction of soot, a common problem with fossil kerosenes. The reduction of sooting tendency, together with benefits from other properties reported in previous works [17,18], make FAME blends worthy of consideration as viable alternatives to aviation fuels.

Finally, the compatibility of three types of materials present in modern airplanes with blends of DFAME and Jet A-1 will be detailed. The fabrication and properties of these blends have been described previously [17,18]. Polymer materials have been tested for traction, resistance, hardness, and mass and dimensional variations. Metals (silver and copper) have been tested for corrosion, and the blends of jet fuel have been tested (before and after immersion of the composites) for acidity, density, viscosity, colour, and Fourier transformed infrared absorption (FT-IR) to evaluate possible compositional changes.

4.2 METHODS USED IN THE TESTING OF FAME COMPONENTS IN JET-A1

4.2.1 Chemical Materials

Refined coconut oil was purchased from Acros Organics (CAS No. 8001-31-8), refined linseed oil (CAS No. 8001-26-1)

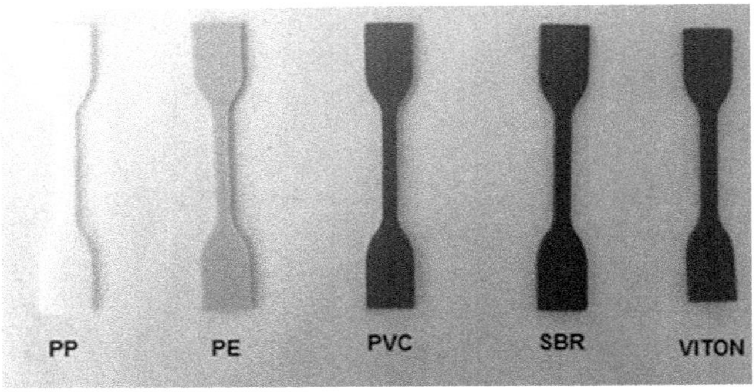

FIGURE 4.1 Samples of polymers [26].

from Fischer Scientific, and crude palm kernel oil and crude babassu oil were supplied by Combustibles Ecologicos Biotel SL. Camelina seeds were supplied by the Laboratory of Fuels & Petrochemistry of the Gomez-Pardo Foundation. Two different kerosene fuels of fossil origin were used to prepare the blends with DFAME: K1 was a straight-run atmospheric distillation cut kerosene, subjected to hydrotreating to remove sulphur compounds, without any additives; K2 was a commercial Jet A-1 with additives, used for civil aviation. Both samples of fossil kerosene were obtained from the Spanish Compañía Logística de Hidrocarburos (CLH). The DFAME were blended at 5, 10, and 20% vol. of DFAME, with the two types of fossil kerosene described above.

All reagents were of commercial grade and were used without any further purification. Dichloromethane (Panreac) was of technical grade. Methanol (Scharlau) was of synthesis grade. *p*-Toluenesulphonic acid (Scharlau), used as a catalyst for the esterification reaction of the crude oils, was also of synthesis grade. Molecular sieve (Scharlau) was of technical purity. The catalytic solution was sodium methoxide 25%wt. in methanol (Acros Organics). Magnesium silicate Magnesol D-60 was kindly supplied by The Dallas Group of America.

4.2.2 Aircraft Materials

The first group of materials tested in this work were polymers, which form an important part of the fuel system of the plane and are under different states of load, pressure, and temperature during their useful life. They are used in the deposits, tanks, valves, joints, and other components of the airplane fuel system. All polymers were purchased from Grupo de Empresas y Talleres (Madrid). Tested polymers were thermoplastics such as polyvinyl chloride (PVC), polypropylene (PP), and polyethylene (PE), and elastomeric polymers such as styrene butadiene rubber (SBR) and VITON, which is a trademark patented by Dupont. To be able to carry out the study, the samples were made according to the standard of tensile test for polymers ASTM D638 (Fig. 4.1 and Table 4.5). They were built with a manual press, model 6051000 CEAST.

The second group of tested materials were metals, silver and copper, both used in the design of the engine. These materials are used to fight corrosion in the components of the fuel system. In aviation, copper oxide is mainly used to form an anticorrosive layer on the underlying layers of metal. It improves their properties at low temperatures, making it suitable for cryogenic applications.

The compatibility test with metals is based on the determination of the corrosion caused by the fuel that is in contact with the metals.

Aviation fuel contains potential corrosive compounds that are limited by the specifications, like mercaptans (maximum 0.002%wt. in Jet A-1) or organic acids. Also, traces of compounds such as sodium, potassium, and other alkaline metals present in the fuel can cause corrosion in the turbine engine. In an attempt to tackle the problem of corrosion, aircraft companies use aluminium to build the fuel deposits, and steel covered by copper and silver to create a protection layer. These two metals were tested for corrosion with our biofuel jet samples.

The third and final group of materials tested were samples of composite materials. Composites have good tensile strength and resistance to compression, making them suitable for use in the manufacture of aircraft parts. The tensile strength of the materials comes from their fibrous nature. The good resistance to compression can be attributed to the adhesive and stiffness properties of the base matrix system. It is the role of the resin to maintain the fibres as straight columns and to prevent them from buckling. In particular, ribs and wings from the airplane provided by Airbus Company were tested. These samples included a main part composed of carbon fibre with a titanium screw electrically isolated, yellow paint on the surface, and phenylene polysulfide in the joints to seal the structure (Fig. 4.2).

4.2.3 Extraction of Camelina Seeds

Camelina seeds are very small dark-yellow grains of elliptical shape (2.09 mm × 1.14 mm). Whole seeds of camelina were dried overnight at 105°C in an oven, and the seeds were extracted with dichloromethane in a 1 L Soxhlet extractor for 24 h. After elimination of the solvent, the oil extracted amounts only to 6%wt. of the dried seeds, and since higher oil yields have been reported for this seed in the literature [27], the seeds were gently ground in an iron mortar, and the extraction procedure was repeated afterwards. The amount of extracted oil from the ground seeds increased to 31.4%wt., closer to the reported values.

4.2.4 Esterification and Transesterification

Transesterification of babassu, camelina, coconut, lineseed, and palm kernel oils was carried out following the experimental procedure previously described in the literature [28]. Refined coconut oil has an acidity index of 1.19 mg KOH/g and a water content of 411 ppm, and the previous esterification step was unnecessary with this feedstock. However, crude palm kernel oil has an acidity index of 4.31 mg KOH/g and a water content of 442 ppm,

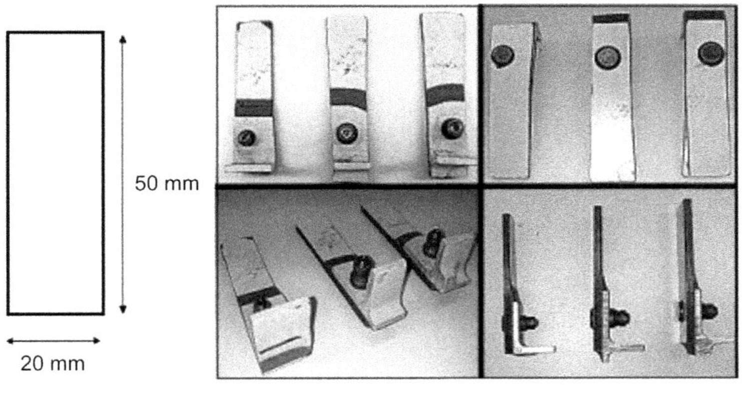

FIGURE 4.2 Samples of composites [26].

and therefore the esterification step was pertinent for reducing the acidity to 0.14 mg KOH/g. The FAME profiles of the coconut and palm kernel biodiesels obtained with this procedure are reported in Table 4.1. Crude babassu and camelina oils also needed the previous esterification step because their acidity numbers were 7.96 and 3.96 mg KOH/g oil, respectively. After esterification, their acidity numbers decreased to 0.24 and 1.58 mg KOH/g oil, respectively. Also after the esterification step, babassu oil was dried over a 4 Å molecular sieve (8%wt. of sieve based in the weight of oil) due to its high water content (984.1 ppm), which could have potentially caused the formation of soaps. It was filtered prior to transesterification. Camelina oil was essentially dry due to the previous drying of the seeds. The FAME profiles of the babassu and camelina biodiesels obtained with this procedure are also reported in Table 4.1. The FAME profile of the coconut, palm kernel, and babassu biodiesels shows that the main constituent is methyl laurate, followed by methyl caprylate and methyl caprate in much smaller amounts. The quantity of FAME higher than C16 does not reach high values, and therefore these feedstocks are appropriate for use as DFAME.

The camelina and linseed oils were different feedstocks. The FAME profile of these feedstocks is unsaturated because in this case the amount of unsaturated FAME reached more than 80%wt. with more than 30%wt. of methyl linolenate (C18:3). The good cold behaviour of unsaturated FAME [29] makes these biodiesels (without distillation) appropriate for the production of biofuels. Table 4.1 shows the FAME profiles of the pure DFAME from babassu (BDFAME100), camelina (CAMFAME100), coconut (CDFAME100), linseed (LINFAME100), and palm kernel (PDFAME100) oils.

4.2.5 Distillation

The DFAME fractions of the babassu, coconut, and palm kernel biodiesel fuels were obtained by fractional distillation using a 41 cm long × 3.5 cm of Vigreux column. The distillation was carried out at 2 torr (2.67 hPa) using a rotary vacuum pump, and the boiling point range of the DFAME was converted to atmospheric equivalent temperature (AET) of distillation using the procedure recently described [30]. The babassu DFAME (BDFAME100) distilled between 47°C and 124°C at 2 torr, which transformed to AET was between 273°C and 359°C, with a yield of 73.3%wt. with respect to the babassu biodiesel fuel, leaving 19.4%wt. yield of a bottom fraction (babassu bottom) that was also characterized. This boiling range is around 100°C higher than that of fossil kerosene, which ranges between 175–185°C and 240–275°C. This situation also happens with biodiesel, which has a much higher distillation temperature than fossil diesel. However, the ASTM D1655 standard fixes a final boiling point temperature for fossil kerosene of 300°C (maximum), and therefore the babassu DFAME produced here surpasses this point at around 60°C. The coconut DFAME (CDFAME100) distilled between 47°C and 114°C at 2 torr (2.67 hPa), which transformed to AET was between 273°C and 349°C with a yield of 81.8% wt. based on the coconut FAME, leaving 12.6% wt. yield of a bottom fraction (coconut bottom) that was also characterized. This boiling range was around 100°C higher than that of fossil kerosene, which ranges between 175–185°C and 240–275°C; therefore, the coconut DFAME produced here surpassed this point at around 50°C. The palm kernel DFAME (PDFAME100) distilled between 35°C and 113°C at 2 torr (2.67 hPa), which transformed to AET was between 259°C and 348°C, again around 100°C higher than the distillation range of fossil kerosene. Thus, the palm kernel DFAME produced here did not meet standards. The yield of palm kernel DFAME was 40.8% based on the palm kernel FAME, leaving a 58.6% yield of a bottom fraction (palm kernel bottom) that was also analyzed.

TABLE 4.1 FAME Profiles of the Biodiesels and DFAME

FAME, (%wt.)[a]	Boiling Point[b] (°C)	Coconut		Palm Kernel		Babassu		Camelina	Linseed
		FAME	DFAME	FAME	DFAME	FAME	DFAME	FAME	FAME
C8:0	193^{760}	8.3	17.3	Traces	3.6	3.91	13.3	–	–
C10:0	224^{760}	3.0	7	Traces	3.5	3.3	11.3	–	–
C12:0	262^{766}	55.5	66.7	56.6	90.8	20.9	69.3	0.84	–
C14:0	295^{751}	14.9	8.9	15.0	2.1	31.3	5.58	0.41	–
C16:0	416^{747}	6.4	0.1	8.5	–	18.0	0.61	5.83	5.9
C18:0	442^{747}	2.5	–	1.7	–	5.51	–	2.97	4.1
C18:1	219^{20}	7.5	Traces	17.1	–	12.5	–	15.8	19.1
C18:2	215^{20}	1.9	–	1.1	–	0.21	–	18.9	17.2
C18:3	$182^{4.0}$	–	–	–	–	3.45	–	34.0	53.5
C20:0	215.5^{13}	–	–	–	–	0.19	–	1.54	–
C20:1	–	–	–	–	–	0.72	–	16.0	–
C22:0	240^{13}	–	–	–	–	–	–	2.2	–
C22:1	$174^{0.266}$	–	–	–	–	–	–	1.4	–
Iodine value, IV	–	9.9	0	16.8	0	20.7	0	148.9	186.5
Mean formula	–	$C_{13.5}H_{26.8}O_2$	$C_{12.1}H_{24.2}O_2$	$C_{14.5}H_{28.8}O_2$	$C_{12.8}H_{25.5}O_2$	$C_{15.1}H_{29.8}O_2$	$C_{12.2}H_{24.4}O_2$	$C_{19.2}H_{35.0}O_2$	$C_{18.9}H_{33.5}O_2$
Molecular weight (g/mol)	–	220.876	201.950	235.584	211.051	242.980	202.860	298.153	292.403
Stoichiometric air/fuel ratio	–	12.004	11.760	12.167	11.891	12.240	11.775	12.508	12.409
C[c] (% wt)	–	73.29	72.06	74.11	72.65	74.50	72.13	77.45	77.52
H[c] (% wt)	–	12.22	12.09	12.31	12.19	12.34	12.11	11.82	11.54
O[c] (% wt)	–	14.49	15.84	13.58	15.16	13.17	15.77	10.73	10.94

[a] *All compositions were normalized to 100%wt.*
[b] *Superscripts are the distillation pressures in mmHg.*
[c] *Calculated from the mean formula.*

The boiling points of FAME of interest are also reported in Table 4.1. The boiling points of methyl laurate and methyl myristate are, respectively, 262^{766} and 295^{751}°C, and that of methyl palmitate $415-418^{747}$°C (superscripts are the distillation pressures in torr). The coconut DFAME produced here was mainly composed of the two former, with no significant amounts of the latter, as can be corroborated by the ester profile. Although the AET distillation curve for biodiesel simulated from the individual FAME boiling points presented lower values than the experimental one [30], in this case the opposite was obtained. In fact, it is surprising that although coconut DFAME produced here was 17.0%wt. of C8:0 FAME (b. p. 192.9^{760}°C) and 7.0%wt. of C10:0 FAME (b. p. 224^{760}°C), the distillation started at 273°C, thus indicating the formation of an azeotropic mixture of maximum boiling point. In the case of palm kernel DFAME, the distillation started earlier (259°C), almost coinciding with the boiling point of methyl laurate (b.p. 262^{766}°C), but also finished at 348°C, although no higher FAME than C14:0 could be found in the ester profile, also indicating the formation of azeotropic mixtures of maximum boiling point. This fact was again opposite to the finding that the AET distillation curve simulated from the FAME individual boiling points was lower than the experimental one [30]. The camelina FAME (CAMFAME100) and linseed FAME (LINFAME100) were used without previous distillation.

4.2.6 Blending

The DFAME were blended at 5, 10, and 20% vol. of DFAME, with two types of fossil kerosene: (1) a straight-run atmospheric distillation cut (hydrotreated) in the boiling range of kerosene, without any additives in it (K1), and (2) commercial jet fuel Jet A-1 (K2). To prepare the blends of fossil kerosene (K1 and K2) and 5, 10, and 20% vol. of FAME, pipettes, beakers, and volumetric flasks with the maximum precision (DIN 12600) were used. These amounts were established in order to evaluate a progressive incorporation of these renewable fuels into the aviation jet fuels.

In kerosene K2, additives are used to prevent the formation of oxidation deposits in aircraft engines and fuel systems, to inhibit steel corrosion, to improve fuel lubricity and oxidation stability, to increase the electrical conductivity of the fuel, to avoid microbiological contamination, to prevent ice formation in fuel systems, and to assist in detecting fuel leaks. The ASTM D1655 standard fixes the allowed additives in jet fuels [1].

Only the specifications considered essential among those required by the standard ASTM D1655 were tested. Of course, all these standard specifications and the fit-for-purpose properties included in the *Handbook of Aviation Fuel Properties* published by the Coordinating Research Council (CRC) [31] should also be met before any test flight.

The following tables list the values of the properties measured for blends of palm kernel (PDFAME100) and babassu DFAME (BDFAME100) with kerosene without additives K1 (Table 4.2), for blends of coconut DFAME(CDFAME100) and camelina FAME (CAMFAME100) with kerosene without additives K1 (Table 4.3), and for blends of palm kernel (PDFAME100) and babassu DFAME (BDFAME100) with kerosene with additives Jet A-1 K2 (Table 4.4).In these tables, the standard procedures, equipment used, and experimental errors are also shown.

4.2.7 Experimental Setup, Conditions, and Calibration of the Smoke Point

Smoke point is defined as the height in millimetres of the highest flame produced without smoking soot breakthrough when the fuel is burned in a specific test lamp [32]. Numerous studies [33–35] have defined the sooting

TABLE 4.2 Properties of the Blends of Palm Kernel (PDFAME) and Babassu (BDFAME) DFAME With Fossil Kerosene Without Additives (K1)

		Palm Kernel DFAME (PDFAME)						Babassu DFAME (BDFAME)							
% DFAME		0	5	10	20	100	Equipment	0	5	10	20	100	Equipment	Method	Experimental Error
Colour and aspect		Clear	Clear	Clear	Clear	Clear	Visual	Clear	Clear	Clear	Clear	Clear	Visual	ASTM D1500	± 0.5 u.c
Acidity (mg KOH/g)		–	–	–	–	–	–	–	0.022	0.033	0.034	0.033	Manual	EN ISO 14104	± 0.02
Water content (mg/kg)		–	–	–	–	–	–	595.2	202.0	197.2	95.35	343.5	Karl-Fischer 831KF	ASTM D1774	< 0.005
Elemental composition	C (%)	85.90	84.79	84.17	82.73	71.84	LECO CHN-600	85.90	84.98	84.79	82.97	73.38	LECO CHNS-932	ASTM D5291	± 0.71
	H (%)	13.88	13.10	13.82	13.13	12.51		13.88	13.16	13.11	12.95	11.55			± 0.18
	O (%)	0.22	2.11	2.01	4.14	15.65		0.22	1.68	1.95	3.93	14.86			–
	N (%)	–	–	–	–	–		0.02	0.02	0.02	0.02	0.11			± 0.11
	S (%)	–	–	–	–	–		0.17	0.16	0.13	0.13	0.10			± 0.01 + 0.06 · mean
Density at 15°C (kg/m^3)		793.0	800.0	804.2	812.0	–	Hydrometer	793.0	805.9	809.2	816.6	874.5	DigitalDM48	ASTM D1298	± 0.3
Higher heating value (MJ/kg)		47.40	45.31	45.08	44.01	37.57	LECO AC300	47.70	45.63	45.16	44.66	37.41	LECO AC300	ASTM 240	± 0.25
Lower heating value (MJ/kg)		44.44	42.51	42.13	41.20	34.89	–	44.44	42.82	42.36	41.89	34.93	–	ASTM 240	–
Energy density (GJ/m^3)		35.23	34.01	33.88	33.45	–		35.23	34.51	34.28	34.21	30.55			
Viscosity at −20°C (mm^2/s)		3.26	3.30	3.49	3.85	7.16	Cannon-Fenske Proton 4378	–	–	–	–	–	Cannon-Fenske Proton 4378	EN ISO 3104	± 0.20%
Viscosity at 40°C (mm^2/s)		–	–	–	–	–		–	1.15	1.46	1.24	2.13		EN ISO 3104	± 0.20%
Flash point (°C)		12.0	16.0	21.0	21.0	–	Herzog 1088	12.0	27.5	30.0	27.1	50.0	Petrotest PMA4	EN ISO 3679	± 1.2
Freezing point (°C)		− 42.5	− 42.5	− 43	− 41	− 15.3	M. Belenguer 528.01	− 42.5	–	–	–	–	ATPM	ASTM D2386	± 3.0

(Continued)

TABLE 4.2 (Continued)

	Palm Kernel DFAME (PDFAME)						Babassu DFAME (BDFAME)							
% DFAME	**0**	**5**	**10**	**20**	**100**	**Equipment**	**0**	**5**	**10**	**20**	**100**	**Equipment**	**Method**	**Experimental Error**
Smoke point (mm)	19.00	21.00	22.50	24.17	–	Analis47551	19.00	18.33	20.33	23.33	22.80	Analis47551	ASTM D1322	± 0.5
Cloud point (°C)	–	–	–	–	–	–	–	− 28	− 28	− 31	− 7	ATPM	ASTM D2500	< 3.0
Pour point (°C)	–	–	–	–	–	–	–	− 38	− 33	− 30	− 23	ATPM	ASTM D2500	< 3.0
Cold filter plugging point (°C)	–	–	–	–	− 5.0	ISL ATPEM 11/86	–	–	–	–	–	–	EN ISO 116	± 1.0
Copper strip corrosion, class	1a	1a	1a	1a	1a	M. Belenguer 534.01	1a	1a	1b	1b	1a	M. Belenguer 534.01	ASTM D130	–
Oxidative stability (h)	–	–	–	–	–	–	–	> 8	> 8	> 8	> 8	Rancimat Methrom 743	EN 14112	0.09 mean + 0.16

TABLE 4.3 Properties of the Blends of Coconut (CDFAME) and Camelina (CAMFAME) FAME With Fossil Kerosene Without Additives (K1)

		Coconut DFAME (CDFAME)						Camelina FAME (CAMFAME)							
% DFAME		0	5	10	20	100	Equipment	0	5	10	20	100	Equipment	Method	Experimental Error
Colour and aspect		Clear	Clear	Clear	Clear	Clear	Visual	Clear	Clear	Clear	Clear	Clear	Visual	ASTM D1500	± 0.5 u.c.
Water content (mg/kg)		47.6	53.8	80.5	113.7	278.9	Karl-Fischer 831KF	–	–	–	–	–	–	EN ISO 12937	± 6.7
Elemental composition	C (%)	85.90	84.84	83.44	82.14	72.35	LECO CHNS-932	85.90	85.29	84.23	83.86	77.09	LECO CHNS-932	ASTM D5291	± 0.71
	H (%)	13.88	13.68	13.37	13.46	12.12		13.88	12.99	12.61	12.45	11.09			± 0.18
	O (%)	0.22	1.48	3.19	4.40	15.54		–	1.65	3.11	3.64	11.74			–
	N (%)	–	–	–	–	–		0.02	0.02	0.02	0.02	0.02			± 0.11
	S (%)	–	–	–	–	–		0.17	0.05	0.03	0.03	0.06			± 0.01 + 0.063 mean
Density at 15°C (kg/m^3)		802.0	805.0	807.0	812.0	867.0	Pyknometer	793.0	806.7	810.8	819.8	889.3	Digital DM48	EN ISO 12185 EN ISO 3675	± 0.3
Density at 23°C (kg/m^3)		–	–	–	–	870.0	Pyknometer	–	–	–	–	–	–	EN ISO 3675	± 0.3
Higher heating value (MJ/kg)		47.40	45.32	44.47	43.94	37.66	Parr 1351	47.70	45.30	45.00	44.80	40.00	LECO AC-300	ASTM 240	± 0.25
Lower heating value (MJ/kg)		44.44	42.39	41.62	41.06	35.06	–	44.44	42.52	42.30	42.14	37.71	–	ASTM 240	–
Energy density (GJ/m^3)		35.63	34.13	33.58	33.34	30.40		35.23	34.30	34.30	34.55	33.46			
Viscosity at −20°C (mm^2/s)		3.26	3.64	3.81	4.18	8.26	Cannon-Fenske Afora 5353/2B	–	3.10	3.20	3.80	6.96	Estimated	ASTM D341	± 0.20 %
Viscosity at 40°C (mm^2/s)		–	–	–	–	2.27		–	1.259	1.308	1.454	4.340	Cannon-Fenske	EN ISO 3104	± 0.20 %
Viscosity at 100°C (mm^2/s)		–	–	–	–	–	–	–	0.680	0.705	0.766	1.708		ASTM D445	± 0.20 %

(Continued)

TABLE 4.3 (Continued)

	Coconut DFAME (CDFAME)						Camelina FAME (CAMFAME)							
% DFAME	0	5	10	20	100	Equipment	0	5	10	20	100	Equipment	Method	Experimental Error
Smoke point (mm)	19.00	18.00	20.67	22.00	–	Analis47551	19.00	19.67	21.67	23.67	–	Analis47551	ASTM D1322	± 0.5
Freezing point (°C)	–	–	–	–	–	–	− 42.5	–	–	–	–	–	ASTM D2500	< 3.0
Cloud point (°C)	–	–	–	–	–	–	–	− 26	− 26	− 22	− 5	ATPEM	ASTM D2500	< 3.0
Pour point (°C)	–	–	–	–	–	–	–	− 40	− 32	− 31	− 5	ATPEM	ASTM D2500	< 3.0
Cold filter plugging point (°C)	− 29.5	− 33.0	− 33.0	− 31.0	− 10.0	ISLFPP 5Gs	–	− 31	− 29	− 26	− 5	ISLFPP 5Gs	EN ISO 116	± 1.0
Lubricity (mm), HFRR test	0.64	0.35	0.28	0.24	0.30	PCS	–	–	–	–	–	–	EN ISO 12156	± 1.0 μm
Oxidative stability (h)	16.0	17.2	18.6	18.6	22.6	Rancimat 743Methrom	–	> 8	> 8	> 8	1.4	Rancimat 743Methrom	EN 14112	± 0.09 mean value +0.16

TABLE 4.4 Properties of the Blends of Palm Kernel (PDFAME) and Babassu (BDFAME) DFAME With Fossil Kerosene With Additives Jet A-1 (K2)

		Palm kernel DFAME (PDFAME)					Babassu DFAME (BDFAME)							
% DFAME		0	5	10	20	Equipment	0	5	10	20	100	Equipment	Method	Experimental Error
Colour and aspect		Clear	Clear	Clear	Clear	Visual	Clear	Clear	Clear	Clear	Clear	Visual	ASTM D1500	± 0.5 u.c.
Acidity (mg KOH/g)		–	–	–	–	–	0.011	0.011	0.011	0.034	0.033	Manual	EN ISO 14104	± 0.02
Water content (mg/kg)		–	–	–	–	–	–	138.6	113.7	346.4	595.3	Karl-Fischer	ASTM D1774	<0.005
Elemental composition	C (%)	84.12	84.47	84.17	82.57	LECO CHN-600	84.12	85.81	84.13	78.35	73.38	LECO CHNS-932	ASTM D5291	± 0.71
	H (%)	14.67	14.24	13.97	14.11		14.67	13.25	13.08	12.64	11.55			± 0.18
	O (%)	1.22	1.29	1.86	3.32		–	0.11	0.13	0.15	0.11			–
	N (%)	–	–	–	–		–	0.10	0.10	0.20	0.10			± 0.11
	S (%)	–	–	–	–		–	0.73	2.56	8.66	14.86			± 0.01 + 0.06 mean
Density at 15°C (kg/m^3)		791.0	802.3	805.5	811.8	Hydrometer	791.0	804.0	807.7	814.9	874.5	DigitalDM48	ASTM D1298	± 0.3
Density at 23°C (kg/m^3)		–	–	–	–	–	–	804.7	807.0	814.7	874.2	DigitalDM48	ASTM D1298	± 0.3
Higher heating value (MJ/kg)		46.04	45.68	45.17	44.21	LECO AC-300	46.04	45.95	45.54	44.75	37.41	LECO AC-300	ASTM D240	± 0.25
Lower heating value (MJ/kg)		42.90	42.64	42.18	41.19	–	42.90	43.12	42.74	42.04	34.93	–	ASTM D240	–
Energy density (GJ/m^3)		33.93	34.21	33.98	33.44		33.93	34.67	34.52	34.26	30.55			
Flash point (°C)		43.0	43.5	45.0	45.5	Herzog 1088	43	43	44	46	50	Petrotest PMA4	EN ISO 3679	± 1.2
Freezing point (°C)		−62.0	−60.0	−48.3	−41.5	M.Belenguer 528.01	−62.0	–	–	–	–	–	ASTM D2386	± 3.0
Cloud point (°C)		–	–	–	–	–	–	−39	−38	−33	−7	ATPM	ASTM D2500	<0.3

(Continued)

TABLE 4.4 (Continued)

	Palm kernel DFAME (PDFAME)					Babassu DFAME (BDFAME)							
% DFAME	0	5	10	20	Equipment	0	5	10	20	100	Equipment	Method	Experimental Error
Pour point (°C)	–	–	–	–	–	–	−43	−42	−32	−23	ATPM	ASTM D2500	<0.3
Smoke point (mm)	23.33	24.33	25.67	26.67	Analis 47551	23.33	23.67	24.67	26.67	22.80	Analis 47551	ASTM D1322	± 5.0
Copper strip corrosion, class	1a	1a	1a	1a	M.Belenguer 534.01	1a	1a	1a	1a	1a	M.Belenguer 534.01	ASTM D130	–
Viscosity at −20°C (mm^2/s)	3.42	3.51	3.67	4.06	Cannon-Fenske Afora 5354/2B	–	–	–	–	–	–	ASTM D445	± 0.20 %
Viscosity at 40°C (mm^2/s)	–	–	–	–	–	–	1.27	1.31	1.32	2.13	Cannon-Fenske	ASTM D445	± 0.0013 mean (0.2 %)
Oxidative stability (h)	–	–	–	–	–	–	>8	>8	>8	>8	Rancimat Methrom 743	EN 14112	± 0.09 mean + 0.16

tendency of fuels and blends as inversely proportional to their smoke point. Consequently, a minimum value is required in jet fuel standards (min. 25 mm). A standard smoke point lamp as specified by ASTM D1322 was used for the smoke point measurements. A steel frame (painted black) surrounding the apparatus was used to control air flow during the measurements. For each fuel, the smoke points were measured three times, averaged, and the standard deviation obtained. Previously, the method was calibrated in accordance with ASTM D1322 using two standard reference fuel blends. The first blend, composed of 15% vol. toluene and 85% vol. 2,2,4-trimethylpentane (iso-octane), led to a smoke point of 25 mm, and the second blend, composed of 40% vol. toluene and 60% vol. 2,2,4-trimethylpentane, led to a smoke point of 15 mm. Comparison of the measured smoke points and those reported in the standard led to a correction factor of 1.006. The ambient temperature and relative humidity of the test site were monitored to make sure that fluctuations of environmental conditions did not affect the smoke points of the blends. The ambient temperature ranged from 21°C to 25°C, and the relative humidity ranged from 47% to 57%. All tests are performed at atmospheric pressure (around 93.5 kPa).

4.2.8 Procedures to Test Materials Compatibility

Different tests were conducted depending on the group of materials.

4.2.8.1 Polymers

Each of the compatibility tests with the polymeric materials was performed with a full immersion of the five samples in the fuels studied. The tests were conducted in order to assess the degree of degradation of the polymer by determining the tensile properties, toughness, discoloration, and variation of the linear dimensions of the specimens. These tests were carried out in sealed vessels and subjected to a temperature of 35°C for a period of 28 days (672 h) in a furnace (EN 400 from NÜVE). The tests, standards, and equipment needed for each test are listed in Table 4.5.

TABLE 4.5 Polymer Tests, Metal Tests, and Characteristics Tested in the Composites After Contact With Fuel Blends

Test	Standard	Equipment	Experimental Error
Tensile test	UNE-ISO 37:2013	INSTRON 3360	± 6.22% (*stress*) ± 7.12% (*strain*)
Hardness test	ASTM D2240-05	Durometer Shore A, model Macro-IRHD-2 (Hildebrand)	± 3.11%
Mass and size variation	–	Balance, model AE 200 (Mettler) and a Vernier caliper	± 0.0002*g* ± 0.01 mm
Silver strip corrosion test	ASTM D7671	Thermostatic bath (Bretteville laboratory)	–
Copper strip corrosion test	ASTM D130	Thermostatic bath (Bretteville laboratory)	–
Acidity	UNE-EN ISO 660	–	± 0.02 mg $\frac{KOH}{g}$
Density	ASTM D4052	Densimeter DMA 48 (Anton Paar GmbH)	± 0.3 kg/m^3
Kinematic viscosity at −20°C	ASTM D445	Ostwald-Cannon-Fenske viscometer	± 0.20%
Colour measurement	DEF STAN 91-91	Colorimeter AF 650 (Lovibond)	± 0.5 *u.c.*

4.2.8.2 *Metals*

The compatibility test with metals is based on the determination of the corrosion caused by the fuel in contact with the metals. In an attempt to address the problem of corrosion, aircraft companies use aluminium to build the fuel deposits and steel covered by copper and silver to create a protection layer.

- Silver strip corrosion test: A silver strip was completely immersed in a volume of fuel for 4 h at a temperature of 50°C. Then the strip was compared with reference strips in standard ASTM D7671 [36].
- Copper strip corrosion test: Similar to the previous test, the copper strips were immersed in a volume of fuel for 4 h at a temperature of 50°C and compared with reference strips in standard ASTM D130 [37].

4.2.8.3 *Composites*

The compatibility with these materials was studied via a different approach. Instead of measuring the changes in the composites caused by the immersion in the fuel samples, the effect of composite contact with the fuel was measured. The samples were immersed in different blends for 90 days (2160 h) at a temperature of 50°C using a furnace (EN 400 NÜVE). The fuel degradation was measured by analyzing the oxidation and nitration bands obtained after undergoing compatibility tests measured with FT-IR. Four samples were taken, the first one at the beginning of the test and the others every 30 days. The rest of the characteristics measured before and after the immersion are summarized in Table 4.5.

4.3 APPLICATION AND TESTING OF SPECIFIC FAME BLENDS

4.3.1 Oxidation Stability

Stability to oxidation at the operating temperatures found in jet aircraft is an important performance requirement. For this measurement, the Rancimat method (EN 15751) specified in the EN 14214 European standard for biodiesel was used. This method uses a temperature of 110°C for the test. In 2015, a new test at 120°C was approved (EN 16568), and this temperature is closer to the temperature of the JFTOT procedure, presented below. All the blends of babassu DFAME and fossil kerosenes K1 and K2 showed induction periods greater than 8 h, and so did the pure babassu DFAME. The blends of coconut DFAME and fossil kerosene K1 also demonstrated an induction period greater than 8 h, while the pure coconut DFAME was 22.6 h. This agrees with the strongly saturated character of the FAME constituents of these DFAME. On the other hand, camelina FAME had a low induction period of 1.4 h due to its highly unsaturated character, though the blends of camelina with fossil kerosene K1 demonstrated induction periods greater than 8 h.

Although some of the tested oils (babassu, coconut, and palm kernel) led to methyl esters that could be suitable for blending with aviation kerosene, a large commercial scale would probably make them environmentally unsustainable [38], and therefore, different production alternatives (eg, via microbial processes) should be explored. However, publication of these results has stirred up a controversy over the use of these FAME as blend components for jet fuel. According to Wilson et al. [19]: 'It is possible to produce FAME blends that will pass the standard test (they refer to ASTM D3241 Standard Test, JFTOT procedure) but that would just open a new line of research to determine if the method provides suitable results with such a fuel'. Of note, the aim of this book is precisely that – to open new lines of research in the field of aviation biofuels.

On the other hand, considering that the Rancimat method is not an approved method for measuring the oxidation stability of jet fuels, in this test, the fuels were heated at

110°C (to pass the test the induction time should be higher than 8 h). In comparison, in the ASTM 3241 test (JFTOT procedure) the duration of heating at 260°C (325°C for SPK) is only 2.5 h. Therefore, the Rancimat temperature is lower, but the test time is almost four times longer. The inverse relation between temperature and induction time, together with the evidence that a lower temperature assures better reproducibility of the test [39], makes the Rancimat an appropriate test for comparing the oxidation stability of FAME-blended aviation fuels. Other methods such as the rapid small-scale oxidation method [40] (which is not approved for jet fuels either) could be another option.

4.3.2 Fuel Metering and Aircraft Range

4.3.2.1 Density

This test is relevant for mass-volume relationships in most commercial transactions, and for the empirical assessment of the heating value, a low density meaning a low heating value per unit volume [1]. All the blends fitted within the limits fixed in the standard (775–840 kg/m^3) (Fig. 4.3), with a neat increase when the amount of DFAME increased. It is notable that the K1 fossil kerosene showed different values with the digital method and the hydrometer method, and that both densities converged at the same value (812 kg/m^3) at 20% vol. of coconut and palm kernel DFAME. Of course, the pure DFAME did not meet the density standard range. Also, the presence of additives did not affect the densities of the palm kernel DFAME blends.

4.3.2.2 Lower Heating Value

This test provided data about the energy obtainable from the fuel. The performance of any jet aircraft depends on the availability of a minimum amount of energy. A reduction of this minimum amount of energy brings about an increase in fuel consumption and a reduction of the flight distance [1]. Only the blends of babassu DFAME at 5% vol. with fossil kerosenes K1 and K2 reached the minimum value of 42.8 MJ/kg, and the 10% vol. blend of babassu with Jet A-1closely approached the specification with a value 42.74 MJ/kg, but not the pure babassu (34.93 MJ/kg) or camelina (37.71 MJ/kg) DFAME, nor any camelina blend (Tables 4.2–4.4). Any blend of fossil kerosene with coconut and palm kernel DFAME reached this minimum value of 42.8 MJ/kg, less the pure coconut (35.06 MJ/kg) or palm kernel (34.89 MJ/kg) DFAME. Only the 5% vol. blends closely approached the specification, with

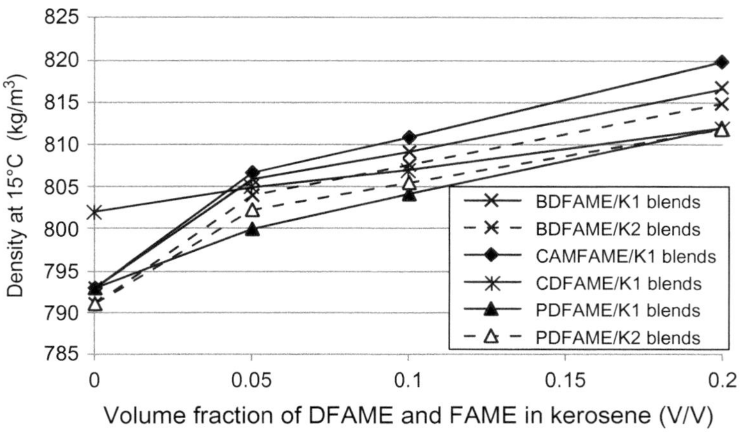

FIGURE 4.3 Density at 15°C of fossil kerosenes (K1 and K2) and blends with babassu DFAME (BDFAME), palm kernel DFAME (PDFAME), coconut DFAME (CDFAME), and camelina FAME (CAMFAME).

values higher than 42.4 MJ/kg. The lower heating value (*LHV*) showed a linear relationship to the mass fraction of DFAME, thus assuring the absence of enthalpy excess in the blending.

Tables 4.2–4.4 show the oxygen content, the density of energy on a volume basis, and the smoke point for these blends and for Jet A-1. The *LHV*, the density (*d*), and the energy density (*ED*) for the already ASTM-approved renewable jet fuels are as follows: FT-SPK, $LHV = 44.2$ MJ/kg, $d = 759$ kg/m^3, and $ED = 33.55$ GJ/m^3; HEFA, $LHV = 44.3$ MJ/kg, $d = 770$ kg/m^3, and $ED = 34.11$ GJ/m^3 [6]. The *LHV* of these DFAME blends is lower than that of Jet A-1, but as their density is also higher, the *ED* (GJ/m^3 fuel) is always higher (or of the same magnitude, except for pure babassu and coconut DFAME) than Jet A-1. It is also higher than that of the biokerosenes approved by the ASTM, FT-SPK, and HEFA. This important fact could make these blends as suitable for commercial use as other biokerosenes, but with the additional presence of oxygen that is always beneficial for fuels as explained above. Since fuels are injected in a volumetric dosage, the fact that the same volume of DFAME blend injected produces the same amount of energy as Jet A-1 could overcome some drawbacks of these new fuels [21,23].

Wilson et al. [19] stated that ASTM D240 is not an appropriate test for measuring the *LHV* of aviation turbine fuels, and they proposed using ASTM D4809 instead. On the contrary, we believe that none of these methods are appropriate for oxygenated fuels. Both standard methods, ASTM D240 and ASTM D4809, use the same equation to calculate the *LHV*:

$$LHV(\text{net, } 25^\circ\text{C}) = HHV(\text{gross, } 25^\circ\text{C}) - 212.2H \tag{4.1}$$

where *LHV* is the net heat of combustion at constant pressure (*LHV*) in kJ/kg, *HHV* is the gross heat of combustion at constant volume (higher heating value) in kJ/kg, and *H* is the mass % of hydrogen in the sample. Eq. (4.1) was derived from simple combustion stoichiometry and considers the thermodynamic conversion from constant volume to constant pressure. However, this equation does not consider the possibility of any oxygen content in the fuel. A similar combustion/thermodynamic approach leads, in the case of oxygenated fuels, to the following equation:

$$LHV(\text{net, } 25^\circ\text{C}) = HHV(\text{gross, } 25^\circ\text{C}) - 213.65H - 0.77O \tag{4.2}$$

where *O* is the mass % of oxygen in the fuel sample. Eq. (4.2) (and not Eq. (4.1)) was used in our previous studies [17,18]. It is clear that the values obtained with Eq. (4.2) are always lower by 20–30 kJ/kg, and therefore, they are more conservative than those obtained by the standard official method.

4.3.3 Kinematic Viscosity

Kinematic viscosity is a critical performance parameter for aviation jet fuels. The kinematic viscosity of jet fuel is related to pumpability at the operating temperature (often below −20°C), and also the ability to lubricate the pump.

The kinematic viscosities at −20°C were measured for the 5, 10, and 20% vol. blends of coconut and palm kernel DFAME with fossil kerosenes [18], and are shown in Fig. 4.4. The kinematic viscosity of the same blends of camelina FAME and fossil kerosene cut K1 were measured at 40°C and 100°C, and extrapolated to −20°C according to the literature [41]. All the above blends showed small nonlinearities with the volume fraction of DFAME and camelina FAME, and they all also met the ASTM D1655 specification (<8 mm^2/s).

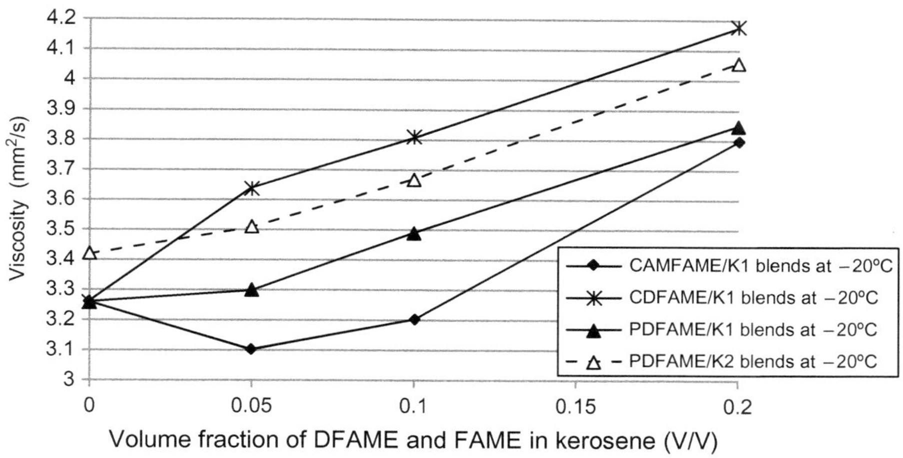

FIGURE 4.4 Kinematic viscosity of blends of fossil kerosenes (K1 and K2) with palm kernel distilled FAME (PDFAME), coconut distilled FAME (CDFAME), and camelina FAME (CAMFAME).

4.3.4 Fluidity at Low Temperature

4.3.4.1 Freezing Point

This property is extremely important for jet fuels, and should be low enough to avoid interference with the jet fuel flow through the filters at the low temperatures observed at high flying altitudes [1]. The test used to evaluate the cold flow behaviour of kerosene is the manual freezing point method (ASTM D2386) [42], which measures the temperature of apparition of the first crystals, with values of −40°C for Jet A and −47°C for Jet A-1. The use of pure DFAME was discarded because their FAME profiles anticipated high crystallization points. For this property, only the 5 and 10% vol. blends of palm kernel DFAME and K2 met the specification of Jet A-1 (−47°C). On the other hand, all the blends met the specification of Jet A (−40°C), although by a very narrow margin. The freezing point of the blends of palm kernel DFAME and fossil kerosene without additive K1 remained almost constant at around −43°C, independent of the amount of DFAME added. However, the freezing point of the blends of this same DFAME and fossil kerosene with additive K2 increased with the volume fraction of DFAME following a linear trend in the studied range of 0–20% vol. of DFAME (Fig. 4.5):

From this it can be inferred that at low DFAME fractions, the presence of additives in the fossil kerosene lowered the freezing point, but at higher fractions (20% vol.), the presence of additives in the fossil kerosene did not have any influence. Similar trends can be found in the literature for the cold flow behaviour of blends of biodiesel and fossil diesel, although these authors extended their study to ranges from 0 to 100% biodiesel, reaching second-order polynomial regressions [43].

4.3.4.2 Cloud Point and Pour Point

In our own experience, the freezing point test has low reproducibility, and the results depend to a great extent on the ability of the operator. Thus, we have checked the cold flow behaviour of the blends of babassu and camelina with fossil kerosene by the cloud point and the pour point test methods.

The cloud point test (ASTM D2500) [44] measured the temperature at which the solution became opaque to light by formation of the first

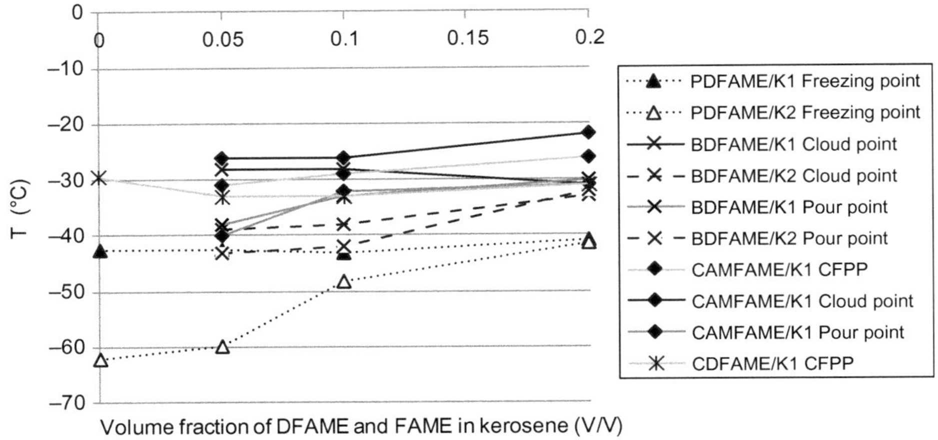

FIGURE 4.5 Cold flow behaviour: Cloud point, freezing point, pour point, and cold filter plugging point of fossil kerosenes (K1 and K2) and blends with babassu distilled FAME (BDFAME), palm kernel distilled FAME (PDFAME), and camelina FAME (CAMFAME).

crystals. If it is assumed that this property is similar to the freezing point, only the blends with 5 and 10% vol. of babassu DFAME with Jet A-1 would meet the specification of Jet A fuel (−40°C), but not that of Jet A-1 (−47°C). It is interesting that the presence of additives in Jet A-1 K2 lowered the cloud point at 5 and 10% vol. of babassu DFAME (Fig. 4.5), but at 20% vol. of the latter, the influence of these additives disappeared and both blends with K1 and K2 had the same value (−32 ± 1°C). This same effect could be observed with the pour point (−32 ± 1°C). For all the blends, the cloud point increased following a linear trend with the volume fraction of FAME, either derived from babassu or from camelina.

The pour point test (ASTM D97) [45] measured the temperature at which the solution ceased to flow, and it was usually lower than the cloud point. The pour point of all the babassu and camelina blends also showed a linear trend with the volume fraction of FAME. Similar relationships can be found in the literature for the cold flow behaviour of blends of biodiesel and fossil diesel [46].

4.3.4.3 Cold Filter Plugging Point

For the blends of coconut DFAME and K1, the automatic cold filter plugging point (CFPP) method was examined, as, in our opinion, this is more reliable than the manual freezing point method. CFPP is not a specific test for jet fuels, but it does give an indication of the cold flow properties of the blends [47]. The CFPP revealed essentially no change in the range of 0–20% vol. of coconut DFAME. Moreover, the experimental error in the CFPP measurement was ±1°C, and the temperature interval between filtrations in the tester was also 1.0°C. Thus, the maximum difference observed of 3.5°C in the measurement of the blends was of the same order of magnitude of this error. The automatic CFPP method was also used for the blends of camelina FAME and fossil kerosene cut K1. The CFPP was between the cloud point (higher) and the pour point (lower), and its variation was almost parallel to that of the cloud point (Fig. 4.5).

Wilson et al. stated [19] that 'only freeze point tests, such as ASTM D2386, D5972, or D7153, are acceptable tests for determining

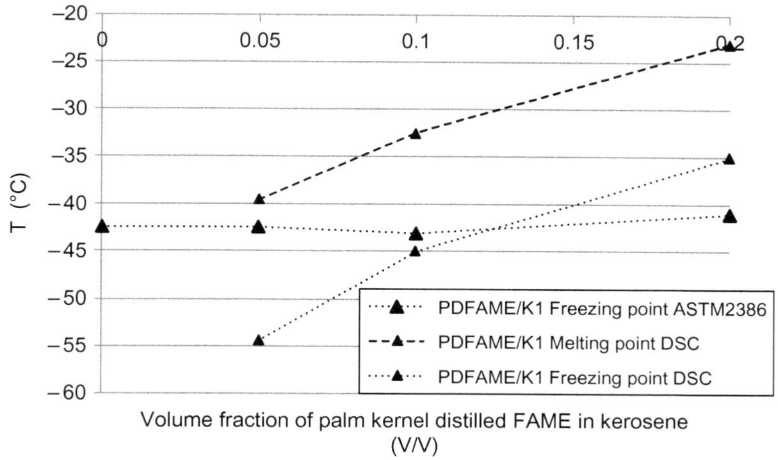

FIGURE 4.6 Freezing points of the blends of palm kernel distilled FAME (PDFAME) with fossil kerosene without additives (K1), measured after ASTM D2386, and using Differential Scanning Calorimetry.

conformance to ASTM D1655. The term *freeze-point* actually describes the recovery temperature of the fuel, which is the temperature at which the fuel will be in the liquid phase regardless of how cold it has been'.

In our opinion, this last sentence describes the 'melting point' of the jet fuel, not the 'freezing point'. To clarify this point, additional differential scanning calorimetric (DSC) analyses were made on the palm kernel DFAME blends with kerosene K1, going from ambient to low temperature (−90°C) and coming back to ambient temperature. A hysteresis always appeared between the freezing point (temperature where the first crystals appear) and the melting point (temperature where all crystals disappear and the fuel return to liquid phase), the freezing point always being lower than the melting point.

Moreover, the freezing point was measured following ASTM D2386 [42] for the blends of palm kernel DFAME with fossil kerosene K1, which showed almost no change from pure K1 to a blend with 20% vol. of biofuel (−42 ± 1°C), while only the pure biofuel showed a substantial increase to −15.3°C. This clearly indicates the difficulties of the measuring procedure [17,18], and that led us to use the cloud point [44], the pour point [45] and the CFPP [47] to check the cold flow behaviour of those blends more precisely. On the contrary, both the DSC-measured freezing and melting points (Fig. 4.6) showed the expected linear increase with the increasing concentration of biofuel (with the above-mentioned hysteresis in between) in agreement with other authors [27,43,46].

4.3.5 Fuel Lubricity: High Frequency Reciprocating Rig (HFRR) Test

Lubricity is defined as the effectiveness of the fuel to lubricate the sliding parts in the aircraft/engine fuel systems. The failure of a jet fuel to meet this specification could cause reduction in the pump flow, or worse, severe mechanical failure leading to in-flight engine shut down. The sensitivity of the aircraft parts to jet fuel lubricity varies with their construction materials, and the intrinsic lubricity of the jet fuel varies with its composition. The general physico-chemical properties of the jet fuel renders it a poor lubricating material, aggravated by the hydrodesulfurization process intended for reducing the sulphur content to

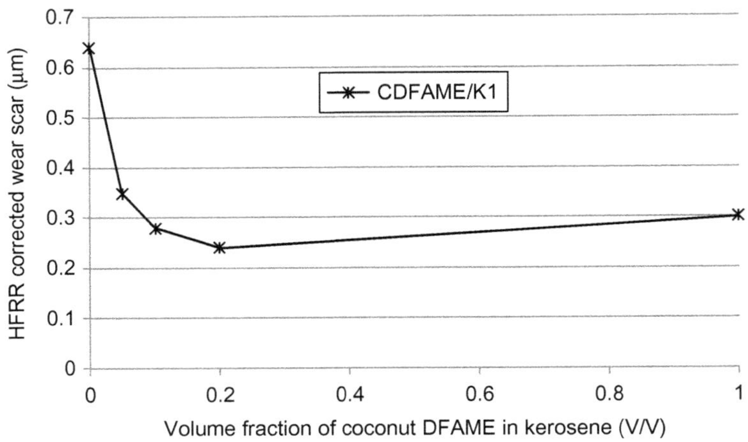

FIGURE 4.7 Fuel lubricity of coconut distilled FAME (CDFAME) blends with fossil kerosene K1.

ultra-low sulphur levels, which are now required for all petroleum fuels [1]. Although the ASTM D1655 standard does not specify any limit for jet fuel lubricity, most modern aircraft fuel systems have been designed to operate on low lubricity fuel (wear diameter scar up to 0.85 mm). All the blends of coconut DFAME and K1 met this limit, and these blends also showed a synergistic effect because the wear scar of the 10 and 20% vol. blends have lower values than those of the pure components (Fig. 4.7). This synergistic effect was observed in a previous work, related to blends of castor oil biodiesel and fossil diesel [48].

4.3.6 Fuel Handling: Flash Point

The flash point is defined as the lowest temperature at which a liquid produces enough vapour to ignite in the presence of a source of ignition, and it is one of the most significant properties of flammable liquids in industrial processes when evaluating process safety [49]. This test gives an indication of the maximum temperature for fuel shipment, storage, and handling without serious fire hazard. ASTM D1655 fixes a minimum flash point for jet fuels of 38°C, but military aircraft regulations establish a higher minimum value of 60°C for JP-5 [1]. All the blends of palm kernel or babassu DFAME and K2 met the ASTM D1655 specification, but not the JP-5 one, though neither of the DFAME blended with K1 met the JP-5 or Jet A-1 specification (Fig. 4.8). This proved that for DFAME concentrations below 80%, the flash point of blends of fossil kerosene and FAME were strongly influenced by the flash point temperature of the fossil kerosene [49]. In any case, the DFAME content always contributed to an increase in the flash points of the blends.

4.3.7 Fuel Cleanliness and Contamination: Acidity and Water Content

Water in a jet fuel onboard an aircraft represents a threat to flight safety and can cause long-term problems such as wear, corrosion, and plugging of filters and other parts [1]. FAME is in general a highly hygroscopic material that when blended with fossil fuels increases the tendency of the latter to absorb water [48]. Thus, in the blends of coconut DFAME and K1, the water content increased with the volume fraction of DFAME (Tables 4.2–4.4), as was expected, because this water came from the 'bio' fraction. All the blends of babassu DFAME with fossil kerosenes K1 and K2 met the acidity specification

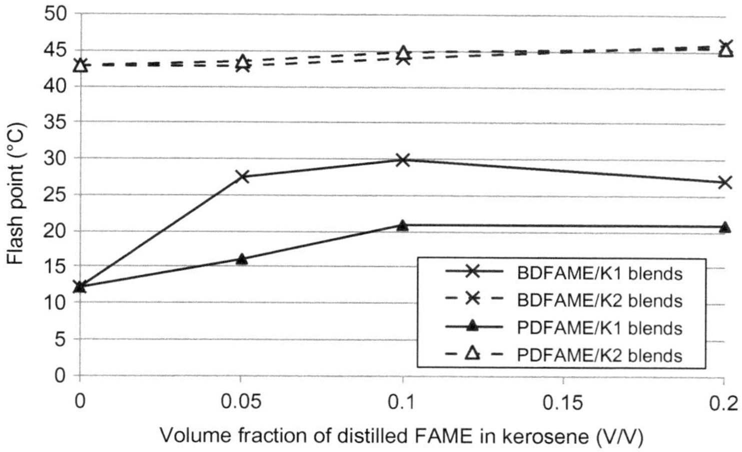

FIGURE 4.8 Flash point of fossil kerosenes (K1 and K2), and blends with babassu distilled FAME (BDFAME) and palm kernel distilled FAME (PDFAME).

(acidity lower than 0.10 mg KOH/g). In the blends of babassu DFAME with K1 and K2 the water content increased with the volumetric fraction of DFAME. The observed results could not be correlated since the storing conditions of the blends were not exactly the same, but all the blends of babassu DFAME met the requirement of ASTM D2709 (max. 0.05% vol) [50].

4.3.8 Colour

Colour could be a useful indicator of jet fuel quality. Colour changes especially should be investigated to determine their cause. Jet fuel colour ranges from colourless (water white) to straw/pale yellow. All the blends tested were clear and colourless (Tables 4.2–4.4).

4.3.9 Fuel Gauging Performance and Other Performance Properties

Aircraft fuel tanks are filled by means of hoses with positive displacement sensors, and once filled, the tanks have pressure sensors that give information of the mass charged in the tank. In both cases, the measurement depends on the density of the fuel, but all the blends described here satisfy the ASTM D1655 requirements for this property. During flight, the measurement of fuel in the tank is carried out by means of capacitive-type sensors that depend on the dielectric constant of the fuel. The scarce literature data about the value of the dielectric constant for blends of fossil diesel and biodiesel give values between 2.0 for diesel and 3.1 for biodiesel [51]. Thus, in case the performance of these blends could be monitored in flight, the necessary adjustment of these instruments should be done.

4.3.10 Payload Range of FAME Blends

Blakey et al. [22] recently published work comparing the performance of conventional Jet A-1 fuel with alternative fuels. In this paper, the most important features were the payload (kg) versus range (km) for three different types of aircrafts. The same study has been carried out here for all the blends of DFAME with fossil kerosenes with and without additives, and a common type of aircraft, the Boeing 747 200B. Figs. 4.9 and 4.10 show this payload (kg) versus range (km) study for the blends of babassu DFAME with K1 and K2. As can be seen, there is no practical difference between using K1 or K2 fossil kerosene. The only

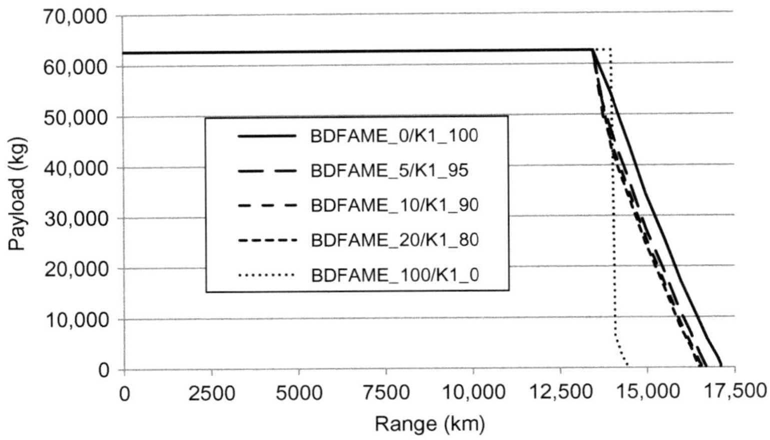

FIGURE 4.9 Payload (kg) versus range (km) for the blends of babassu distilled FAME (BDFAME) with fossil kerosene K1.

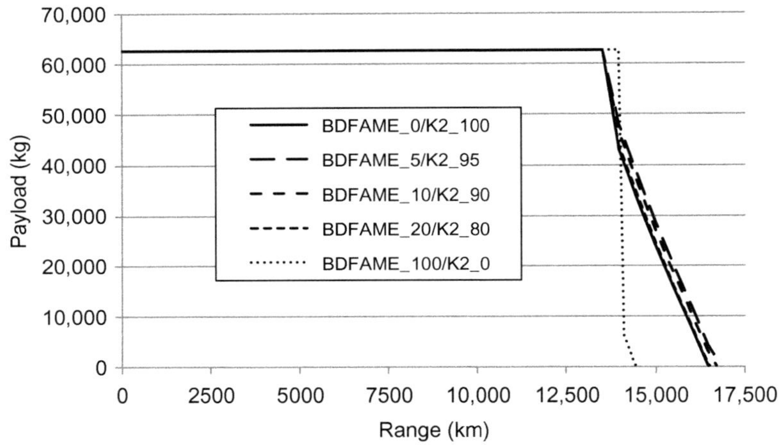

FIGURE 4.10 Payload (kg) versus range (km) for the blends of babassu distilled FAME (BDFAME) with fossil kerosene K2.

remarkable difference was observed when using pure babassu DFAME, which clearly reduced flight range due to the *LHV*, and also the slightly better range of the pure K1 fossil kerosene. In this study, the most favourable Breguet equation model was used for the calculus, which lightly expanded the range when compared with other more conservative models like Raymer and PIANO [22].

4.3.11 Sooting Tendency

Paraffinic hydrocarbons offer the most desirable combustion characteristics of cleanliness for jet fuels. On the contrary, aromatics, although they help avoid leakages from seal swelling, generally have less appropriate combustion characteristics for aircraft turbine fuels because their diffusion flames tend to produce much more soot [1].

This parameter is defined as the height in millimetres of the highest flame produced without smoking soot breakthrough when the fuel is burned in a specific test lamp. Previous studies [34,35] have shown that the sooting tendency of fuels and blends is inversely proportional to their smoke point. Consequently, a minimum value is required in jet fuel standards. All the

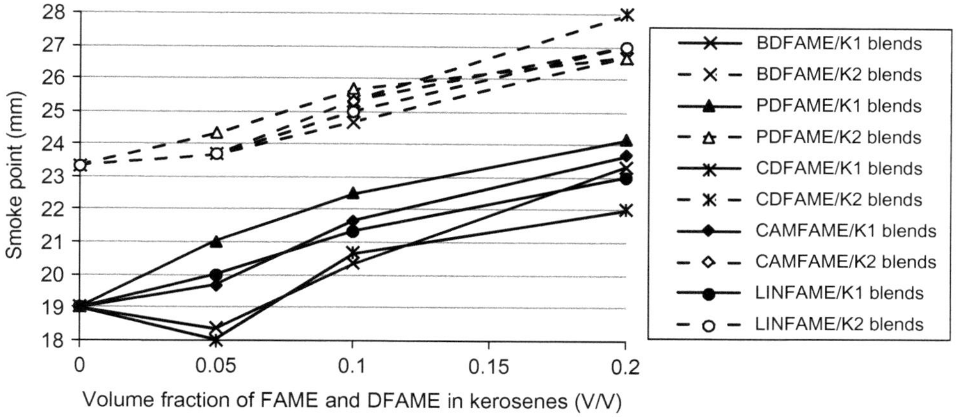

FIGURE 4.11 Smoke points of fossil kerosenes (K1 and K2) and blends with babassu distilled FAME (BDFAME), palm kernel distilled FAME (PDFAME), coconut distilled FAME (CDFAME), camelina FAME (CAMFAME), and lineseed FAME (LINFAME).

blends of palm kernel DFAME with Jet A-1 (K2) met this specification (min. 25 mm), but in the blends with K1, only the blend with 20% vol. of PDFAME met the specification. For all blends, the smoke point varied with the volume fraction of DFAME (Fig. 4.11) and seemed to follow a linear relationship, with increasing smoke points for higher FAME fractions. This agrees with the explanation given above because the saturated FAME were similar to the paraffinic hydrocarbons but very different to the aromatic hydrocarbons, and the increased volume fraction of the FAME linearly increased the smoke point of the blends.

All the blends of babassu DFAME with Jet A-1 (K2) met this specification. As shown in Fig. 4.11, none of the tested fossil kerosenes met the smoke specification. The commercial Jet A-1 (named K2 here) showed a smoke point value close to the limit specified ($SP > 25$ mm) and also similar to the smoke point of other commercial jet fuels such as JP-8, measured by other researchers under similar conditions [52]. However, kerosene K1, directly obtained from straight-run atmospheric distillation cut, was far from meeting specification, thus requiring significant, and therefore expensive, use of additives. Blending with selected and distilled (termed DFAME here) or non-distilled FAME is an effective alternative to increase the smoke point. The paraffinic molecular structure of saturated FAME, enhanced by the inclusion of two oxygen atoms per molecule, together with the absence of aromatic hydrocarbons, favoured the increase of the smoke point of fossil kerosene. In fact, the blends of coconut and palm kernel DFAME, camelina, and linseed methyl esters with Jet A-1 (K2) at 10% vol. met this specification. Only babassu DFAME did not meet it, although it was close. Blends at 20% met the specification in all cases. In almost all instances, the smoke point of the blends increased approximately linearly with the volume fraction of the DFAME. In the case of kerosene K1, a linear extrapolation indicated that around 30% of most of the DFAME would be enough to reach the specification limit. Based on the mean standard deviation (0.425 mm), the described trends can be considered significant.

The inverse correlation between smoke point and sooting tendency of hydrocarbons was formulated by Calcote and Manos [53]. From this correlation, they defined the Threshold Sooting Index (*TSI*), which arbitrarily ranks the fuels from 0 to 100, with 0 being the lowest sooting value (assigned to ethane) and 100 the highest (assigned to naphthalene). The use of *TSI* permitted the use of all the literature data to interpret molecular structure effects and predict the sooting tendency of fuels under internal combustion engine conditions, and especially, of jet fuels in turbine combustors. It was defined as follows:

$$TSI = a\left(\frac{MW_f}{SP}\right) + b \tag{4.3}$$

where a and b are constants for any given experimental setup, MW_f is the molecular weight of the fuel, and SP is the measured smoke point. The inclusion of the molecular weight in the *TSI* correlation was justified by Calcote and Manos [53] because an increase in molecular weight requires more oxygen to diffuse into the flame to consume a unit volume of the fuel, which translates into an increased flame height. However, in the case of oxygenated fuels, the molecular weight used in the *TSI* correlation is not a precise measure of the volumetric Air-To-Fuel ratio required by the flame for stoichiometric combustion. For a generic fuel with a mean formula $C_nH_mO_p$, this requirement is exactly quantified by $n + m/4 - p/2$, which is not proportional to $MW_f = 12n + m + 16p$, especially in the case of oxygenates. Based on this evidence, a new Oxygen Extended Sooting Index (*OESI*) was recently proposed [14]:

$$OESI = a'\left(\frac{n + \frac{m}{4} - \frac{p}{2}}{SP}\right) + b' \tag{4.4}$$

Similar to *TSI*, *OESI* permits the comparison of different sets of data obtained in different laboratories or different burners by scaling and adjusting the proportionality constants a' and b' of each data set. In this case, these constants were determined assigning the sooting index values proposed by Olson et al. [54] to toluene and iso-octane (pure reference hydrocarbons used for the smoke point calibration in ASTM D1322), and the sooting index values obtained by mole-fraction weighting to the blends of both. This mixing rule was proposed by Gill and Olson [35], who proved that the sooting tendency of blends in diffusion flames can be weighted proportionally to the molar fraction of their components. The identification of the *TSI* values proposed by Olson et al. [54] with the *OESI* values obtained here guarantees that the latter sooting index values remain in the same scale (0–100) referred to for ethane-naphthalene. The values finally obtained for the application of Eq. (4.4) were $a' = 43.588$ mm and $b' = -5.7177$.

The translation of the formerly presented smoke point results to sooting tendency, quantified with the *OESI*, leads to the results presented in Fig. 4.12. These results are not exactly inversely proportional to those presented in Fig. 4.11 because the chemical compositions of the DFAME were slightly different. However, these results provide a quantification of the ability of DFAME to reduce the sooting tendency of kerosene fuels. All tested DFAME showed significant reductions of sooting tendency. Among them, that made from palm kernel oil was shown to be the most effective (24% reduction when blended with K1 at 20% vol.), and non-distilled methyl esters (from camelina and linseed oils) showed, in general, lower effectiveness than distilled DFAME in reducing sooting tendency.

Distilled methyl esters have shorter carbon chains, and therefore higher oxygen fraction and lower carbon–carbon bond fraction, which more greatly contributes to the formation of soot nuclei than those with higher concentrations of carbon–hydrogen and carbon–oxygen bonds [14]. The latter conclusion is not observed from the smoke point data (Fig. 4.11), demonstrating that quantification through the *OESI* reveals some valuable additional information about the sooting tendency of the blends [15].

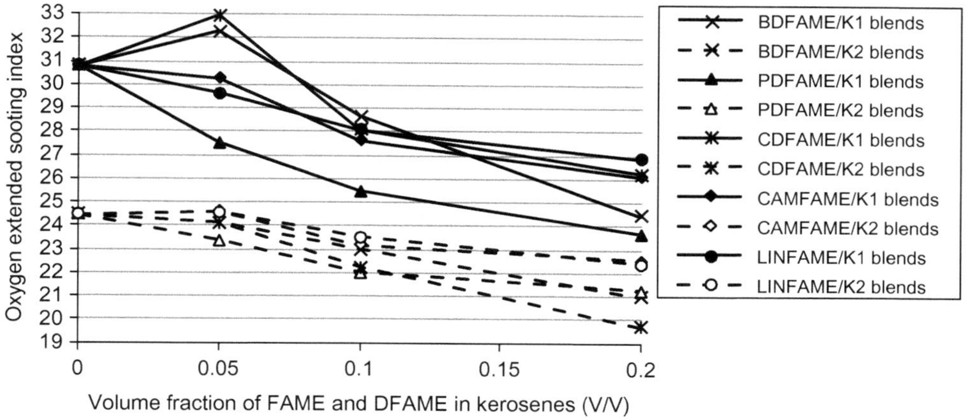

FIGURE 4.12 Oxygen Extended Sooting Index (OESI) of fossil kerosenes (K1 and K2) and blends with babassu distilled FAME (BDFAME), palm kernel distilled FAME (PDFAME), coconut distilled FAME (CDFAME), camelina FAME (CAMFAME), and linseed FAME (LINFAME).

4.4 MATERIAL COMPATIBILITY WITH DFAME

4.4.1 Polymers

4.4.1.1 Tensile Stress Test

This test is used to determine the stress–strain properties of traction for vulcanized rubbers, elastomers, and thermoplastics. The properties that can be determined are the tensile strength, the elongation at break, the effort at a given elongation, the elongation at a given effort, and the effort and elongation at yield point. Traction means the effort that the material suffers under the application of two forces acting in opposite directions. The tensile test of a material consists of subjecting a specimen of standardized dimensions to an increasing axial force tensile until the specimen breaks. During the test, the elongation of the specimen due to tensile force (F) is measured, thereby obtaining the chart tensile force (F) – elongation (ΔL). But in order to make comparisons between the test results with materials of different dimensions, we used the strain diagram (σ) – deformation (ε). Table 4.6 shows the tensile stress at maximum load (MPa) and the strain (elongation) at maximum load (%), and it specifies the differences found between the polymer samples tested with Jet A-1 and BDFAME20 and another sample without contact with the fuel. According to this tensile test three important conclusions can be inferred:

- The behaviour of the samples tested with Jet A-1 or BDFAME20 were similar.
- The two polymer groups displayed different behaviours: the thermoplastic polymers (PVC, PP, and PE), after contact with the fuel, tended to increase the strain at maximum load, and the elastomeric polymers (SBR and VITON) had the opposite effect. This means that elastomers allowed the inclusion of fuel molecules between polymeric chains, thus lowering the strain, but thermoplastics did not.
- The most affected parameter was the deformation on strain. For example, in the case of VITON, it was reduced nearly 43%, while the PVC sample reached an increase of 50.6%. Although VITON has been designed to resist most chemicals, the fuel molecules greatly reduced its strain, whereas the chlorine presence in PVC enhanced this property after fuel immersion.

TABLE 4.6 Results of the Tensile Test

	Jet A-1		BDFAME20	
	Difference[a]		Difference[a]	
Standard UNE-ISO 37:2013	Δ	(%)	Δ	(%)
PVC				
Stress at maximum load (MPa)	12.78	2.80	− 16.50	− 3.61
Strain at maximum load (%)	1.68	50.60	1.68	50.60
PP				
Stress at maximum load (MPa)	− 31.20	− 9.63	− 40.76	− 12.58
Strain at maximum load (%)	0.84	12.54	1.67	25.08
PE				
Stress at maximum load (MPa)	− 75.32	− 21.01	− 95.29	− 26.58
Strain at maximum load (%)	3.34	50.15	6.67	100.15
SBR				
Stress at maximum load (MPa)	− 1.16	− 12.75	− 0.44	− 4.86
Strain at maximum load (%)	− 18.34	− 8.94	− 15.00	− 7.32
VITON				
Stress at maximum load (MPa)	0.46	11.39%	− 0.02	− 0.50%
Strain at maximum load (%)	− 35.01	− 42.86	− 26.67	− 32.66

[a]*Difference between the samples without and within immersion in the fuel.*

4.4.1.2 *Hardness Test*

This test method permits hardness measurements based on initial indentation, indentation after a specified period of time, or both. The method is based on the penetration of a specific type of indentor when forced into the material under specified conditions. The indentation hardness is inversely related to the penetration and is dependent on the elastic modulus and viscoelastic behaviour of the material. The geometry of the indentor and the applied force influence the measurements such that no simple relationship exists between the measurements obtained with one type of durometer and those obtained with another type of durometer or other instruments used for measuring hardness. No simple relationship exists between indentation hardness determined by this test method and any fundamental property of the material tested. Table 4.7 summarizes the results of the hardness test:

- The thermoplastic polymers showed higher hardness than the elastomeric polymers.
- While samples tested with BDFAME20 always showed a change less than 2%, samples of PP and VITON tested with Jet A-1 had greater changes. The biggest change was found in the case of VITON, which showed an increase of hardness of 10.88%.

TABLE 4.7 Results of the Hardness Test

	Jet A-1		Difference		BDFAME20		Difference	
	Before	**After**	**Δ**	**(%)**	**Before**	**After**	**Δ**	**(%)**
PVC								
Hardness	98.16	98.22	0.06	0.06	98.16	98.42	0.26	0.26
PP								
Hardness	98.06	95.57	− 2.49	− 2.54	98.06	98.58	0.52	0.53
PE								
Hardness	98.52	98.54	0.02	0.02	98.52	97.76	− 0.76	− 0.77
SBR								
Hardness	78.76	78.91	0.15	0.19	78.76	77.54	− 1.22	− 1.55
VITON								
Hardness	81.31	90.16	8.85	10.88	81.31	82.56	1.25	1.54

- Although the performance of the samples showed no important changes in the hardness, comparing the behaviour between samples tested with Jet A-1 and samples tested with BDFAME20, remarkable differences were observed. PP has a chemical structure based in hydrocarbons like Jet A-1, but the ester structure of BDFAME20 affected it in a different way. The PP sample tested with Jet A-1 reduced its hardness, while the one tested with BDFAME20 increased it. In the case of the PE and SBR samples, the opposite happened. An increase of the hardness was observed in the test with Jet A-1, while the samples in BDFAME20 reduced its hardness.

4.4.1.3 *Dimensional Linear Variation*

The thermoplastic polymers PVC, PP, and PE did not show any change in size during the tests, irrespective of the fuel used (see Table 4.8).A remarkable difference between SBR and VITON was observed, with the width of the VITON increasing by 11.47% compared to the Jet A-1 test. The dimensional reduction that occurred in the test with SBR in Jet A-1 was not observed in the testing of VITON, which increased both in length and thickness. In relation to the behaviour between samples tested with Jet A-1 and BDFAME20, no significant changes were found.

4.4.1.4 *Mass Variation*

Similarly, the thermoplastic polymers PVC, PP, and PE did not show any mass variation during the tests. Related to the elastomeric polymers, a major mass increase was observed during the tests with VITON and Jet A-1 and BDFAME20 reaching 18.16% and 25.82%, respectively (Table 4.9). This happened by absorption of the fuel during the test and was higher in the case of BDFAME20.

4.4.1.5 *Solubility of Elastomers in the Fuels*

The Hansen solubility parameters (HSP) are physicochemical parameters, and are widely used to estimate the type of interactive forces responsible for compatibility between materials. The basis of the HSP is the assumption that the cohesive energy (E) might be divided

TABLE 4.8 Results of the Dimensional Linear Variation Test

	Jet A-1		Difference		BDFAME20		Difference	
	Before	**After**	**Δ (mm)**	**(%)**	**Before**	**After**	**Δ (mm)**	**(%)**
SBR								
L[a] (mm)	74.46	74.32	−0.14	−0.19	74.62	74.86	0.24	0.32
HW[b] (mm)	12.22	12.12	−0.10	−0.82	12.17	12.18	0.01	0.08
LW[c] (mm)	3.85	3.72	−0.14	−3.51	3.79	3.72	−0.08	−1.98
T[d] (mm)	1.83	1.83	0.00	0.00	1.91	1.95	0.04	2.09
VITON								
L[a] (mm)	74.67	82.59	7.93	10.61	74.58	84.63	10.05	13.47
HW[b] (mm)	12.29	13.70	1.41	11.47	12.39	13.94	1.55	12.51
LW[c] (mm)	3.81	4.13	0.32	8.41	3.83	3.96	0.13	3.26
T[d] (mm)	2.00	2.16	0.16	7.75	2.00	2.18	0.18	8.75

[a] *L, Length.*
[b] *HW, Higher Width.*
[c] *LW, Lower Width.*
[d] *T, Thickness.*

TABLE 4.9 Results of the Mass Variation Test

	Jet A-1		Difference		BDFAME20		Difference	
	Before	**After**	**Δ (g)**	**(%)**	**Before**	**After**	**Δ (g)**	**(%)**
SBR								
Mass (g)	2.26	2.12	− 0.04	− 1.89	2,33	2.34	0.01	0.56
VITON								
Mass (g)	2.00	2.36	0.36	18.16	2.01	2.53	0.52	25.82

into three parts corresponding to atomic dispersion (E_d), molecular dipolar interactions (E_p), and hydrogen-bonding interactions (E_h). Similarly, the total solubility parameter can be divided into three components corresponding to the above-mentioned different types of molecular interactions: dispersion (δ_d), polar (δ_p), and hydrogen-bonding (δ_h) [55].

$$E = E_d + E_p + E_h \tag{4. 5}$$

$$\delta_T^2 = \delta_d^2 + \delta_p^2 + \delta_h^2 \tag{4. 6}$$

For elastomers, fuel compatibility is estimated by the difference of solubility between the elastomer and the fuel. In practice, the extent of compatibility is measured by the volume swelling of the elastomer [56].

TABLE 4.10 $\Delta\overline{\delta}$ Values for the Elastomers Immersed in Jet A-1 and BDFAME20

Materials		δ_d	δ_p	δ_h	$\Delta\overline{\delta}_{Jet\text{-}A1}$	$\Delta\overline{\delta}_{BDFAME20}$
Elastomers	SBR	16.60	0.50	0.00	0.43	1.50
	Viton	15.38	10.49	2.47	10.47	9.90
Fuels	BDFAME	15.96	2.11	5.48	–	–
	Jet-A1	16.53	0.30	0.37	–	–
	BDFAME20	16.40	0.69	1.48	–	–

The mean HSP of the FAME are the average values obtained from those of the individual FAME by the van Krevelen and Stefanis-Panyiotou methods [57]. The solubility parameters for SBR are $\delta_d = 16.6$, $\delta_p = 0.5$, and $\delta_h = 0.0$, according to Guo et al.'s work [58], and those for VITON are $\delta_d = 15.38$, $\delta_p = 10.49$, and $\delta_h = 2.47$, according to Evans and Hardy's work [59]. According to reference [57], polarity (δ_p) and hydrogen-bonding forces (δ_h) decrease with carbon chain length for saturated methyl ester, while dispersion (δ_d) is almost the same. Meanwhile, dispersion (δ_d), polarity (δ_p), and hydrogen-bonding forces (δ_h) of unsaturated esters (C18:1) are relatively higher than their saturated counterparts (C18:0). The HSP of DFAME and blends of DFAME and Jet A-1 could be calculated for the three solubility components with Eq. (4.7):

$$\overline{\delta}_X = \Sigma_i x_i \delta_X \tag{4.7}$$

where x_i is the weight fraction of each component present in DFAME or the blend. The difference between the solubility parameter of fuel (F) and elastomer (E) could then be determined using the $\Delta\overline{\delta}$factor [57]:

$$\Delta\overline{\delta} = \sqrt{(\delta_{dF} - \delta_{dE})^2 + (\delta_{pF} - \delta_{pE})^2 + (\delta_{hF} - \delta_{hE})^2} \tag{4.8}$$

where δ_{iF} is the solubility parameter for the fuel and δ_{iE} is the solubility parameter for the elastomer. For compatibility between fuel and elastomer, the difference between the solubility parameters should be small. The calculation results of $\Delta\overline{\delta}$ for SBR and VITON immersed in Jet A-1 and BDFAME20 are summarized in Table 4.10. It could be offered as an explanation of the much larger dimensional swell of VITON as compared to SBR, as observed in Tables 4.8 and 4.9.

4.4.2 Metals

A positive result for this test (ASTM D130) indicates that the jet fuel will not corrode copper or copper-based alloys in the different parts of the fuel system and turbine [1]. All the blends of palm kernel DFAME with K1 and K2 passed this test with the highest value (1a). Most of the blends of babassu DFAME with K1 and K2 passed this test with the highest value (1a). Only the 10% and 20% vol. blends of babassu with fossil kerosene cut K1 show values of 1b (Tables 4.2–4.4).

In the silver strip corrosion test every sample was given a level of '0' according to the scale of the standard ASTM D7671. In accordance with the results obtained in the tests it can be said that both metals, silver and copper, show a perfect compatibility with the DFAME blends.

TABLE 4.11 Results of the Acidity, Density, and Viscosity Tests

	Before	After	Difference	
Acidity	**TAN[a] (mg KOH/g)**	**TAN[a] (mg KOH/g)**	**Δ (mg KOH/g)**	
Jet A-1	0.01	0.06	0.05	
CDFAME20	0.02	0	− 0.02	
BDFAME20	0.03	0	− 0.03	
PDFAME20	0.01	0.05	0.04	
	Before	**After**	**Difference**	
Density	**Density (kg/m^3)**	**Density (kg/m^3)**	**Δ (kg/m^3)**	**(%)**
Jet A-1	788.4	788.5	0.1	0.01
CDFAME20	812	814.5	2.5	0.31
BDFAME20	814.9	814.7	0.2	− 0.02
PDFAME20	811.8	814.8	3.0	0.37
	Before	**After**	**Difference**	
Kinematic Viscosity at −20°C	**Viscosity (mm^2/s)**	**Viscosity (mm^2/s)**	**Δ (mm^2/s)**	**(%)**
Jet A-1	1.03	1.07	0.04	4.01
CDFAME20	1.19	1.38	0.19	15.75
BDFAME20	1.32	1.32	0.00	− 0.13
PDFAME20	1.27	1.39	0.12	9.11

[a] *TAN, Total acid number.*

4.4.3 Composites

The results of the acidity, density, and viscosity tests are shown in Table 4.11. Regarding the acidity, no sample exceeded the maximum given by the standard ASTM D1665, which is 0.1 mg KOH/g. It is remarkable that the acidity even decreased when the composite was submerged in CDFAME20 and BDFAME20. The density was also in the range of the standard, which was 775–840 kg/m^3, and all the samples had more density than the Jet A-1. This is usually the case with FAME.

Values of kinematic viscosity showed slight changes after the test, and all of them were far below the maximum given by the standard ASTM D1655 (8 mm^2/s). As for the colour test, the results showed no change from the initial values, demonstrating that there was no fuel degradation during testing. In addition, all samples met the standard for aviation fuel DEF STAN 91-91, which requires the kerosene used in aviation to be clear and colourless. Through FT-IR testing, the oxidation and nitration bands were measured to study the degradation of the composites, but spectra did not show any trace of oxidation compounds [26].

4.5 CONCLUSIONS

Coconut and palm kernel oils were transesterified with methanol by the classical homogeneous base catalysis method with good yields. The FAME was subjected to fractional distillation at vacuum, and the low boiling point fractions were blended with two types of fossil kerosene, a straight-run atmospheric distillation cut (hydrotreated, K1), and commercial Jet A-1 (K2). The blends of palm kernel DFAME and Jet A-1 met some specifications selected for study of the ASTM D1655 standard – smoke point, density, flash point, viscosity at −20°C, and freezing point – but did not comply with the *LHV* by a narrow margin (−0.16 MJ/kg for 5% vol. and −0.52 MJ/kg for 10% vol. of PDFAME). On the other hand, the blends of fossil kerosene K1 and palm kernel DFAME only met the density parameter. The blends of coconut DFAME and K1 met the following specifications: density, viscosity at −20°C, and lubricity. It was especially noticeable that all the blends of 5% vol. of DFAME and fossil kerosenes K1 and K2 did not meet the *LHV* by a narrow margin of less than 1.0 MJ/kg.

Babassu and camelina oils were also transesterified with methanol by the classical basic homogeneous catalytic method. The babassu FAME was distilled at vacuum and the major low boiling point fraction (babassu DFAME) was blended with the two types of fossil kerosene, fossil kerosene cut, and Jet A-1. The blends of babassu DFAME with fossil kerosene Jet A-1 met the following ASTM D1655 specifications: colour, acidity, density, *LHV* (5% and 10% vol. blends), flash point, cloud point (not included in the standard), copper strip corrosion, oxidation stability, and smoke point. The blends of babassu DFAME and fossil kerosene cut K1 met the following specifications of the ASTM D1655 standard: colour, acidity, density, copper strip corrosion, oxidation stability, and *LHV* (only the 5% vol. blend). The camelina FAME was blended with fossil kerosene cut K1, avoiding the previous distillation step, and these blends were shown to meet the following specifications of the ASTM D1655 standard: colour, density, viscosity at −20°C, and oxidative stability.

DFAME produced from transesterification and distillation favoured the reduction in soot emissions (and thus particulate emissions) as a consequence of the presence of molecular oxygen. This fact should be considered, in addition to other benefits and drawbacks studied previously, when comparing this type of DFAME with other renewable options, such as HEFA. The sooting tendency was quantified in this work with an *OESI*, based on smoke point measurements. Among the tested FAME (derived from babassu, camelina, coconut, linseed, and palm kernel oils), *OESI* results have shown that the DFAME made from palm kernel oil was the most effective one, and non-distilled methyl esters (from camelina and linseed oils) showed lower effectiveness than distilled DFAME to reduce soot tendency. This conclusion was not evident when smoke point results were directly observed, demonstrating that the quantification through the *OESI* revealed additional information about sooting tendency.

After studying material compatibility, there was no different performance of the materials using Jet A-1 than using the blends studied. In addition there was a clear distinction between how the polymer responded to a test depending on whether it was an elastomer or a thermoplastic polymer. Moreover, VITON was the material with a bigger change in its properties once it was mixed with the fuel. Regarding the metals, both silver and copper showed no corrosion after the tests, and all the tests done with the composites had results according to their standard. Therefore, all the DFAME samples evaluated showed a good compatibility with the materials used: polymers, metals, or composites.

Previous publications from the authors of this chapter have aroused some controversy over the use of FAME as blending components with fossil kerosenes. Many of the arguments given are based on the evidence that the methods used in the previous works are not approved for jet fuels in the standard methods. The authors of the criticism conclude that the use of FAME in any amount is, thus, inappropriate, therefore discouraging any research in this field and any revision of the standard methods. However, some of the standard methods are not updated for the consideration of oxygenated components. This is the case for *LHVs*, for which an equation that does not consider the possibility of any oxygen content in the fuel is still used. Other methods have been shown to have low accuracy, such as the methods for measuring the freezing point, and some alternative methods, such as the CFPP or DSC for cold flow properties, or Rancimat, for oxidation stability of oil-derived kerosenes, may provide better reproducibility, at least in the case of oxygenated fuels.

Acknowledgements

We wish to express our thanks to the Laboratory of Fuels and Petrochemistry of the Gómez-Pardo Foundation (Technogetafe Scientific Park, Universidad Politécnica de Madrid, UPM) for its help in carrying out this work – especially Ana-María Al-Lal, Francisco Sanz, and Miguel Hernández – and for the supply of camelina seeds. Thanks also to all the staff of the School of Mines and Energy Laboratory (UPM), especially Alberto Llamas, David Bolonio, and Emilio García. Thanks to Miguel Cobo, formerly with the company Combustibles Ecológicos Biotel SL, for the gift of palm kernel and babassu oils, and to the Dallas Group of America for the gift of Magnesol D-60. We also thank the graduate and postgraduate students Diego Koss, María Rodríguez de la Rubia (UCLM), Emilia Mangue, Pedro Martín Sanz, Iván Cruz Plata, Plácida Esono Eyenga, and Sergio Alonso Lázaro (UPM) for their technical contribution to this work. We would also like to mention the contribution of the Airbus Company, who provided the composite materials of an airplane necessary for the tests, and of the JEMAC SL company that cut the composite samples for the immersion tests. Special thanks to the publishers of Global NEST Journal for allowing us to reproduce Figs. 4.1 and 4.2 for this chapter; they were originally published in this journal (reference [26]).

References

[1] ASTM International, ASTM D1655-15c. Standard Specification for Aviation Turbine Fuels, ASTM International, West Conshohocken, PA, 2015.

[2] ASTM International, ASTM D7566-15a. Standard Specification for Aviation Turbine Fuel Containing Synthesized Hydrocarbons, ASTM International, West Conshohocken, PA, 2015.

[3] M. Balakrishnan, E.R. Sacia, S. Sreekumar, G. Gunbas, A.A. Gokhale, C.D. Scown, et al., Novel pathways for fuels and lubricants from biomass optimized using life-cycle greenhouse gas assessment, Proc. Natl. Acad. Sci. U.S.A. 112 (25) (2015) 7645–7649.

[4] ASTM International, ASTM D4054-14. Standard Practice for Qualification and Approval of New Aviation Turbine Fuels and Fuel Additives, ASTM International, West Conshohocken, PA, 2014.

[5] <http://www.caafi.org/information/fuelreadinesstools.html#ASTMD4054>.

[6] M.M. Bomgardner, Flying the green skies with biofuels, Chem. Eng. News 90 (24) (2012) 18–20.

[7] T.D. Hong, T.H. Soerawidjaja, I.K. Reksowardojo, O. Fujita, Z. Duniani, M.X. Pham, A study on developing aviation biofuel for the tropics: production process—experimental and theoretical evaluation of their blends with fossil kerosene, Chem. Eng. Process.: Process Intensif. 74 (2013) 124–130.

[8] D. Chiaramonti, M. Prussi, M. Buffi, D. Tacconi, Sustainable bio kerosene: process routes and industrial demonstration activities in aviation biofuels, Appl. Energy 136 (2014) 767–774.

[9] B.G. Harvey, W.W. Merriman, A. Thomas, Koontz high-density renewable diesel and jet fuels prepared from multicyclic sesquiterpanes and a 1-hexene-derived synthetic paraffinic kerosene, Energy Fuels 29 (4) (2015) 2431–2436.

[10] R.O. Dunn, Alternative jet fuels from vegetable oils, Trans. ASAE 44 (6) (2001) 1751–1757.

[11] P. Dagaut, S. Gail, Kinetics of gas turbine liquid fuels combustion: jet A1 and biokerosene, Proc. ASME Turbo Expo 2 (2007) 93–101.
[12] D.M. Korres, D. Karonis, E. Lois, M.B. Linck, A.K. Gupta, Aviation fuel JP-5 on a biodiesel engine, Fuel 87 (2008) 70–78.
[13] A.W. Wagutu, S.C. Chabra, C.L. Thoruwa, T.F. Thoruwa, R.L.A. Mahunnah, Indigenous oil crops as a source for production of biodiesel in Kenia, Bull. Chem. Soc. Ethiop. 23 (3) (2009) 359–370.
[14] E.J. Barrientos, M. Lapuerta, A.L. Boehman, Group additivity in soot formation for the example of C-5 oxygenated 4 hydrocarbon fuels, Combust. Flame 160 (2013) 1484–1498.
[15] A. Llamas, M. Lapuerta, A.M. Al-Lal, L. Canoira, Oxygen extended sooting index of FAME blends with aviation kerosene, Energy Fuels 27 (11) (2013) 6815–6822.
[16] Alternative jet fuels cut particulate emissions. Chemical & Engineering News 2011, November 21st, 2.
[17] A. Llamas, A.M. Al-Lal, M. Hernandez, M. Lapuerta, L. Canoira, Biokerosene from babassu and *camelina* oils: production and properties of their blends with fossil kerosene, Energy Fuels 26 (2012) 5968–5976.
[18] A. Llamas, M.J. García-Martínez, A.M. Al-Lal, L. Canoira, M. Lapuerta, Biokerosene from coconut and palm kernel oils: production and properties of their blends with fossil kerosene, Fuel 102 (2012) 483–490.
[19] G.R. Wilson III, T. Edwards, E. Corporan, R.L. Freerks, Certification of alternative aviation fuels and blend components, Energy Fuels 27 (2013) 962–966.
[20] ASTM International, ASTM D3241-14be1 Standard Test Method for Thermal Oxidation Stability of Aviation Turbine Fuels (JFTOT), ASTM International, West Conshohocken, PA, 2014.
[21] S. Blakey, L. Rye, C.W. Wilson, Aviation gas turbine alternative fuels: a review, Proc. Combust. Inst. 3 (2011) 2863–2885.
[22] S. Blakey, C.W. Wilson, M. Farmery, R. Midgley, Fuel effects on range versus payload for modern jet Aircraft, Aeronaut. J. 115 (1172) (2011) 627–634.
[23] C.J. Chuck, J. Donnelly, The compatibility of potential bioderived fuels with Jet A-1 aviation Kerosene, Appl. Energy 118 (2014) 83–91.
[24] S.S. Gill, A. Tsolakis, K.D. Dearn, J. Rodríguez-Fernández, Combustion characteristics and emissions of Fischer–Tropsch diesel fuels in IC engines, Prog. Energ. Combust. Sci. 37 (2011) 503–523.
[25] S. Pinzi, D. Leiva, G. Arzamendi, L.M. Gandia, M.P. Dorado, Multiple response optimization of vegetable oils fatty acid composition to improve biodiesel physical properties, Bioresour. Technol. 102 (2011) 7280–7288.
[26] A. Al-Lal, A. Llamas, D. Bolonio, F. Sanz-Pérez, M. Lapuerta, L. Canoira, Airplane materials compatibility with blends of fossil kerosene Jet A1 with biokerosenes from babassu, palm kernel and coconut oils, Global NEST J. 16 (6) (2014) 1066–1075.
[27] M.P. Dorado, Raw materials to produce low-cost biodiesel, in: A. Nag (Ed.), Biofuels Refining and Performance, McGraw-Hill, New Delhi, 2008, p. 107.
[28] L. Canoira, M. Rodríguez-Gamero, E. Querol, R. Alcántara, M. Lapuerta, F. Oliva, Biodiesel from low-grade animal fat: production process assessment and biodiesel properties characterization, Ind. Eng. Chem. Res. 47 (2008) 7997–8004.
[29] R.O. Dunn, M.O. Bagby, Low-temperature properties of triglyceride-based diesel fuels: transesterified methyl esters and petroleum middle distillate/ester blends, J. Am. Oil Chem. Soc. 72 (8) (1995) 895–904.
[30] M. Lapuerta, L. Canoira, J. Raez, Improved method for determining the atmospheric distillation curve of biodiesel fuels from reduced pressure, Ind. Eng. Chem. Res. 50 (2011) 7041–7048.
[31] Coordinating Research Council (CRC), Handbook of Aviation Fuels Properties, CRC, Alpharetta, GA, 2004, CRC Report 635.
[32] ASTM International, ASTM D 1322-15e1. Standard Test Method for Smoke Point of Kerosine and Aviation Turbine Fuel, ASTM International, West Conshohocken, PA, 2015.
[33] S.T. Minchin, Luminous stationary flames: the quantitative relationship between flame dimensions at the sooting point and chemical composition, with special reference to petroleum hydrocarbons, J. Inst. Pet. Technol. 17 (88) (1931) 102–120.
[34] A.E. Clarke, T.G. Hunter, F.H. Garner, The tendency to smoke of organic substances on burning, J. Inst. Pet. 32 (1946) 627–642.
[35] R.J. Gill, D.B. Olson, Estimation of soot thresholds for fuel mixtures, Combust. Sci. Technol. 40 (5–6) (1984) 307–315.
[36] ASTM International, ASTM D7671-10e1. Standard Test Method for Corrosiveness to Silver by Automotive Spark–Ignition Engine Fuel–Silver Strip Method, ASTM International, West Conshohocken, PA, 2010.
[37] ASTM International, ASTM D130-12. Standard Test Method for Corrosiveness to Copper from Petroleum Products by Copper Strip Test, ASTM International, West Conshohocken, PA, 2012.

[38] J. Fargione, J. Hill, D. Tilman, S. Polasky, P. Hawthorne, Land clearing and the biofuel carbon debt, Science 319 (5867) (2008) 1235–1238.
[39] M. Lapuerta, J. Rodríguez-Fernández, A. Ramos, B. Alvarez, Effect of the test temperature and anti-oxidant addition on the oxidation stability of commercial biodiesel fuels, Fuel 93 (1) (2012) 391–396.
[40] European Committee for Standardization. EN 16091:2011. Liquid petroleum products. Middle distillates and fatty acid methyl ester (FAME) fuels and blends. Determination of oxidation stability by rapid small scale oxidation method. European Committee for Standardization 2011, Brussels, Belgium.
[41] ASTM International, ASTM D341-09. Standard Practice for Viscosity Temperature Charts for Liquid Petroleum Products, ASTM International, West Conshohocken, PA, 2015.
[42] ASTM International, ASTM D2386-15e1, Standard Test Method for Freezing Point of Aviation Fuels, ASTM International, West Conshohocken, PA, 2015.
[43] H. Tang, S. Salley, K.S. Ng, Fuel properties and precipitate formation at low temperatures in soy-, cottonseed- and poultry–fat based biodiesel blends, Fuel 87 (2008) 3006–3017.
[44] ASTM International, ASTM D2500-11, Standard Test Method for Cloud Point of Petroleum Products, ASTM International, West Conshohocken, PA, 2011.
[45] ASTM International, ASTM D97-12, Standard Test Method for Pour Point of Petroleum Products, ASTM International, West Conshohocken, PA, 2012.
[46] R.O. Dunn, Effect of minor constituents on cold flow properties and performance of biodiesel, Prog. Energ. Combust. Sci. 35 (2009) 481–489.
[47] ASTM International, ASTM D6371-05, Standard Test Method for Cold Filter Plugging Point of Diesel and Heating Fuels, ASTM International, West Conshohocken, PA, 2010.
[48] L. Canoira, J. García- Galeán, R. Alcántara, M. Lapuerta, R. García-Contreras, Fatty acids methyl esters (FAME) from castor oil: production process assessment and synergistic effects in its properties, Renewable Energy 35 (2010) 208–217.
[49] J. Lazzús, Prediction of flash point temperature of organic compounds using a hybrid method of group contribution + neural network + particle swarm optimization, Chin. J. Chem. Eng. 18 (2010) 817–823.
[50] ASTM International, ASTM D2709-96 e1. Standard Test Method for Water and Sediment in Middle Distillate Fuels by Centrifuge, ASTM International, West Conshohocken, PA, 2011.
[51] R.A.Vazquez-Gay, Design, construction and calibration of an equipment to measure the dielectric constant of lubricants diluted with different fuels. BSc Thesis No. 10-10-200930, Universidad De Castilla-La Mancha, ETS Ingenieros Industriales, Ciudad Real, Spain.
[52] Y. Yang, A.L. Boehman, R.J. Santoro, A study of jet fuel sooting tendency using the threshold sooting index (TSI) model, Combust. Flame 149 (2007) 191–205.
[53] H.F. Calcote, D.F. Manos, Effect of molecular structure on incipient soot formation, Combust. Flame 49 (1983) 289–304.
[54] D.B. Olson, J.C. Pickens, R.G. Gill, The effects of molecular structure on soot formation II: diffusion flames, Combust. Flame 62 (1985) 43–60.
[55] C.M. Hansen, Hansen Solubility Parameters: A User's Handbook, second ed., CRC, Boca Raton, FL, 2007.
[56] M. Kass, T. Theiss, C. Janke, S. Pawel, J.T. Chapin, E. Yang, et al., Compatibility of elastomers with test fuels of gasoline blended with ethanol, Sealing Technology 12 (2012) 7–12.
[57] M.M. Batista, R. Guirardello, M.A. Krahenbuhl, Determination of the solubility parameters of biodiesel from vegetable oils, Energy Fuels 27 (2013) 7497–7509.
[58] R. Guo, A.G. Talma, R.N. Datta, W.K. Dierkes, J.W.M. Noordermeer, Solubility study of curatives in various rubbers, Eur. Polym. J. 44 (2008) 3890–3893.
[59] K.M. Evans, J.K. Hardy, Predicting solubility and permeation properties of organic solvents in Viton glove material using Hansen's solubility parameters, J. Appl. Polym. Sci. 93 (2004) 2688–2698.

CHAPTER

5

Aviation Biofuels Through Lipid Hydroprocessing

A.K. Sinha, M. Anand and S.A. Farooqui

CSIR – Indian Institute of Petroleum, Dehradun, Uttarakhand, India

5.1 INTRODUCTION

Lipids are long-chain organic compounds predominantly consisting of carbon, hydrogen, and oxygen linkages, classified mainly as glycerides (triglycerides, diglycerides, monoglycerides), free fatty acids, and their derivatives, including phospholipids. These are naturally occurring, and as these compounds are produced by living organisms through their own metabolism, are one of the most abundant usable sources of renewable carbon.

Biofuels can be produced by processing lipids by various routes, including transesterification [1–4], hydrothermal [4,5] reactions, or hydroprocessing [1–4]. All routes have their own limitations and advantages, and in time, with technological developments, the most economical and sustainable route to drop-in biofuel production by lipid processing will prevail [2,3,6–13].

All over the globe, researchers have hydroprocessed various sources of lipids such as soybean oil [14–16], sunflower oil [17–20], palm oil [21,22], rapeseed oil [23–25], castor oil [26,27], tall oil [28], jatropha oil [8,11,29–37], pomace oil [38], and fresh and waste cooking oil [9,39–42] either directly or by co-processing [8,14,18,19,22,25,35,43,44] with gas oil [7–9,11,13,14,18–20,24,25,28–30, 39–41,45]. To understand the reaction mechanisms and intermediate compounds for lipid hydroprocessing, researchers have studied model compounds such as tristerian (glycerol tristearate C_{18}) [46], triolean (glucerol trioleate C_{18}) [46], tricaprylin (octanoic acid triglyceride C_8) [47], caprylic acid (octanoic acid C_8) [47], stearic acid [48,49], and oleic acid [50,51] – as well as actual feedstocks such as Jatropha – over different catalysts in hydrogen and inert atmospheres [31–33,45,47–50,52]. Fixed-bed as well as batch reactors and semi-batch reactors have been explored for performing these reactions. Both mono-functional [28,30,47] and bi-functional [8,14,29,31,32] catalysts have been utilized, depending on target reactions, such as hydrotreatment or hydrocracking reactions. Hydrotreatment targets only removal of hetroelements in hydrocarbons such as 'O, N, S' and require a catalyst with hydrogenation

Biofuels for Aviation.
DOI: http://dx.doi.org/10.1016/B978-0-12-804568-8.00005-6

functionality such as Pd, Pt, Pt-Re, or sulfided NiW, NiMo, CoMo supported over a nonacidic support like γ-Al_2O_3 or activated carbon [11,13,20,28,30,46]. Hydrocracking reactions target production of a wide range of hydrocarbon distillates and require bi-functional catalysts with hydrogenation functionality along with acidic functionality incorporated onto supports such as zeolites, silica-alumina, silicoaluminophosphates, titanosilicates, etc. [8,14,29]. These bi-functional catalysts perform hydrogenation/dehydrogenation, cracking, isomerization, cyclization, and aromatization reactions during lipid hydroprocessing.

Lipids are mostly comprised of triglyceride molecules, which are bulky in size, and these molecules need to access the active sites on the catalyst surface through its porous structure. Moreover, lipids on hydrotreatment produce water, which highlights other important criteria for catalyst selection (ie, hydrothermal stability). The catalyst's activity, stability, life, and regenerability are the four most important criteria for catalyst preparation and selection. Researchers have modified the conventional catalyst properties and have also developed new nonconventional catalysts for hydroprocessing reactions. To achieve the desired properties, variations in the composition of catalyst supports have been studied as have changes in active metal composition, metal loading, use of different kinds of active metals, and additives, etc. [53–55]. Apart from conventional hydroprocessing catalyst supports, many materials such as clays [56], carbon [11,13,20,28,30,46,57,58], oxides [59–62] like SiO_2, MgO, ZrO_2, TiO_2, and mixed oxide derivatives, such as TiO_2–ZrO_2, TiO_2–Al_2O_3, SiO_2–TiO_2 [57,63–69], zeolites like Na–Y, USY, HY, SAPO [27,34,70–72], mesoporous materials like MCM-41 [73–77], HMS [78], SBA-15 [79,80], and nano-hydroxyapatite [37] have been researched as support materials. Mesoporous aluminosilicate MCM-41 with a combination of large surface area, uniform pore size distribution, and mild acidity and high stability have been reported to have increased hydrodesulfurization (HDS), hydrodenitrogenation (HDN), and hydrocracking performance over conventional alumina- and zeolite-based catalysts [78]. TiO_2–ZrO_2-, TiO_2–Al_2O_3-, and SiO_2–TiO_2-supported catalysts have all shown synergetic effects with increased activity and stability. A unique characteristic of mesoporous materials is the three-dimensional structure with increased surface areas and porosities combined with higher thermal stability. As such, they have been studied for hydroprocessing reactions and for lipid hydroprocessing [23,27,29,30,32,37].

Kinetics and pathways for various reactions occurring during lipid hydroprocessing have been studied, as well as the extent of their effects on the yields of various biofuel components. Major reactions occurring during lipid hydroprocessing are hydrocracking and hydroisomerization reactions, with simultaneous depropanation (C_3H_8 elimination) and hydrodeoxygenation (H_2O elimination), decarboxylation (CO_2 elimination), and decarbonylation (CO elimination) reactions [8,14,19,29,31–33, 45,52,80]. The above-mentioned primary reactions are also accompanied by unavoidable gas phase and side reactions, such as water gas shift, methanation, oligomerization, cyclization, and aromatization, which have also been studied. Conditions for minimization of these reactions have been reported in the literature [31–33,45,51,52].

This chapter describes academic and technological advances for the processing of lipids obtained from various sources via hydroprocessing routes to produce biofuels suitable for aviation. The effects of different catalytic systems, operating parameters, reaction pathways, and kinetics involved in the hydroprocessing of lipids are also described. A path forward for the commercial realization of a future aviation biofuel industry via hydroprocessing of lipids is described at the end of the chapter.

5.2 EFFECT OF CATALYSTS ON LIPID HYDROPROCESSING

5.2.1 Composition of Lipids and Their Effects

Lipids from various vegetable oils are comprised of triglycerides formed from fatty acids (Table 5.1) ranging from C_{14} to C_{22} carbons. These oils have high viscosities that range between 49 cP for sunflower oil to 970 cP for castor oil, with boiling ranges >350°C and densities of around 0.88–0.96 g/mL. Having high viscosities and densities, along with voluminous structure, selective hydroprocessing of these lipid triglyceride molecules to particular products is challenging. The process parameters, reactor type, and catalyst properties need to be tailored for maximization of the yield of the desired product and for harnessing the potential of these renewable oils.

In addition, the chemical composition, along with the degree of unsaturation available in these lipids, plays a crucial role in deciding the nature of reactions during lipid hydroprocessing. If renewable diesel range is the targeted product, then lipids from corn, soybean, sunflower, camelina, jatropha, castor, rice bran,

TABLE 5.1 Fatty Acid Composition of Lipids From Various Sources

	Fatty Acid (wt%) [5,31,81]								
Fatty Acid	**Myristic**	**Palmitic**	**Palmitoleic**	**Stearic**	**Oleic**	**Linoleic**	**Linolenic**	**Archidic**	**Behenic**
Crop Oil	**14:00**	**16:00**	**16:01**	**18:00**	**18:01**	**18:02**	**18:03**	**20:00**	**22:00**
Corn		12.2	0.1	2.2	27.5	57	0.9	0.1	
Peanut	0.1	11.6	0.2	3.1	46.5	31.4		1.5	3
Soybean	0.1	11		4	23.4	53.2	7.8	0.3	0.1
Sunflower	0.2	6.8	0.1	4.7	18.6	68.2	0.5	0.4	
Camelina[a]		7.8		3	16.8	23.1	31.2		
Jatropha		19.5		7.9	45.4	27.3			
Palm	0.9–1.5	39.2–45.8	0.0–0.4	3.7–5.4	37.4–44.1	8.7–12.5	0.0–0.6	0.0–0.5	
Castor[b]		1		1	3	4.2	0.3	0.3	
Rape seed	4.3	1.7	63.7	15.4	14.3				
Algal		51		2	39	7			
Olive	0.1–1.2	7–16		1–3	65–80	4.0–10.0		0.1–0.3	
Rice bran	0.4–1	12–18	0.2–0.4	1–3	40–50	29–42	0.5–1		
Sesame		7–9		4–5	40–50	35–45		0.4–1	
Neem	0.2–2.6	13.6–16.2		14.4–24	49–62	2.3–15.8		0.8–3.4	4.2–5.3
Karanj		3.7–7.9		2.4–8.9	44.5–71.3	10.8–18.3		2.2–4.7	
Mahua		20–25		20–25	41–51	10–14		0–3.3	
Sal		4.5–8.6		34.2–44.2	41.4–42.2	2–3		6.3–12.2	

[a] *12% C20:1 and 2.8% C22:1.*
[b] *89.5% Ricinoleic acid ($C_{18}H_{34}O_3$).*

and lipids from some indigenous trees (neem, karanj, mahua, sal) containing increased percentages of C_{18} range acids (Table 5.1), would undergo hydrodeoxygenation and isomerization reactions. Alternatively, cracking, isomerization, and hydrodeoxygenation reactions would be targeted for obtaining aviation kerosene from these feedstocks. Lipids comprised of more C_{16} and lower-range fatty acids such as palm, algal, or coconut oil would give increased yield of $C_{12}-C_{15}$ hydrocarbons in the aviation kerosene range under less severe conditions by mild cracking along with deoxygenation reactions. The feedstock's properties and desired products determine the catalyst properties, such as acidity, hydrogenation/dehydrogenation ability, and morphological properties.

The presence of unsaturation in lipids favours the formation of reaction intermediates that promote cyclization, aromatization, and isomerization reactions along with cracking. An increased amount of single unsaturation (oleic and palmitoleic) in jatropha, algal, rapeseed, and olive oils promotes cyclization and aromatization reactions in the presence of a suitable catalyst and processing conditions. Fatty acids with two and three double bonds (linoleic and linolenic acid) in soybean, sunflower, jatropha, and camelina promote oligomeric reactions along with aromatization, cyclization, isomerization, and cracking reactions, leading to products comprised of greater carbon numbers than those present in the fatty acid feedstock. Researchers have reported that the occurrence of oligomerization, cyclization, and aromatization reactions over these lipid sources, and production of cyclics and aromatics during the hydroprocessing reactions, reduces the overall hydrogen requirement and provides an economic advantage [33,34,82].

5.2.2 Lipid Hydroprocessing Over Various Catalytic Systems

In the literature, researchers have reported lipid hydroprocessing over various catalytic systems, and have classified the liquid products either based on the carbon number distribution ($<C_8$ as naphtha, C_9-C_{15} as kerosene, $>C_{15}$ as diesel) or fractional cut ranges obtained by simulated distillation (IBP-135°C as naphtha, 135–260°C as kerosene, 260-FBP as diesel). Lipids from various sources have been processed either directly or co-processed after mixing with gas oil or other hydrocarbon streams from the refinery. The following section describes more about lipid co-processing and direct processing over various catalytic systems and under different operating conditions.

5.2.2.1 Co-Processing Lipids With Refinery Streams

Lipids have been co-processed with refinery streams of mainly gas oil (both light and heavy) so that the process of production can be carried out in the existing refinery infrastructure and additional capital costs avoided. Researchers have co-processed jatropha [8], soybean [14,43], sunflower [18,19], waste cooking oil [44], and rapeseed oil [25] under hydrotreating conditions and over a nonacidic sulfided metal-based catalyst system, Ni-Mo/Al_2O_3. Very few studies have been done on the co-processing of lipids with gas oil (heavy and vacuum) over hydrocracking catalysts [18,43]. Table 5.2a details lipid co-processing (10–30% lipid in gas oil) at 350°C and 370°C temperatures, 50 bar H_2 pressure, and 1–5 h^{-1} space velocity over hydrotreating catalyst. Table 5.2b describes lipid co-processing over hydrocracking catalysts. The main aim of these hydrotreating studies was to evaluate the extent of conversion and selectivity for competing deoxygenation and desulfurization reactions; whereas over hydrocracking catalysts, the objective was to maximize the yield of middle distillates of aviation kerosene.

Oxygen present in the lipids is similar to heteroatoms (such as sulfur) in fractionated

TABLE 5.2 Yield of Different Products Obtained by Co-processing of Lipids From Various Sources With Refinery Gas Oil Over (a) Hydrotreating, and (b) Hydrocracking Catalysts

a. Ni-Mo/Al_2O_3 Catalyst at 50 bar Pressure

	350°C (10–15% Lipid)					**370°C (20–30% Lipid)**			
Feed (Space Velocity)	**Jatropha ($2\ h^{-1}$) [8]**	**Soybean[a] ($4\ h^{-1}$) [34,43]**	**Sunflower ($5.2\ h^{-1}$) [19]**	**Rapeseed ($2\ h^{-1}$) [25]**	**Cooking ($1\ h^{-1}$) [44]**	**Soybean[a] ($2\ h^{-1}$) [14,43]**	**Sunflower ($5.2\ h^{-1}$) [19]**	**Rapeseed[a] ($2\ h^{-1}$) [25]**	**Cooking ($1\ h^{-1}$) [44]**
Naphtha (IBP-150)	0	0	0.09	0	0	0	2	0	1
Kerosene (150–250)	9	5	0.05	32.4	5	4.5	1	34.4	8
Diesel (250-FBP)	91	95	99.86	67.6	95	95.5	97	65.6	91
HDS %	96	89	34		92%		48		96%
C_{17}/C_{18}		1.6	0.49			0.75	0.9		
C_{15}/C_{16}		1.15	0.5			1	0.75		
CO			0.08				0.66		
CO_2			0.025				0.07		
Density, g/cc	0.839	0.846		0.835	0.8508			0.831	0.84

b. Hydrocracking Catalyst at 1000–1500 Nl/L H_2 per Feed Ratio

Feed [Reference]	**Soybean[b] [43]**	**Sunflower[c] [18]**	**Soybean[b] [43]**
% Lipid	25	30	40
Naphtha (IBP-150)	0	10	0
Kerosene (150–250)	35	26	30
Diesel (250-FBP)	65	64	70
HDS%	84		
C_{17}/C_{18}	1.6		1.58
C_{15}/C_{16}	1.5		1.48
Density, g/cc		0.769	0.818

[a] *380°C.*
[b] *370°C, $2\ h^{-1}$.*
[c] *350°C, $1.5\ h^{-1}$.*

products from crude oil sources. These heteroatoms need to be removed before further processing or utilization of products in any downstream processes. Deoxygenation reactions such as hydrodeoxygenation ($-H_2O$), decarboxylation ($-CO_2$), and decarbonylation ($-CO$) reactions compete with hydrodesulfurization ($-H_2S$) reactions for active sites. It was anticipated that a negative effect of these renewable feedstocks on the catalyst and extent of desulfurization reaction would be observed. However, Kumar et al. [8] demonstrated that co-processing lipid sources (jatropha oil) increased the percentage hydrodesulfurization (%HDS) achieved compared to that achieved when processing gas-oil separately. Bezergianni et al. [44] and Huber et al. [19] also demonstrated that on increasing the percentage of lipid (cooking and sunflower oil, respectively) in gas oil, the %HDS achieved was increased (Table 5.2). High %HDS was observed for jatropha (96%) and soybean (89%) oils, whereas only 34% was observed in the case of sunflower oil, which may be attributed to operation at low severity – that is, at high space velocities (5.2 h^{-1}) as compared to other co-processed lipids (Table 5.2).

Renewable diesel range (250-FBP) products were mainly produced during co-processing reactions because cracking reactions were kept to a minimum – only hydrotreatment reactions were targeted (Table 5.2a). High throughputs with operation at 4 and 5 h^{-1} space velocity for co-processing of soybean [14,43] and sunflower [6] oils have been reported, respectively. In comparison, waste cooking [44], rapeseed [25], and jatropha [1] oils have been co-processed at lower space velocities of 1–2 h^{-1}. An increase in the kerosene range (150–250°C) products was observed for these lipid sources at lower space velocity (1–2 h^{-1}) due to milder cracking reactions. Rapeseed oil instead showed 32% yield of kerosene range fractions, which were mainly due to co-processing with a light gas oil mixture that contained 28% kerosene range in the feed itself (Table 5.2a). Overall <10% kerosene range products could be obtained at both 350°C and 370°C. Upon increasing the percent of lipid fraction in gas oil (Table 5.2), there was a marginal increase in cracked products yield (naphtha and kerosene), which may be attributed to the increased acidity of the reaction media due to the increased formation of acidic intermediates at high lipid concentrations and higher temperatures.

Unlike Ni-Mo/Al_2O_3 (low-acidity) catalysts, hydroprocessing of lipids from soybean and sunflower sources over hydrocracking catalyst gave high yields of desirable kerosene range hydrocarbons (25–35% yield) at 370°C and 2 h^{-1} space velocity (Table 5.2b). As a consequence, lower diesel yield was obtained due to cracking of these diesel range hydrocarbons into lower-range distillates.

The ratios of decarboxylation and decarbonylation (carbon rejection) reactions to hydrodeoxygenation (hydrogen addition) reactions (C_{17}/C_{18} and C_{15}/C_{16}) have also been reported (Table 5.2), which indicates an increase in carbon rejection reactions at higher temperatures (370°C) as compared to lower temperatures (350°C) for sunflower oil at 5 h^{-1} [19]. A large increase in the concentration of carbon monoxide (eight times) at high temperatures (370°C) due to decarbonylation during co-processing supports this conclusion. Increased carbon rejection reactions were observed for hydrocracking catalysts (higher C_{17}/C_{18} and C_{15}/C_{16} ratios) as compared to hydrotreating (Table 5.2a and b).

The density of products obtained after co-processing of various lipid sources decreases only a little on increasing temperatures to 370°C, and on increasing % lipid fraction in feed (Table 5.2), which is due to formation of lighter cracked products. The density of products obtained upon co-processing sunflower oil over hydrocracking catalytic systems was comparatively lower due to increased cracking reactions (naphtha formation), even at lower temperature (350°C). On average, the density of products obtained over hydrocracking catalytic systems was lower compared to products formed over hydrotreating catalytic systems (Table 5.2)

5.2.2.2 *Direct Processing of Lipid Sources*

Lipids have been processed directly under both hydrotreating and hydrocracking conditions in the presence of noble metal and

TABLE 5.3 Influence of Temperature Variation on Product Distribution During Hydrotreatment of Lipids From Various Sources

Lipid Source	Castor [26]		Palm [21]			Sunflower [19]			Soybean [14,15]		Jatropha Oil [31]			Cooking [35,36]		
Catalyst System	5% Pd/C		$NiMo/Al_2O_3$			$NiMo/Al_2O_3$			$NiMo/Al_2O_3$		$CoMo/Al_2O_3$			Commercial HDT Catalyst		
Pressure, bar (Space Velocity, h^{-1})	25		50 (1)			50 (5.2)			50 (2)	92	80 (4)			81(1)		
Temperature	300	340	300	330	420	300	350	420	380	400	320	340	360	330	350	398
Naphtha	0	0				5.4	6.3	12	0	2.5	1	1.5	3.35	0	0	5
Kerosene	0	0	0.8	1	12.8	0.4	1	5	3	7.5	2.5	3	7.13	5	5	15
Diesel	40	100	95	93.7	41.6	93.2	91.8	76	97	90	85	88	85	95	95	80
C_{15}/C_{16}		0[a]	0.2	0.2	0.2	1	1	0.8		1.9				0.8	1.2	1.4
C_{17}/C_{18}	28[a]	9.7	0.2	0.2	0.2	0.4	0.7	0.8		2.5				1.1	1.3	1.8
m (kerosene)	0	0				0.2	0.05	0.3		0.5				0.1	0.1	0.6
Conversion	100	100	100	100	100	100	100	100		100	92.2	95.1	98.5	92.5	92	96.5

[a]*Methyl Stearate.*

sulfided base metal catalyst systems supported over acidic and nonacidic supports. Tables 5.3 and 5.4 describe the processing of various lipid sources at different temperatures over hydrotreating and hydrocracking catalytic systems, respectively.

5.2.2.2.1 HYDROTREATING OF LIPIDS

Castor, palm, and sunflower oils were processed by Meller et al. [26], Sirifa et al. [21], and Huber et al. [19] at 300°C (Table 5.3). There was a negligible amount of kerosene range products observed, with predominantly renewable diesel product. Palladium supported over carbon was used as a catalyst for processing castor oil [26] at a low pressure of 25 bar, while palm and sunflower oils were processed using sulfided Ni-Mo supported over alumina as the catalyst at 50 bar pressure. Complete conversion of triglycerides could be observed for all three feedstocks under these conditions. Although complete conversions could be observed at 25 bar pressure and 300°C in the case of castor oil over Pd/C catalyst, only 40% yield of diesel range compounds was observed, unlike the NiMo(S)-alumina catalyst, where a >90% diesel yield was obtained. Sirfa et al. [26] noted the formation of many oxygenated intermediate compounds along with a 28% yield of methyl stearate. Similar oxygenated intermediates were also observed at reduced severity by Kubicka et al. [23] at lower temperatures, by Anand et al. [31] at reduced pressures, and by Huber et al. [3,19] at higher space velocities.

Upon increasing the temperature to 340°C, methyl stearate disappeared and an increase in the yield of diesel range molecules was observed in the case of castor oil (Table 5.3). On the contrary, there was only a slight reduction in diesel yield and an increase in kerosene and naphtha products yield upon increasing the temperature to 350°C. On further increasing the temperature to 420°C, a drastic decrease in diesel yield was observed; the highest decrease was observed in the case of

TABLE 5.4 Product Distribution for Hydroprocessing of Lipids from Various Sources Over Different Hydrocracking Catalyst Systems at $1-1.5\ h^{-1}$

a. Sulfided

Lipid Source [Reference]	Soybean [16]		Algal [29]	Jatropha [8]		Jatropha [34]								Cooking [9]		
Pressure, Bar	40		50	50		70								130		
Catalyst System	NiMo/HY		NiMo/Meso-ZSM-5	NiW/SiO_2-Al_2O_3	$CoMo/Al_2O_3$	NiMo/ MSP-1		NiMo/ MSP-2		NiW/ MSP-1		NiW/ MSP-2		Commercial HC catalyst		
Temperature, °C	390	410	410	360	360	400	450	400	450	400	450	400	450	350	370	390
Naphtha						17.3	37	5.6	30.5	14	40	11	28.5	0	1.8	1.2
Kerosene	49.1	55	78.5	19.7 ($<C_{15}$)	40.2 ($<C_{15}$)	8	28.5	8	34.5	3.5	24.5	4	37.5	5.3	8.2	19.9
Diesel	18.2	5		80.8	49.2	74.7	34.5	86.4	35	82.5	35.5	85	34	66.7	65	61.0
C_{15}/C_{16}				1.3	0.8	0.5		0.3								
C_{17}/C_{18}				2.2	0.7	0.7		0.7								
i/n			2.5	1.1	0.3	0.8	2.3	2.5	3.1	1.5	3.5	0.8	3.8			
Conversion	98	100	98	100	100	100	100	100	100	100	100	100	100	72	75	82

b. Nonsulfided

Lipid Source	Jatropha [30]				Castor [27]				
Pressure, Bar	65 (H_2/N_2: 85/15)+				30				
Catalyst System	Pt/H-ZSM-5	Pt/USY	Pt/CNT	Pt/H-ZSM-5*	25% Ni/SAPO-11	25% Ni/H-Beta	25% Ni/ZSM-5	25% Ni/USY	25%Ni/USY-APTES-MCM-41
Si/Al	23	6.3		23			38		
Temperature	270				300				
Naphtha	8.0	0.0	0.0	18.2	1.2	97.7	96.8	79.4	13.8
Kerosene	3.8	1.4	13.3	6.0	3.2	1.5	2	19.2	80.3
Diesel	88.4	98.6	86.7	75.8	95.4	0	0	0.8	5.4
C_{15}/C_{16}	0.0	0.8	40.9	1.9					
C_{17}/C_{18}	0.4	0.5	0.7	0.1					
i/n					0.1	0	0	5.3	4.4
Conversion		31.2	13.6	14.2	99	99	99	99	99

+ cat:jatropha:H_2O ratio(1:1:9; *1:10:9). CNT, carbon nanotubes; HY, Y zeolite (H-form); Meso-ZSM-5, hierarchical mesoporous H-ZSM-5 zeolite; MSP, hierarchical mesoporous SAPO-11(MSP-1 (Si/Al: 0.4) & MSP-2 (Si/Al: 0.27)

palm oil (Table 5.3). Increase in cracking and isomerization reactions were observed at higher temperatures, which led to increased kerosene and naphtha yields. A greater reduction in diesel yield was observed in the case of palm oil, not only because of operation at reduced space velocity (palm: 1 h^{-1}; sunflower: 5.2 h^{-1}), but also due to its more saturated composition. Soybean oil was also processed over $NiMo/Al_2O_3$ catalyst at 50 and 90 bar pressures, and on increasing temperature from 380°C to 400°C, a decrease in diesel yield and an increase in kerosene and naphtha yield were also reported (Table 5.3).

Jatropha [31] and cooking [40,41] oils were processed at a high pressure of 80 bar over $CoMo/Al_2O_3$ and commercial hydrotreating catalytic systems. At lower temperatures – that is, 320°C (jatropha oil) and 330°C (cooking oil) – only 92% conversion was observed with lower diesel yield (85%) in the case of $CoMo/Al_2O_3$ with jatropha oil (Table 5.3), as compared to that over commercial hydrotreating catalyst with cooking oil. Lighter cracked products – that is, naphtha and kerosene (3.5% yield) – could be observed even at 4 h^{-1} space velocity in the case of $CoMo/Al_2O_3$ systems, whereas only 5% kerosene was observed at 330°C and 1 h^{-1} space velocity (high severity) in the case of cooking oil (Table 5.3). Diesel yield increased (88%) upon increasing the temperature to 340°C for jatropha oil due to an increase in conversion, whereas it remained almost constant in the case of cooking oil (Table 5.3). On further increasing temperatures to 360°C (jatropha) and 398°C (cooking oil) at 80 bar pressure, there was a decrease in diesel yield and an increase in cracked products yield, and increased formation of kerosene range hydrocarbons (Table 5.3).

Anand et al. [31] demonstrated formation of acidic intermediates upon processing lipid feedstocks over nonacidic $Co\text{-}Mo/Al_2O_3$ catalyst, which justified the increase in cracking reactions under less severe hydrotreating conditions. An increase in decarbonation selectivity (CO, CO_2 elimination) over hydrodeoxygenation selectivity (H_2O elimination) was observed for all feeds at higher temperatures (Table 5.3), except in the case of palm oil, wherein low decarbonation selectivity (low C_{15}/C_{16} and C_{17}/C_{18} ratio) was observed. This indicated that heavier lipids such as castor, jatropha, soybean, sunflower, and cooking oils favoured the carbon rejection mechanism at higher temperatures, whereas hydrodeoxygenation for the removal of oxygen was a preferred mechanism in the case of lighter lipid sources such as palm oil (Table 5.3). These facts indicate that the consumption of hydrogen in the case of palm oil would be greater compared with other lipid sources towards deoxygenation reactions.

In summary, oxygenated intermediates were produced at reduced reaction severity (low hydrogen partial pressures, low reaction temperatures, and reduced residence time, that is, high space velocity). Upon increasing the reaction severity, these intermediates converted to deoxygenated products and increased temperatures led to cracking and isomerization reactions. The $CoMo/Al_2O_3$ catalytic system induced acidity in the reaction media due to formation of acidic intermediates and led to increased cracking reactions and formation of kerosene range molecules even at reduced severity. Lighter lipid sources such as palm oil favoured hydrodeoxygenation reactions over carbon-rejection reactions, whereas heavier lipid sources deoxygenated primarily by carbon-rejection mechanisms at higher temperatures. Generally, hydrotreating cannot directly produce desirable yields of aviation kerosene. But it is possible to convert the diesel – selectively produced from hydrotreating vegetable oil in a first reactor – into aviation kerosene by mild hydrocracking in a second reactor (a two-step process).

5.2.2.2.2 HYDROCRACKING OF LIPIDS

Hydrocracking of lipids has been studied in the literature over a wide range of catalytic systems, such as conventional sulfided catalysts and their combinations with various hierarchical mesoporous supports, or the nonconventional nonsulfided catalytic systems such as Ni and Pt metal-based catalysts used in reduced conditions over different zeolites, silicoaluminophosphates, and their combinations. Conventional catalysts were evaluated between 40–130 bar (hydrogen pressures) and at temperatures ranging between 360°C and 450°C, whereas nonconventional catalytic systems were evaluated at lower pressures 30–55 bar (hydrogen pressures) and temperatures between 270°C and 300°C.

5.2.2.2.3 CONVENTIONAL SULFIDED CATALYSTS

Lipids from soybean, algal, jatropha, and waste cooking oils have been processed over Ni-Mo, Ni-W, Co-Mo-based catalytic systems where Mo and W are the active metals providing the required hydrogenation/dehydrogenation ability to the catalyst along with Ni or Co as the promoters, all loaded over an acidic support (Table 5.4a). Nearly complete conversion was observed for all lipid feedstocks except waste cooking oil lipids, with a gradual increase in conversion from 72% to 82% upon increasing temperature observed at 130 bar (Table 5.4a). The conversion calculations as defined by Bezergianni et al. [9] consider boiling fractions above 360°C as unconverted feedstock. Considering the reaction products boiling above 360°C as unconverted lipid from cooking oil, reduced conversions over the commercial hydrocracking catalytic system are inferred.

Cheng et al. [16], Kumar et al. [8], and Verma et al. [29,34] instead studied the conversion of lipids of different sources, and actual concentrations of unconverted feed (measured with chromatography) have been used for conversion calculations. High renewable diesel yield of around 60–80% was obtained at temperatures below 400°C and pressures between 50 and 130 bar (Table 5.4a) for jatropha and cooking oil lipids over Ni-Mo- and Ni-W-based catalysts [8,16,29,34]. Upon further increase of the temperature, the yield of the diesel fraction decreased and more cracked products (ie, kerosene and naphtha) were observed. Comparatively, over a Co-Mo/Al_2O_3 catalyst, low diesel yields were observed (Table 5.4a), along with a 40% yield of cracked ($<C_{15}$) products. Similar observations of increased cracked product yields over a Co-Mo/Al_2O_3 catalytic system were observed by Anand et al. [31,33]. Increased cracking observed over a Co-Mo catalyst supported on low acidity Al_2O_3 indicated the formation of acidic intermediates in the reaction media that were catalysing the cracking reactions. Moreover, for a Co-Mo catalyst, C–O bond breaking was preferred – that is, hydrodeoxygenation was preferred over decarbonation mechanisms (0.8 (C_{15}/C_{16}); 0.7 (C_{17}/C_{18})) – whereas Ni-W/SiO_2-Al_2O_3 catalytic systems preferred decarbonation reactions over hydrodeoxygenation reactions (1.3 (C_{15}/C_{16}); 2.2 (C_{17}/C_{18})) due to the acidic SiO_2-Al_2O_3 support, which catalyses C–C bond cracking and leads to the formation of CO and CO_2 [8].

Mesoporous silicoaluminophospate SAPO-11 (MSP)-based catalytic systems supported with Ni-Mo and Ni-W active metals showed similar yield patterns with increased aviation kerosene and naphtha yield upon increasing temperature from 400°C to 450°C (Table 5.4a). A corresponding decrease in diesel yield was also observed. There was drastic improvement in isomer-to-normal hydrocarbon (i/n) ratios upon increasing the temperature, indicating increased isomerization selectivities over SAPO-11-based catalytic systems at elevated temperatures (Table 5.4a). A maximum aviation kerosene yield of 37.5% was observed for jatropha oil lipids over Ni-W/MSP systems at 450°C temperature and 70 bar pressure, whereas for soybean oil lipids, 49% yield of aviation kerosene was observed over NiMo/HY zeolite catalyst at reduced temperatures and pressures (390°C and 40 bar). Increasing temperature to 410°C, the yield of aviation kerosene further increased over NiMo/HY catalyst (Table 5.4a). Similar observations were made for algal oil lipids, with a maximum aviation kerosene yield of 78% (at 410°C) over hierarchical mesoporous zeolite (ZSM-5)-supported NiMo catalyst [29], which indicated increased lipid cracking ability of these zeolites over other catalytic systems.

Fig. 5.1 shows the product yield distribution for hydrocracking of various vegetable oils over different catalysts. For jatropha oil hydrocracking over mesoporous silica-alumina, hierarchical mesoporous H-ZSM-5 and titanosilicate (MTS) supports loaded with sulfided NiMo, NiW, and CoMo active metals gave the highest yield of naphtha, followed by that of aviation kerosene and diesel, with aviation kerosene ranging between 30–35% at 400–420°C temperatures and 80 bar pressure. However, over CoMo/MTS systems at a reduced temperature of 360°C, cracking of jatropha oil was low, with 80% renewable

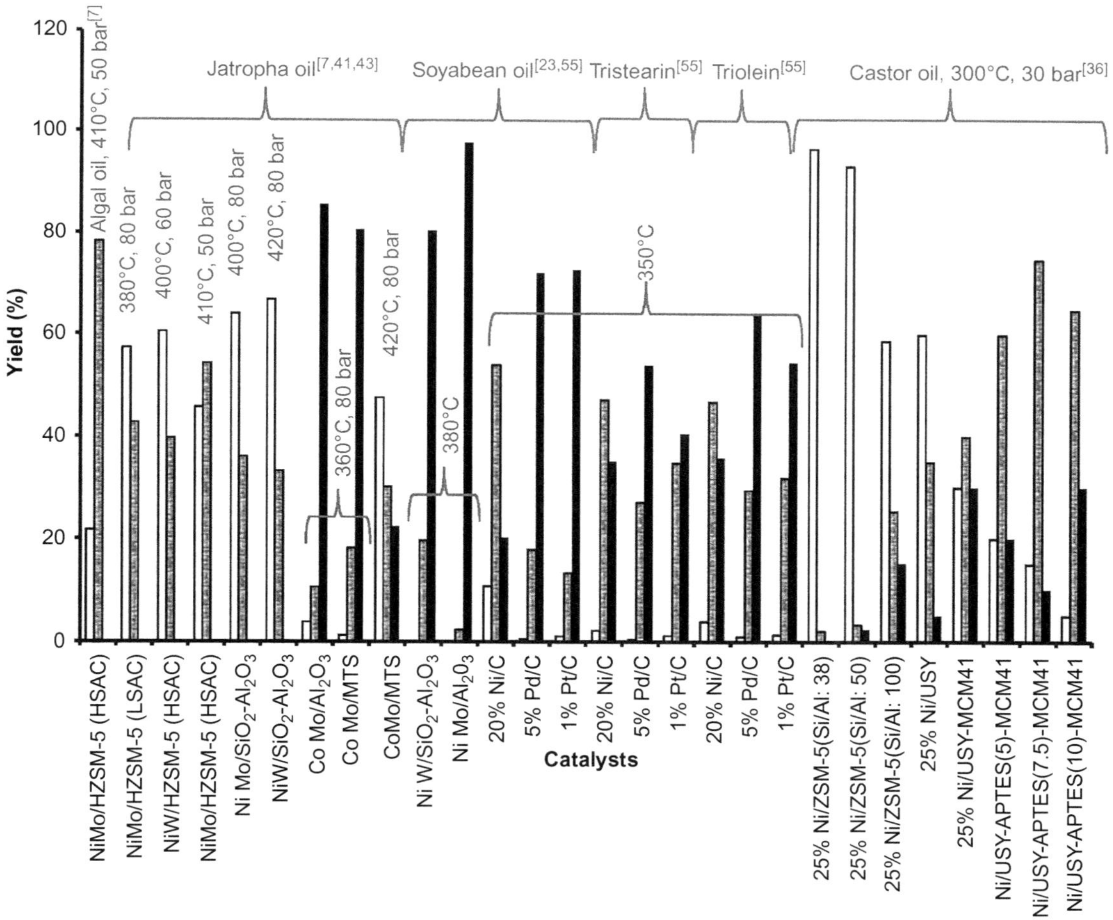

FIGURE 5.1 Product yield distribution for hydroprocessed vegetable oils over various catalysts (□ Lighter distillate; ▩ Middle distillate; ■ Heavier distillate).

diesel as the major product, followed by aviation kerosene (~20%) and a negligible yield of naphtha (Fig. 5.1). Increase in acidic intermediates formed [8,31,33], in addition to higher reaction severity, at higher temperature, led to cracking and isomerization reactions, even over low-acidity supports like MTS, as was also observed for CoMo-alumina (Table 5.3) [31]. NiMo/HZSM-5(HSAC), which has hierarchical porous, high surface area, crystalline HZSM-5 zeolite with framework-confined mesoporosity as support, furnished the highest yield of aviation kerosene (~60%) with negligible diesel, from jatropha oil. The same catalyst gave the highest reported yield of aviation kerosene (~80%) when algal oil was used as the feed (Fig. 5.1) [7].

5.2.2.2.4 NONCONVENTIONAL NONSULFIDED CATALYSTS

Jatropha and castor oil lipids were processed over reduced Pt and Ni catalysts on various supports (Table 5.4b). At low temperatures (270°C) conversions were low (15–30%), even at a high hydrogen partial pressure of 55 bar [28], but nearly complete (99%) conversions were achieved for castor oil at slightly higher temperature (300°C) and a low pressure of 30 bar, indicating that slight changes in reaction temperature have a strong influence on the conversion of lipids [27]. There was increased formation of cracked naphtha range products observed by Murata et al. [30] over a Pt/H-ZSM-5 catalytic system for jatropha feeds, whereas at similar conversion levels Pt/carbon nanotubes gave increased formation of kerosene range molecules (Table 5.4a). This may be attributed to the lower acidity of the carbon nanotube support for Pt as compared to that for H-ZSM-5. Ultrastable Y (USY) zeolites gave two times the conversion (31%) compared to other Pt-loaded catalytic systems (Table 5.4b), indicating Pt/USY as a better catalyst at lower temperature (270°C) for increased diesel production (98.6%). Upon increasing temperature to 300°C, 25% Ni/USY showed complete conversions of castor oil, and a naphtha yield of 79.4% and kerosene yield of 19.2% with negligible diesel yield (Table 5.4b). A similar yield pattern was also observed for H-Beta and USY catalytic systems with 25% Ni loading, and increased naphtha yield was reported. These observations indicate that such highly acidic catalytic systems with 25% Ni loading promoted cracking and isomerization reactions even at 300°C and a low pressure of 30 bar, whereas conventional sulfided Ni-Mo-, Co-Mo-, and Ni-W-based catalyst systems promoted these reactions at higher temperatures, >360°C (Table 5.4a). However, a major limitation for commercial success of the Ni/Zeolite-based catalysts is that they deactivate much more rapidly than conventional sulfided catalysts supported on moderately acidic silica-alumina supports.

Liu et al. [27] also demonstrated for castor oil lipids that a catalytic system with moderate acidity increased the desired aviation kerosene yield to as high as 80% (Table 5.4b) with increased i/n ratios. The yield of aviation range kerosene increased nearly six times using a combination of USY, APTES ((3-aminopropyl)-tri-ethoxysilane), and MCM-41 (Fig. 5.1). Addition of MCM-41 to USY zeolite blocked the strong acid sites, which resulted in greater yields of bio-aviation fuel range compounds and reduced naphtha range compounds (Table 5.4b) [27] (Fig. 5.1). Addition of APTES increased the mesoporosity of the system, which in turn also improved the diffusion (mass transfer) of bulkier triglyceride lipid molecules into the pores of the active sites. Moderate increase in APTES content (5 to 7.5) increased the aviation kerosene yield, but on further increase of the APTES content (7.5 to 10) the yield of cracked products (ie, naphtha and kerosene range molecules) started decreasing. This decrease in cracked products yield was attributed to blockage of active sites/zeolitic pores caused due to excessive use of APTES

[27], in addition to negligible moderate acidity shown by high APTES content systems. Liu et al. [27] demonstrated that upon increasing the Si/Al ratio, for 25% Ni/ZSM-5 catalyst, the Lewis acidity decreased, which in turn decreased the yield of cracked lighter distillate products (Fig. 5.1), with gradual increase in middle and heavier distillates. Reduced secondary cracking reactions due to the reduction of strong acidity led to an increase in aviation kerosene yields upon increasing the Si/Al ratio.

Soybean oil was hydroprocessed over Pt-, Pd-, and Ni-based catalysts supported on carbon at 300°C (Fig. 5.1) [55]. Negligible secondary cracking responsible for formation of naphtha range product, and mild primary cracking, led to 15–20% kerosene range middle distillates for the Pt- and Pd-based catalysts. On the contrary, lipid hydroprocessing over 20% Ni supported over carbon yielded mostly kerosene range products (50%) (Fig. 5.1) under similar operating conditions. Studies by both Morgan et al. [46] and Liu et al. [47] over lipids from soybean and castor oils showed Ni-supported catalytic systems suitable for the production of aviation kerosene range hydrocarbons. Carbon-supported catalyst systems are inherently nonacidic in nature and do not catalyse cracking reactions. The higher yield of middle distillate kerosene range products (Fig. 5.1) observed over Ni-based catalyst systems indicated formation of acidic intermediates which promoted cracking of lipids at lower temperature (300°C).

Model triglyceride compounds, both saturated (tristearin, $C_{57}H_{110}O_6$) and unsaturated (triolein, $C_{57}H_{104}O_6$), were also hydroprocessed over Ni-, Pt-, and Pd-based systems supported on carbon [46]. Similar results with increased cracking (kerosene 46–47%) were observed over Ni-based catalyst as compared to Pt- and Pd-based systems (kerosene, 27–34%), along with increased conversion of 81–85% for Ni/C as compared to 22–47% in the case of Pt and Pd systems (Fig. 5.1). Product patterns observed over both saturated and unsaturated lipid molecules were similar over all the catalytic systems, with an exception of increased formation of internal alkenes in the case of triolein feedstocks, which were probably the result of reduced hydrogenation functionalities at 300°C [46].

Ni-based catalytic systems show a promising future for increased production of kerosene range products, but further experimentation and analysis is necessary for determining the catalyst life and properties of these products vis-a-vis specifications for aviation kerosene. Apart from these catalytic systems, various newer conventional and nonconventional catalytic systems are being developed for the production of kerosene range hydrocarbons, but their utilization is subject to the products meeting the ASTM 7566 or Jet A-1 test specifications for aviation kerosene. Ni-Mo and Ni-W catalytic systems supported over acidic mesoporous supports such as SiO_2-Al_2O_3 or ZSM-5 have shown desirable catalytic performance, with properties of kerosene range products meeting the ASTM 7566 specifications. These acidic mesoporous supports, supported with active metals with strong hydrogenation/dehydrogenation ability at 400–450°C temperatures and high hydrogen pressures (80–150 bar), yield the desired isomerization and deoxygenation of lipids to produce aviation range hydrocarbons. Continuous pilot-scale runs and catalytic evaluations have resulted in better understanding of these systems and catalytic life cycles for scale-up and commercial-scale productions. Reduction in reaction temperatures and pressures, with increased productivity of desired aviation kerosene fuel without compromising the quality of the fuel would result in reduced operating costs and increased profit margins, leading to rapid commercialization of these processes.

5.2.2.3 *Isomerization Selectivity*

The products from vegetable oil hydroprocessing can be used as liquid transportation fuels

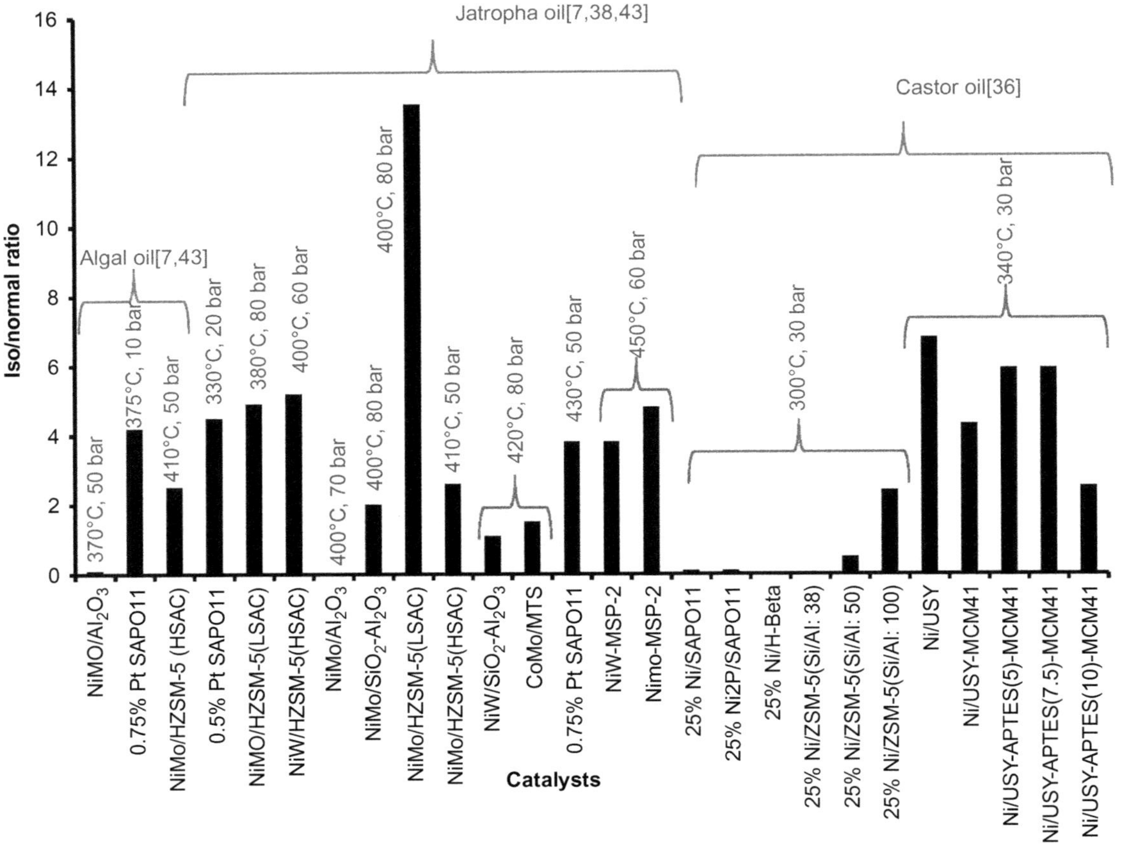

FIGURE 5.2 Isomer/Normal ratio in middle distillate (kerosene range) product for hydroprocessed vegetable oils over various catalysts.

only if they meet required specifications, such as those defined by ASTM. Isomer-to-normal hydrocarbon ratio is a very important criteria for meeting the desired properties of all liquid transportation fuels. High i/n ratio is necessary for increasing the octane number of gasoline, and for reducing the freeze point of aviation kerosene or pour point of renewable diesel.

Isomerization reactions can occur during lipid hydroprocessing at the acidic sites on the catalyst surface. Generally, alumina support-based catalysts have low acidity and hence do not favour isomerization reactions. On the other hand, acidic supports such as silica-alumina and zeolites provide the required acidity for these reactions. Fig. 5.2 shows i/n hydrocarbon ratios for kerosene range hydrocarbons obtained by the hydroprocessing of various lipids over different acidic and non-acidic catalytic systems. Alumina-based catalytic systems showed almost negligible i/n ratio for algal- and jatropha-based hydroprocessed products at 370°C and 400°C, indicating a negligible effect of temperature on i/n ratio over these NiMo/Al_2O_3-based catalytic systems. Verma et al. [29] hydroprocessed algal and jatropha oil over various hierarchical structured and porous catalytic systems – H-ZSM-5-, SAPO-11-, and

SiO_2-Al_2O_3-supported with both sulfided and noble metal-based hydrogenation/dehydrogenation functionalities. They demonstrated that crystalline HZSM-5 (low surface area crystalline (LSAC)) promoted more isomerization reactions as compared to semi-crystalline HZSM-5 (high surface area semi crystalline (HSASC)). Medium-strength acid sites (Brönsted acidity) were responsible for desirable isomerization reactions, whereas strong acidic sites (Lewis acidity) promoted undesired cracking reactions. Liu et al. [27] processed castor oil lipids over 25% Ni/ZSM-5 catalytic systems with varying Si/Al ratios at 300°C and 30 bar, wherein similar results were reported, in other words, increasing isomerization reactions with increasing Si/Al ratio, which was ascribed to lower Lewis acidity (with increasing Si/Al ratio) because of the decrease in extra-framework Al content. Acidic site distribution results shown by both Verma et al. [29] and Liu et al. [27] indicate a decrease in Lewis acidity (strong acid sites) upon increasing the Si/Al ratio. A decrease in the yield of lighter cracked products with increasing Si/Al ratio [27] also supported the fact that cracking was favoured over Lewis acidic sites and isomerization reactions over Brönsted acid sites. Novel mixed catalytic systems (USY, MCM-41, APTES) were also studied by Liu et al. [27] for hydroprocessing of castor oil, and i/n hydrocarbon ratios of 5–6 were obtained, which are comparable to the values obtained over most of the other catalysts, such as NiMo/HZSM-5 (Fig. 5.2).

Over NiMo catalytic systems similar i/n (2.2) were observed for algal oil lipids as well as jatropha oil lipids at 410°C for HZSM-5 (HSASC) supports. Upon increasing the temperature from 380°C to 400°C over HZSM-5 (LSAC) supports, the i/n hydrocarbon ratio increased three times, whereas such a drastic increase was not observed for any other catalytic systems, indicating NiMo/H-ZSM-5 (LSAC) systems were better-suited for isomerization reactions at higher temperatures. Higher isomerization selectivity (i/n = 4.5) was reported for Ni-W/H-ZSM-5 (HSASC) as compared to Ni-Mo/H-ZSM-5 (HSASC) (i/n = 2.5), indicating the role of the stronger hydrogenation/dehydrogenation ability of 'W' over 'Mo' supported catalysts for isomerization reactions. However, a similar observation was not reported by Verma et al. [34] in the case of hierarchical mesoporous SAPO-11-supported NiMo and NiW catalytic systems. A slight decrease in the i/n ratio for NiW (i/n = 3.8) was reported as compared to that for NiMo (i/n = 4.8) catalytic systems (450°C, 60 bar). These differences could be due to the differences in the acidities of the two catalysts. On further increasing the hydrogenation/dehydrogenation ability of the catalytic systems by using noble metal-supported catalysts (0.75% Pt/SAPO-11), a similar i/n ratio (i/n = 4) could be obtained at slightly reduced temperatures and pressures (430°C, 50 bar). Even upon further reducing the temperature to 330°C and the pressure to 20 bar, the isomerization selectivity (i/n = 4) was maintained, indicating that isomerization reactions were favoured over stronger hydrogenation/dehydrogenation functionalities. Different i/n ratios were obtained for algal and jatropha lipids, indicating that isomerization selectivities are also dependent on feed composition.

5.3 KINETICS, REACTION MECHANISMS, AND PATHWAYS

The reaction mechanism and the pathway followed for lipid conversion into various hydrocarbons mainly depend on the type of catalyst system chosen for the conversion. Lipids from various sources mostly contain triglycerides, along with some diglycerides, monoglycerides, and free fatty acids. These

a.

Bond Saturation
$TG^{=} + H_2 \rightarrow TG$
Depropanation
$TG + 3H_2 \rightarrow 3R_XCOOH + CH_3CH_2CH_3$
Hydrodeoxygenation
$R_XCOOH + 3H_2 \rightarrow R_XCH_3 + 2H_2O$
Decarboxylation
$R_XCOOH \rightarrow R_XH + CO_2$
Decarbonylation
$R_XCOOH + H_2 \rightarrow R_XH + H_2O + CO$

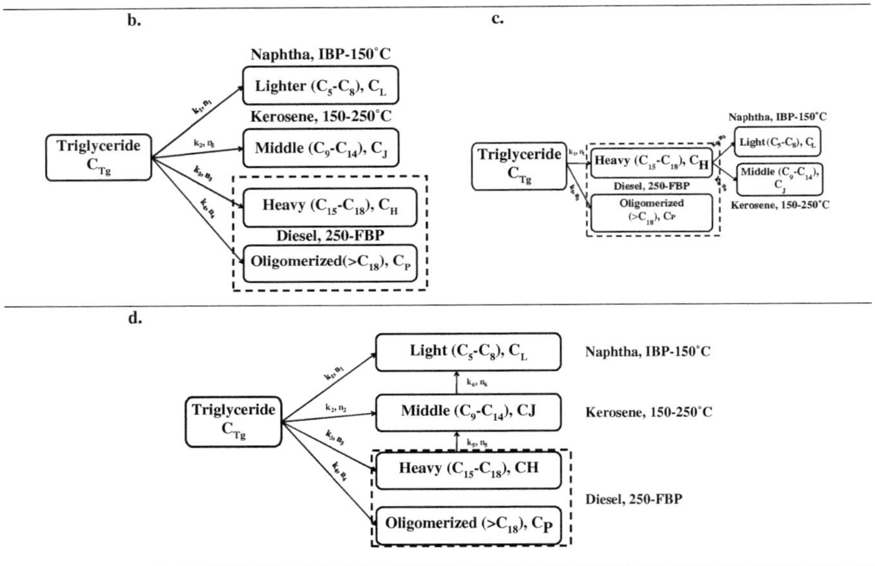

SCHEME 5.1 Reaction schemes for the hydroconversion of lipids into transportation fuels.

molecules contain double bonds (Table 5.1) along with heteroatoms (mainly oxygen), which need to be removed prior to use.

Over a hydrotreating catalytic system, such as CoMo or NiMo supported over low acidity Al_2O_3, the unsaturated lipid molecules are primarily saturated and deoxygenated to yield a corresponding saturated hydrocarbon [21,23,45]. During deoxygenation, the glycerol linkage of the triglyceride molecule is broken to produce a propane molecule along with three corresponding carboxylic acid molecules (Scheme 5.1a) [7]. In the case of a multifunctional catalyst with cracking/isomerization functionality and hydrogenation/dehydrogenation ability, cracking and isomerization reactions are observed [7,31,32]. In addition, cyclization and aromatization reactions may also be observed if conjugated double bonds are present [5]. Gas phase compositions of products obtained over hydrotreating and hydrocracking catalysts for palm and jatropha oil (Table 5.5) indicate that 30–60% of the gaseous products is propane, which is

TABLE 5.5 Molar Composition of Gaseous Products from the Hydroprocessing of Lipids at 50 bar and 1 h^{-1} Over Both Hydrocracking and Hydrotreating Catalytic Systems

Process	Hydrotreating			Hydrocracking		
Temperature, °C	360			360	400	
Lipid [Reference]	Palm [21]	Jatropha [8]		Jatropha [8]	Jatropha [34][a]	
Catalyst	$NiMo/Al_2O_3$	$NiMo/Al_2O_3$	$CoMo/Al_2O_3$	NiW/SiO_2-Al_2O_3	NiMo/MSP-1	NiW/MSP-2
Gas HC – C_3	1.8	0.53	1.72	1.54	1.43	1.51
C_3	1.7	0.69	0.74	0.83	1.09	0.66
CO_2	1.3	0.49	0.55	2.15	4.35	1.47
CO	0.1	2	1.8	5.98	2.25	0.93

[a]*70 bar; HC, hydrocarbons; MSP, hierarchical mesoporous SAPO-11 (MSP-1 (Si/Al: 0.4) & MSP-2 (Si/Al: 0.27)).*

predominantly produced from cleavage and hydrogenation of the triglyceride molecule (C—O bond).

The oxygen present in these complex molecules can be removed either by deoxygenation reactions (ie, hydrodeoxygenation, H_2O) or by decarbonation reactions (CO and CO_2). Following the deproponation reaction, the intermediate acid molecule produced undergoes a deoxygenation reaction to yield hydrocarbon products. The depropanation and deoxygenation reactions can either occur simultaneously or consecutively – that is, depropanation to produce acid followed by deoxygenation of acidic molecules [23]. The route by which the lipid molecules or the acid intermediate molecules are deoxygenated (ie, the deoxygenation pathways) also depends on the processing conditions chosen and catalyst functionality (acidity/hydrogenation ability). Strong hydrogenation catalytic systems over nonacidic supports – that is, hydrotreating catalysts such as sulfided CoMo, NiMo supported over Al_2O_3 – favour hydrogenation reactions leading to the breakage of C—O bonds. These bonds are present in the acid linkages and upon reaction form hydrocarbons as the major product with H_2O as the main by-product, along with small quantities of CO and CO_2; whereas hydrocracking catalysts with acidic supports such as silica-alumina, silicoaluminophospates, zeolites, etc., favour C—C bond breaking. This leads to comparatively more CO_2 and CO formation than H_2O. Accordingly, increased molar yields of CO and CO_2 are observed over hydrocracking catalysts compared to hydrotreating catalysts (Table 5.5).

The produced deoxygenated hydrocarbons undergo further cracking and isomerization reactions to yield lower boiling point range hydrocarbons. Anand et al. [31] and Sharma et al. [32] studied various lumped kinetic models for hydroconversion of jatropha oil lipids over CoMo catalyst systems supported over Al_2O_3 and MTS. They considered all the possible reactions and the reaction products formed during the hydroconversion of lipids. Triglycerides in lipid feed were considered as a single entity and the liquid hydrocarbon products produced after hydroprocessing were lumped according to their corresponding carbon ranges – that is, gasoline (C_5–C_8), kerosene (C_9–C_{14}), gasoline + kerosene ($<C_{15}$), diesel (C_{15}–C_{18}), and oligomerized product ($>C_{18}$). Rate equations were framed for the models and reaction

TABLE 5.6 Rate Constants and Activation Energies for Hydroprocessing of Lipids and Co-processing of Lipids With Gas Oil at Various Temperatures

Lipid (Jatropha Oil) – Co-Mo				
Catalyst System	**Titanosilicates [32]**	**Alumina [31]**		
Temperature, °C	**320**	**320**	**360**	**Activation Energy, kJ/mol**
$k_{Triglyceride}$, h^{-1}	14.4	17	24	26
$k_{Naphtha}$, $C_5–C_8$, h^{-1}	0.04	0.3	1.1	83
$k_{Kerosene}$, $C_9–C_{15}$, h^{-1}	0.1	0.3	2	127
k_{Diesel}, $>C_{15}$, h^{-1}	1.2	17	20.9	47

Co-processing Lipids – NiMo/Al_2O_3					
Temperature, °C	**325**				**350**
% Lipid	**25% Soya [43]**	**10% Soya [43]**	**10% Jatropha [8]**	**5% Jatropha [8]**	**20% Jatropha[a] [35]**
k_{HDS}, s^{-1}	4.8	4.7	4.9	4.4	–
k_{HDO}, s^{-1}	4.3	4.1	4.3	4.2	–
E_{HDS}, kJ/mol	15	17.9	12.7	12.9	–
E_{HDO}, kJ/mol	136.6	140.2	129.3	132.3	148.4

[a]*Co-processing with N-hexadecane/4,6-dimethyldibenzothiophene.*

parameters were estimated for different reaction pathways based on the experimental results (Table 5.6). Titanosilicate-based catalyst systems showed reduced rates for triglyceride conversion at 320°C as compared to alumina-supported catalyst systems. The marginally higher acidity of MTS as compared to the alumina-supported catalyst [32], led to direct cracking of the lipid molecules into naphtha and kerosene (Scheme 5.1b), whereas in the case of the CoMo/Al_2O_3 system at 320°C, the triglycerides were initially deoxygenated to yield diesel range products, which were further cracked into lower range naphtha and kerosene range hydrocarbons (Scheme 5.1c). The conversion pathways for lipids over these catalyst systems were different, due to alternative metal support interactions and the different physicochemical properties of these systems. The MTS system catalysed an alternative pathway for the formation of lower range hydrocarbons directly from the triglyceride at lower temperatures. The rate of formation of primary cracked products (ie, gasoline and kerosene range compounds over MTS catalyst) were lower when compared to the Al_2O_3 system (Table 5.6). There was also increased conversion and corresponding diesel rate of formation in the case of Al_2O_3 catalyst systems as compared to MTS systems, indicating CoMo/Al_2O_3 systems were better-suited for the formation of diesel range compounds at lower temperature (320°C).

Upon further increase of the temperature to 360°C in the case of the CoMo/Al_2O_3 catalyst, a shift in the reaction pathway was observed from only deoxygenation reactions to direct cracking into naphtha and kerosene range products, along with internal conversion of diesel range products into cracked kerosene range products (Scheme 5.1d). There was an increased (four to six times) rate of formation of naphtha, kerosene along with a four-unit increase in the rate of formation for diesel range hydrocarbons upon increasing temperature (360°C) as compared to those obtained at lower temperature, 320°C (Table 5.6). The substantial cracking and change in the reaction pathway observed upon increasing temperature (320–360°C) was attributed to acidic intermediates produced over the $CoMo/Al_2O_3$ catalyst system, which promoted cracking reactions at 360°C. Reaction pathways were affected by the acidic intermediates produced during the hydroconversion of lipids over different catalysts. These intermediates promoted cracking reactions and changed the type of reactions that occurred. Over MTS these promoted mild cracking reactions along with oligomerization reactions at lower temperatures of 300°C and 320°C.

Upon co-processing the lipids from soy and jatropha oils a marginal increase in the rate of the HDS reaction was observed upon increasing the lipid concentration in gas oil (Table 5.6). A decrease in activation energy required for HDS reactions over $NiMo/Al_2O_3$ catalytic systems also supports the above conclusion. Reduction in the activation energy upon increasing the lipid concentration may be attributed to different feed/catalyst interactions that occur in the presence of lipids (which produces acidic intermediates). Activation energy on the order of 40 kJ/mole for the conversion of triglycerides were observed in the case of the MTS systems [32], compared to only 26 kJ/mole in the case of the Al_2O_3 system [31]. This indicated that different conversion pathways were followed over these catalytic systems at lower temperatures (<320°C). Very high activation energy for hydrodeoxygenation reactions was observed (130–150 kJ/mole) while co-processing lipids over $NiMo/Al_2O_3$ catalyst systems as compared to direct processing of lipids, indicating direct processing of lipids is advantageous due to the absence of competing HDS reactions.

A very high activation energy of around 127 kJ/mole was required for the conversion of jatropha oil lipids into aviation fuels over the $CoMo/Al_2O_3$ catalyst system. A catalyst system with lower activation energy for the formation of aviation kerosene would therefore be beneficial in terms of increasing yields as well as reducing the cost of production of these fuels.

5.4 PATH FORWARD AND CHALLENGES

Aviation kerosene is a global commodity, and due to various global drivers, the aviation sector is at the brink of needing to develop alternative fuels. Many organizations such as DARPA, IATA, NATO, and the European Union (EU) have already taken enormous interest in the development of an alternative aviation fuel, funding projects that will bring green jet fuel to commercial flights and to strategic applications. It is expected that by 2050 biofuels will reduce aviation emissions by 80% [83,84]. United Airlines signed an agreement with AltAir to buy 15 million gallons of Bio-jet fuel over a period of three years starting in 2014. This fuel is for use in flights departing from their Los Angeles hub. AltAir Fuels will retrofit idle portions of a Los Angeles petroleum refinery to produce 30 million gallons of advanced biofuels and chemicals per year from nonedible natural oils and agricultural wastes using a two-step process. This renewable jet fuel is expected to achieve at least a 50% reduction in greenhouse gas emissions on a lifecycle basis [84].

In light of the above (both fuel demand and environmental issues), it can be said that aviation fuel from renewable sources has great potential and could provide a clear alternative solution to fossil fuels. For the commercial production of hydrocarbons from vegetable oil (lipid), major challenges with respect to hydroprocessing technology are as follows: (1) high hydrogen requirement, (2) high exothermicity of the reaction, (3) continuous feed supply for large production, and (4) feed pretreatment issues.

High hydrogen requirement: A hydrocracking unit is typically the largest hydrogen consumer in the refinery, and hydroprocessing of vegetable oil requires much more (nearly two to three times) hydrogen compared to a typical hydrocracker. In addition to typical hydrocracking reactions, other reactions like hydrodeoxygenation and saturation of a high percentage of unsaturated components in some vegetable oils like jatropha are the main reason for such high hydrogen consumption. The hydrogen requirement may also increase as unsaturated hydrocarbons form due to unwanted cracking reactions becoming saturated and requiring extra hydrogen, which adds additional cost to the process. For commercial viability, purification and recycling of the unused hydrogen is necessary to make the process economical. Even though simple water washing or amine washing is generally done to recycle unused hydrogen, it cannot remove certain impurities, specifically CO, which may adversely affect catalyst performance. New technologies like membrane purification give better hydrogen purity compared to conventional systems. Impurities such as CO, CO_2, and methane can be removed using such membrane systems to obtain the desired (99.9%) purity of hydrogen in the recycle stream.

High exothermicity of the reaction: The conversion of renewable feedstocks such as plant, animal, and algal oil triglycerides and free fatty acids into aviation fuel and other hydrocarbons is highly exothermic. Therefore, quench hydrogen requirements, reactor bed temperature controls, and high deactivation rates for the catalyst are major concerns for commercialization of the processes. The high exothermicity of these reactions also leads to excess cracking and coke formation reactions in the catalyst pores, further leading to high pressure drops, low catalyst life, and hence, a less economical process.

Continuous feed supply for large production: To make the process competitive, it is necessary to have a continuous supply of feedstock to be converted into fuel, while taking care to avoid food-versus-fuel issues and deforestation. Besides waste cooking oils, cultivation of oilseed bearing trees (afforestation) could be one of the ways to increase the feed supply in an environment-friendly way. Government support in terms of policy making and funding is a must for advancing renewable aviation fuel technology.

Feed pretreatment issue: Most of the vegetable oil contains significant quantities of impurities (such as metals, Na, K, Ca, etc.) and large amount of phosphorus, which need to be completely removed prior to conversion into aviation fuel so as to maintain catalyst life. To increase catalyst life it is important to pretreat the feed as it contains metals that may cause catalyst poisoning. The metals in vegetable oil are generally present along with phospholipids. To remove these metals and phospholipids, a 'water degumming' process and guard bed catalysts are required.

Overcoming the above challenges can make the renewable aviation fuel from the hydroprocessing of lipids competitive with fossil-based aviation fuel. Fuel coming from renewable sources has marginally higher calorific value as well as low sulfur and aromatics content, which results in lower emissions compared to fossil fuels.

5.5 CONCLUSION

It is clear that there is a great need for alternative sources of aviation fuel. Due to stringent specifications for aviation fuels, higher investment in terms of engine/aircraft modifications, and complications in terms of fuel certification strategy, etc., alternative aviation fuels that are drop-in by nature are the most promising solutions to meeting increased fuel and environmental demands.

Hydroprocessing routes to the production of alternative aviation fuels have become a well-established technology, though economically not yet cost-competitive due to the higher cost of animal- and plant-derived triglycerides/lipids. Among different catalysts reported and discussed in this chapter, the sulfided mesoporous catalysts with moderate acidity and higher surface area are the best. Since these catalysts are similar to currently used hydrocracking catalysts, they are easier to retrofit in the current refinery infrastructure for large-scale production, even under co-processing conditions. As discussed above, if challenges like regular feedstock availability, control of exothermicity of the reaction by using modern strategies (liquid quench, heat integration), pretreatment of feedstock to increase the catalyst life, and recycle gas purification are met, and government support to farmers/suppliers to streamline these alternative feedstocks are provided, it is expected that aviation fuel could be produced from animal- and plant-based oils (lipids) competitively.

References

[1] I. Kubickova, D. Kubicka, Utilization of triglycerides and related feedstocks for production of clean hydrocarbon fuels and petrochemicals: a review, Waste Biomass Valor. 1 (2010) 293–308.

[2] M. Stocker, Biofuels and biomass-to-liquid fuels in the biorefinery: catalytic conversion of lignocellulosic biomass using porous materials, Angew. Chem. Int. Ed. Engl. 47 (2008) 9200–9211.

[3] G.W. Huber, A. Corma, Synergies between bio- and oil refineries for the production of fuels from biomass, Angew. Chem. Int. Ed. Engl. 46 (2007) 7184–7201.

[4] J. Calero, D. Luna, E.D. Sancho, C. Luna, F.M. Bautista, A.A. Romero, et al., An overview on glycerol-free processes for the production of renewable liquid biofuels, applicable in diesel engines, Renew. Sust. Energy Rev. 42 (2015) 1437–1452.

[5] L. Li, E. Coppola, J. Rine, J.L. Miller, D. Walker, Catalytic hydrothermal conversion of triglycerides to non-ester biofuels, Energy Fuels 24 (2010) 1305–1315.

[6] C.E. Wyman, Alternative fuels from biomass and their impact on carbon dioxide accumulation, Appl. Biochem. Biotechnol. 45–46 (1994) 897.

[7] A.K. Sinha, M. Anand, B.S. Rana, R. Kumar, S.A. Farooqui, M.G. Sibi, et al., Development of hydroprocessing route to transportation fuels from non-edible plant-oils, Catal. Surv. Asia 17 (2013) 1–13.

[8] R. Kumar, B.S. Rana, R. Tiwari, D. Verma, R. Kumar, R. K. Joshi, et al., Hydroprocessing of jatropha oil and its mixtures with gas oil, Green Chem. 12 (2010) 2232–2239.

[9] S. Bezergianni, A. Kalogianni, Hydrocracking of used cooking oil for biofuels production, Bioresour. Technol. 100 (2009) 3927–3932.

[10] J.O. Metzger, Production of liquid hydrocarbons from biomass, Angew. Chem. Int. Ed. Engl. 45 (2006) 696–698.

[11] Y. Liu, R. Sotelo-Boyas, K. Murata, T. Minowa, K. Sakanishil, Hydrotreatment of Jatropha Oil to Produce Green Diesel over Trifunctional Ni–Mo/SiO2–Al2O3 Catalyst, Chem. Lett. 38 (2009) 552–553.

[12] A.L. Mascarelli, Gold rush for algae, Nature 461 (2009) 460–461.

[13] A.A. Lappas, S. Bezergianni, I.A. Vasalos, Production of biofuels via co-processing in conventional refining processes, Catal. Today 145 (2009) 55–62.

[14] R. Tiwari, B.S. Rana, R. Kumar, D. Verma, R. Kumar, R.K. Joshi, et al., Hydrotreating and hydrocracking catalysts for processing of waste soya-oil and refinery-oil mixtures, Catal. Commun. 12 (2011) 559–562.

[15] B. Veriansyah, J.Y. Han, S.K. Kim, S.A. Hong, Y.J. Kim, J.S. Lim, et al., Production of renewable diesel by hydroprocessing of soybean oil: effect of catalysts, Fuel 94 (2012) 578–585.

[16] J. Cheng, T. Li, R. Huang, J. Zhou, K. Cen, Optimizing catalysis conditions to decrease aromatic hydrocarbons and increase alkanes for improving jet biofuel quality, Bioresour. Technol. 158 (2014) 378–382.

[17] P. Simacek, D. Kubicka, I. Kubickova, F. Homola, M. Pospisil, Premium quality renewable diesel fuel by hydroprocessing of sunflower oil, Fuel 90 (2011) 2473–2479.

[18] S. Bezergianni, A. Kalogianni, I.A. Vasalos, Hydrocracking of vacuum gas oil-vegetable oil mixtures for biofuels production, Bioresour. Technol. 100 (2009) 3036–3042.

[19] G.W. Huber, P.O. Connor, A. Corma, Processing biomass in conventional oil refineries: production of high quality diesel by hydrotreating vegetable oils in heavy vacuum oil mixtures, Appl. Catal. A Gen. 329 (2007) 120–129.

[20] M. Krár, S. Kovács, D. Kalló, J. Hancsók, Fuel purpose hydrotreating of sunflower oil on CoMo/Al_2O_3 catalyst, Bioresour. Technol. 101 (2010) 9287–9293.
[21] A. Sirifa, K. Faungnawakiji, V. Itthibenchapong, N. Viriya-empikul, T. Charinpanitkul, S. Assabumrungrat, Production of bio-hydrogenated diesel by catalytic hydrotreating of palm oil over $NiMoS_2$/γ-Al_2O_3 catalyst, Bioresour. Technol. 158 (2014) 81–90.
[22] A. Vonortas, D. Kubika, N. Papayannakos, Catalytic co-hydroprocessing of gasoil–palm oil/AVO mixtures over a NiMo/γ-Al2O3 catalyst, Fuel 116 (2014) 49–55.
[23] D. Kubicka, P. Simacek, N. Zilkova, Transformation of vegetable oils into hydrocarbons over mesoporous-alumina-supported CoMo catalyst, Top. Catal. 52 (2009) 161–168.
[24] P. Šimácek, D. Kubicka, G. Šebor, M. Pospíšil, Hydroprocessed rapeseed oil as a source of hydrocarbon-based biodiesel, Fuel 88 (2009) 456–460.
[25] J. Walendziewski, M. Stolarski, R. Łużny, B. Klimek, Hydroprocesssing of light gas oil—rape oil mixtures, Fuel Process. Technol. 90 (2009) 686–691.
[26] E. Meller, U. Green, Z. Aizenshtat, Y. Sasson, Catalytic deoxygenation of castor oil over Pd/C for the production of cost effective biofuel, Fuel 133 (2014) 89–95.
[27] S. Liu, Q. Zhu, Q. Guan, L. He, W. Li, Bio-aviation fuel production from hydroprocessing castor oil promoted by the nickel-based bifunctional catalysts, Bioresour. Technol. 183 (2015) 93–100.
[28] B. Rozmyslowicz, P. Maki-Arvela, S. Lestari, O.A. Simakova, K. Eranen, I.L. Simakova, et al., Catalytic deoxygenation of tall oil fatty acids over a palladium-mesoporous carbon catalyst: a new source of biofuels, Top. Catal. 53 (2010) 1274–1277.
[29] D. Verma, R. Kumar, B.S. Rana, A.K. Sinha, Aviation fuel production from lipids by a single-step route using hierarchical mesoporous zeolites, Energy Environ. Sci. 4 (2011) 1667–1671.
[30] K. Murata, Y. Liu, M. Inaba, I. Takahara, Production of synthetic diesel by hydrotreatment of jatropha oils using Pt − Re/H-ZSM-5 catalyst, Energy Fuels 24 (2010) 2404–2409.
[31] M. Anand, A.K. Sinha, Temperature-dependent reaction pathways for the anomalous hydrocracking of triglycerides in the presence of sulfided Co-Mo-catalyst, Bioresour. Technol. 126 (2012) 148–155.
[32] R.K. Sharma, M. Anand, B.S. Rana, R. Kumar, S.A. Farooqui, M.G. Sibi, et al., Jatropha-oil conversion to liquid hydrocarbon fuels using mesoporous titanosilicate supported sulfide catalysts, Catal. Today 198 (2012) 314–320.
[33] M. Anand, M.G. Sibi, D. Verma, A.K. Sinha, Anomalous hydrocracking of triglycerides over CoMo-catalyst–influence of reaction intermediates, J. Chem. Sci. 126 (2014) 473–480.
[34] D. Verma, B.S. Rana, R. Kumar, M.G. Sibi, A.K. Sinha, Diesel and Aviation Kerosene with desired Aromatics from Hydroprocessing of Jatropha oil over Hydrogenation Catalysts Supported on Hierarchical Mesoporous SAPO-11, J. Apcata 490 (2015) 108–116.
[35] P. Rizo-Acosta, M.T. Linares-Vallejo, J.A. Munoz-Arroyo, Co-hydroprocessing of a mixture: vegetable oil/n-hexadecane/4,6-dimethyldibenzothiophene for the production of sustainable hydrocarbons. A kinetic modelling, Catal. Today 234 (2014) 192–199.
[36] J. Guo, G. Xu, F. Shen, Y. Fu, Y. Zhang, Q. Guo, Catalytic conversion of jatropha oil to alkanes under mild conditions with a Ru/La$(OH)_3$ catalyst, Green Chem. 17 (2015) 2888–2895.
[37] K. Fan, J. Liu, X. Yang, L. Rong, Hydrocracking of jatropha oil over Ni-$H_3PW_{12}O_{40}$/nano-hydroxyapatite catalyst, Int. J. Hydrogen Energy 39 (2014) 3690–3697.
[38] F. Pinto, F.T. Varela, M. Goncalves, R.N. Andre, P. Costa, B. Mendes, Production of bio-hydrocarbons by hydrotreating of pomace oil, Fuel 116 (2014) 84–93.
[39] S. Bezergianni, S. Voutetakis, A. Kalogianni, Catalytic Hydrocracking of Fresh and Used Cooking Oil, Ind. Eng. Chem. Res. 48 (2009) 8402–8406.
[40] S. Bezergianni, A. Dimitriadis, A. Kalogianni, P.A. Pilavachi, Hydrotreating of waste cooking oil for biodiesel production. Part I: effect of temperature on product yields and heteroatom removal, Bioresour. Technol. 101 (2010) 6651–6656.
[41] S. Bezergianni, A. Dimitriadis, T. Sfetsas, A. Kalogianni, Hydrotreating of waste cooking oil for biodiesel production. Part II: effect of temperature on hydrocarbon composition, Bioresour. Technol. 101 (2010) 7658–7660.
[42] S. Bezergianni, A. Dimitriadis, A. Kalogianni, K.G. Knudsen, Toward hydrotreating of waste cooking oil for biodiesel production. Effect of pressure, H_2/oil ratio, and liquid hourly space velocity, Ind. Eng. Chem. Res. 50 (2011) 3874–3879.
[43] B.S. Rana, R. Kumar, R. Tiwari, R. Kumar, R.K. Joshi, M.O. Garg, et al., Transportation fuels from co-processing of waste vegetable oil and gas oil mixtures, J. Biombioe 56 (2013) 43–52.
[44] S. Bezergianni, A. Dimitriadis, G. Meletidis, Effectiveness of CoMo and NiMo catalysts on co-hydroprocessing of heavy atmospheric gas oil-waste cooking oil mixtures, Fuel 125 (2014) 129–136.
[45] S. Melis, S. Mayo, B. Leliveld, Vegetable oil co-processing in diesel hydrotreaters, Biofuels Technol. 1 (2009) 43–47.
[46] T. Morgan, D. Grubb, E.S. Jimenez, M. Crocker, Conversion of triglycerides to hydrocarbons over supported metal catalysts, Top. Catal. 53 (2010) 820.

[47] L. Boda, G. Onyestyak, H. Solt, F. Lonyi, J. Valyon, A. Thernesz, Catalytic hydroconversion of tricaprylin and caprylic acid as model reaction for biofuel production from triglycerides, Appl. Catal. A Gen. 374 (2010) 158–169.

[48] S. Lestari, P. Maki-Arvela, H. Bernas, O. Simakova, R. Sjoholm, J. Beltramini, et al., Catalytic deoxygenation of stearic acid in a continuous reactor over a mesoporous carbon supported Pd catalyst, Energy Fuels 23 (2009) 3842–3845.

[49] M. Snare, I. Kubickova, P. Maki-Arvela, K. Eranen, D. Y. Murzin, Heterogeneous catalytic deoxygenation of stearic acid for production of biodiesel, Ind. Eng. Chem. Res. 45 (2006) 5708–5715.

[50] M. Ahmadi, A. Nambo, J.B. Jasinski, P. Ratnasamy, M.A. Carreon, Decarboxylation of oleic acid over Pt catalysts supported on small-pore zeolites and hydrotalcite, Catal. Sci. Technol. 5 (2015) 380–388.

[51] J.G. Immer, H.H. Lamb, Fed-batch catalytic deoxygenation of free fatty acids, Energy Fuels 24 (2010) 5291–5299.

[52] B. Donnis, R.G. Egeberg, P. Blom, K.G. Knudsen, Hydroprocessing of bio-oils and oxygenates to hydrocarbons. Understanding the reaction routes, Top. Catal. 52 (2009) 229–240.

[53] Y. Okamoto, M. Breysse, G.M. Dhar, C. Song, Effect of support in hydrotreating catalysis for ultra clean fuels, Catal. Today 86 (2003) 1.

[54] G.M. Dhar, B.N. Srinivas, M.S. Rana, M. Kumar, S.K. Maity, Mixed oxide supported hydrodesulfurization catalysts—a review, Catal. Today 86 (2003) 45.

[55] M. Breysse, P. Afanasiev, C. Geantet, M. Vrinat, Overview of support effects in hydrotreating catalysts, Catal. Today 86 (2003) 5–16.

[56] S.K. Maity, B.N. Srinivas, V.V.D.N. Prasad, A. Singh, G.M. Dhar, T.S.R.P. Rao, Recent advances in basic and applied aspects of industrial catalysis, Stud. Surf. Sci. Catal. 113 (1998) 579.

[57] R. Prins, V.H.J.D. Beer, G.A. Somorjai, Structure and Function of the Catalyst and the Promoter in Co—Mo Hydrodesulfurization Catalysts, Catal. Rev. Sci. Eng. 31 (1989) 1–41.

[58] T.M. Sankaranarayanan, M. Banu, A. Pandurangan, S. Sivasanker, Hydroprocessing of sunflower oil-gas oil blends over sulfided Ni-Mo-Al-zeolite beta composites, Bioresour. Technol. 102 (2011) 10717–10723.

[59] L.C. Caero, A.R. Romero, J. Ramirez, Niobium sulfide as a dopant for Mo/TiO2 catalysts, Catal. Today 78 (2003) 513–518.

[60] M.C. Barrera, M. Viniegra, J. Escobar, M. Vrinat, J.A.D.L. Reyes, F. Murrieta, et al., Highly active MoS_2 on wide-pore ZrO_2-TiO_2 mixed oxides, Catal. Today 98 (2004) 131.

[61] S.K. Maity, M.S. Rana, B.N. Srinivas, S.K. Bej, G.M. Dhar, T.S.R.P. Rao, Characterization and evaluation of ZrO_2 supported hydrotreating catalysts, J. Mol. Catal. A Chem. 153 (2000) 121.

[62] G.M. Dhar, F.E. Massoth, J. Shabtai, Catalytic functionalities of supported sulfides: I. Effect of support and additives on the CoMo catalyst, J. Catal. 85 (1984) 44–52.

[63] S. Damyanova, L. Petrov, M.A. Centeno, P. Grange, Characterization of molybdenum hydrodesulfurization catalysts supported on ZrO_2-Al_2O_3 and ZrO_2-SiO_2 carriers, Appl. Catal. A Gen. 224 (2002) 271.

[64] M.S. Rana, E.M.R. Capitaine, C. Leyva, J. Ancheyta, Effect of catalyst preparation and support composition on hydrodesulfurization of dibenzothiophene and Maya crude oil, Fuel 86 (2007) 1254.

[65] M.S. Rana, S.K. Maity, J. Ancheyta, G.M. Dhar, T.S.R.P. Rao, MoCo(Ni)/ZrO_2-SiO_2 hydrotreating catalysts: physico-chemical characterization and activities studies, Appl. Catal. A Gen. 268 (2004) 89.

[66] F.E. Massoth, G.H. Dhar, J. Shabtai, Catalytic functionalities of supported sulfides: II. Effect of support on Mo dispersion, J. Catal. 85 (1984) 53.

[67] F.P. Daly, H. Ando, I.L. Schmitt, E.A. Sturm, Hydrodesulfurization over a $TiO_2 \cdot ZrO_2$-supported $CoO \cdot MoO_3$ catalyst, J. Catal. 108 (1987) 401–408.

[68] M.S. Rana, B.N. Srinivas, S.K. Maity, G.M. Dhar, T.S. R.P. Rao, Origin of Cracking Functionality of Sulfided (Ni) CoMo/SiO_2–ZrO_2 Catalysts, J. Catal. 195 (2000) 31.

[69] W. Zhaobin, X. Qin, G. Xiexian, E.L. Sham, P. Grange, B. Delmon, Titania-modified hydrodesulfurization catalysts: II. Dispersion state and catalytic activity of molybdena supported on titania-alumina carrier, Appl. Catal. 75 (1991) 179.

[70] D. Li, A. Nishijima, D.E. Morris, Zeolite-supported Ni and Mo catalysts for hydrotreatments: I. catalytic activity and spectroscopy, J. Catal. 182 (1999) 339.

[71] W.J.J. Welters, G. Vorbeck, H.W. Zandbergen, L.J.M. V.D. Ven, E.M.V. Oers, J.W.D. Haan, et al., NaY-supported molybdenum sulfide catalysts: I. catalysts prepared via impregnation with ammonium heptamolybdate, J. Catal. 161 (1996) 819.

[72] R. Zarchin, M. Rabaev, R. Vidruk-Nehemya, M.V. Landau, M. Herskowitz, Hydroprocessing of soybean oil on nickel-phosphide supported catalysts, Fuel 139 (2015) 684–691.

[73] A. Wang, Y. Wang, T. Kabe, Y. Chen, A. Ishihava, W. Qian, et al., Hydrodesulfurization of Dibenzothiophene over Siliceous MCM-41-Supported Catalysts: II. Sulfided Ni–Mo Catalysts, J. Catal. 210 (2002) 319.

[74] T. Klimova, M. Calderon, J. Ramirez, Ni and Mo interaction with Al-containing MCM-41 support and its effect on the catalytic behavior in DBT hydrodesulfurization, Appl. Catal. A Gen. 240 (2003) 29.

[75] A. Wang, Y. Wang, T. Kabe, Y. Chen, A. Ishihara, W. Qian, Hydrodesulfurization of Dibenzothiophene over Siliceous MCM-41-Supported Catalysts: I. Sulfided Co–Mo Catalysts, J. Catal. 199 (2001) 19.

[76] U.T. Turaga, C. Song, MCM-41-supported Co-Mo catalysts for deep hydrodesulfurization of light cycle oil, Catal. Today 86 (2003) 129.

[77] A. Corma, A. Martinez, V.M. Soria, J.B. Monton, Hydrocracking of Vacuum Gasoil on the Novel Mesoporous MCM-41 Aluminosilicate Catalyst, J. Catal. 153 (1995) 25.

[78] T. Chiranjeevi, P. Kumar, M.S. Rana, G.M. Dhar, T.S. R.P. Rao, Physico-chemical characterization and catalysis on mesoporous Al-HMS supported molybdenum hydrotreating catalysts, J. Mol. Catal. A Chem. 181 (2002) 109.

[79] G.M. Dhar, G.M. Kumaran, M. Kumar, K.S. Rawat, L.D. Sharma, B.D. Raju, et al., Physico-chemical characterization and catalysis on SBA-15 supported molybdenum hydrotreating catalysts, Catal. Today 99 (2005) 309.

[80] B.A. Alwan, S.O. Salley, K.Y.N. Simon, Hydrocracking of DDGS corn oil over transition metal carbides supported on Al-SBA-15: effect of fractional sum of metal electronegativities, Appl. Catal. A Gen. 485 (2014) 58–66.

[81] T.H. Applewhite, Fats and fatty oils, in: M. Grayson, D. Eckorth (Eds.), Kirk-Othmer, Encyclopedia of Chemical Technology, 9, John Wiley and Sons, New York, NY, 1979, pp. 795–831.

[82] S.K. Kim, J.Y. Han, H. Lee, T. Yuma, Y. Kim, J. Kim, Production of renewable diesel via catalytic deoxygenation of natural triglycerides: Comprehensive understanding of reaction intermediates and hydrocarbons, Appl. Energy 116 (2014) 199–205.

[83] Sustainable Aviation CO2 Road-MAP 2012. <http://www.sustainableaviation.co.uk/>, March 2012 (accessed August 2015).

[84] Committee on Climate Change, report on "Meeting the UK aviation target – options for reducing emissions to 2050". <https://www.theccc.org.uk/archive/aws2/Aviation%20Report%2009/21667B%20CCC%20Aviation%20AW%20COMP%20v8.pdf>, 2009.

CHAPTER

6

Low-Carbon Aviation Fuel Through the Alcohol to Jet Pathway

K.P. Brooks[1], L.J. Snowden-Swan[1], S.B. Jones[1], M.G. Butcher[1], G.-S.J. Lee[1], D.M. Anderson[1], J.G. Frye[1], J.E. Holladay[1], J. Owen[2], L. Harmon[2], F. Burton[2], I. Palou-Rivera[2], J. Plaza[3], R. Handler[4] and D. Shonnard[4]

[1]Pacific Northwest National Laboratory (PNNL), Richland, WA, United States, [2]LanzaTech, Skokie, IL, United States, [3]Kopius Energy Solutions LLC, Seattle, WA, United States, [4]Michigan Technological Institute, Houghton, MI, United States

ACRONYMS AND ABBREVIATIONS

A4A	Airlines for America
AEO	Annual Energy Outlook
AFRL	Air Force Research Laboratory
ASTM	American Society for Testing and Materials
ATJ	alcohol to jet
BOF	basic oxygen furnace
CAAFI	Commercial Aviation Alternative Fuels Initiative
CAFE	corporate average fuel economy
EIA	US Energy Information Administration
EISA	Energy Independence and Security Act
EPA	US Environmental Protection Agency
FAA	Federal Aviation Administration
FAME	fatty acid methyl esters
F–T	Fischer–Tropsch
FT-SPK	Fischer–Tropsch hydroprocessed synthesized paraffinic kerosene
GDP	gross domestic product
GHG	greenhouse gas
HEFA	hydroprocessed esters and fatty acids
HEFA-SPK	synthesized paraffinic kerosene from hydroprocessed esters and fatty acids
IATA	International Air Transport Association
JIG	Joint Inspection Group
LCA	life cycle assessment
LOHAS	lifestyles of health and sustainability
LTO	landing/take-off
MBM	market-based measure
MJ	megajoule
OECD	Organization for Economic Co-operation and Development
OPEC	Organization of the Petroleum Exporting Countries
PM	particulate matter
RFS2	Renewable Fuel Standard (programme)
RIN	Renewable Identification Number
SIP	synthetic isoparaffins
SKA	synthetic kerosene containing aromatics
SPK	synthetic paraffin kerosene
SwRI	Southwest Research Institute
VFA	volatile fatty acid

Biofuels for Aviation.
DOI: http://dx.doi.org/10.1016/B978-0-12-804568-8.00006-8

6.1 INTRODUCTION

The purpose of this chapter is to review the opportunities for production of alternative jet fuel components from alcohol. This includes a review of potential conversion technologies and their state of development, a survey of the market and economic drivers, and an assessment of competing technologies. Given the lack of public information, the production costs for the various conversion technologies were not assessed.

Concerns about climate change and energy supply are driving the production of more sustainable fuels and chemicals [1a]. Renewable fuels such as bioethanol are a promising alternative to fossil-based transport fuels, and extensive research and development has been carried out to deploy technologies for their commercial production [1b].

The International Air Transport Association (IATA), comprising the top 250 airlines in the world and representing 83% of global air traffic (IATA 2015), has committed to carbon-neutral growth by 2020, stating that it intends to reduce emissions by 50% in 2050 from its 2005 levels. This is the key driver for the world's development of sustainable jet fuel solutions. In order to achieve carbon-neutral growth, new aviation fuel technologies must be commercialized. The most significant piece of the sustainable jet fuel puzzle is the development of competitively priced, sustainable aviation fuels.

In addition to reducing the aviation sector's carbon footprint, sustainable aviation fuel is perhaps the best solution to reduce dependence on fossil fuels and offset the risks associated with jet fuel price volatility. While significant improvements have been made in operational efficiency in recent years, airlines need low-carbon fuel options in order to continue to grow as an industry. While this is due to various reasons, including possible forthcoming regulations that would limit aircraft emissions, airlines are also particularly vulnerable to petroleum price swings. For example, from 2001 to 2012, the price of crude oil increased by 262%. Alternative fuels may provide better price stability compared to conventional fuels. Not only is the price of fuel the airlines' largest cost component, it remains its most volatile. During that time, 25 airlines were bankrupted or stopped operations largely due to fuel costs [1c].

Because of this, fuel represents nearly 40% of an airline's total operating costs [2]. This is a significant driver for sustainable aviation fuel. Meanwhile, in an effort to control price volatility and as a matter of national security, governments are also pushing for sustainable aviation fuel use in their fleets. For example, the US Navy has demonstrated alternative fuels with their aircraft and has announced a goal of supplying half of its total energy consumption from alternative sources by 2020 (Fig. 6.1).

According to IATA 'a 3–6% share in global commercial aviation fuel demand should be achievable by 2020'. This translates to almost 5 billion gallons per year of sustainable aviation fuel, a rather significant market.

In order to support these market drivers, there has been a plethora of activities by various aviation stakeholders, including, but not limited to, airlines, aircraft manufacturers, engine manufacturers, government organizations with the

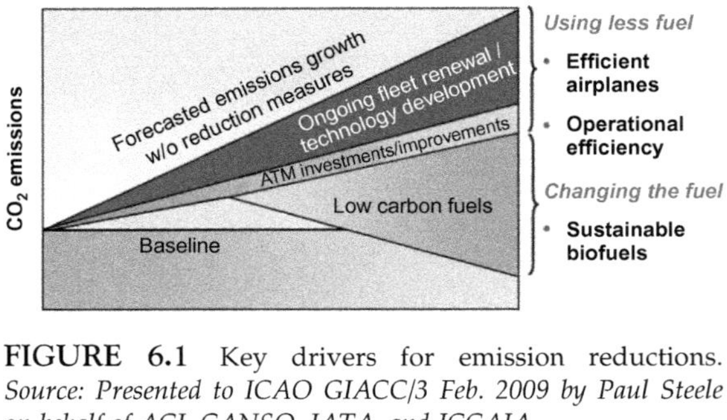

FIGURE 6.1 Key drivers for emission reductions. *Source: Presented to ICAO GIACC/3 Feb. 2009 by Paul Steele on behalf of ACI, CANSO, IATA, and ICCAIA.*

military playing a key role, technology developers, feedstock providers, nongovernmental organizations, as well as corporations looking to improve their sustainability footprint. In addition to the technology development and partnership announcements in the space, numerous workshops, conferences, and road mapping exercises evaluating the challenges and opportunities for achieving aviation industry goals have occurred, and while many are global exercises, some are region and country specific. Meanwhile, thousands of flights have already demonstrated the use of alternative jet fuel in commercial flights. Airlines involved include Lufthansa, Qatar Airways, United Airlines, Virgin Atlantic, British Airways, among others.

With this in mind, we provide a new look at producing jet fuel from alcohols. Our own research in this space starts with waste gases rich in carbon monoxide, such as co-produced in steel production. LanzaTech has demonstrated at large scale in Asia the conversion of such gases to ethanol. Working with its partner, Pacific Northwest National Laboratory (PNNL), we have developed new routes that can either produce cyclic hydrocarbons or selectively produce only synthetic paraffinic kerosene (SPK), which match current needs. The key aspect of this focus is producing the intermediate alcohol from a stream that today is not valorized and which contributes to greenhouse gases (GHGs). This provides a new paradigm, as feedstock costs are the largest operating cost in producing renewable fuels. In addition to our own process we will examine the other alcohol to jet (ATJ) options forthcoming.

6.2 BACKGROUND

6.2.1 Types of Jet Fuel

Many types of jet fuel are used for commercial and military aviation. Some fuels used more extensively in past years have since become less preferable to fuels with more desirable properties. The following sections describe the most widely used conventional kerosene-grade turbine aviation fuels. This serves to highlight the types of fuels and properties that the ethanol to jet blend components will need to meet. Select specifications for each fuel are given in Appendix A.

6.2.1.1 *Jet A*

Jet A became the standard jet fuel for all United States and many international commercial airlines in the 1960s when it was selected over the more highly flammable JP-4 for passenger safety reasons [3]. Now it is normally available only in the United States, where it is used in civil aviation for domestic flights and international flights originating in the United States because of its relatively low cost. It typically has a flash point of 38°C and a freeze point maximum of −40°C [4].

6.2.1.2 *Jet A-1*

Jet A-1 is available and used worldwide, including in the United States. It is virtually the same as Jet A in its basic properties, but with agreement between the supplier and purchaser, it may require one or more additives not usually required for Jet A (eg, static dissipater, icing inhibitor, or antioxidant). The flash point minimum is the same as for Jet A, but Jet A-1 has a lower freezing point maximum of −47°C [4]. The lower freezing point makes Jet A-1 more suitable for long international flights, especially on polar routes during the winter [5].

6.2.1.3 *JP-8*

JP-8 is currently the dominant military jet fuel grade for NATO air forces [6]. Military aircraft have been using JP-8 since the 1980s [3]. JP-8 is essentially Jet A-1, but with the addition of a static dissipater additive, corrosion inhibitor/lubricity improver, and fuel system icing inhibitor, and it may also contain antioxidant and metal deactivators [7]. Like Jet

A-1, it is specified to a maximum freezing point of −47°C and a flash point of 38°C. The US Air Force is in the process of converting the continental US operations from military specification fuel (JP-8) to commercial specification jet fuel (Jet A) with specified additives to meet its performance requirements [8]. The impact this will have on future US global military operations is yet to be seen.

6.2.1.4 JP-9 and JP-10

JP-9 and JP-10 are specialty fuels designed for applications, such as air-launched cruise missiles, that require very clean-burning fuel, superior volumetric energy content, and low-temperature performance [5,9]. JP-9 has been replaced by JP-10 in most military applications [9,10]. JP-10 is a high-density synthetic missile fuel composed exclusively of C_{10} molecules [11,12]. It is tailored for operational use with high density as a primary requirement [13]. JP-10 is higher priced than kerosene jet fuel, because of a lower volumetric demand, and it is more expensive to make [14]. The Navy is exploring the potential of using JP-10 as a blending component with conventional fuels for applications in renewable missile or air-breathing rocket fuels [11].

6.2.1.5 Jet Fuels Containing Hydrocarbons From Nonconventional Sources

Sasol[1] semisynthetic fuel is a blend of conventionally produced kerosene and a synthetic isoparaffinic kerosene alone, or combined with Sasol heavy naphtha #1, and is specified in Defense Standard 91-91 [4]. Sasol fully synthetic fuel is a blend of up to five synthetic streams specified in Defense Standard 91-91 (ASTM Specification D1655, Annex A1). These two semisynthetic jet fuels are said to have composition, properties, and performance that are essentially identical to conventional jet fuels [15], and are considered acceptable fuels from nonconventional sources.

6.2.2 Composition of Turbine Jet Fuels

Conventional aviation turbine fuel consists of refined hydrocarbons derived from conventional sources including crude oil, natural gas liquid condensates, heavy oil, shale oil, and oil sands [4]. There is no standard formula for jet fuels because the exact composition of a jet fuel depends on the crude oil from which it is refined [16]. Kerosene-type jet fuels consist of hundreds of different carbons ranging from 8 to 16 [5]. The major categories are *n*-paraffins, isoparaffins, cycloparaffins, and aromatics. Each effect on fuel properties is shown in Table 6.1 and Fig. 6.2.

The proportion of these components depends on the source of the feedstock and refining processes used to produce the finished fuel. Between 75% and 85% of the fuel by volume is made up of the paraffins (normal-, iso-, and cyclo-paraffins, the latter also referred to as naphthenes), and the remainder from aromatics (less than 25 vol.%) and olefins [17]. Some aromatics are required to ensure proper seal swelling, but above that amount they are the least desirable component. Differences in the proportions of these compound classes can result in differences in the bulk properties of the fuels [5], leading to adjustments in refining processes, blending, or use of additives to provide the required fuel properties.

In addition to its primary function as an energy source for the jet engine, jet fuel also functions as hydraulic fluid and coolant. Fuel additives are added in small quantities to provide specific performance properties such as enhanced fuel handling and maintenance [4]. Jet A in the United States typically has no additives, but it might include an antioxidant, while Jet A-1 typically includes a static

[1]Sasol Limited, Johannesburg, South Africa

TABLE 6.1 Standard Specifications Used to Control Turbine Jet Fuel Quality

Class	Characteristics	Jet Fuel Properties
Paraffins and isoparaffins	High heat release per unit weight [13], burns cleanly [16], isoparaffins offer potential for low-temp fluidity not offered by *n*-paraffins	Most desirable for combustion
Cycloparaffins (naphthenes)	Lower hydrogen-to-carbon ratio, lower heat release per unit weight, greater density, lower freeze point than the paraffins [13]	Next most desirable
Olefins	Good combustion characteristics, but reactive	Limited to 1% or less because of gum formation
Aromatics	Burns with a smoky flame, releases more of its chemical energy as thermal radiation than other hydrocarbons (ASTM D1655-13)	Least desirable with regard to combustion characteristics
Heteroatoms	Sulphur, nitrogen, oxygen [3]	Limited allowance

dissipater and possibly an antioxidant [5]. JP-8 will contain a static dissipater, corrosion inhibitor/lubricity improver, fuel system icing inhibitor, and possibly an antioxidant and metal deactivator [7]. Greater use of additives is the principal difference between commercial and military jet fuels, with US military jet fuels containing three or more additives [5].

Renewable routes offer the potential to provide the most desirable hydrocarbon components to the mixture that petroleum refineries can produce. ASTM approvals to date have been on those components that are most desirable for combustion quality, isoparaffin-rich SPK.

6.2.3 Approaches to Producing Alternative Fuels

To help ensure an affordable and stable supply of environmentally progressive aviation fuels, the Federal Aviation Administration (FAA), in partnership with airlines, airports, and manufacturers, has launched the Commercial Aviation Alternative Fuels Initiative (CAAFI). With support from CAAFI, and based on the positive results from a series of test flights from 2008 to 2011, the

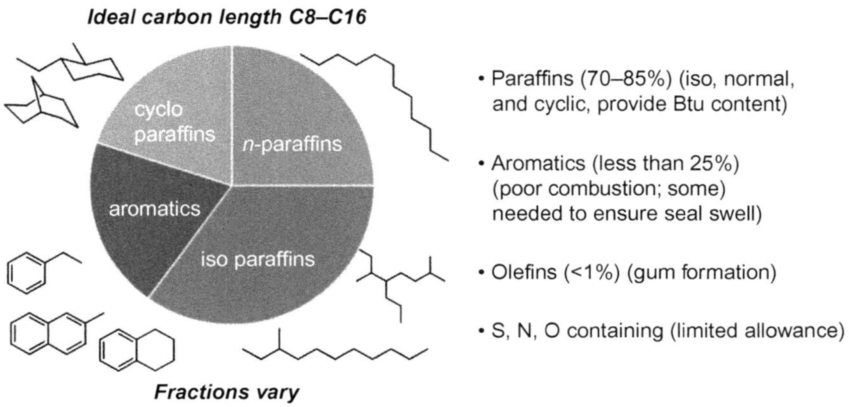

FIGURE 6.2 Hydrocarbon classes in jet fuel.

aviation industry was given approval to use biofuels in passenger flights in the US as defined under ASTM Standard D7566. Some of these initial flights are shown in Appendix B.

The fuels used for these demonstrations were developed from the Fischer–Tropsch (F–T) and the hydroprocessed esters and fatty acids (HEFA) processes. The F–T process is a commercially available technology for converting natural gas and coal to liquid fuels. In this process, the fuels are reformed to produce syngas, a mixture of carbon monoxide and hydrogen. Using the F–T process, the syngas reacts to produce a distribution of straight-chain alkanes. Jet fuel range hydrocarbons are possible with this process, but with a high capital cost that has been a barrier to deployment, as associated capital investments can exceed $100,000 per barrel per day production capacity [18]. The HEFA process uses oil feedstocks from palm and soy, for example, and performs similar processes to those used in the petrochemical industry to produce jet fuel range hydrocarbons. The process includes hydrocracking of the large molecules, hydrodeoxygenation, and hydrogenation to obtain paraffinic hydrocarbons. While this process is less capital cost intensive than the F–T process, it is over 80% dependent on the feedstock cost [18]. This dependency can open up opportunities for technologies that utilize alternate feedstocks.

The F–T and HEFA processes, as well as most of the ATJ processes that are the subject of this chapter, produce SPK, which consists of paraffin hydrocarbons that meet the requirements of ASTM Specification D7566. Due to their production chemistries, SPKs do not contain the levels of aromatics and cycloparaffins required to meet standard specifications. As a result, they are blended up to 50% with petroleum-based jet fuel to provide these additional compounds. When producing alternative fuels using ATJ processes, this blending eliminates the need for aromatization of the paraffins. These processes are discussed in Sections 6.3 and 6.4.

If a drop-in fuel is required with no blending of petroleum-based jet fuel, the alternative fuel must contain the levels of aromatics and cycloparaffins required to meet ASTM D7566-13. This fuel is called synthetic kerosene with aromatics (SKA). Process steps beyond those typically used in SPK production may be needed to achieve the levels of aromatics required for SKA jet fuel. These additional process steps are discussed in Section 6.5.

6.2.4 Standard Specifications

For all new fuel production processes, there is a rigorous evaluation process by ASTM before it can be used commercially. This covers use in the United States, but the approval agencies in other regions follow ASTM.

Because of the complexity of aviation turbine fuels, the specifications have evolved over time into guidelines of performance more than composition [4]. For conventional jet fuels, if results from standard performance property tests fall within specified ranges, they are assumed to have acceptable composition to provide the performance required. This is not the case for new synthetic jet fuels containing hydrocarbons from nonconventional sources. While at least some of these new synthetic fuels may have composition and performance properties that are similar to conventional fuels, limited experience with them requires case-by-case evaluation. This is highlighted in Table 6.2, where conventional fuels require only reporting certain properties, whereas fuels mixed with synthetic hydrocarbons from sources such as FT-SPK or HEFA-SPK, described in Section 6.2.4, must meet specific standards.

It is important to note that while initial requirements are more stringent for these new fuels, once the requirements of ASTM D7566-13 are met and the fuel is certified and released, it is regarded as D1655 aviation turbine fuel [19], and any re-certification is done in accordance with Specification D1655. Further details can be found in Appendix A.

Synthetic jet fuel has been produced and qualified for blending with petroleum jet fuels by ASTM through F–T, HEFA, and most recently, synthetic isoparaffins (SIP) (Farnesene). The ATJ pathway to produce SPK-type fuel has recently been submitted to the ASTM for review [20] and is expected to be approved for use by 2016. The ATJ pathway is of interest to CAAFI and FAA, both for a 50% blend as an SPK and eventually as a 100% replacement for jet fuel (ie, SKA). This route shares many similarities to the approved F–T production route as highlighted below (Fig. 6.3).

TABLE 6.2 Example of Increased Specifications for Nonconventional Fuel Sources

	Jet A Standard Specifications		
Fuel Property	**Conventional Jet A or Jet A-1 (ASTM D1655-13)**	**Conventional Jet A or Jet A-1 Blended With 50% FT-SPK or HEFA-SPK (ASTM D7566-13)**	**100% FT-SPK or HEFA-SPK Blend Component for Jet A or Jet A-1 (ASTM D7566-13)**
COMPOSITION			
Acidity (total mg KOH/g)	≤0.10	≤0.10	≤0.015
Total aromatics (volume %)	≤25	≥8≤25	≥8≤25
Aromatics (mass %)	report	report	≤0.5
Cycloparaffins (mass %)	report	report	≤15
Paraffins, (mass %)	report	report	report
Carbon and hydrogen (mass %)	report	report	≥99.5
NON-CARBON COMPOSITION			
Sulphur (total mass %)	≤0.3	≤0.3	≤0.3
Sulphur (mg/kg)	report	report	≤15
Sulphur, mercaptan, mass percent	≤0.003	≤0.003	≤0.003
Nitrogen (mg/kg)	report	report	≤2
Water (mg/kg)	report	report	≤75
Phosphorus (mg/kg)	report	report	report
Metals (mg/kg)	report	report	≤0.1 per metal
Fatty acid methyl esters (mg/kg)			For HEFA-SPK ≤5
Halogens (mg/kg)	report	report	≤1

6.3 APPROACHES TO PRODUCING SYNTHETIC PARAFFIN KEROSENE (SPK) JET FUEL FROM ETHANOL

Technologies for converting alcohols and other oxygenates to SPK jet fuel are being developed by several companies. A partial list includes BYOGY,[2] Cobalt,[3] Gevo,[4] Swedish Biofuels,[5] Terrabon,[6] UOP,[7] LanzaTech,[8] Zeachem,[9] and Vertimass[10] [21]. While these companies use a variety of alcohols and oxygenated intermediates, this section focuses on the conversion of ethanol to SPK jet fuel. The pathways can be categorized according to the chemistry involved: (1) production through ethylene as an intermediate, (2) production through a propylene intermediate, (3) production through a higher alcohol intermediate, or (4) production through a carbonyl intermediate.

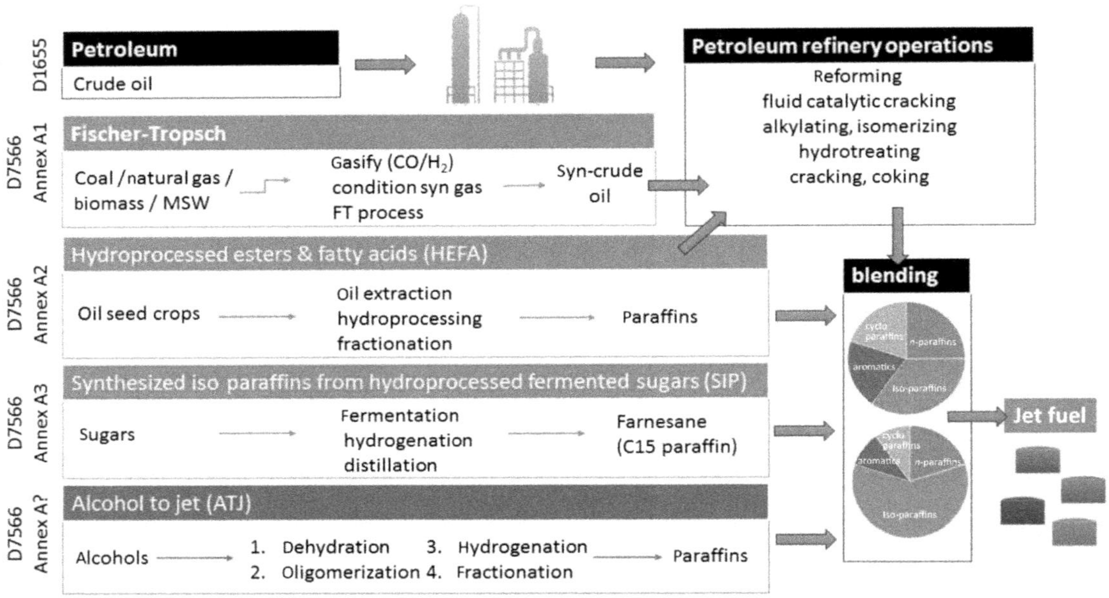

FIGURE 6.3 ASTM pathways for synthetic kerosene.

[2]BYOGY Renewables, Inc., San Jose, CA.

[3]Cobalt Technologies, Mountain View, CA.

[4]Gevo, Inc., Englewood, CO.

[5]Swedish Biofuels, Lidingo, Sweden.

[6]Terrabon, Inc., Houston, TX.

[7]UOP, Des Plaines, IL.

[8]LanzaTech Inc., Skokie, IL.

[9]Zeachem, Inc., Lakewood, CO.

[10]Vertimass, LCC, Irvine, CA.

CASE STUDY: LANZATECH PATHWAY TO CERTIFICATION WITH ASTM

LanzaTech has been involved in navigating the certification process for new fuels since 2010. The ASTM certification process for emerging or new fuels is guided by individuals from the FAA and Southwest Research Institute (SwRI). The process is a collaborative effort amongst a variety of stakeholders, including fuel producers, research laboratories that perform fuel testing, and engine and aircraft manufacturers, amongst others. This process can take several years, and requires different quantities of fuel, based on the type of testing, but these tests generally require larger volumes as testing progresses. Fuel producers are required to provide fuel samples and testing results for an ASTM research report that supports final decision making. LanzaTech is currently involved in the (ATJ) ASTM Taskforce, which is working through the qualification process for jet fuel produced from alcohols. The specific steps within this process are illustrated below:

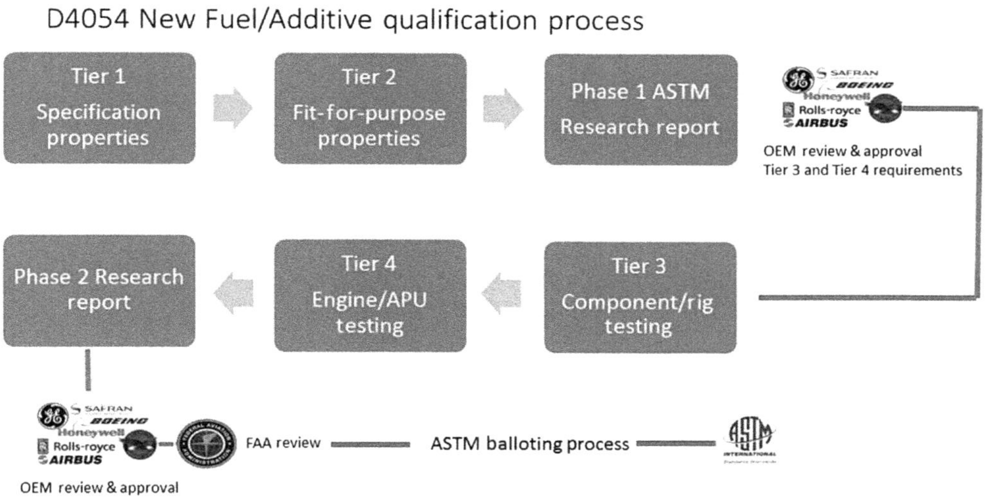

To date, several sustainable aviation fuel pathways have been certified for use, including:

1. Fischer–Tropsch hydroprocessed synthesized paraffinic kerosene (F–T)
2. Synthesized paraffinic kerosene from HEFA
3. Synthesized isoparaffins from hydroprocessed fermented sugars (SIP, farnesane)

A taskforce consisting of fuel producers and engine and airframe Original Equipment Manufacturers (OEMs) was created in 2010 to help certify the ATJ production pathway. A ballot for the production of SPK from an isobutanol feedstock is targeted to take place during the fall of 2015 with the issuing of the modified ASTM D7566 spec. LanzaTech has been actively working with the OEMs and FAA to achieve approval of the ethanol feedstock shortly after this in 2016.

(*cont'd*)

To aid this approval, LanzaTech has collaborated with Virgin Atlantic Airlines and HSBC to install a demonstration plant at its Freedom Pines facility in the United States. LanzaTech has a commercial licence for the PNNL ATJ conversion technology being employed, and has installed the largest ethanol to jet fuel demonstration plant, completing key process scale-up milestones, including employing commercial catalysts and equipment. Project partners have guided the fuel testing plan, helping to ensure smooth approval through ASTM. LanzaTech and Virgin Atlantic will conduct a flight demonstration in 2016 with sustainable aviation fuel produced from the Freedom Pines Facility. The initial feedstock for the ethanol is steelmaking off-gases from the Beijing Shougang LanzaTech New Energy Technology Ltd. demonstration plant in China. This will be a first-of-its-kind demonstration of sustainable aviation fuel produced from waste gases.

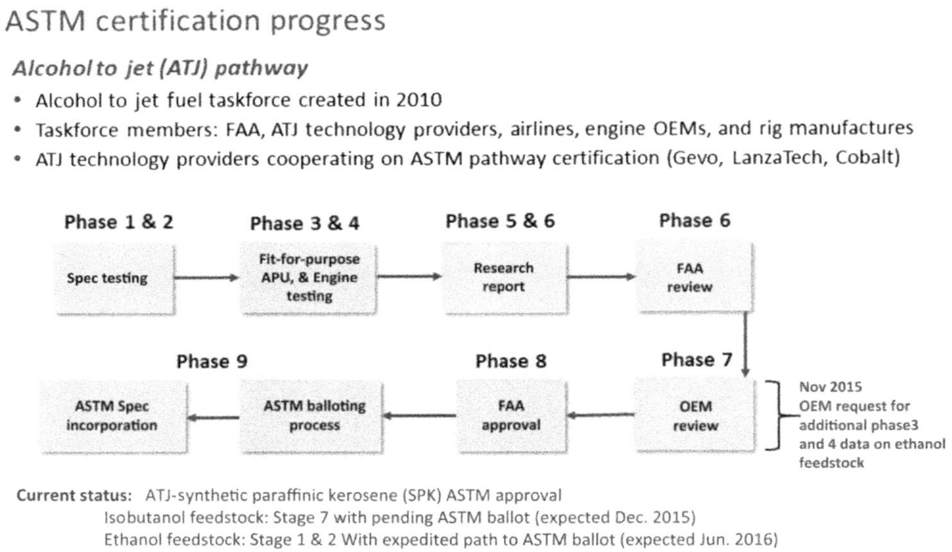

6.3.1 Ethylene Intermediate

One approach to producing jet fuel is to dehydrate the ethanol to ethylene as an intermediate product. Ethanol dehydration is a well-developed and commercially practiced process. Braskem[11] opened an ethanol dehydration plant using bio-derived ethanol to produce 'green polyethylene' in Brazil that has been operating since 2010. Ethylene monomer

[11]Braskem, Sao Paulo, Brazil.

is produced through dehydration of sugar cane-derived ethanol [22]. High ethylene yields are achievable. For example, one reported process achieves 99.7% ethylene selectivity at 240°C using an HZSM-5 catalyst [23]. Hence, no technology development is required for this initial step.

There are a variety of pathways to producing jet fuel from ethylene. Although ethylene can be readily produced from ethanol, oligomerization of this intermediate product to jet-range paraffins can be challenging. The tradeoffs of the available approaches are discussed in detail below. In all cases, following the oligomerization process, the products are hydrotreated to produce primarily jet range paraffin components.

6.3.1.1 Direct Oligomerization of Ethylene

The most straightforward route for converting ethylene to jet and diesel fuels relies on a direct oligomerization approach. Fig. 6.4 shows an example of the direct ethylene oligomerization to jet fuel. Homogeneous-type catalysts have long been used to produce long-chain alpha olefins for chemical use [24]. These catalysts have been used to selectively target more valuable olefins such as hexene and octene rather than producing a wide array of products. The specific olefins generated are also used for co-monomers for the polymerization of linear, low-density, polyethylene as well as plasticizers, detergents, surfactants, and lubricants [25]. In homogeneous catalysis, the reactants, products, and catalyst are all in the same phase and often all dissolved in the same solvent. To be commercially viable catalysts for industrial processes, they must have a high turnover frequency. This allows them to be added in very small quantities. If not, the process becomes expensive and separation of product from the catalyst can be difficult. As a result, heterogeneous catalysts are generally preferred.

One of the earliest heterogeneous olefin oligomerization catalysts is supported on polyphosphoric acid. While this catalyst is inexpensive, it has struggled with low productivity, poor thermal stability, poor flexibility to control the product composition, and corrosion issues. As a result, recent studies have focused

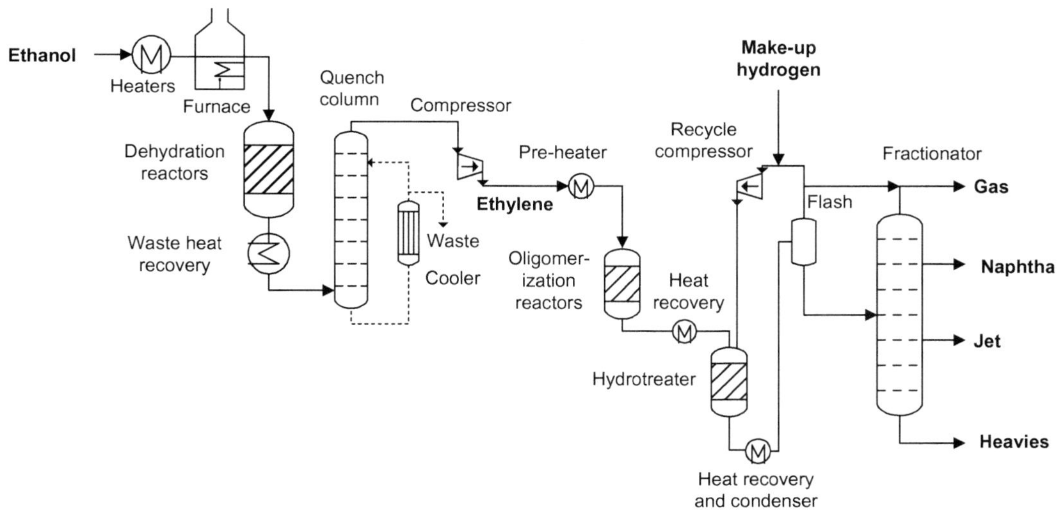

FIGURE 6.4 Process flow diagram of direct oligomerization of ethylene.

on ethylene oligomerization with zeolites and acidic mesoporous catalysts. These acidic catalysts oligomerize olefins through carbocation intermediates. Because of the greater stability of tertiary carbocations, they generate highly branched hydrocarbons, which results in an improved product with lower melting point. Ethylene oligomerization using ZSM-5 zeolite was investigated to study the role of Bronsted acid sites in the formation of higher hydrocarbons [26]. The oligomerization of olefins is dependent on the acidity of ZSM-5 zeolite, and can be increased by dealuminating for different periods according to the acidic ion-exchange method to produce ZSM-5 with various Si/Al ratios. The performance of dealuminated zeolite was tested for ethylene oligomerization. The result demonstrated that the dealumination of ZSM-5 led to slightly higher ethylene conversion and jet fuel range production, as seen in Table 6.3. This table illustrates low conversions and selectivity toward short olefins rather than jet fuel. As a result, these processes would require significant recycle to obtain the $C_{10}-C_{16}$ range olefins.

One way to extend the hydrocarbon chain length and improve jet fuel yield is to incorporate active metals (ie, nickel) into zeolites, and amorphous or mesoporous silica-alumina catalysts. In the latter case, conversion of ethylene typically reaches 97–99%, but the yields of $C_{10}+$ oligomers have been reported at no more than ~30–35% by mass [27,28].

TABLE 6.3 Direct Ethylene Oligomerization With ZSM-5 and Its Derivatives [26]

Catalyst[a]	Ethylene Conversion (%)	Selectivity of Hydrocarbons		
		C_3-C_4	C_5-C_{10}	$C_{11}-C_{15}$
ZSM-5	14	0.3	93.9	5.8
DZ-1	21.7	6.4	85.9	7.3
DZ-2	18.9	17.2	76.4	6.3
DZ-3	16.4	18.4	74.2	7.4

[a] *DZ-1, 2, and 3 are ZSM-5 catalyst treated with 1 N HCl for 12, 24, and 36 h dealumination time, respectively.*

Following the oligomerization step, the alpha olefins are hydrogenated to saturation at high hydrogen partial pressures. Hydrotreating is a well-developed technology in the petrochemical industry, and these olefins require relatively modest levels of hydrogen, pressure, and temperature conditions to achieve near 100% conversions to paraffins [29].

In summary, although this direct ethylene to jet process is the simplest available, its selectivity is not high and significant recycle would be required. Furthermore, the economics of these reactions have not been favourable; as a result, progress on this process has been minimal. Therefore, it has not been developed beyond the laboratory.

6.3.1.2 Conversion of Ethylene to Intermediate Olefins for Oligomerization

Because oligomerization of ethylene is difficult, alternative pathways have been developed that first convert the ethylene to an intermediate olefin prior to oligomerization to jet fuel-length olefins. This two-step process consists of conversion of ethylene to a mixture of C_4-C_{10} and is only slightly more complicated than the direct ethylene process. These higher olefins are more easily oligomerized to jet fuel than ethylene itself. Butene and hexene are considered here, although these processes generally use a range of intermediates.

Butene Intermediate. The first step in converting ethylene to butene produces olefins, primarily in the C_4 and C_6 range. One possible catalyst is the heterogeneous nickel-exchanged AlMCM-41. The test results for this catalyst at 150°C and 3.5 MPa for a reaction time of 1 h are shown in Table 6.4 [30,31]. Both linear and branched olefins are produced by this type of catalyst [32].

Using a homogeneous catalyst based on titanium tetrabutoxide and triethylaluminum, conversions of ethylene in excess of 95% have been achieved with product selectivity primarily to butene (~80%) and hexene (16–18%) [33]. Separation of this catalyst is required following the reaction. This is achieved by hydrolysing the catalyst with water to produce a filterable TiO_2 precipitate.

Once butenes are produced, they can be oligomerized selectively to jet fuel range oligomers. A C_{12} hydrocarbon stream can be produced via a subsequent trimerization of *n*-butene with Ni-doped HZSM-5 [34,35]. Conversions of 77.5 wt% with selectivities of 50.5 wt% for C_{12} have been achieved. The remaining products are primarily C_8 with a small fraction of C_{16}. This product has excellent cold-flow properties. Using a zirconium-based methylaluminoxane homogeneous catalyst, near 100% conversion is possible with a selectivity of 75% jet fuel range olefins [30]. Isobutene oligomerizes to C_{12} more easily than other isomers with selectivities of more than 90% [36]. As with the direct ethylene conversion process, the final step is to hydrotreat the α-olefins to produce a paraffinic product.

These results indicate that butene intermediates are a viable approach to producing jet fuel. Single pass yields can be as high as 70% and higher yields can be achieved with adequate recycle. The processes for converting both ethylene to butene and subsequently butene to higher hydrocarbons are commercially available at full scale. For the former reaction, homogeneous catalysts are generally used because heterogeneous catalysts such as AlMCM-41 are not yet commercially available. The latter reaction is commonly done in the petrochemical industry using feedstocks between C_2 and C_4 with a heterogeneous phosphoric acid catalyst.

Hexene Intermediate. The use of 1-hexene as an intermediate between ethylene and jet fuel is another way to overcome the difficulties of the direct ethylene to jet range hydrocarbons with a similar level of added complexity as the butene route described above. 1-Hexene can be efficiently generated from ethylene with the aid of a variety of homogenous catalysts via trimerization [37]. This technology has a turnover frequency of more than 17,000, which is high enough to make it economical for industrial processes, and is well developed. A new plant has been developed using Chevron Phillips Chemical's proprietary homogeneous chromium-based catalyst technology to convert ethylene to 1-hexene with 'exceptional product purity' [38]. Selectivities of 93 wt% have been

TABLE 6.4 Catalytic Behaviour of Ni-MCM-41 Catalysts for Ethylene Oligomerization [30,31]

Catalyst Average Nickel Loading	Activity (g/gh)	Selectivity (%)					TOF[a] (h^{-1})
		C4	C6	C8	C10	C12+	
Ni1.4-MCM-41(3.5)	89	48	31	14	7	0	13,330
Ni2.0-MCM-41(3.5)	150	45	33	15	5	2	15,720
Ni5.6-MCM-41(3.5)	149	42	37	16	4	1	5580
Ni7.5-MCM-41(3.5)	126	46	35	15	4	0	3520
Ni1.4-MCM-41(10)	113	44	35	15	5	1	16,920
Ni2.0-MCM-41(10)	158	40	33	16	7	4	16,560

[a] *TOF (turn over frequency) = (activity/28)/(Ni wt%/58.7).*

achieved [39]. An example process flow diagram is shown in Fig. 6.5.

Once 1-hexene is produced, conversion of 1-hexene to jet fuel can be performed with a homogeneous Cp_2ZrCl_2/methylaluminoxane catalyst at ambient temperature and pressure, resulting in 100% conversion of 1-hexene with greater than 80% selectivity to a mixture of the dimer and trimer that can be separated to produce jet and diesel fuels. The dimer can be used for jet fuel. The trimer is very desirable for diesel due to a high cetane number [40]. Once again, a mild hydrotreating is required as a final step.

6.3.2 Propylene Intermediate

It is possible to convert the ethanol to propylene as an intermediate rather than to ethylene, as shown in Fig. 6.6. Propylene is more easily oligomerized than ethylene. Because the second intermediate is removed, it is a simpler process than described in the previous section. It does however produce lower yields, requiring higher recycles than the butene intermediate pathway. With increased temperatures, the conversion of ethanol increases, but selectivity becomes dominated by the dehydration reaction to ethylene rather than propylene production.

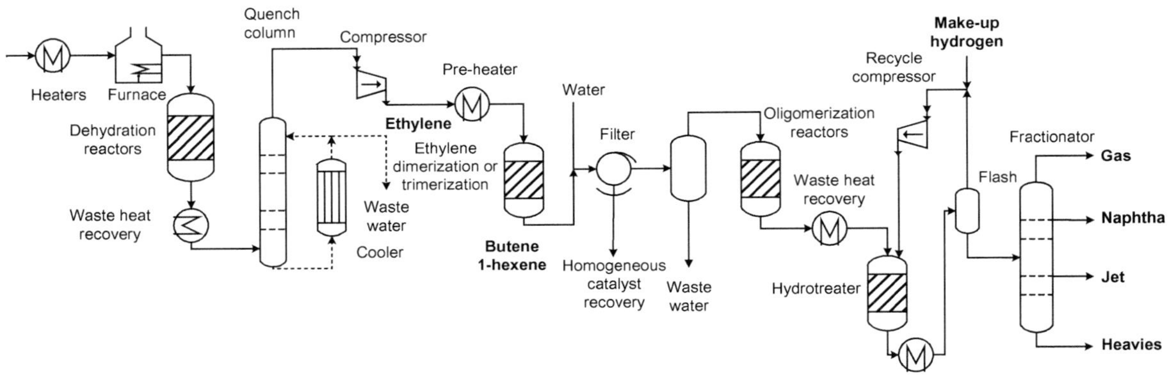

FIGURE 6.5 Process flow diagram of ethylene to butene or 1-hexene for oligomerization.

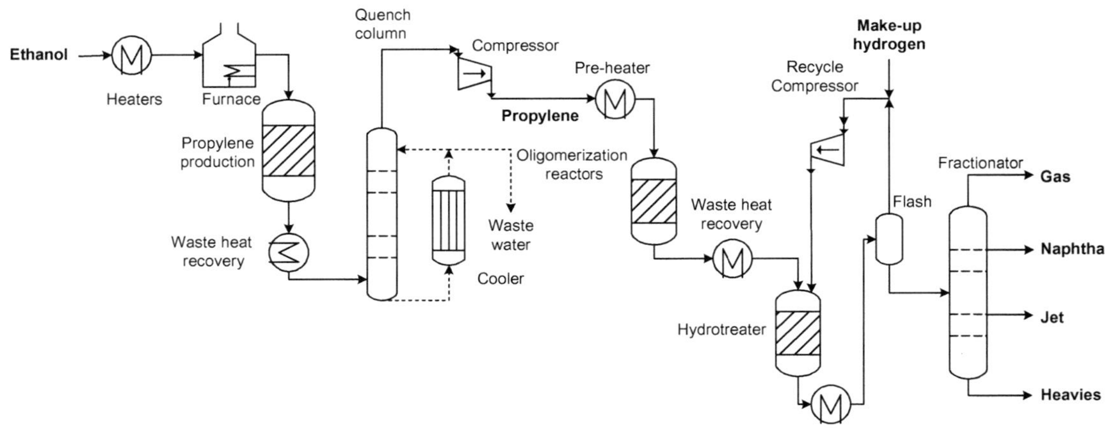

FIGURE 6.6 Process flow diagram of ethanol to jet via propylene intermediate.

An HZSM5-type zeolite containing alkaline earth metals can achieve yields of only 32% [41]. In addition, butene is formed. Both of these intermediates can then be oligomerized to jet fuel constituents. Inoue [42] has also reported similar yields for this reaction with an La-modified HZSM5 zeolite with a SiO_2/Al_2O_3 ratio of 280. To enhance the propylene production, the strength of the zeolite acidity needs to be reduced. The modification with alkaline earth metals, Sc, and La, and the dealumination of HZSM5 zeolite, seem to help with propylene production. The proposed mechanism for propylene production involves first ethanol dehydration and ethylene oligomerization/cracking to propylene in a 1-reactor process [42]. In spite of these modifications to improve the catalyst, these processes generate low-molecular-weight alkanes, which are not useful in the production of jet fuel.

Nickel ion-loaded mesoporous silica at 673K can achieve 68% conversion of ethylene with a propylene selectivity of 48%. A similar process is claimed with ethanol to propylene. The proposed reaction mechanism involves dimerization to butene followed by metathesis of butene and ethene to yield propene. As a result of this mechanism, butene is a major by-product (~40% selectivity) that can likewise be converted to jet fuel [43].

Once propylene is produced, catalysts are available for the oligomerization reaction, although their selectivity to jet fuel is not as high as butene oligomerization. Partial substitution of aluminium atoms in ZSM-5 zeolite by zirconium atoms results in a catalyst with moderate acidity, medium pore size, and improved catalytic properties. Documented yields to jet fuel have been less than 35% [44]. To improve the selectivity toward jet fuel, a less reactive catalyst, solid phosphoric acid, is preferred. The carbon number distribution during catalysed propene oligomerization is limited, resulting in a kerosene-rich product, especially during tetramer mode of operation. UOP has licensed more than 250 catalytic condensation units using a proprietary solid phosphoric acid catalyst to polymerize propylene to nonene and the propylene tetramer [45]. Propylene conversion is about 90% to the propylene trimer and tetramer with selectivities of 60% and 30%, respectively [46]. The catalyst's drawbacks are its shorter lifetime and lower structural stability.

6.3.3 Higher Alcohol Intermediate

Carbon–carbon coupling of alcohols is commonly regarded as the Guerbet reaction, named after its inventor, Marcel Guerbet [47]. The overall Guerbet reaction for ethanol coupling to butanol involves dehydration of two molecules of ethanol (Eq. 6.1).

$$2CH_3CH_2OH \rightarrow CH_3CH_2CH_2CH_2OH + H_2O \quad (6.1)$$

At least two pathways have been proposed for the Guerbet reaction mechanism. The most commonly accepted route includes dehydrogenation of ethanol to acetaldehyde, self-adolization of the acetaldehyde, dehydration, and hydrogenation to form butanol [48,49]. Note that hydrogen used for hydrogenation in this pathway is derived through hydrogen transfer reactions. A second proposed pathway involves direct condensation of two ethanol molecules, shown to be the dominant pathway using hydroxyapatite catalyst [50]. Formed butanol can subsequently undergo additional Guerbet reactions with other butanol molecules or ethanol. Butanol and higher alcohols can then be dehydrated to mixed olefins, which can be used to make jet range fuel via the same oligomerization reactions discussed in the previous sections.

A variety of homogeneous, homogeneous/heterogeneous, and heterogeneous catalyst systems have been studied for the Guerbet reaction [48]. Homogeneous systems have generally used KOH or NaOH. These must be neutralized and disposed of, making the process expensive. Favourable activity and selectivity have been reported for various basic oxide catalysts, including MgO, Mg/Al mixed oxides, hydroxyapatites, and others [48].

(6.2)

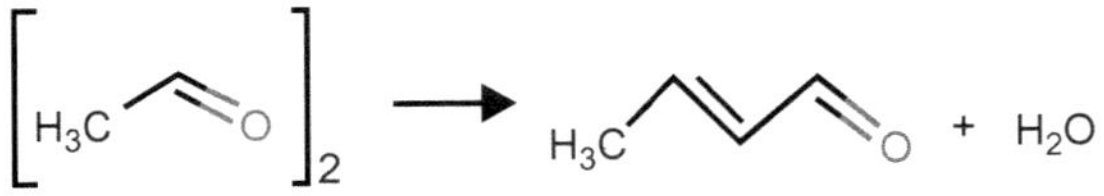

(6.3)

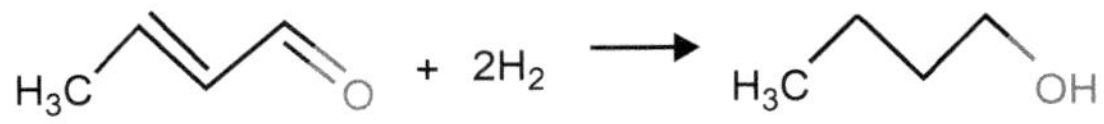

(6.4)

While higher alcohols undergo the Guerbet reactions more easily, ethanol is more resistant to the dehydrogenation reaction, [51]. Rittonen et al. [52] reported 25% ethanol conversion with 80% selectivity to 1-butanol over alumina-supported nickel catalysts. Tsuchida et al. [53] reported 20% conversion and 70% selectivity using hydroxyapatite. The heterogeneous catalysts for this reaction are not commercially available and the base-catalysed reaction has waste disposal issues. As a result, an improved catalyst is needed to achieve reasonable conversions of ethanol to higher alcohols before it can be used on a large scale.

Abengoa[12] is developing Guerbet chemistry to catalytically convert ethanol to butanol for sale into the value-added chemicals market [54]. Butenes obtained from butanol dehydration can easily be oligomerized to jet fuel range components (Section 6.4.3).

6.3.4 Carbonyl Intermediate

Jet fuel can also be produced from alcohols through a carbonyl intermediate via hydroformylation or 'oxo synthesis'. This approach is being developed by Swedish Biofuels. Hydroformylation is an important industrial process, producing aldehydes for final products as well as important intermediates in the production of numerous bulk and specialty chemicals [55]. Companies that employ hydroformylation processing include BASF, Exxon, Shell, Sasol, Dow Chemical, and Mitsubishi [55].

The general process configuration is shown in Fig. 6.7. The sequence is similar to that described in Section 6.3.2, but also includes an oxo synthesis step and a second dehydration step to produce higher olefins than ethylene for further processing, thus making it more complicated than many of the other processes discussed previously.

Hydroformylation is a homogeneously catalysed industrial process in which alkenes are reacted with syngas (CO and H_2) to produce aldehydes, Eq. (6.5), resulting in the addition of a formyl group (H-C = O) and a hydrogen atom to the carbon double bond of the alkene.

$$\text{Alkene} + \text{CO} + \text{H}_2 \rightarrow \begin{matrix}\text{Aldehyde}\\ \text{(linear and branched)}\end{matrix} \tag{6.5}$$

Industrial processes use either Co- or Rh-based homogeneous complexes, with most relying on Rh-based systems [56]. Cobalt-based catalysts are active at high pressure (100–300 atm), while Rh-based catalysts are active at atmospheric pressure. Golubkov [57] suggests hydroformylation of ethylene can be performed with Co-carbonyl catalyst at 160°C and 30 MPa, with Co catalyst modified by phosphorus compounds at 175°C and 7.5 MPa, or with Co-Rh catalyst at 90°C and 2 MPa. Researchers have demonstrated propanal formation at such mild conditions as 100°C and 250 psig with high product yield [58].

The aldehydes produced in hydroformylation, such as propanal, are then hydrogenated

[12]Abengoa, S.A., Seville, Spain.

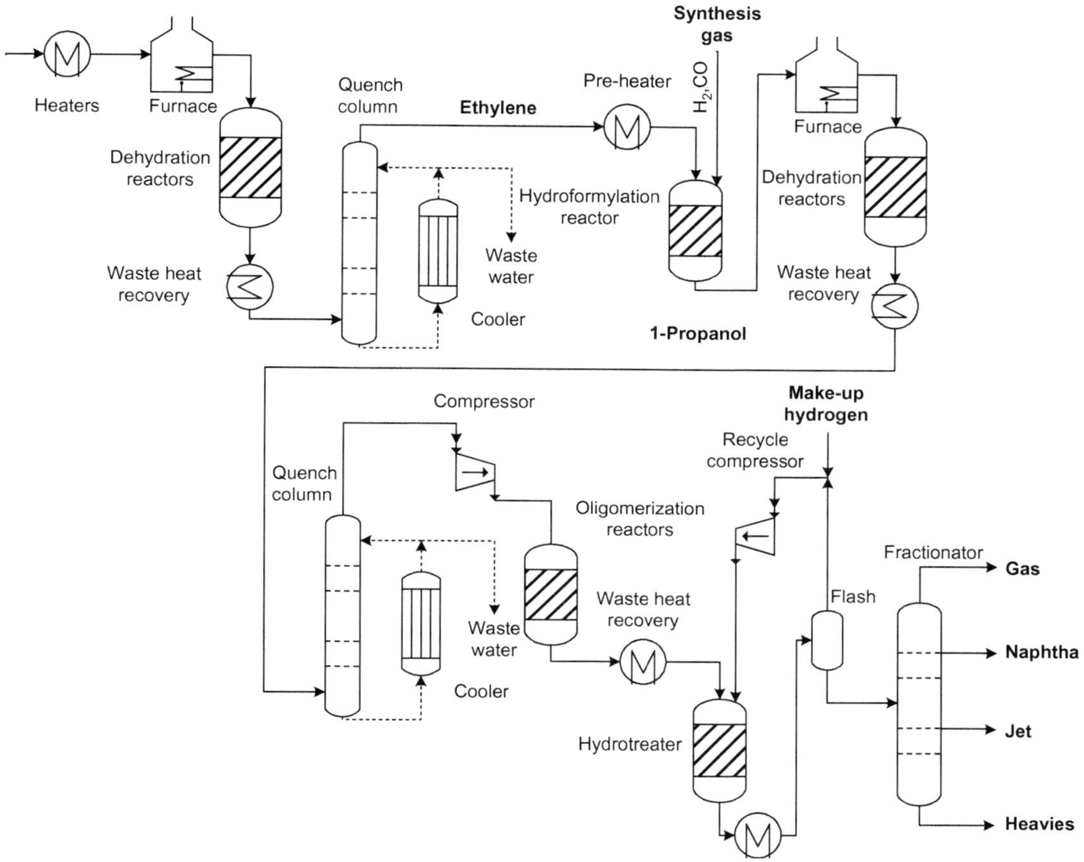

FIGURE 6.7 Process flow diagram of ethylene conversion with hydroformylation process.

into an alcohol, Eq. (6.6), for further processing to final fuels.

$$\text{Aldehyde} + H_2 \rightarrow \text{Alcohol} \qquad (6.6)$$

This reaction occurs at reactor conditions of 150±50°C and 1–2 MPa in the presence of Ni catalyst [57]. Certain catalyst systems enable hydroformylation and hydrogenation in one reactor, directly producing an alcohol in a single step.

The resulting olefin mixture can then be converted to jet fuel with processes similar to those described in Section 6.3.2. Oligomers falling within the jet fuel carbon number fraction are separated and hydrogenated into jet fuel. The advantage of this approach is that olefins higher than ethylene are produced, which are more easily oligomerized than ethylene. The disadvantage of this approach is that a source of renewable, inexpensive syngas (eg, biomass and/or waste gas) is needed for the hydroformylation step.

6.3.5 Summary of Ethanol Processes

A summary of the evaluated ethanol processes based on publically available information is provided in Table 6.5. The processes are compared based on the following: fraction of ethanol converted to jet fuel, process simplicity, required feedstocks, cost of the catalyst, and the development level of the technology. In some cases, two

TABLE 6.5 Comparison of Technologies for Conversion of Ethanol to Jet Fuel

Process Used	Ethanol Converted to Jet Fuel[a]	Process Simplicity[b]	Minimum Added Feedstock[c]	Low Catalyst Cost[d]	Well-Developed Process[e]	Company Developing Technology[f]	Notes
ETHYLENE INTERMEDIATE							
Direct to jet	F	G	G	G	F	None	Produces primarily low MW oligomers
Butene intermediate	G	G	G	F	F/E	None	Selective to good jet fuel (only published data is for homogeneous catalysis)
Hexene intermediate	F	F	E	G	F/E	None	Only published data is for homogeneous catalysts, process control issues
Propylene intermediate	F	G	G	G	F/E	None	Low ethanol-to-propylene Conversion
Higher alcohol intermediate	F	G	F	F	F	None	Guerbet reaction obtains low yields with ethanol
Carbonyl intermediate	G	F	F	F	G	Swedish Biofuels	Requires syngas and expensive catalysts

[a]*Ethanol Converted to Jet Fuel: (E) Excellent, >90%; (G) Good, 70–90%; (F) Fair, 30–70%; (P) Poor, <30% (fraction ethanol to jet fuel).*
[b]*Process Simplicity: (E) Excellent, <5; (G) Good, 5–10; (F) Fair, 10–15; (P) Poor, >15 (major unit operations).*
[c]*Minimum Added Feedstock: (E) Excellent, Only ethanol; (G) Good, H_2 for mild hydrotreating; (F) Fair, Other gases.*
[d]*Low Catalyst Cost: (E) Excellent, Simple zeolite; (G) Good, Metal-doped or highly de aluminated catalyst; (F) Fair, Precious metal or not developed.*
[e]*Well-Developed Technology: (E) Excellent, Industrial; (G) Good, Pilot-scale; (F) Fair, Laboratory-Scale; (P) Poor, Not yet demonstrated.*
[f]*'None' is used to denote cases where no company could be identified through this review.*

scores are given since the first step is not well developed but the second step is an industrial process. The table also provides examples of companies that are currently developing the technologies. As can be seen from the table, each technology has its own advantages and disadvantages. Simple processes tend to have low jet fuel yields. Improved processes generally result in more-complex additional feedstocks or more-complicated catalysts. In most cases, significant technology development is still required to achieve commercialization and ultimately cost-competitiveness.

6.3.6 Higher Alcohols

Other alcohols besides ethanol are further along in the development of a process to produce jet fuel. Normal butanol is primarily produced from petrochemicals using propylene feedstock and hydroformylation. It can also be made by fermentation of sugars using the Clostridia class of bacteria based on the well-known acetone-butanol-ethanol (ABE) fermentation process. Isobutanol is also produced using alcoholic fermentation with genetically modified yeast cells [59]. These higher alcohols have higher energy content and lower water solubility than ethanol. With rather minor changes, an ethanol plant can be repurposed to produce butanol [60], but with a lower volumetric yield of alcohol. In addition, butanol and other higher alcohols have an advantage in that they dehydrate at lower temperatures and pressures than ethanol, and their derivatives easily oligomerize with higher yields toward jet fuel.

CASE STUDY: ETHANOL TO JET-LANZATECH APPROACH

LanzaTech has developed a novel carbon recycling technology that captures and reuses waste gas emissions for fuel and chemical production. Proprietary microbes consume waste gases to make alcohol, much like yeast consuming sugars to make beer in a brewery. This alcohol is then be converted to jet fuel through a series of catalytic steps in the ATJ pathway.

LanzaTech's microbes can use a variety of gas streams, including industrial emissions that would otherwise enter our atmosphere. The alcohol feedstock used for this project came from captured steel mill emissions in China.

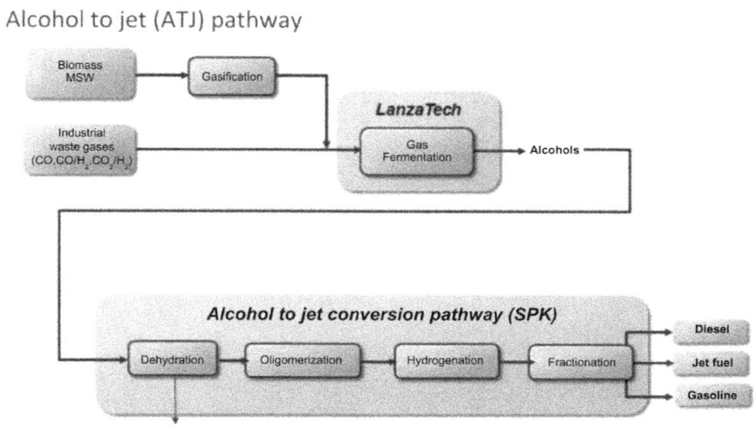

A novel route to synthetic jet fuel

LanzaTech's ATJ pathway is the process pioneered by PNNL and scaled up by LanzaTech and partners. It involves the conversion of ethanol to a synthetic paraffinic jet fuel blendstock (50/50 blend with conventional fuel) using an ethylene intermediate. The project will produce fuel to support testing and pathway certification. Key data will also be collected via a world-first test flight using steel mill gas-based ethanol as a feedstock.

Certification will enable the fuel to be used commercially throughout the aviation sector. The fuel is being produced at LanzaTech's Freedom Pines Biorefinery in Soperton, GA.

Image of LanzaTech's Freedom Pines Biorefinery plant in Soperton, GA.

In making steel, impurities are removed from iron ore and carbon is added. In fact, steel is essentially an iron alloyed with carbon. Carbon monoxide is produced as a side product. That CO is converted into CO_2 in the atmosphere. The magic of the LanzaTech/PNNL jet fuel process is that it starts with a waste gas, produced as a co-product of steel, and using a combination of biotechnology and catalysis converts that gas into a liquid fuel that exceeds the standards required to propel jet aircraft.

In the bigger picture this is the epitome of recycling. We think about recycling aluminium or plastic. Now we should add carbon recycling. Carbon is part of our everyday life. It is the backbone of food, chemicals, materials, and fuels. The ability to take carbon that is already in the cycle prevents us from needing to take carbon from the ground. There are many sources of carbon that today are underutilized. Most of it is also highly

(*cont'd*)

distributed. A central goal of technology development should be to consider this resource – some wet, some dry, some gaseous – and develop energy-efficient means for converting (recycling) that carbon into the fuels and chemicals we need in our world.

LT/PNNL Sample – Key Properties are Confirmed

Property		ASTM Test Method	ASTM D7566	ATJ-SPK (July 2015)
Hydrogen Content, mass%		D7171	n/a	15.4
Freeze point, °C		D5972	≤ -40	< -75
Flash point, °C		D93	≥ 38	44
Density at 15°C, kg/L		D4052	0.775 – 0.840 (Jet A) 0.751 - 0.770 (SPK)	0.763
Viscosity	-20°C, cSt		< 8	4.42
	-40°C, cSt		< 12 target	9.30
Heat of combustion, MJ/kg		D4809	≥ 42.8	43.89
Thermal Stability (325°C)		D3241	2/25	1/0 (pass)
Hydrocarbon Type Analysis				
Aromatics, volume %		D6379	≤ 0.5	<0.01 (GC)
Paraffins, mass%		D2425	report	97.45 (GC)
Distillation		D86		
10%			205 max	170
Final Boiling Point			300 max	263
T90-T10, °C			>22	68

(cont'd)

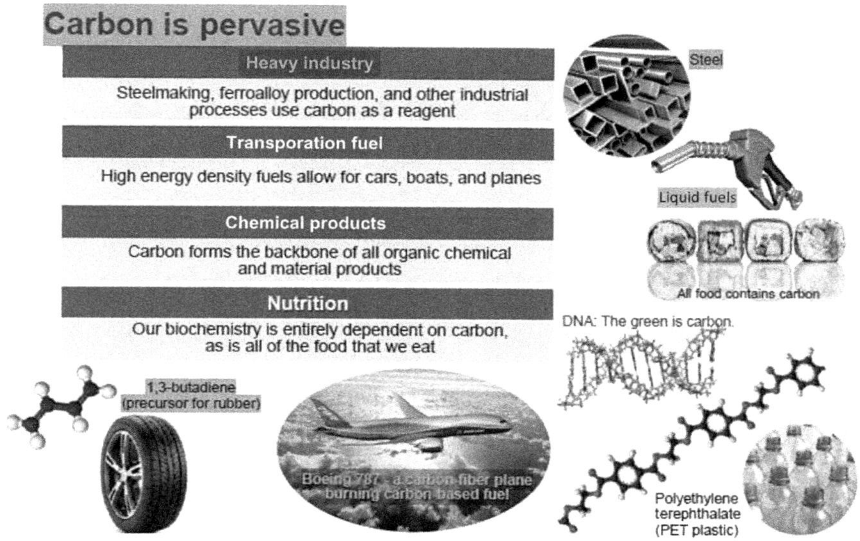

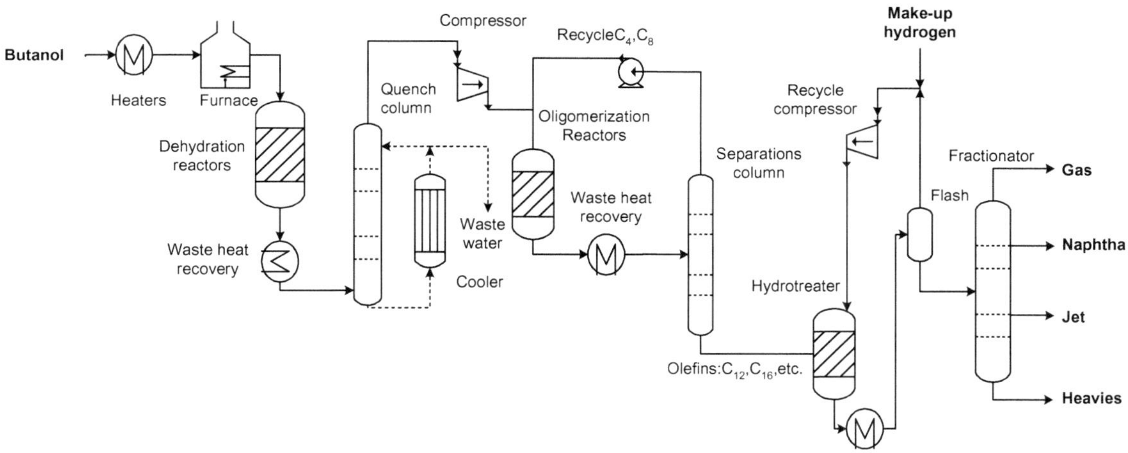

FIGURE 6.8 Process flow diagram of the butanol to jet fuel process.

UOP, Gevo, and Cobalt/US Navy based their ATJ technology on butanol. In all three cases, the alcohols follow the basic process of dehydration, oligomerization, and hydrogenation to produce jet fuel in a reasonably simple process (Fig. 6.8). The conversion of butanol to butene has a conversion rate of 98–99% using gamma alumina, Mg-modified gamma alumina, or ZSM-5 catalysts.

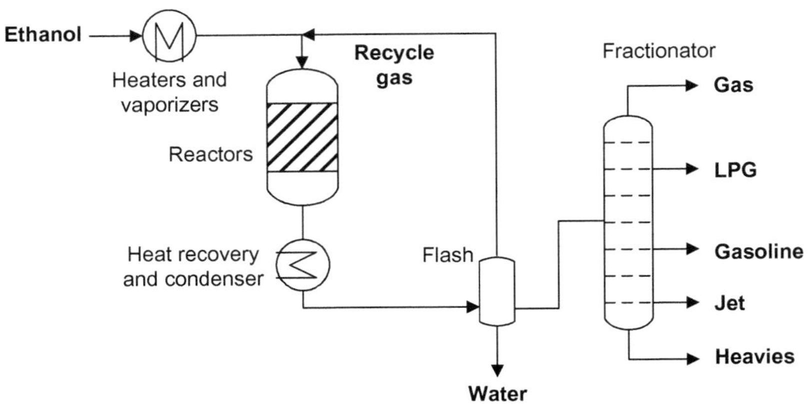

FIGURE 6.9 Single-step process for aromatic production.

The butenes can then be converted to higher olefins using similar processes as described in Section 6.3. Unreacted butene and its dimer C_8 from oligomerization can be recycled or sent to a separate dimerization reactor. The final olefin product is hydrotreated to produce SPK jet fuel. Both Gevo and Cobalt/Navy produce a highly branched jet fuel. The Gevo process is based on isobutanol from fermented sugars [59], while the Cobalt process is based on the use of *n*-butanol and Ziegler–Natta chemistry [61].

6.4 ADJUSTING THE AROMATIC CONTENT OF THE PRODUCT

As noted in Section 6.3, conventional jet fuel can have an aromatic content of up to 25 wt%. Paraffinic hydrocarbons alone cannot meet all jet fuel specifications, but can be blended with conventional jet fuel. Producing a 100% replacement fuel (SKA-type fuel) requires a certain amount of aromatic material. This can be made directly from ethanol or through an olefinic intermediate. It is an advantage of alcohol conversion technologies that both aromatics and paraffin production is possible. However, making aromatics as a blending agent with conventional jet fuel is not desired, as aromatics provide poor combustion characteristics. Blending conventional fuels with renewable aromatics leads to a jet fuel that can easily exceed aromatic content limits.

6.4.1 Ethanol to Aromatics

Catalytic production of hydrocarbons from ethanol has been reported by many researchers. In most cases, this reaction is performed with H-ZSM-5 zeolite catalysts [62]. Aromatics can be produced directly over high-temperature zeolites. This process consists of a single-step heterogeneous catalysis and is similar to the commercial ExxonMobil Methanol-to-Gasoline process, as shown in Fig. 6.9. The advantages of this process are that it is simple, well-established, and may have no hydrogen demand. However, the selectivity to the jet fuel range is poor. Costa et al. [63] reported yields of C_9 aromatics in the 2–11 wt% range, and C_{10} aromatics in the 2–7 wt% range. Furthermore, the high temperatures required to produce longer-chain hydrocarbons in the jet fuel range generally produce more aromatics (~60%), while lower temperatures produce fewer aromatics (~20%) but shorter chain hydrocarbons. This is shown in Table 6.6.

TABLE 6.6 Products from Zeolite at Different Temperatures [64]

Temperature	Product
175–250°C	Primarily diethyl ether.
280°C	Dehydration to ethylene begins.
300–360°C	Oligomerization, dehydrocylization, hydrogenation, cracking with carbon numbers between C_2 and C_{12}. Liquid product is ~70% aromatic, 10% paraffinic, and 10% naphthalene, and balance is mostly naphthenic with a small amount of olefins.
Above 360°C	Cracking predominates, increased gas, decreased liquid products.

6.4.2 Olefins to Aromatics

Aromatics can also be formed from the oligomerization-derived olefins. This reaction is similar to some of the chemistry that drives the reaction from ethanol to aromatics described in Section 6.4.1 [65]. In a zeolite system, yields to C_9+ aromatics are 10 wt% or less [66]. A two-step scheme, as described in Ref. [67], would take the aromatics derived from the zeolite system and alkylate them with light olefins to create distillate range material ($C_{10}+$) with a yield of 30 wt%.

6.4.3 Biomass Liquefaction to Hydrocarbons

Aromatic blendstocks can also be derived from the direct liquefaction of biomass. Direct liquefaction processes include fast pyrolysis, catalytic fast pyrolysis, hydropyrolysis, and hydrothermal liquefaction. These processes are discussed in further detail in other chapters, but briefly, pyrolysis-type processes involve making contact between dried, ground biomass and a heat carrier at approximately 500°C and atmospheric pressure. Residence times are short (on the order of 1–2 s), and the produced vapours are rapidly cooled to prevent further reaction. Organic liquid yields for uncatalysed pyrolysis are feedstock-dependent and can range from 40 to 65 wt% on dry biomass [68]. The oxygen content of the bio-oil is 35–40 wt%. Catalytic pyrolysis involves making contact between catalysts, such as zeolites, and pyrolysis vapours prior to cooling. The resulting bio-oil has a much lower oxygen content, with the exact amount dependent upon operating conditions. Hydropyrolysis requires contact between the biomass and a catalyst in the presence of hydrogen at approximately 300 psig. Hydrothermal liquefaction differs significantly from the pyrolysis-type processes. Wet feed, typically a slurry of 10–20 wt% solids, is processed at 350°C and 3000 psia. High pressure is needed to maintain water in the condensed state at the reaction temperature. Residence times are longer than pyrolysis processes, typically 20–30 min [69].

Each process generates a bio-oil that is rich in aromatic and naphthenic compounds, but also contains a significant amount of oxygenated species. Hydrodeoxygenation and hydrocracking are necessary to produce a hydrocarbon fuel product in the desired boiling range.

Envergent is a joint venture between Ensyn (a fast pyrolysis vendor) and UOP (a petroleum refining process vendor that also performs hydrotreatment) with the goal of commercializing pyrolysis and hydrotreating for liquid transportation fuel production [70]. GTI is in the process of licensing hydropyrolysis [71]. The National Advanced Biofuel Consortium has recently concluded studies regarding hydrothermal liquefaction [72].

6.5 MARKET DRIVERS FOR ETHANOL TO JET FUEL

This section identifies potential drivers for considering jet fuel production from low-cost renewable ethanol rather than selling ethanol directly as a commodity chemical. To make this comparison, the value of ethanol relative to jet fuel in terms of its properties, price, and market size are evaluated. Changes to the market that make ethanol-derived jet fuel blendstocks more attractive to produce than petroleum-based fuels are discussed. The influence of government taxes and incentives both within the United States and abroad is also addressed.

6.5.1 The Need for Alternative Jet Fuel

There is significant interest within the aviation industry for affordable, high-quality alternative fuels [73]. High fuel prices, fuel price volatility, the desire to reduce GHG emissions, and the fact that some governments are considering or are implementing carbon emissions penalties for aviation make using low-carbon fuels attractive and beneficial [74]. There are three primary reasons that aviation has a strong need for non-petroleum-based jet fuel: cost, climate change, and national security.

6.5.1.1 Fuel Cost

Airlines are particularly vulnerable to petroleum price swings. For example, from 2001 to 2012, the price of crude oil increased by 262%. As a result, fuel now accounts for nearly 40% of an airline's total operating costs [2]. The risks associated with fuel price fluctuations inhibit the airline industry's growth. Due to the competitive nature of the airline industry, airlines may not always be able to pass their costs of higher gas prices to their consumers by increasing fares. To address potential fuel increases, hedging is used to minimize this risk. Such hedging can be risky and costly itself. As a result, in a fluctuating market, the airline companies instead may be reluctant to expand their market and may even reduce capacity to address possible fuel price changes. Additionally, they may lease or purchase fewer new aircraft in preparation for possible fuel upswings. Having a steady fuel price would be good for both airline companies and consumers.

6.5.1.2 Climate Change

Concerns about climate change have brought increased consumer and governmental pressure for the industry to be environmentally sustainable. The airline industry has increased engine efficiency and reduced emissions per passenger mile. For example, Boeing's 787 jet is reported as producing 20% lower emissions per passenger than similar-sized planes [75]. The industry is now looking to alternatives to further reduce CO_2 emissions. The use of renewable fuels and improved aircraft together are expected to play an important role in making this possible. Aviation biofuels are cleaner, with around an 80% reduction in CO_2 life cycle emissions compared with fossil jet fuel [76]. In an effort to increase the rate of development of renewable jet fuel, both the United States government and industry have set targets, created organizations to further this work, and funded demonstrations.

The environmental profile of jet fuel produced from ethanol can be assessed with standardized environmental life cycle assessment (LCA) methods.[13] Arriving at a complete understanding of the environmental impacts, such as GHG emissions or energy demand, requires an investigation into each step of the value chain to understand the implications of material processing and transport as the

[13]International Organization for Standardization (2006) ISO 14044: *Environmental Management—Life Cycle Assessment—Requirements and Guidelines.*

feedstocks are transformed into higher-value products. For the ATJ process envisioned by LanzaTech and collaborators, this value chain involves the following: (1) feedstock preparation, either by growing, harvesting, transporting, and gasifying biomass, or by capturing carbon-containing industrial waste gases; (2) ethanol production via fermentation; (3) ethanol transport if needed; (4) upgrading to jet fuel; (5) jet fuel transport; and (6) jet fuel blending and combustion as a transportation fuel. Inputs of materials and energy must be tracked at each of these life cycle stages, and the embodied environmental impacts associated with these inputs can be quantified with life cycle inventory databases and LCA software modelling packages that utilize transparent and accepted impact assessment methods. A clear carbon accounting process is also important in the ultimate quantification of GHG emissions, as carbon moves into and out of the fuel life cycle in different locations and in different forms.

Initial LCA studies of jet fuel have been conducted by Michigan Technological University researchers in collaboration with LanzaTech and PNNL. In this study, ethanol was produced by fermentation of carbon-containing gases from either basic oxygen furnace (BOF) waste exhaust streams, or a variety of biomass streams that were grown, harvested, transported, and gasified to a syngas blend of CO and H_2.[14] Ethanol was then upgraded to a mix of jet fuel and other hydrocarbon fuels using catalytic technology (provided by PNNL) under a range of operating parameters. Abbreviated results are shown below in Table 6.7 to illustrate the general LCA results for cumulative GHG emissions of LanzaTech-PNNL jet fuel and emissions reductions compared to traditional fossil-derived jet fuel, after allocation of emissions between final fuel products was performed on the basis of energy content. This data serves to illustrate a few important points surrounding our initial modelling and evaluation of these conversion systems. Presentation of stage-by-stage results depends on important assumptions surrounding carbon accounting: here, we assume

TABLE 6.7 Abbreviated Results to Illustrate General LCA Results

	BOF, EU	BOF, US	Corn Stover	Switchgrass	Forest Residue
Ethanol production (g CO_2eq/MJ ethanol)	25.2−71.4 = −46.2	31.4−71.4 = −40	8.0−71.4 = −63.4	11.4−71.4 = −60.0	1.5−71.4 = −69.9
Jet/Gasoline production stage (g CO_2eq/MJ fuel)	13	13	13	13	13
Jet/Gasoline combustion (g CO_2eq/MJ fuel)	72.9	72.9	72.9	72.9	72.9
Jet/Gasoline production, complete life cycle (g CO_2eq/MJ fuel)	29−35	36−42	10−17	14−22	2−12
Jet/Gasoline reduction over fossil jet fuel	60−65%	55−60%	80−90%	75−85%	85−95%

[14]Handler, RM, Shonnard, DR, Griffing, EM, Lai, A, Palou-Rivera, I. Life cycle assessments of LanzaTech ethanol production: Anticipated greenhouse gas emissions for cellulosic and waste gas feedstocks. *Submitted to Industrial and Engineering Chemistry.*

that carbon-containing waste gases or gasified biomass streams entering the LanzaTech ethanol fermentation process contain carbon that was either prevented from release to the atmosphere or sequestered from the atmosphere during biomass growth, therefore large carbon credits appear for net gas absorption during the ethanol production stage, making per MJ GHG emissions results for ethanol production appear large and negative. Small differences are apparent at the ethanol production stage due to different requirements for procuring biomass feedstock types. Jet fuel production has a relatively small GHG emissions footprint, which is variable due to the ability to tailor the process to internally recycle a variable amount of small-chain carbon molecules for hydrogen and heat, or separate these as an additional co-product. Utility consumption during the entire process is an important source of GHG emissions, but is low enough to result in favourable life cycle GHG emissions for the final fuel products, achieving at least 50% reduction in GHG emissions in all scenarios investigated.

6.5.1.3 Other Air Pollutants

In addition to reducing CO_2 life cycle emissions, alternative jet fuels that are low in aromatics will reduce air toxic emissions as well. The levels of aromatics found in petroleum-based jet fuel (up to 25%) have been shown to increase the production of particulate matter (PM) emissions in jet engine exhaust. The sulphur compounds (up to 0.3%) result in a proportional increase in sulphuric oxides and sulphuric acid. In contrast, the SPK alternative fuels contain negligible levels of aromatics and sulphur. Alternative fuel emissions tests have been performed with F–T and FAME and compared to conventional petroleum fuels (Jet A, Jet A-1, and JP-8) in commercial jet engines (See Fig. 6.10). PM reductions of as much as 50% were achieved [77]. Similar PM reductions are anticipated with ATJ alternative fuels.

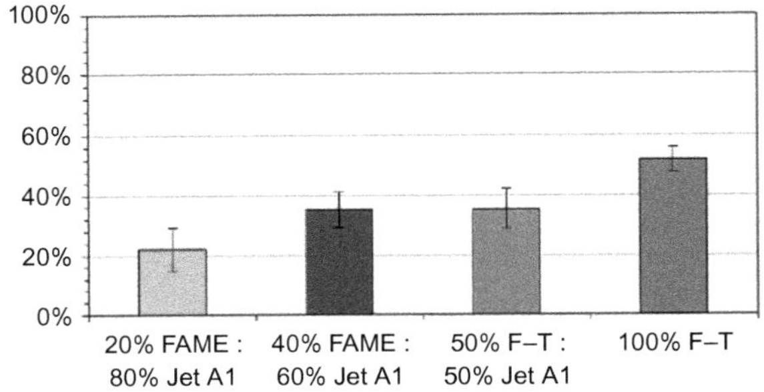

FIGURE 6.10 Overall particulate matter (PM) number emissions reduction over the LTO cycle achieved using candidate alternative fuels. *Source: P. Lobo, D.E. Hagen, P.D. Whitefield, Comparison of PM emissions from a commercial jet engine burning conventional, biomass, and fischer–tropsch fuels, Environ. Sci. Technol. 45 (2011) 10744–10749.*

6.5.1.4 National Security

Another reason for aviation's need for bio-jet fuel is energy and national security. Jet fuel constitutes the largest share of the armed services' fuel demand, accounting for 73% of total purchases [78]. With the US importing much of its petroleum, national security could be compromised. As a result, the US Department of Defense (DoD) has focused on powering its jets with domestically produced alternatives. Over-reliance on oil is believed to put US troops at risk of supply disruptions during military and humanitarian missions, and adopting advanced 'drop-in' fuel is expected to help the nation achieve its broader national security objectives [12].

Although it remains a concern, national security has recently become less of a driving force for producing fuel domestically, primarily due to the introduction and proliferation of hydraulic fracturing, which has made it possible to tap large reserves of shale oil and gas on state and private lands [79]. In its annual long-term global energy forecast, ExxonMobil Corp. foresees that oil production from North American shale oil fields will exceed that of every member of the Organization of the

Petroleum Exporting Countries (OPEC), except Saudi Arabia [80], by 2014.

6.5.1.5 Government Interest in Alternative Jet Fuel

As the largest user of jet fuel in the United States military, the Air Force has made plans to convert one-half of the petroleum-based jet fuel that it uses to non-petroleum-based fuel by the year 2016. In an effort to reach this goal, they demonstrated bio-jet fuel with a B-52 in 2006. Similarly, the US Navy has demonstrated alternative fuels with their aircraft and has announced a goal of supplying half of its total energy consumption from alternative sources by 2020. In addition to the military, the FAA has set a target of 1 billion gallons per year of aviation biofuel capacity by 2018. For every $0.25 increase in the price of a gallon of oil, the DoD incurs over $1 billion in additional fuel costs [12]. Additionally, the US Department of Agriculture, Airlines for America, Inc. (A4A), and Boeing signed a resolution formalizing their commitment to work together on the 'FARM to FLY' initiative to accelerate availability of sustainable biofuel in the United States.

6.5.1.6 Industry Interest in Alternative Jet Fuel

The commercial aviation industry has also set goals to increase its use of renewable jet fuel. The IATA, comprised of the top 240 airlines in the world and representing 84% of global air traffic [81], has a stated target of 10% of the aviation fuel market (approximately 200 million barrels per annum) to be from renewable sources by 2017 [82]. The recently executed agreement for United Airlines and World Fuel Services to purchase a minimum of 15 million gallons of advanced biofuels from AltAir Fuels for use at United's Los Angeles International Airport hub and other locations further demonstrates a significant commitment by the airlines to use alternative jet fuel [83].

In addition, the Sustainable Aviation Fuel Users Group was formed in Sep. 2008 with the support and advice of the world's leading environmental organizations, such as the Natural Resources Defense Council and the Roundtable on Sustainable Biomaterials. The group is focused on accelerating the development and commercialization of sustainable aviation fuels.

Companies like Boeing, the world's largest airplane maker, are working with fuel developers from around the world to find alternative fuels that will shrink jet flight's substantial environmental footprint without requiring an overhaul of the world's existing airplane fleet [75]. Boeing has a timeline to develop 'carbon neutral' machines that use biofuels, and less fuel overall, by 2050 [73]. Boeing estimates that biofuels could reduce flight-related GHG emissions by 60–80% [75]. Airbus, along with the European Commission, leading European airlines, and fuel producers, has launched an industry-wide sustainable fuel initiative called the 'European Advanced Sustainable Aviation Fuel Flightpath' [84]. The initiative aims to replace 4% of European aviation fuel with sustainable alternatives by the year 2020.

The global airline industry is committed to a 50% reduction in emissions by 2050 from its 2005 levels. In 2009, IATA adopted its Aviation Carbon-Neutral Growth (CNG2020) Strategy, which has the goal of becoming carbon neutral by the year 2020, cutting net emissions in half by 2050 compared to 2005 levels, and improving fuel efficiency by 1.5% annually to 2020 [81].

The 38th Assembly of the United Nations International Civil Aviation Organization (ICAO) committed to developing a global market-based measure (MBM) that is essential to enabling the civil aviation industry to achieve carbon-neutral growth by 2020. Such a global MBM is expected to complement airline industry progress towards its strategy to manage aviation's climate change impact, and it strengthens the resolution passed by IATA in 2013 asking governments to develop a global mandatory carbon off-setting scheme [81].

A wide variety of airlines have demonstrated alternative jet fuels in commercial flights. These include Lufthansa, Qatar Airways, United Airlines, Virgin Atlantic, British Airways, and Porter Airlines. Organizations such as CAAFI have been created to approve ASTM fuel specifications for bio-jet fuel.

Over 1500 commercial flights have been demonstrated using a blend of alternative and petroleum jet fuels [85]. A partial list of the airlines, the amount of biofuel used, distance travelled, and the number of flights and references are provided in Tables B1 in Appendix B. These demonstrations illustrate the commitment that the airline industry has to alternative jet fuel. To further drive this work forward, organizations such as CAAFI have been developed with industry and government partners. These organizations have worked with ASTM International to successfully certify F–T, HEFA, and farnesane jet fuels (2009, 2011, and 2014, respectively). As discussed in Section 6.2.4, the ATJ pathway is nearing approval.

6.5.2 Inherent Improvement in Energy Density From Ethanol to Jet Fuel

An underlying incentive for converting ethanol to jet fuel is the resulting increase in fuel product energy density. Jet fuel primarily consists of paraffins in the C_{10}–C_{14} range. These paraffins are a blend of straight and branched alkanes as well as cyclic compounds. Ethanol, in contrast, is a partially oxygenated hydrocarbon. As a result, the gravimetric and volumetric energy density of jet fuel is significantly higher than that of ethanol. A comparison of densities of ethanol, constituents of jet fuel, and the ASTM jet fuel standards is shown in Table 6.8. As can be seen in the table, the energy densities of the jet fuel both gravimetrically and volumetrically are approximately 60% higher than that of ethanol. A high-energy-density material is critical for aviation applications. As a result, this energy density improvement makes jet fuel a higher-value commodity when used in combustion applications.

TABLE 6.8 Comparison of Energy Densities of Ethanol and Jet Fuel Constituents

Compound	Gravimetric Energy Density (MJ/kg) (LHV)	Volumetric Energy Density (MJ/L) (LHV)
Ethanol	26.8	21.1
Decane	44.2	32.3
Dodecane	46.4	34.8
Tetradecane	47.3	36.1
ASTM Standards for both petroleum and synthesized hydrocarbon jet fuel (ASTM D1655-13 and D7566-13)	> 42.8	> 33.2

6.5.3 Government Impacts/Incentives

For bio-jet fuels to succeed in the market, favourable government policies are likely required. Policies that penalize carbon emissions and incentivize lower carbon-emitting fuels already are contributing to establishment of the bio-jet market in the United States. This section discusses the influence of carbon taxes imposed on liquid fuels and the influence of production credits for approved biofuels.

6.5.3.1 Carbon Taxes

Imposition of carbon taxes or emission allowances increases the cost of transportation modes that rely on fossil-based fuels. In the case of jet fuels, the European Union (EU) recently added airline emissions to its emissions allowance trading system. This system issues credits to the airlines, based on the establishment of baseline flight operations in Europe. Then, over time, the number of annual emission credits granted to each airline is gradually reduced, forcing the affected airline to either reduce emissions or buy additional emissions credits on the market to maintain or grow their businesses. Currently, the portion of commercial airline flights occurring over EU airspace is subject to an EU-wide emission allowance cost reflected in the cost of purchasing emissions allowance credits of roughly $9.70 per ton of CO_2 ([86], based on a Euro-to-dollar exchange rate of 1.37). This value is generally viewed as temporarily low due to the deep recession in Europe, and is expected to increase as economic activity increases and emission allowance credits become scarcer. Several global businesses account internally for the expected cost of carbon. These values vary widely, from prices below the $9.70 per ton reported above to as much as $60 per ton [87].

If a similar structure is enacted in the United States, at the 2012 average carbon price reported above, the aggregate cost of carbon for all domestic US flights (any airline) would be roughly $3 billion annually, estimated based on 2012 figures [88,89], or roughly 29 cents per gallon of added cost, based on current fuel costs [90]. These are new costs airlines and airfreight transporters would face. Thus, bio-jet fuels would be able to reduce these costs by roughly 30% if blended at the expected 1:1 ratio with fossil jet fuel. These estimates depend on a number of factors, including the ultimate carbon tax rate and mitigative actions commercial airlines and airfreight transporters can take, such as fare increases.

6.5.3.2 Renewable Fuel Credits

Renewable fuel credits are production incentives paid to producers or fuel blenders as a tax credit (blender's tax credit) or as a component of the price received for the fuel by the producer/blender. The US Environmental Protection Agency (EPA) has developed the Renewable Identification Number (RIN) as a mechanism to value the production of federally qualifying biofuels. The RIN system enables the market to incentivize biofuel production reliably when market fuel prices alone cannot cover the cost of production faced by refiners. In the case of bio-jet fuel, each gallon of qualifying product is assigned a number of RIN credits based on its fossil fuel equivalence value. Typical bio-jet drop-in fuel has an equivalence value of 1.6, meaning that for each gallon produced, 1.6 credits are assigned. Depending on the market value of the credits, they can be treated as a direct incentive to the producer by offsetting the cost per gallon by the amount of the credit price. A recent study [91] indicated that a plausible credit price could be $2.25 per gallon for isoparaffinic kerosene. This incentive applies in addition to the blender's tax credit of $1.00 per gallon for advanced biofuel production. Thus, it may be reasonable to expect $3.25 per gallon in total federal incentives for bio-jet fuel production in the near term while tax credits are in force. Tax credits are subject to renewal by Congress, and are not expected to be permanent.

Renewable fuel credits are also available for ethanol. When comparing the incentives possible by just producing and selling ethanol to those that could be available if the ethanol is converted to jet fuel, the tax benefit would be determined by the difference in the credits available for each fuel and the additional cost of the jet fuel production. The relative cost of production for ethanol versus jet would then determine at what point to take the RINs incentive.

6.5.4 Marketing Advantages

The use of sustainable fuels can be a marketing tool to help differentiate an airline or air cargo company from its competitors. According to Roper ASW's [92] Green Gauge study, 30% of Americans closely follow the environmental records of large companies [92]. This segment is sometimes referred to as the 'lifestyles of health and sustainability' (LOHAS) market, estimated by the National Marketing Institute (NMI) to be worth $290 billion in the United States in 2008 [93]. Sustainable jet fuel could influence market sectors such as eco-tourism and alternative transportation, which make up 20% of the LOHAS market (NMI). Results have shown that the benefits are usually much greater when such a company is enjoying a 'first-mover advantage', being the first in their region or industry to make the change [94]. One company that is making an alternative fuel change is United Parcel Service. They have set a goal of one billion alternative fuel miles by 2017, and are engaging external initiatives such as the A4A and 'Future of Fuels' Working Group at Businesses for Social Responsibility to reach that goal [95,96]. In 2015, Colorado biofuel producer Red Rock Biofuels announced that FedEx had agreed to purchase 3 million gallons of low-carbon fuel per year.

The use of sustainable fuels can also strengthen a company's image with employees and shareholders. Approximately half of employers surveyed in a KPMG survey who had included green energy in their portfolio believed that employee motivation, and therefore the 'war for talent', was a major driver for such a change [97].

The primary concern for any company is the financial impact of using sustainable fuels. In spite of the benefits of sustainable energy and fuel, consumers continue to be unwilling to pay for 'green travel'. In a 2009 survey, only 9% of consumers indicated that they would be willing to pay more to use travel service suppliers that offer eco-friendly options for travellers. This is corroborated with the statistic that only 3% of passengers have purchased a carbon offset when booking travel [98]. Studies have shown that there can be a large disparity between consumer attitudes toward renewable energy and their behaviours. Market research findings were consistently favourable to green energy; however, marketing results generally have shown much lower levels of actual response to green power offerings [99]. In spite of these findings, several recent studies have shown a correlation between environmental performance and financial performance. According to a Thomson Reuters study, 82% of investors evaluate environmental, social, and governance criteria as part of their investment decision because they believe these actions affect share price [100].

Part of the challenge will be to create consumer awareness and understanding of the benefits of alternative fuels. Efforts to educate the public are already underway. Organizations such as A4A, CAAFI, and IATA provide information about progress on the development of alternative jet fuels. Websites targeting green consumer products are providing data on green airlines as well (eg, wespire.com, greenopia.com). Armed with this information, stakeholders can make educated decisions regarding sustainable jet fuels.

6.5.5 Markets and Prices

This section will consider the price differential between ethanol and jet, competition from higher alcohols, and the opportunities for low-cost ethanol.

6.5.5.1 Ethanol and Jet Prices

Fuel supply and demand and the price differential between ethanol and jet fuel also provide incentives. Fig. 6.11 shows ethanol and jet

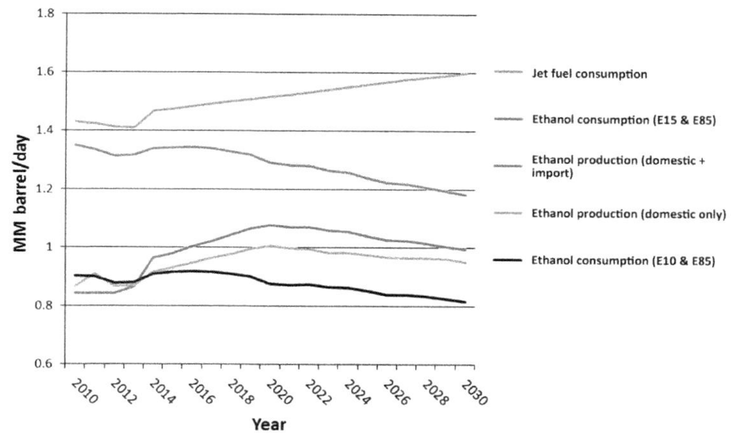

FIGURE 6.11 Projected ethanol supply and demand [101].

fuel volume projections based on data from the US Energy Information Administration's (EIA's) Annual Energy Outlook (AEO) analysis [101]. The portion of total US refinery output of kerosene-type jet fuel going to military applications is historically between about 7% and 10%. Ethanol consumption values include ethanol used in gasoline and in E85. E85 is assumed to contain 74% by volume ethanol. Two cases are considered for gasoline ethanol content, E10 (10% by volume ethanol) and E15 (15% by volume ethanol). In the case of E10, ethanol supply exceeded consumption in about 2013 (ie, the so-called 'blend wall' was hit). In an effort to recognize the practical limits on ethanol blending, the EPA has proposed reducing the 2014 renewable fuel mandate to 15.21 billion gallons from the 18.15 billion gallons originally set forth in the 2007 Energy Independence and Security Act (EISA) Renewable Fuel Standard (RFS2) programme. This corresponds to 13.01 billion gallons of corn ethanol, down from the 2013 mandate of 13.8 billion gallons and significantly reduced from the 14.4 billion gallons originally set forth in the RFS2 [102–104]. This is the first time the EPA has proposed lowering the total renewable and advanced fuel requirements since implementation of the RFS2 in 2007. The EPA has also proposed a cellulosic biofuel standard of 17 million gallons for 2014, up from 6 million for 2013, but much less than the originally intended 1.75 billion gallons in RFS2.

As shown in Fig. 6.12, gasoline consumption is projected to steadily decrease in the future. This is because of increased fuel efficiency due to the corporate average fuel economy (CAFE) standards for light duty vehicles implemented by the EISA. At the same time, E85 growth is assumed to remain relatively flat over the projection period, due to market penetration challenges such as infrastructure compatibility. These two factors lead to a surplus in ethanol supply of approximately 0.2 MM barrels per day over the long term if gasoline continues to contain a maximum of 10% ethanol. Clearly, this extra supply of ethanol could be a resource of feedstock for jet fuel production. Assuming a 50% ethanol to jet conversion yield, this surplus would equate to about 4% of the jet fuel demand in 2022. On the other hand, if the gasoline ethanol content is increased to 15% throughout the projection, there is a shortfall of ethanol necessary to meet the gasoline need (Fig. 6.11). Gasoline currently can be blended with up to 15% ethanol in light-duty 2001 and newer vehicles; however,

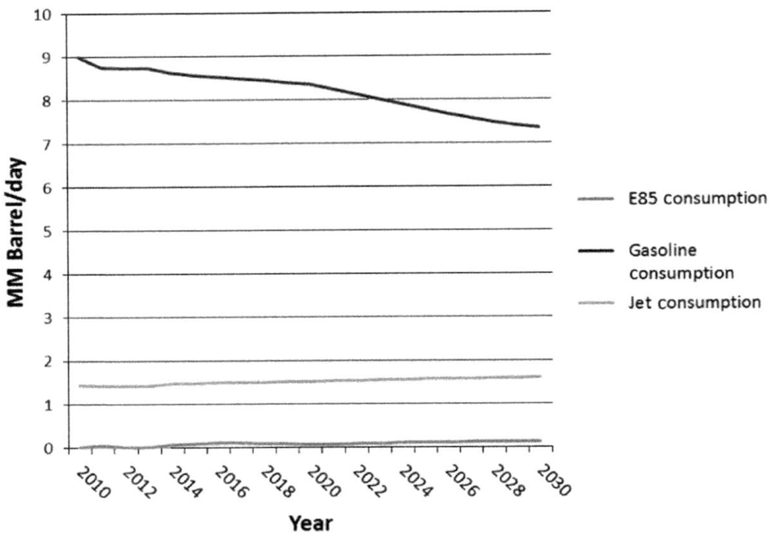

FIGURE 6.12 Fuel consumption for gasoline, jet, and E85 (Reference case) [101].

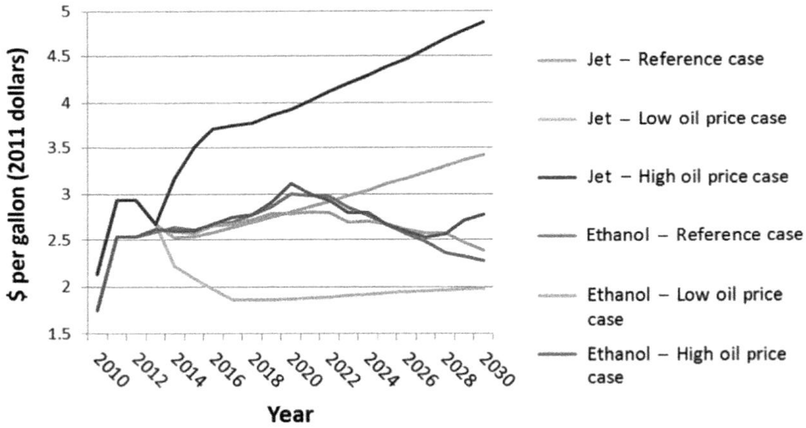

FIGURE 6.13 Projected jet and ethanol wholesale price [101].

E15 is only available in a few states in the midwest United States at filling stations that also offer E85 [105]. Actual widespread adoption of E15 will depend on the cost-competitiveness of blending with ethanol compared to other gasoline-blending components in the future [106].

Historical and projected wholesale prices for fuel-grade ethanol and conventional jet fuel from EIA's AEO report [101] are shown in Fig. 6.13. The AEO projection scenarios shown include the Reference case, corresponding to 'business as usual' (2.5% annual US gross domestic product (GDP) growth), and also low and high oil price cases to illustrate the effect of oil price uncertainty on these projections. The main assumptions behind each of these scenarios are listed in Table 6.8. In all cases, ethanol price reaches a high around 2022, coinciding with the RFS2 2022 target. After this, ethanol price declines, as gasoline demand (ie, E10 and E15) continues to drop and E85 demand levels off (Table 6.9).

While ethanol prices shown in Fig. 6.13 are for fully refined fuel ethanol for the wholesale market, refiners are likely to use an intermediate type of ethanol feedstock (eg, higher water content) adequate for conversion in hydrocarbon fuel. Not only would this ethanol intermediate be less refined, and therefore less expensive to make than fuel grade, it would not have the transfer costs associated with a wholesale-priced product. Therefore, a more useful comparison for rough economic analysis is the price difference between an intermediate grade of ethanol and jet fuel. Fig. 6.14 shows the price difference between wholesale jet fuel and ethanol intermediate (plotted y-values equal the wholesale jet price minus the ethanol intermediate price). Several cases are shown, corresponding to the three EIA scenarios (see Table 6.8), and assuming an ethanol intermediate price of 50% and 75% of the projected wholesale fuel grade ethanol price. As shown, the future price of oil has a large effect on the incentive to take ethanol to jet. As an example, in the year 2022, the price differential ranges from \$0.20 per gallon (low oil price; intermediate ethanol at 75% wholesale price) to \$2.66 per gallon (high oil price; intermediate ethanol at 50% wholesale price).

Biofuels Digest [107] reports that alcohol for jet fuel will need to be produced at \$1.25–\$1.50 per gallon to make jet fuel affordable at commercial scale (nominally a

TABLE 6.9 EIA's AEO Modelling Assumptions for Reference, Low Oil Price, and High Oil Price Cases [101]

	Reference Case	Low Oil Price Case	High Oil Price Case
Oil price cost in 2040	\$163/bb	\$75/bb	\$237/bb
Liquid fuel demand and GDP in non-OECD nations		Lower than reference	Higher than reference
OPEC share of liquid production in 2040	40–43% (remains constant)	49%	37–40%
Jet price growth rate	1.2%	− 1.2%	2.4%
Ethanol price growth rate	− 0.10%	− 0.30%	0.90%

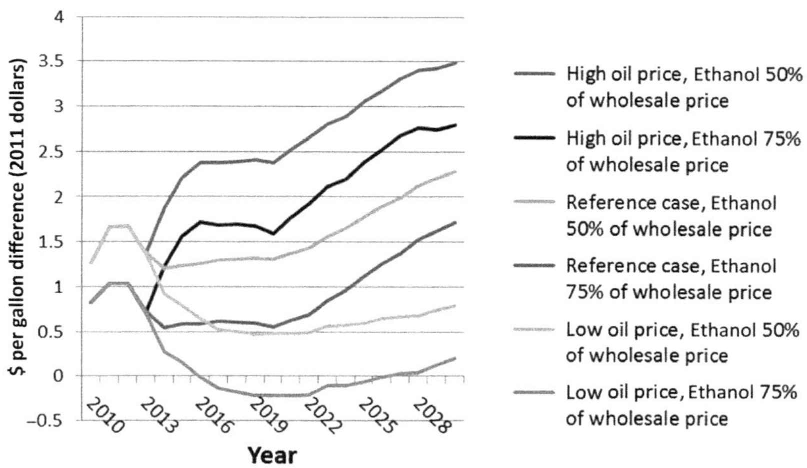

FIGURE 6.14 Projections of cost difference between wholesale jet price and ethanol intermediate price (shown for 50% and 75% of wholesale price) [101].

50 million gallon per year plant). The 2011 average wholesale price for ethanol in the United States was $2.54 per gallon (EIA), based on domestically grown corn ethanol. Price estimates for second-generation ethanol vary substantially. Algenol has reported a $1.00 per gallon ethanol feedstock production cost for jet fuel production [108]. LanzaTech has reported ethanol production costs of $0.60 per gallon from waste feedstocks and $1.71 per gallon from biomass [109]. Refined ethanol from biochemical conversion of lignocellulosics (corn stover) is estimated at $2.15 per gallon (2007 dollars) [110]. Ethanol from thermochemical conversion of wood chips is estimated at $2.05 per gallon (2007 dollars) [111]. Reduced capital and operating costs to produce nonrefined ethanol, as well as leveraging low-cost feedstocks, suggest that the $1.25–$1.50 per gallon raw ethanol price may be feasible. Production economics estimates for ethanol to jet fuel are not yet available in the open literature as the technology is still in the pre-commercialization phase.

The EIA projections of fuel prices and ethanol supply and demand suggest that there could be a substantial feedstock resource and feasible price differential for future ethanol to jet fuel production for all cases, except in the Low Oil Price case. While the EIA projections provide a rough estimate of potential market for ethanol to jet fuel, they depend on many assumptions and are inherently uncertain. Along with price swings historically associated with OPEC-produced oil (EIA 2015[15]), recent developments in North America will also have an effect. For example, the surge in shale oil and gas production is expected to launch the United States ahead of all but one of the OPEC nations [112], which could significantly impact the global supply balance. Future fuel supply, demand, and prices are interdependent and will be determined according to economic conditions, policy, and politics, both nationally and globally.

CASE STUDY: LANZATECH RESEARCH WITH PNNL AND IMPERIUM

The basis of the technology conversion of ethanol to hydrocarbons was generated by PNNL and John Plaza, CEO of Imperium Aviation Fuels, in 2010. The initial research initiative undertaken by PNNL and initially co-funded by Imperium and Battelle, the operator of PNNL, was to assess the viability of converting ethanol into hydrocarbon fuels for aviation from biomass-sourced alcohols. The belief that the ethanol 'blend wall' issues were likely going to create oversupply of ethanol for US Markets, and significant availability of bio-sourced ethanol, justified the research into a technology pathway for building higher value molecules from ethanol to jet fuel. This development allowed PNNL researchers to demonstrate that ethanol could be used as a feedstock for preparing an all-hydrocarbon fuel blend that met or exceeded standardized international specifications for jet fuel. The initial fuel produced was a mixture of all the hydrocarbon components, including aromatics.

In 2011, DOE issued a call for proposals for a new aviation biofuel research project. PNNL and Imperium jointly responded with LanzaTech as the lead applicant to the DOE. The research was focused on the ability to produce the ethanol from biomass, and then the minimally processed ethanol would be converted into biomass-based jet fuel. The proposal

[15]http://www.eia.gov/todayinenergy/detail.cfm?id = 22852

(*cont'd*)

was successful with DOE providing up to $4 million in grant funding for the project, of which about $2 million went to PNNL over a three-year period that began in 2012. In the DOE-funded work, PNNL adapted its technology so it produced solely isoparaffinic kerosene.

Five years later, it is clear that waste streams to ATJ is the clear path forward for success for the conversion of ethanol to hydrocarbon-based fuels. The key to sustainable alternative aviation fuel is abundant, low-cost feedstock. For the ATJ pathway, low-cost ethanol sourced from companies such as LanzaTech will enable ATJ jet fuel to compete with conventional Jet A-1 over the long term. For the ATJ pathway, most of the cost of production is tied to the feedstock, so utilizing a low-cost ethanol sourced from companies such as LanzaTech will enable ATJ jet fuel to compete with conventional Jet A-1 over the long term.

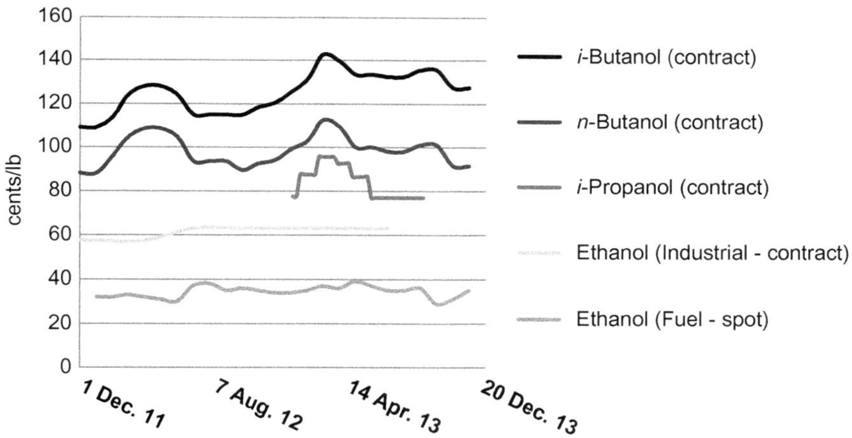

FIGURE 6.15 Ethanol and higher alcohol price history, US gulf coast [113].

6.5.5.2 Ethanol and Higher Alcohols

As described in Section 6.3.6, conversion to jet fuel from propanol through propylene, and butanol via butene, is simpler and better developed than from ethanol through ethylene. However, the market share and pricing for higher alcohols put them at a disadvantage when compared with ethanol. Fig. 6.15 shows that current pricing for higher alcohols is significantly greater than ethanol for both the fuel and chemicals markets. This reflects the fact that the existing market for higher alcohols is almost entirely for chemicals.

n-Butanol for chemical production is approximately 2.5 billion pounds per year [114] and is well-established. Butanol fuel production, however, is a nascent industry with Gevo announcing an increase in isobutanol production at their Luverne, MN, plant to a range of 750,000 to 1 million gallons in 2016.[16] Gevo's ATJ is being produced at its demo biorefinery in Silsbee, TX, using isobutanol

[16]http://ir.gevo.com/phoenix.zhtml?c = 238618&p = irol-newsArticle&ID = 2089814

produced at its Luverne, MN, fermentation facility. The company has successfully flown test flights with the US Air Force, US Army, and US Navy. Gevo also intends to begin test flights with the commercial aviation industry, including Alaska Airlines, following receipt of ASTM International certification. Fuel butanol prices will certainly be lower than chemical prices as purification needs can be relaxed. However, the large existing production capacity and market for fuel ethanol, coupled with low butanol capacity and higher prices in the chemical market, suggest that the incentives to build additional butanol capacity to produce a lower value product (namely fuel) may be low. Ultimately, the choice of ethanol versus higher alcohols as a jet fuel intermediate will depend upon the speed at which ethanol conversion technology can be developed versus development of sufficient butanol production capacity coupled with production at a competitive price.

6.6 CONCLUSIONS AND RECOMMENDATIONS

There is significant interest from both government and private sectors to displace petroleum-based jet fuel in favour of alternative bio-based fuels. Even with the expansion of fossil hydrocarbon fuel production in the United States, there still remains a strong interest for renewable transportation fuel. This increased interest results from concerns relative to climate change, oil price volatility, and the need to reduce our dependence on foreign oil in the future. As a result, demonstrations of alternative fuels, commitments to increased usage, and increased grants for research and development have been made. There is also significant interest from the public in increasing the use of alternative fuels. While this interest may provide some marketing advantages for companies that employ alternative fuels, surveys, and test programs demonstrate that consumers are unwilling to pay higher prices for these alternatives. As a result, favourable economics are critical to alternative fuel usage.

When the cost and market size of ethanol and jet fuel are compared, based on the AEO Reference case, it appears that ATJ will be economical if the production costs for the intermediate alcohol are substantially reduced from wholesale and the alcohol conversion step(s) is also inexpensive. DOE analyses of cellulosic ethanol suggest a wholesale price of \$2.05–2.15 per gallon [110,111]. However, alternative ways of producing ethanol (eg, using waste gases) could provide a more favourable price differential for ethanol to jet. Both Algenol and LanzaTech claim to be able to produce raw ethanol at the cost necessary for this purpose. The incentive to produce jet fuel blend components rather than produce finished ethanol also appears to be viable given the expected excess supply of ethanol over that required for gasoline blending. As a result, the market for ATJ is unconstrained and will accommodate a wide range of both ethanol and nonethanol intermediate-based technologies. Additionally, there is little need to produce SKA fuels (direct drop-in replacements), and research can instead emphasize 50% and less blends of SPK.

Options for converting ethanol to jet range components economically are still in development. Existing processes are available for homogeneous-type catalysis, but processing costs make these more applicable for the higher value chemicals market. Improved processes with better catalysts, likely heterogeneous, are needed to meet the required yields and make these processes economical. It is a particular advantage that they can be tailored to make both linear and cyclic compounds. The conversion of higher alcohols to jet fuel blend components presents less of a technical hurdle. However, limited infrastructure

TABLE 6.10 Summary of Drivers for Ethanol to Jet Fuel

Driver	Issue	How Ethanol to Jet Addresses Issue
Cost	Crude oil is historically vulnerable to price swings, deeply affecting airline fuel costs (40% of total).	Domestically produced fuel is less likely to experience such extreme fluctuations.
Climate change	The airline industry has maximized engine efficiency and has restricted renewable options compared to terrestrial transportation.	Bio-jet fuels are one of the only sustainable fuel options for aviation and can provide up to or even exceed 80% reduction in carbon emissions.
National security	Over-reliance on foreign oil is believed to put US troops at risk of supply disruptions.	Domestically produced advanced 'drop-in' fuels are expected to help the nation achieve its broader national security objectives.
Other air emissions	Particulate matter and SO*x* are formed by combustion of petroleum-based fuels.	ATJ is low in sulphur and can be engineered to contain low aromatics (soot formers).
Energy density of fuel	Ethanol has low energy density compared to petroleum fuels.	Ethanol to jet provides a 60% increase in energy density.
Carbon taxes and fuel credits	Future carbon taxes could be imposed on the airlines similar to that recently adopted in the European Union.	Bio-jet fuels could reduce the cost impact of a carbon tax by roughly 30% in a 50% blend with fossil jet fuel. Renewable fuel credits can incentivize production by as much $3.25 per gallon when market fuel prices cannot cover production costs.
Green consumer market	Consumers are increasingly seeking 'green' alternatives.	Airlines' use of bio-based jet fuel can target the 'lifestyles of health and sustainability' (LOHAS) market and provide a marketing advantage.
Ethanol 'Blend Wall'	Gasoline content is limited to 10% and consumption will continue to fall because of corporate average fuel economy (CAFE) standards.	EIA projections suggest there may be a significant surplus of ethanol that could potentially fill 4% of jet demand in 2020.
Price differential between ethanol and jet	There must be an adequate difference between the wholesale jet fuel and intermediate ethanol feedstock on a $ per gallon basis.	EIA projections suggest there is a positive price differential in most cases, except in the low oil price case.
Competition from higher alcohols	Technology availability and market share.	The ethanol fuel market is well established and much larger than the chemicals market for higher alcohols.
Flexibility to produce paraffins or cyclic components	Not all alternative/biofuel routes allow for production of paraffins or cyclics.	Ethanol to jet has a low hydrogen demand and greater flexibility to produce isoparaffins or cyclic components in synthetic jet fuel.

and more lucrative chemical markets may slow entry into ATJ. Future work should focus on quantifying yields, capital and operating requirements, and external energy requirements (power, natural gas) for ethanol to jet, which are beyond the scope of this chapter.

Given sufficiently low production cost, ethanol supplies and existing infrastructure suggest that it is a viable intermediate for the production of alternative jet fuel components. Table 6.10 summarizes how ethanol addresses the issues related to ATJ.

Acknowledgements

The authors would like to thank the Office of Energy Efficiency and Renewable Energy within the US Department of Energy (DOE), the Federal Aviation Administration (FAA) and the Defense Advanced Research Projects Agency (DARPA) for support of research for alternative jet fuel.

References

[1a] N. Stern., The Economics of Climate Change, Cambridge University Press, Cambridge, UK, 2007.

[1b] S.N. Naik, V.V. Goud, P.K. Rout, A.K. Dalai, Production of first and second generation biofuels: a comprehensive review, Renew. Sustain. Energy Rev. 14 (2010) 578–597.

[1c] AsiaOne News, More airlines fold as fuel prices soar: IATA. <http://news.asiaone.com/News/Latest+News/Business/Story/A1Story20080708-75407.html>, 2008 (accessed 08.07.08).

[2] MASBI, Fueling a sustainable future for aviation. Midwest Aviation Sustainable Biofuels Initiative, 2013.

[3] U. Yildirim, S. Abanteriba, Manufacture, qualification and approval of new aviation turbine fuels and additives, Proc. Eng. 49 (2012) 310–315.

[4] ASTM D1655-13, Standard Specification for Aviation Turbine Fuels, ASTM International, West Conshohocken, PA, 2013, pp. 1–7.

[5] Chevron, Aviation fuels technical review, p. 24–29. Available from: <https://www.cgabusinessdesk.com/document/aviation_tech_review.pdf>, 2006.

[6] Shell, Military jet fuel grades and specifications (NATO). Available from: <http://www.shell.com/global/products-services/solutions-for-businesses/aviation/shell-aviation-fuels/fuels/types/military-jet-fuel-grades.html>, 2013.

[7] DOD, Detail specification: turbine fuel, aviation, kerosene type, JP-8 (NATO F-34), NATO F-35, and JP-8 + 100 (NATO F-37), MIL-DTL-83133H w/ Amendment 1, (September 2012), p. 2–18, 2012.

[8] Air Force (US), Conversion from military specification fuel to commercial specification fuel at United States air force bases. Headquarters Department of the Air Force Program Guidance Letter 12-03 4 June 2013.

[9] J. Schmitigal, J. Tebbe, JP-8 and other military fuels. Briefing charts presentation for US Army TARDEC, 6501 E.11 Mile Rd, Warren, MI, 48397-5000, 1 December 2011. <http://www.dtic.mil/dtic/>, 2011.

[10] Purvis Bros, Aviation Fuels Info: Modern US Military Turbine Missile Fuel Specifications, Purvis Brothers, Inc., Mars, PA, 2001. Available from: <http://purvis-bros.com/afi/modernmilturbinemissle.htm>

[11] H.A. Meylemans, L.C. Baldwin, B.G. Harvey, Low-temperature properties of renewable high-density fuel blends, Energy Fuels 27 (2) (2012) 883–885.

[12] J. Lane, Breaking news: US Senate votes 62-37 to back military aviation biofuels. Biofuels Digest. <http://www.biofuelsdigest.com/bdigest/2012/11/28/breaking-news-us-senate-votes-62-37-to-back-military-aviation-biofuels/>, 2012 (accessed 08.11.12).

[13] CRC, *Handbook of Aviation Fuel Properties*. CRC Report No. 530, Coordinating Research Council, Inc., Alpharetta, GA, 1983.

[14] J. Lane, Making superfuels affordable, via biofuels: the JP-10 story. Biofuels Digest. <http://www.biofuelsdigest.com/bdigest/2013/02/05/making-superfuels-affordable-biofuels-and-jp-10/>, 2013a. (accessed 05.02.13).

[15] FAA, Federal aviation safety, September 2011, Bulletin NE-11-56, 'Special Airworthiness Information Bulletin', pp. 1–2. Federal Aviation Administration, Washington, DC, 2011.

[16] US DHHS, Toxicological Profile for JP-5 and JP-8, Chemical and Physical Information, U.S. Department of Health and Human Services, Public Health Service, 1998, pp. 97–102. Available from: <http://www.atsdr.cdc.gov/ToxProfiles/tp121.pdf>.

[17] Blakey, Proceedings of the Combustion Institute 33, 2011. pp. 2863–2865.

[18] CAAFI, Frequently asked questions: are alternative fuels economically viable? Commercial Aviation Alternative Fuels Initiative. <http://www.caafi.org/about/faq.html>, 2013 (accessed 22.11.13).

[19] ASTM D7566-13, Standard Specification for Aviation Turbine Fuel Containing Synthesized Hydrocarbons, ASTM International, West Conshohocken, PA, 2013, pp. 1–13.

[20] ASTM WK41378, Work Item: ASTM WK41378—New Specification for RESEARCH REPORT: Evaluation of Alcohol to Jet Synthetic Paraffinic Kerosenes (ATJ-SPKs). ASTM International, West Conshohocken, PA. Available from: <http://www.astm.org/DATABASE.CART/WORKITEMS/WK41378.htm>.

[21] IATA, Report on alternative fuels, international air transport association, December 2011. International Aviation Transport Association, Geneva, Switzerland, 2011.

[22] Braskem, Braskem Ethanol-to-Ethylene Plant, Brazil. <http://www.chemicals-technology.com/projects/braskem-ethanol/>, 2013 (accessed 17.12.13).

[23] D. Fan, D. Dai, H. Wu, Ethylene formation by catalytic dehydration of ethanol with industrial considerations, Materials 6 (2013) 101–115.

[24] A. Forestière, H. Olivier-Bourbigou, L. Saussine, Oligomerization of monooelfins by homogeneous

catalysts, Oil Gas Sci. Technol.–Rev. IFP 64 (6) (2009) 649–667.

[25] T. Agapie, Selective ethylene oligomerization: recent advances in chromium catalysis and mechanistic investigations, Coord. Chem. Rev. 255 (2011) 861–880.

[26] N.A.S. Amin, D.D. Angorro, Dealuminated ZSM-5 zeolite catalyst for ethylene oligomerization to liquid fuels, J. Nat. Gas Chem. 11 (2002) (2002) 79–86.

[27] J. Heveling, C.P. Nicolaides, M.S. Scurrell, Catalysts and conditions for the highly efficient, selective and stable heterogeneous oligomerisation of ethylene, Appl. Catal. A Gen. 173 (1998) 1–9.

[28] M.D. Heydenrych, C.P. Nicolaides, M.S. Scurrell, Oligomerization of ethene in a slurry reactor using a Nickel(II)-Exchanged Silica–Alumina catalyst, J. Catal. 197 (2001) 49–57.

[29] SRI Consulting, Syntroleum Process by Slurry-Phase F–T Synthesis with an Integrated Power Block PEP Yearbook. Menlo Park, CA, 2007.

[30] M.E. Wright, B.G. Harvey, R.L. Quintana, Highly efficient zirconium-catalyzed batch conversion of 1-Butene: a new route to jet fuels, Energy Fuels 22 (5) (2008) 3299–3302. Available from: http://dx.doi.org/10.1021/ef800380b.

[31] M. Lallemand, A. Finiels, F. Fajula, V. Hulea, Ethylene oligomerization over Ni-containing mesostructured catalysts with MCM-41, MCM-48 and SBA-15 topologies, Stud. Surf. Sci. Catal. 170 (2007) 1863–1869.

[32] V. Hulea, F. Fajula, Ni-exchanged AlMCM-41—An efficient bifunctional catalyst for ethylene oligomerization, J. Catal. 225 (1) (2004) 1–258.

[33] S.H. Mahdaviani, D. Soudbar, M. Parvari, Selective ethylene dimerization toward 1-butene by a new highly efficient catalyst system and determination of its optimum operating conditions in a buchi reactor, Int. J. Chem. Eng. Appl. 1 (3) (2010) 276–281.

[34] X. Zhang, J. Zhong, J. Wang, J. Gao, A. Liu, Trimerization of butene over Ni-doped zeolite catalyst: effect of textural and acidic properties, Catal. Lett. 126 (2008) 388–395.

[35] X. Zhang, J. Zhong, J. Wang, L. Zhang, J. Gao, A. Liu, Catalytic performance and characterization of Ni-doped HZSM-5 catalysts for selective trimerization of *n*-butene, Fuel Process. Technol. 90 (2009) 863–870.

[36] R. Alcantara, E. Alcantara, L. Canoira, M.J. Franco, M. Herrera, A. Navarro, Trimerization of isobutene over Amberlyst-15 catalyst, Reactive Funct. Polym. 45 (2000) 19–27.

[37] M. Parvari, et al., Selective ethylene dimerization toward 1-Butene, Int. J. Chem. Eng. Appl. 1 (2) (2010).

[38] Chevron Phillips Chemical, Chevron Phillips Chemical announces plans for world-scale 1-hexene plant. PlastInfo. <http://www.plastinfo.net/information/news/2544_18.10.2010/>, 2010 (accessed 18.10.10).

[39] IHS, 1-Hexene from ethylene by the phillips trimerization technology. PEP Review 95-1-8, December 1997, IHS Chemical, Englewood, CO. <http://www.ihs.com/products/chemical/technology/pep/reviews/hexene-from-ethylene.aspx>, 1997 (accessed 17.12.13).

[40] B.G. Harvey, H.A. Meylemans, 1-Hexene: a renewable C6 platform for full-performance jet and diesel fuels, Green Chem. 16 (2013) 770–776 on-line, accepted 9 September 2013. Available from: http://dx.doi.org/10.1039/c3gc41554f.

[41] D. Goto, Y. Harada, Y. Furumoto, A. Takahashi, T. Fujitani, Y. Oumi, et al., Conversion of ethanol to propylene over HZSM-5 type zeolites containing alkaline earth metals, Appl. Catal. A Gen. 383 (2010) 89–95.

[42] K. Inoue, Conversion of ethanol to propylene, Catal. Lett. 136 (1–2) (2010) 14–19.

[43] M. Iswamoto, One step formation of propene from ethene or ethanol through metathesis on nickel ion-loaded silica, Molecules 16 (2011) 7844–7863.

[44] C.D. Chang, Propylene oligomerization over silica modified zeolites. EP-493931A1, 8 July 1992, 1991.

[45] UOP, Higher olefins. Des Plaines, IL. Available from: <http://www.uop.com/processing-solutions/petrochemicals/detergents/higher-olefins/>, 2013.

[46] L. Yinong, D. Qinghua, Investigation of propene oligomerization catalyzed by phosphotungstic acid catalysts, China Pet. Process. Petrochem. Technol. 14 (3) (2012) 10–16.

[47] M. Guerbet, C.R. Hebd, Séances, Acad. Sci. 128 (1899) 511–513.

[48] D. Gabriels, W. Hernandez, B. Sels, P. Van Der Voort, A. Verberckmoes, Review of catalytic systems and thermodynamics for the Guerbet condensation reaction and challenges for biomass valorization, Catal. Sci. Technol. 5 (2015) 3876–3902.

[49] J. Kozlowski, R. Davis, Heterogeneous catalysts for the Guerbet coupling of alcohols, ACS Catal. 3 (2013) 1588–1600.

[50] J.F. Scalbert, R. Thibault-Starzyk, D. Jacquot, Morvan, F. Meunier, Ethanol condensation to butanol at high temperatures over a basic heterogeneous catalyst: how relevant is acetaldehyde self-aldolization? J. Catal. 311 (2014) 28–32.

[51] G.R.M. Dowson, M.F. Haddow, J. Lee, R.L. Wingad, D.F. Wass, Catalytic conversion of ethanol into an advanced biofuel: unprecedented selectivity for *n*-butanol, Angew. Chem. Int. Ed. 52 (2013) 9005–9008.

[52] T. Rittonen, E. Toukoniitty, D. Madnani, A. Leino, K. Kordas, M. Szabo, et al., One-pot liquid phase catalytic conversion of ethanol to 1-Butanol over Al2O3, Catalysis 2 (2012) 68–84.

[53] T. Tsuchida, J. Kubo, T. Yoshioka, S. Sakuma, T. Takeguchi, W. Ueda, Reaction of ethanol over hydroxyapatite affected by Ca/P ratio of catalyst, J. Catal. 259 (2008) 183–189.
[54] Abengoa, Abengoa launches a technology development project for catalytic conversion of ethanol to biobutanol. Available from: <http://www.abengoa.es/htmlsites/boletines/en/febrero2013/tecnologia/>, 2013.
[55] R. Franke, D. Selent, A. Börner, Applied hydroformylation, Chem. Rev. 112 (2012) 5675–5732.
[56] J. Pospech, I. Fleischer, R. Franke, S. Buchholz, M. Beller, Alternative metals for homogeneous catalyzed hydroformylation reactions, Angew. Chem. Int. Ed. 52 (2013) 2852–2872.
[57] I. Golubkov, Method for producing hydrocarbons and oxygen-containing compounds from biomass. U.S. Patent 0112739 A1, October 2012, 2005.
[58] R.M. Deshpande, B.M. Bhanage, S.S. Divekar, S. Kanagasabapathy, R.V. Chaudhari, Kinetics of hydroformylation of ethylene in a homogeneous medium: comparison in organic and aqueous systems, Ind. Eng. Chem. Res. 37 (6) (1998) 2391–2396.
[59] C. Machado, Technical characteristics and current status of butanol production and use as biofuel. BIO2010 Seminario Latinoamericano y de Caribe de Biocombustibles, Santiago, Chile, 17, 18 August 2010.
[60] R. Kolodziej, J. Scheib, Bio-isobutanol: the next-generation biofuel, Hydrocarbon Process. 91 (9) (2012) 79–85.
[61] Green Car Congress, Naval Air Warfare Center awards contract to Albemarle for processing Cobalt Technologies' bio *n*-butanol to renewable jet fuel using Alcohol-to-Jet process. <http://www.greencarcongress.com/2012/03/nawcwd-20120320.html>, 2012 (accessed 20.03.12).
[62] S.N. Chaudhuri, C. Halik, J.A. Lercher, Reaction of ethanol over HZSM-5, J. Mol. Catal. 62 (3) (1990) 289–295.
[63] E. Costa, A. Uguina, J. Aguado, P. Hernandez, Ethanol to gasoline process: effect of variables, mechanism and kinetics, Ind. Eng. Chem. Process Des. Dev. 24 (1985) 239–244.
[64] K. Ramasamy, Y. Wang, Thermochemical conversion of fermentation-derived oxygenates to fuels, in: B. Zhang, Y. Wang (Eds.), Chapter 17 in *Biomass Processing, Conversion and Biorefinery*, Nova Science Publishers, Hauppauge, NY, 2013.
[65] N. Viswanadham, S. Saxena, J. Kumar, P. Sreenivasulu, D. Nandan, Catalytic performance of nano crystalline H-ZSM-5 in ethanol to gasoline (ETG) reaction, Fuel 95 (2012) 298–304.
[66] V.R. Choudhary, P. Devadas, S. Banerjee, A.K. Kinage, Aromatization of dilute ethylene over Ga-modified ZSM-5 type zeolite catalysts, Microporous Mesoporous Mater. 47 (2001) 253–267.
[67] B. Umansky, M. Clark, C. Lopez, K. Weiger, Process and system to convert olefins to diesel and other distillates. ExxonMobil Research and Engineering, U.S. Patent Application 2011/0147263 A1, 23 June 2011.
[68] A.V. Bridgwater, Review of fast pyrolysis of biomass and product upgrading, Biomass Bioenergy 38 (2012) 68–94.
[69] D.C. Elliott, Hydrothermal processing, in: R.C. Brown (Ed.), Chapter 7 in Thermochemical Processing of Biomass: Conversion Into Fuels, Chemicals and Power, Wiley-Blackwell, Oxford, 2011, pp. 200–231.
[70] Envergent Technologies, <www.envergenttech.com>, 2013.
[71] T. Marker, et al., Biomass to gasoline and diesel using integrated hydropyrolyis and hydroconversion. Technical Report Available from: <http://www.osti.gov/scitech/biblio/1059031>, 2013.
[72] Y. Zhu, M.J. Biddy, S.B. Jones, D.C. Elliott, A.J. Schmidt, Techno-economic analysis of liquid fuel production from woody biomass via hydrothermal liquefaction (HTL) and upgrading, Appl. Energy 129 (2014) 384–394.
[73] M. Wilson, Flying high with aviation biofuels, Farm Futures (2013). Posted 22 February 2013. Available from: <http://farmfutures.com/blogs-flying-high-aviation-biofuels-5179>.
[74] Renewable Jet Fuels, What's driving the development of renewable fuels? Available from: <http://renewablejetfuels.org/what-we-do/faq>, 2013.
[75] A. Gonzalez, To go green in jet fuel, Boeing looks at algae. The Seattle Times, 30 August 2007, <http://seattletimes.com/html/businesstechnology/2003858756_boeingenergy30.html>, 2007 (accessed 18.12.13).
[76] General Biomass Company, Why aviation biofuels. Available from: <http://www.generalbiomass.com/renewablejetfuel1.htm>, 2013.
[77] P. Lobo, D.E. Hagen, P.D. Whitefield, Comparison of PM emissions from a commercial jet engine burning conventional, biomass, and fischer–tropsch fuels, Environ. Sci. Technol. 45 (2011) 10744–10749.
[78] SAFN, Sustainable aviation fuels northwest: powering the next generation of flight. 2011 Report. Available from: <http://www.safnw.com/wp-content/uploads/2011/06/SAFN_2011Report>, 2011.
[79] USA Today, Big milestone: US producing more oil than it imports. 13 November 2013. <http://www.usatoday.com/story/news/nation/2013/11/13/us-oil-production-exceeds-imports/3518245/>, 2013 (accessed 17.12.13).

[80] Bloomberg News, North American shale to trump most OPEC members in two years. 12 December 2013. <http://www.businessweek.com/news/2013-12-11/north-american-shale-to-exceed-most-opec-nations-output-in-2015>, 2013b (accessed 17.12.13).
[81] IATA, Fact sheet: alternative fuels. International Aviation Transport Association, Geneva, Switzerland. Available from: <http://www.iata.org/pressroom/facts_figures/fact_sheets/pages/alt-fuels.aspx>, 2013.
[82] Jatenergy, The market renewables. Available from: <http://jatenergy.com/the-market-renewables/>, 2013.
[83] J. Lane, Fly the (environmentally) friendly skies: united to commence wide-scale aviation biofuels flights in 2014, Biofuels Digest 5 June 2013 (2013). Available from: <http://www.biofuelsdigest.com/bdigest/2013/06/05/fly-the-environmentally-friendly-skies-united-to-commence-wide-scale-aviation-biofuels-flights-in-2014/>.
[84] Airbus, <http://www.airbus.com/innovation/eco-efficiency/operations/alternative-fuels/>, 2013.
[85] Innovate Washington, Aviation biofuels update. December, 2013.
[86] European Union, The EU emissions trading system. The Publications Office of the European Union, Luxembourg. Available from: <http://ec.europa.eu/clima/publications/docs/factsheet_ets_en.pdf>. http://dx.doi.org/10.2834/55480. 2013.
[87] Carbon Disclosure Project, Use of internal carbon price by companies as incentive and strategic planning tool; A review of findings from CDP 2013 disclosure. A white paper from CDP North America, December 2013. Available from: <http://www.nationaljournal.com/free/document/download/4880-1>, 2013.
[88] DOT, Total passengers on US airlines and foreign airlines US Flights Increased 1.3% in 2012 from 2011. 4 April 2013. Press Release No. BTS-16-13, US Department of Transportation, Washington, DC. Available from: <http://www.rita.dot.gov/bts/press_releases/bts016_13>, 2013a.
[89] EPA, Optional emissions from commuting, business travel and product transport. Office of Air and Radiation, US Environmental Protection Agency, Washington, DC. Available from: <http://www.epa.gov/climateleadership/documents/resources/commute_travel_product.pdf>, 2008.
[90] DOT, Airline fuel cost and consumption (US Carriers—Scheduled), January 2000–October 2013. Research and Innovative Technology Administration, Bureau of Transportation Statistics, US Department of Transportation, Washington, DC. Available from: <http://www.transtats.bts.gov/fuel.asp>, 2013b.
[91] NARA, Western montana corridor, summary report—August 2013. Northwest Advanced Renewables Alliance, Pullman, WA. Available from: <http://wood-symposium.wsu.edu/documents/WMCAtlasBook8-23.pdf>, 2013.
[92] Roper ASW, Green gauge report 2002: Americans Perspective on Environmental Issues—Yes, But. November, 2002.
[93] American Spa, LOHAS market-size data released: A $290 billion opportunity. Taken from National Marketing Institute. <http://www.americanspa.com/spa-news/lohas-market-size-data-released-290-billion-opportunity>. More recent data is available from National Marketing Institute for a fee at <http://www.nmisolutions.com/index.php/research-reports/us-sustainability-report>, 2013 (accessed 30.10.13).
[94] B. Friedman, Buying green power: does it make cents for your company? E-Source, GE-F-7, July 2004.
[95] K. Petrella, UPS reports reduction of greenhouse gas emissions. SustainableBuisness.com. <http://www.sustainablebusiness.com/index.cfm/go/news.viewpressrelease/id/351?utm_source = feedburner&utm_medium=feed&utm_campaign=Feed%3A+SBPressReleases+%28SustainableBusiness.com+Press+Releases%29>, 2013 (accessed 25.07.13).
[96] UPS, UPS corporate sustainability report 2012. United Parcel Service, Atlanta, GA. <http://www.responsibility.ups.com/community/Static%20Files/sustainability/UPS_CSR2012_WEB_072213.pdf>, 2012 (accessed 17.12.13).
[97] C. Hanson, The business case for using renewable energy. World Resources Institute, Instalment 7, December 2005.
[98] G. Dooley, Travelers unwilling to pay more for green travel. Travel Agent Central. <http://www.travelagentcentral.com/green-travel/travelers-unwilling-pay-more-green-travel-17065>, 2009 (accessed 25.08.09).
[99] E. Holt, R. Wiser, Understanding consumer demand for green power. Prepared for the National Wind Coordinating Committee, 1999.
[100] U. Burns, Is the green movement a passing fancy? Bloomberg Business week. 25 January 2009. <http://www.businessweek.com/technology/content/jan2009/tc20090126_136438.htm>, 2009 (accessed 02.12.13).
[101] EIA, Annual energy outlook. US Energy Information Administration, Washington, DC, 2013a.
[102] R. Schnepf, B.D. Yacobucci, Renewable Fuel Standard (RFS): Overview and Issues, Congressional Research Service (CRS), 23 January 2012.

[103] EPA, EPA proposes 2014 renewable fuel standards, 2015 biomass-based diesel volume. November 2013 regulatory announcement, Office of Transportation and Air Quality. EPA-420-F-13-048, 2013a.

[104] EPA, EPA finalizes 2013 renewable fuel standards. August 2013 regulatory announcement, Office of Transportation and Air Quality. US Environmental Protection Agency, Washington, DC. EPA-420-F-13-042, 2013b.

[105] DOE, Alternative fuels data center: E15. Office of Energy Efficiency and Renewable Energy, US Department of Energy, Washington, DC. <http://www.afdc.energy.gov/fuels/ethanol_e15.html>, 2013 (accessed 09.12.13).

[106] EIA, Annual energy outlook assumptions report. US Energy Information Administration, Washington, DC, 2013b.

[107] Biofuels Digest, Race for scale: united, alaska airlines lunch commercial aviation biofuels flights. 7 November 2011.

[108] Advanced Biofuels USA, Call to action for a truly sustainable renewable future. <http://advancedbiofuelsusa.info/economic-jet-a-fuel-from-ethanol-achieved>, 2013 (accessed 18.08.13).

[109] J. Holmgren, Innovative use of industrial waste gases to produce sustainable fuels & chemicals. Presented at the Low Carbon Jet Fuel: The Industrial Roadmap Avalon Air Show, Australia. 27 February. Available from: <http://ussc.edu.au/ussc/assets/media/docs/other/130226_LCF_Holmgren.pdf>, 2013.

[110] D. Humbird, R. Davis, L. Tao, C. Kinchin, D. Hsu, A. Aden, Process design and economics for biochemical conversion of lignocellulosic biomass to ethanol dilute-acid pretreatment and enzymatic hydrolysis of corn stover. Technical Report NREL/TP-5100-47764, May 2011, National Renewable Energy Laboratory, Golden, CO, 2011.

[111] A. Dutta, M. Talmadge, J. Hensley, Process design and economics for conversion of lignocellulosic biomass to ethanol; thermochemical pathway by indirect gasification and mixed alcohol synthesis. Technical Report NREL/TP-5100-51400, May 2011, National Renewable Energy Laboratory, Golden, CO, 2011.

[112] Bloomberg News, Big milestone: US producing more oil than it imports. 12 December 2013. <http://www.businessweek.com/news/2013-12-11/north-american-shale-to-exceed-most-opec-nations-output-in-2015>, 2013a (accessed 17.12.13).

[113] ICIS, ICIS chemical pricing. Available from: <http://www.icispricing.com>, 2013.

[114] B. Bowen, *n*-Butanol ICIS Chemical Business, 23–29 September 2013, p. 47. Accessed through ABI/INFORM Trade & Industry, 2013.

CHAPTER

7

Metabolic Engineering Strategies to Convert Carbohydrates to Aviation Range Hydrocarbons

A. Bergman[1,2] *and V. Siewers*[1,2]

[1]Department of Biology and Biological Engineering, Chalmers University of Technology, Gothenburg, Sweden, [2]Novo Nordisk Foundation Center for Biosustainability, Chalmers University of Technology, Gothenburg, Sweden

7.1 INTRODUCTION

Metabolic engineering enables the creation of modified biological systems with a precise mission – the conversion of biomass and carbon-containing gases into specific biochemicals. This chapter presents recent progress in the development of cell factories capable of producing aviation range biofuels. It is divided into two major parts – the first covering production routes towards fatty acid-based biofuels, such as alkanes, and the second covering metabolic pathways and strategies targeting isoprenoid-based fuels, of which there are commercial examples available. Each section covers natural precursor metabolism, information on central biofuel-forming pathways, metabolic engineering examples to generate efficient cell factories for the discussed fuels, and a discussion on challenges and opportunities in the field. In this introductory section we discuss the advantages of using cell factories for production of biofuels and primary strategies and principles within the field of metabolic engineering.

7.1.1 Why Use Microbes for Fuel Production?

The use of yeast for ethanol formation and molds for production of penicillin are examples of well-known biotechnological industrial processes today [1,2]. Both are illustrations of classical uses of cell factories to convert carbohydrate substrates into specific metabolic products, where the natural substrate-product spectrum of the producing organism is exploited. However, this limits the range of possible products because it is problematic to

Biofuels for Aviation.
DOI: http://dx.doi.org/10.1016/B978-0-12-804568-8.00007-X

culture many organisms on a large scale, and often the compound of interest is present only in low quantities [3]. Progress in genomics and genetic engineering during the last few decades has allowed us to vastly expand the microorganism production portfolio. Today, nearly any metabolic product detected in a life form can be produced in a nonnative host as long as the metabolic pathway and genetic elements within the parent organism are elucidated. This has spurred a great interest in the research fields of synthetic biology and metabolic engineering: to design and create custom-made organisms with high product-to-substrate ratios and low by-product formation [4]. Furthermore, it allows for a sustainable economy, where carbon dioxide-consuming plants and organisms can act as the substrates, or even as cell factories, themselves [5].

A major challenge in the field of metabolic engineering is to develop economically viable alternatives to petroleum-based products, especially within the massive fuel market. Ethanol is a strong biofuel candidate for markets directed towards regular combustion engines, but to be able to power the aviation industry, more energy-dense fuel molecules have to be established that can be used in gas-turbine engines [6]. Jet fuels tend to be comprised of a mixture of hydrocarbons – most commonly linear, branched, and cyclic alkanes – that bring about different attributes to the fuel, for example, low-temperature flow properties, high combustibility, and high-energy content [7]. Hydrocarbons are naturally present in a variety of organisms [8], and there are numerous biological compounds that are of interest as jet fuel candidates, such as alkanes/alkenes and isoprenoid-based fuels.

A great advantage of using fermentation-based technologies instead of chemical conversion of plant-based material is the broad range of substrates that can potentially be utilized. Currently, chemical production of biodiesel relies to a large extent on vegetable oils as substrates, which is questionable both from an environmental as well as a food-versus-fuels point-of-view [9]. The microbial community, on the other hand, has conquered virtually every environmental niche imaginable, and has consequently adapted to metabolize numerous food sources – traits that can also be transferred between organisms by genetic engineering. Thus, the future holds great promise in using current waste products from agriculture and forestry industries to produce next-generation biofuels if microbes can be constructed to catalyse an efficient conversion, for example, yeasts engineered to utilize cellulose and hemi-cellulose as substrates [10]. Additionally, a microbial reaction is catalysed under mild and physiological conditions, which reduces the formation of toxic by-products.

7.1.2 Natural Fuel Molecules and Prospective Production Hosts

There are two major cellular metabolic pathways that give rise to precursor hydrocarbons with relevance for the jet fuel industry – fatty acid and isoprenoid metabolism [11]. Alkanes, alkenes, fatty alcohols, and fatty acid ethyl esters (FAEE) are all prospective fuel molecules that can be derived from natural fatty acid metabolism, which is an essential pathway in all organisms in order to produce lipids (eg, membrane lipids). For example, the majority of cyanobacterial species and certain plants can naturally produce alkanes and alkenes [12,13], and several plants generate biodiesel-like molecules such as wax esters [14]. Farnesene and bisabolene are both prospective aviation range fuels belonging to the group of isoprenoid compounds. Isoprenoid production is also essential in all living organisms, and these molecules constitute the largest group of chemicals produced in biological systems, entailing more than 40,000 compounds [15].

With regards to the different kinds of microbial cell factories to use, cyanobacteria have been discussed as the gold standard, as they can convert light energy to chemical energy by fixating carbon dioxide (instead of converting biomass); many research groups focus on optimizing such microorganisms [16]. However, scaling-up of such processes is troublesome because cyanobacteria need direct access to light [16], and currently yields of biofuel-related compounds remain at a proof-of-principle stage [17]. A large number of the metabolic engineering efforts for biofuel production have gone into research and development with the traditional workhorses of biotechnology that can consume carbohydrates: the yeast *Saccharomyces cerevisiae* [18] and the bacterium *Escherichia coli* [19]. Naturally, these organisms do not produce large amounts of fuel precursors, but are genetically very flexible. Extensive knowledge exists about their biology, there are many tools for modifying them, and they are well known in industry. More work is also going into utilizing certain yeast species that naturally accumulate high quantities of lipids – oleaginous yeasts, of which *Yarrowia lipolytica* has become a model organism [20]. Such yeasts are promising because they would constitute a good platform for fatty acid-based fuel production, but they are less well characterized compared to *S. cerevisiae*, from an applied as well as a fundamental point of view. This chapter mainly focuses on progress made in *S. cerevisiae* and *E. coli*, but to some extent other organisms as well.

7.1.3 General Metabolic Engineering Principles

Genetic engineering has paved the way for the engineering of life forms to make them optimal for production of a certain biochemical compound or protein – the engineering of cell factories. This section gives a summary of the different strategies adopted to optimize microbial production of metabolites.

The metabolic engineering process is depicted in Fig. 7.1. It is inherently an iterative process due to the complexity of biological systems and our incomplete understanding of them. Commonly, an engineering strategy is based on the available knowledge of metabolism of the chosen organism, such as simplified biochemical reaction networks. The strategy is then implemented in the host strain through the use of recombinant DNA technology, such as plasmid cloning strategies and genome engineering tools, a process referred to as strain construction. The created strains are then tested in different cultivation settings and samples are taken to quantify production capacity and detect physiological changes on different levels. In the concluding step, the obtained data are analysed to verify if the initial strategy was successful and detect anomalies that might explain unpredicted behaviour. The obtained knowledge can then be added to the previous model, and the cycle can restart to optimize the production of the target compound. Potentially, joint advances in systems biology and metabolic modelling can help create improved predictive genome-scale metabolic models constrained by experimental data, such as fluxomics, to aid in strain construction strategies [21].

The re-direction of carbon flux in targeted biochemical reaction networks is a central concept in metabolic engineering. Increased flux through a certain pathway can, for example, be achieved by increasing the activity of pathway enzymes occurring naturally in the organism of choice, or by introducing genes from heterologous (nonnative) sources into the host, which can provide the cell with new, desired functions. The activity of a gene and corresponding enzyme of interest can in turn be modulated on many different levels. Increased *transcriptional activity* (generation of messenger

RNA that later on can be translated into a protein) is commonly accomplished either by increasing gene copy number or utilizing a strong promoter (the regulatory DNA sequence upstream of a gene that recruits the transcriptional machinery). After transcription, the *translational efficiency* can also be increased in prokaryotic species by modulating ribosome binding sites to increase the absolute enzyme number [22]. In eukaryotes as well as prokaryotic species, translational efficiency of heterologous genes can be increased by performing so-called codon-optimization, through which a gene sequence is adjusted to match the most

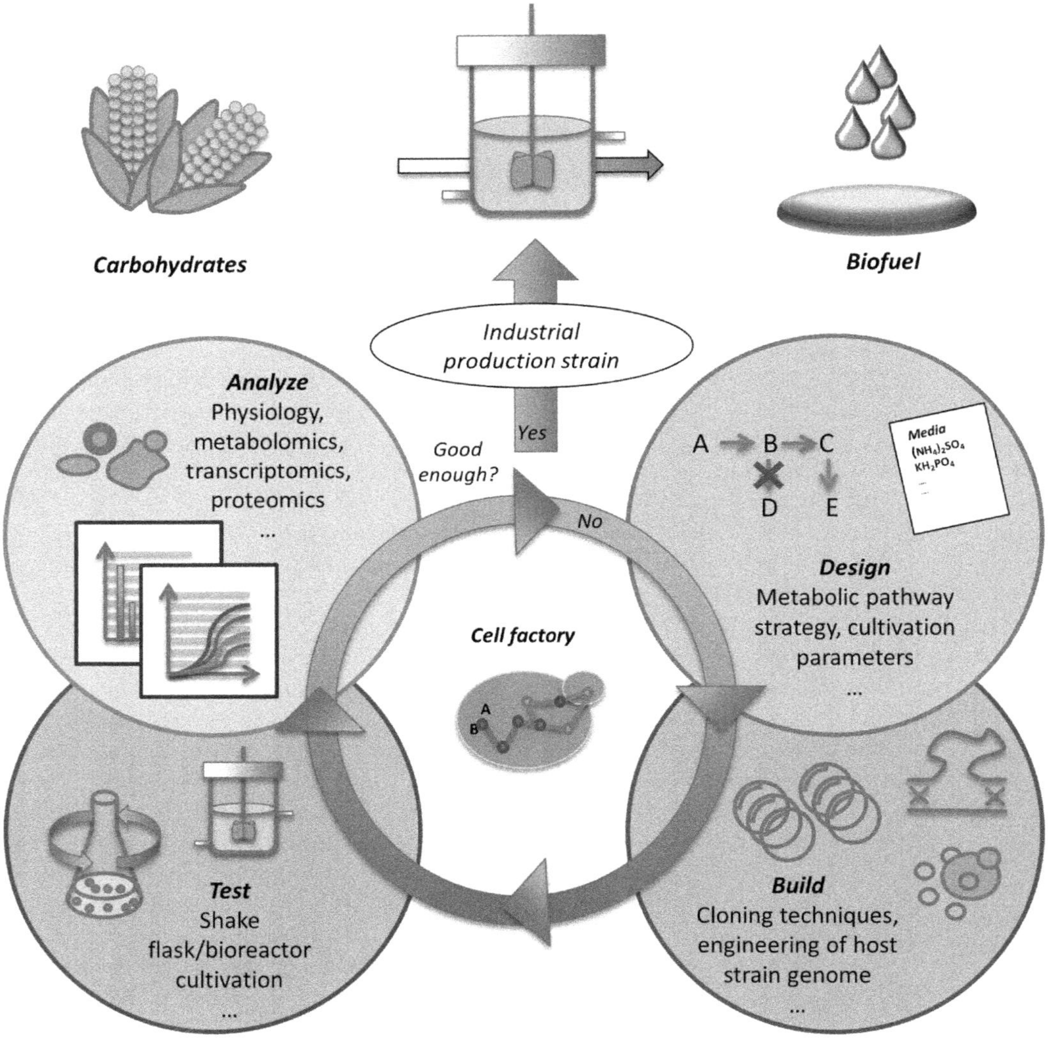

FIGURE 7.1 The iterative metabolic engineering cycle used to construct cell factories.

prevalent codon usage in the receiving organism [23]. The *enzymatic efficiency* can also be modified, for example, by making targeted engineering efforts to the catalytic site of the protein or abolishing post-translational regulation of the enzyme [24].

A metabolic engineering scheme usually targets metabolism in an organism at many different points simultaneously. Since it is beyond the scope of this chapter to describe them in detail, a small summary of different engineering approaches and their purposes is made in Table 7.1. For a more comprehensive summary of metabolic engineering strategies, see the review by Lee et al. [25].

7.2 FATTY ACID-DERIVED AVIATION RANGE BIOFUELS

The possibility to construct cell factories that produce the prospective aviation fuel molecules alkanes and alkenes has become a reality due to the recent advances in elucidation of the genetic and biochemical character of the final converting enzymes in these metabolic pathways. Microbial alkane-formation is dependent on enzymatic conversion of fatty acid species – aliphatic carboxylic acids. Therefore, a critical part in the process of engineering a microbial factory for alkane-based fuels is to engineer the endogenous fatty acid metabolism.

TABLE 7.1 Summary of Commonly Utilized Metabolic Engineering Strategies

Strategy	Example
Optimize/expand substrate utilization	Engineering of cells to co-utilize natural substrates in order to increase productivity, or installing heterologous genes that help cells utilize substrates that could not be used naturally.
Reduce/eliminate by-product formation	Deletion of genes that enable cells to produce nonessential metabolites in order to optimize yield of the desired product, or down-regulating genes that are essential but divert flux from the pathway of interest.
Increase precursor metabolite pool	Up-regulating genes responsible for catalysis of reactions upstream of a terminal (eg, biofuel-forming) enzyme of interest to increase substrate availability and reaction rate of the terminal enzyme.
Rerouting of metabolic pathways	Use of heterologous pathways that can catalyse a certain reaction faster/with fewer intermediate catalytic steps/from another intracellular metabolite/with reduced energy consumption, in order to optimize productivity.
Co-factor engineering	Modifying a metabolic pathway enzyme requiring a certain co-factor to prefer one that is more abundant in the biological system, or engineering the cell to increase production of required co-factors.
Synthetic enzyme scaffolds	Utilizing a protein-based structure that can dock multiple enzymes in order to bring proximity of enzymes in a modular pathway and their respective substrates, thereby increasing reaction rate, and/or to be able to dock additional copies of a slower enzyme onto the scaffold in order to balance the carbon flux.
Pathway optimization	Optimizing expression of the pathway enzymes or construction of a system where the metabolic pathway gene expression is regulated by substrate presence, which can generate flux balance and lead to a more energy efficient system.

Adapted after J.W. Lee, et al., Systems metabolic engineering of microorganisms for natural and non-natural chemicals, Nat. Chem. Biol. 8 (6) (2012) 536–546.

This subsection is divided into four parts – the first discusses key features of fatty acid metabolism in yeast and *E. coli*, the second deals with metabolic engineering strategies to overproduce fatty acids in these hosts, the third summarizes information about the terminal enzymes responsible for natural straight-chain hydrocarbon formation and engineering strategies to produce these molecules in microbes, and the final section discusses challenges and opportunities in the field.

Fatty alcohols and FAEE are also prospective biofuels derived from fatty acids that do not, however, have optimal properties for jet fuel purposes, though research in this field is described in detail in chapter 'The Suitability of Fatty Acid Methyl Esters (FAME) as Blending Agents in Jet A-1' [26]. When the general approach to producing them pertains to the production of alkanes/alkenes, they will be brought into the discussion.

7.2.1 Fatty Acid Biosynthesis and Degradation

The overall reaction principles of fatty acid biosynthesis (FAB) are evolutionarily conserved from bacteria to eukaryotes, but the enzymatic machinery differs from a structural point of view. In short, the synthesis is based on a cyclic system, where aliphatic thioester-intermediates are extended by two carbons at a time through four steps: (1) condensation, (2) reduction, (3) dehydration, and (4) reduction. The mechanisms are illustrated in Fig. 7.2.

The first and committed step of FAB is initiated when the compound acetyl-coenzyme A (acetyl-CoA, Ac-CoA), a two-carbon metabolic intermediate at the junction of catabolic and anabolic pathways, is converted to malonyl-CoA (Ma-CoA) by addition of dissolved CO_2 in a biotin- and ATP-dependent reaction catalysed by acetyl-CoA carboxylase (Acc1 in yeast and AccABCD in *E. coli*, respectively) [27]. The malonyl-moiety is transferred to an acyl-carrier protein (ACP), and then serves as the extender unit in FAB, in which the fatty acid chain is elongated by two carbons at a time.

In eukaryotic organisms such as yeast, the cytosolic fatty acid synthase (FAS) enzyme consists of two distinct polypeptides – encoded by *FAS1* and *FAS2* – containing all enzymatic domains required for FAB, a system called type I FAS [27]. In the bacterial machinery, the enzymes involved in FAB are instead composed of individual proteins, called the type II FAS system [28]. The main yeast FAB-system is located in the cytosol, but yeast also contain a mitochondrial system of type II, which is essential for respiratory growth due to the mitochondrial requirement for lipoic acid [29].

In yeast, cytosolic FAB is initiated through the transfer of Ac-CoA to ACP located within the Fas2-polypeptide, while ACP-bound intermediates in *E. coli* are formed in a dissociated state (as ACP is expressed as an individual polypeptide). The first reaction in the FAB cycle is a condensation reaction between the acetyl and malonyl unit catalysed by a ketoacyl synthase (KS subunit or FabH), during which one carbon is lost in the form of CO_2. In yeast, the acetyl unit is bound to ACP, while in *E. coli* the acetyl-unit remains in CoA form. The chain elongation beyond the first cycle is catalysed in *E. coli* by other enzymes than FabH (ie, FabB/FabF) due to the required change in substrate specificity (CoA to ACP), while the KS module performs all condensation reactions in yeast. After condensation, the produced ketoacyl-ACP is reduced to β-hydroxyacyl-ACP in an NADPH-consuming step by a ketoacyl reductase (KR subunit or FabG). In the next stage, the β-hydroxyacyl-ACP is dehydrated to β-enoylacyl-ACP by enoyl dehydratase (DH subunit or

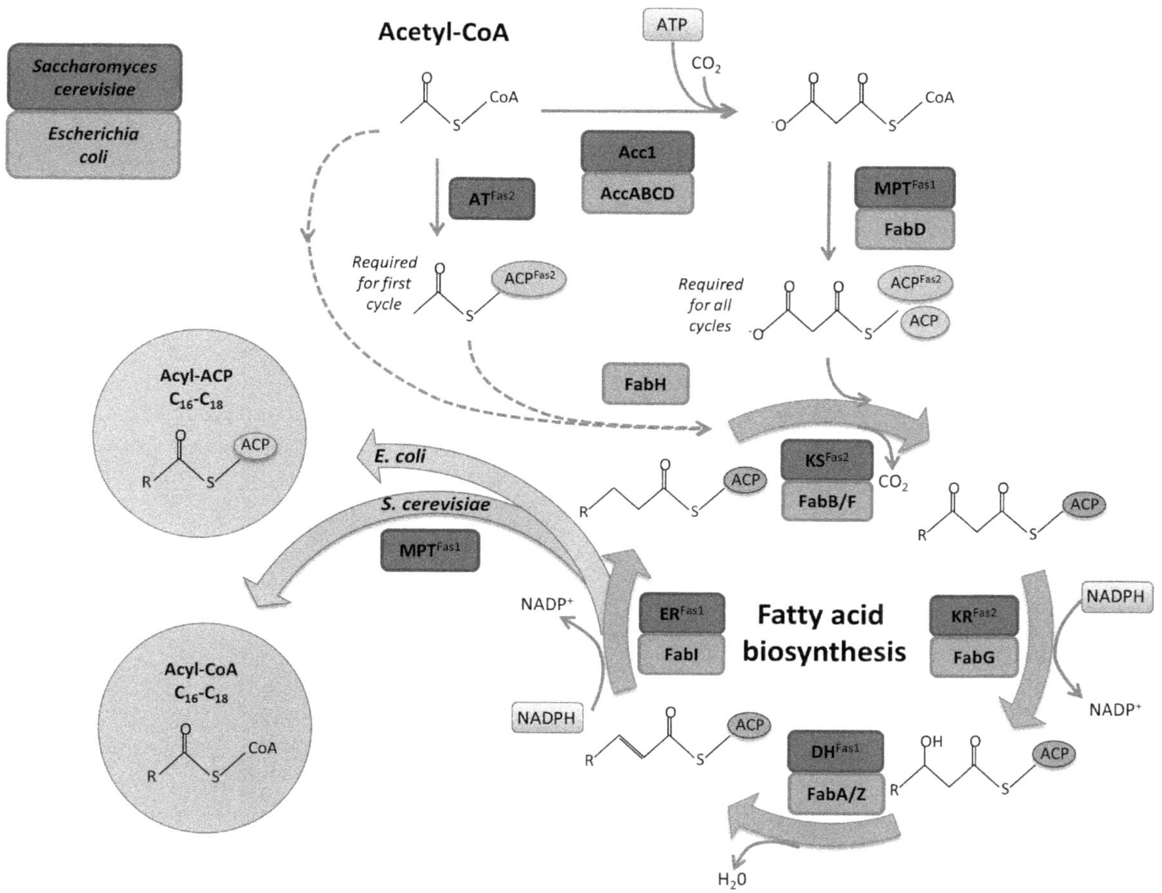

FIGURE 7.2 The reaction principles and enzymatic components of the fatty acid biosynthesis system in *Saccharomyces cerevisiae* and *Escherichia coli*, respectively. Fas1/2, fatty acid synthase subunits; Acc1 and AccABCD, acetyl-CoA carboxylase; ACP, acyl-carrier protein; AT, acetyl transferase; MPT and FabD, malonyl-palmitoyl transacylase; KS and FabB/F/H, ketoacyl synthase; KR and FabG, ketoacyl reductase; DH and FabA/Z, enoyl dehydratase; ER and FabI, enoyl reductase.

FabZ/FabA), and further reduced to an acyl-ACP molecule of C_{n+2} while consuming an additional molecule of NADPH through the action of an enoyl reductase (ER subunit or FabI). This newly formed acyl-ACP then undergoes subsequent elongation cycles to reach a final length of normally 16 or 18 carbons, which constitute the major fatty acid types in both *E. coli* and yeast [27,28]. A distinction is that the end product of FAB in yeast is generated when the Fas2-bound acyl-chain is enzymatically released as acyl-CoA, while the bacterial end product is acyl-ACP. This difference has implications in downstream applications, since biofuel-converting enzymes have different specificities towards these intermediates.

The formed acyl-chains have multiple destinations within cells. A fundamental function is the formation of membrane lipids. This requires that the fatty acids be converted into amphiphilic phospholipids that can form bilayers. A large proportion of the yeast fatty acids formed, around 75–80%, are enzymatically desaturated by individual enzymes, a feature that helps to increase membrane fluidity [30]. The desaturation process in *E. coli* instead takes place during the fatty acid chain synthesis by the activity of the enzymatic players FabA and FabB as opposed to FabF and FabZ, which will produce saturated fatty acid species [31].

Compared to bacteria, the complexity of lipid metabolism in yeast increases due to the distinct feature of eukaryotic cells: compartmentalization, which is the presence of multiple types of membrane-enclosed organelles with different function and physiology. For example, in the endoplasmic reticulum, fatty acid elongation up to C_{26} is executed by the proteins Elo1, 2, and 3 [27], in the lipid droplet triacylglycerol (TAGs) and sterol esters (SEs) are formed and stored [32], and in the peroxisomes β-oxidation (the breakdown of fatty acid chains) takes place. β-Oxidation is a critical system for cells in order to grow on fatty acid substrates, and essentially consists of the reversed reactions compared to FAB: cyclic steps of (1) oxidation, (2) hydration, (3) oxidation, and (4) thiolytic cleavage. The mechanisms in *E. coli* and *S. cerevisiae* are depicted in Fig. 7.3.

FAB demands a lot of energy, and production of one C_{18}-fatty acid will consume nine Ac-CoA, eight ATP, and 16 NADPH. Thus, in order to not waste the cell's energy resources, the process is heavily regulated, where transcriptional, translational [27,28], as well as posttranslational mechanisms are involved [33]. Furthermore, fatty acid synthesis is subject to product inhibition by activated fatty acid species on multiple levels [34–36]. This regulation poses great challenges to the overproduction of fatty acid-derived compounds.

The previously described FAB-system for yeast is valid for both *S. cerevisiae* and *Y. lipolytica*. However, one important distinction between the two is the presence of an ATP citrate lyase (ACL) in *Y. lipolytica*, as well as in other oleaginous yeasts. This enzyme produces cytosolic acetyl-CoA from citrate, which can be exported out of the mitochondrion, thus increasing the precursor supply for FAB [32]. During lipid accumulation – for example, in nitrogen-limited conditions when nucleotides and proteins cannot be produced – the mitochondrial isocitrate dehydrogenase is down-regulated, increasing the flux through ACL and towards TAG-synthesis [37]. ACL is therefore hypothesized to be a critical component adding to the oleaginous character of this and other yeasts.

7.2.2 Engineering Examples to Increase FAB and Lipid Production

There have been extensive efforts to optimize microbial FAB for oleochemical production purposes. This includes increasing precursor supply as well as FAB, altering regulatory mechanisms within the concerned system, and exploring synthetic systems. Representative examples of this are described below (see also Fig. 7.4).

7.2.2.1 Engineering Acetyl-CoA Metabolism

The metabolic intermediate acetyl-CoA is the universal precursor for biosynthesis of fatty acids, and also for isoprenoids, in yeast. Accordingly, multiple efforts to up-regulate the precursor pool of acetyl-CoA in cell factories with the aim to increase production of such compounds have been made [38].

Due to compartmentalization, there exist distinct acetyl-CoA pools in yeast, of which the cytosolic one is responsible for FAB for membrane and storage lipid biosynthesis. An example of cytosolic acetyl-CoA engineering

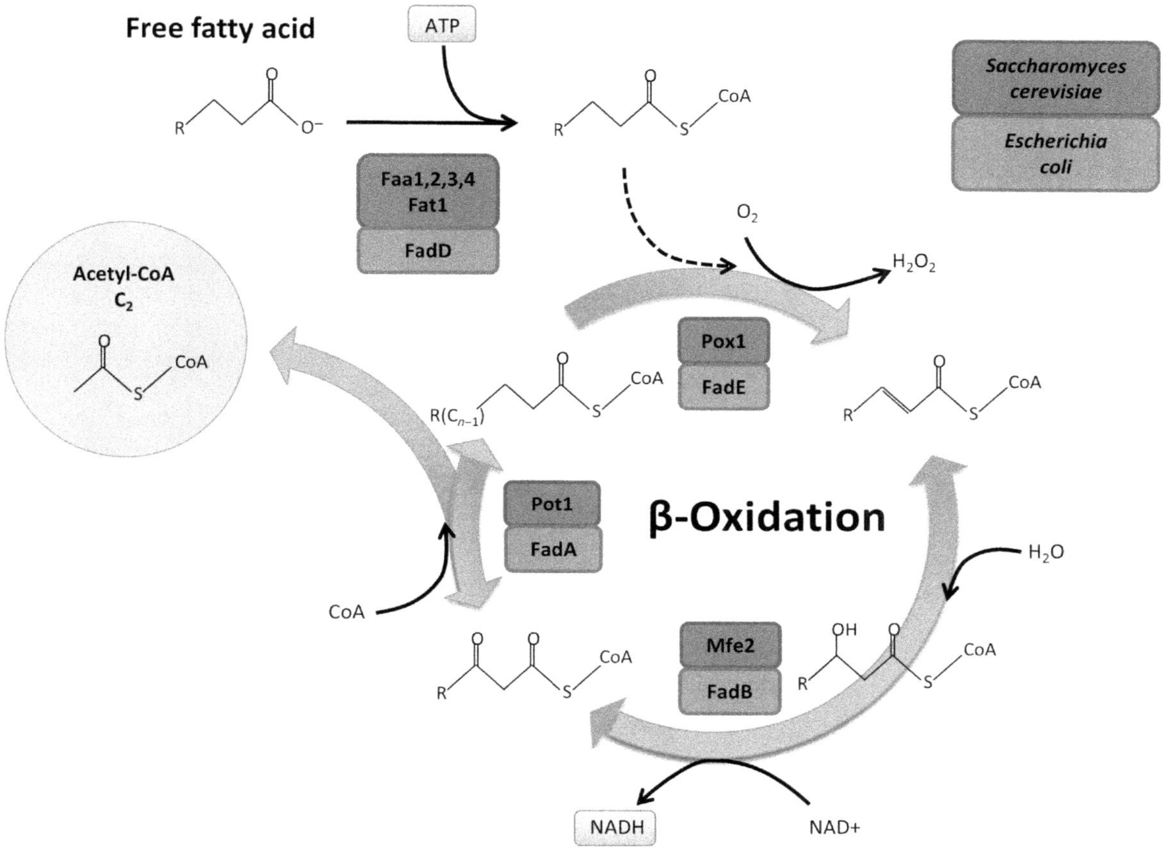

FIGURE 7.3 The reaction principles and enzymatic components of β-oxidation in *Saccharomyces cerevisiae* and *Escherichia coli*, respectively. Faa1/2/3/4 and FadD, fatty-acyl CoA synthase(s); Fat1, very long chain fatty acyl-CoA synthase; Pox1 and FadE, acyl-CoA oxidase; Mfe2 and FadB, enoyl-CoA hydratase/hydroxyacyl-CoA dehydrogenase; Pot1 and FadA, ketoacyl-CoA thiolase.

in yeast was set by Shiba et al., who targeted the pyruvate dehydrogenase (PDH) bypass. By overexpressing the endogenous acetaldehyde dehydrogenase Adh6 and a deregulated, heterologous version of acetyl-CoA synthase (acs) from *Salmonella enterica*, acs^{L641P}, flux towards the cytosolic acetyl-CoA node, and the isoprenoid product amorphadiene was increased fourfold at the expense of ethanol formation [24].

Many heterologous pathway options to produce cytosolic acetyl-CoA in yeast have been considered, in part since the native route is energy demanding. In addition, the native route bypasses the metabolite acetaldehyde, which can be efficiently converted to ethanol, a by-product that severely lowers the theoretical product-on-substrate yield for fatty acid and isoprenoid-based products. For example, pathways based on phosphoketolase [39], ACL [40–42], pyruvate formate lyase [43], acetylating acetaldehyde dehydrogenase (A-ALD) [43], and the PDH complex [44] are all acetyl-CoA-forming pathways used to channel carbon towards the cytosolic acetyl-CoA node in yeast to increase acetyl-CoA-derived product formation. The development of an efficient alternative route will help decrease the amount of carbon lost to ethanol as well as reduce energy costs. In addition, it could complement the deletion of the three isoenzymes encoding pyruvate decarboxylase (Pdc1, 5, 6), creating a strain that is completely deficient in ethanol production, and thus has potential to act as a yeast platform strain with low by-product formation [45].

In *E. coli*, several enzymes intersect the acetyl-CoA metabolism for FAB since no compartmentalization is present. To keep a high acetyl-CoA level, many strategies have aimed to delete metabolic pathways leading to fermentation products, such as acetate [46,47], up-regulating the acetyl-CoA-forming enzymes, such as PDH [48], and fast-channeling the formed acetyl-CoA to downstream pathways, for example, by overexpressing the first enzyme in FAB, acetyl-CoA carboxylase [49].

More details on the subject of engineering acetyl-CoA metabolism in microorganisms are available in a comprehensive review recently published on the topic [38].

7.2.2.2 Fatty Acid and Lipid Overproduction

Fatty acids and lipids are not only valuable as biofuel precursors, but also as food additives, consumer products, etc. This is why attempts and strategies to overproduce these compounds in microorganisms are numerous. Fatty acid-accumulating strains can be used as platform strains for specific applications (eg, for alkane synthesis). Due to the sheer quantity of studies performed to increase yields, and numerous strategic overlaps in these studies, the principle targets of these strategies are described, and a few key studies presented in more detail.

7.2.2.2.1 OVEREXPRESSION OF ENZYMES DIRECTLY RELATED TO FAB

In order to overproduce fatty acids, the strategies described in the previous section to increase acetyl-CoA levels are applicable. The subsequent step is to focus on FAB. A straightforward and beneficial principle in yeast has been to co-overexpress the endogenous acetyl-CoA carboxylase and the FAS-enzymes [50]. The sole overexpression of AccABCD in *E. coli* had no impact on fatty acid yields, possibly due to product inhibition of acyl-ACP, but a small positive effect when a "pull" from the acyl-ACP pool was implemented [51]. Acc1 in *S. cerevisiae* is inhibited by posttranslational phosphorylation, and removal of this regulation increases its activity in metabolic engineering strategies [33,52]. The utilization of heterologous FAS-systems in *S. cerevisiae* has also been reported, for example, functional utilization of a modified human FAS-system [53], or utilization of a bacterial FAS type I system which is assumed to bypass native regulation [54]. Heterologous expression of a FASI-system in *E. coli* has also been reported in an attempt to overproduce fatty alcohols [55]. The terminal enzyme for fatty alcohol production acts upon the intermediate acyl-CoA, which is directly produced by the FASI FAB-system, but requires two additional enzymatic reactions from acyl-ACP in prokaryotic systems. The system was functional in vivo, but failed to increase yields compared to the endogenous system.

7.2.2.2.2 RELEASE OF FREE FATTY ACIDS AS A MEANS TO INCREASE FLUX THROUGH FAB

Due to the previously mentioned repression of the FAB-machinery by their products, acyl-ACP in *E. coli* and acyl-CoA in yeast, a well-explored method to increase yields of fatty acids in *E. coli* is to convert these end products into free fatty acids (FFA), a reaction catalysed by thioesterases. Early attempts to utilize this principle for engineering of fatty acid metabolism showed that overexpression of a native thioesterase, TesA, modified to localize to the cytosol [56], and a thioesterase from *Umbellularia californica* [51], both improved FFA-yields significantly in *E. coli*. Subsequent studies of a broad range of thioesterases in *E. coli* showed a great variety of substrate specificity (C_4 to C_{16}) and activity (up to 2 g/L produced FFA by overexpression of a single thioesterase), suggesting thioesterases as the major determinants of fatty acid chain length, thus providing an indispensable tool for engineering of fatty acid metabolism [57–60]. The thioesterase principle has been shown to work in *S. cerevisiae* as well. Expression of the truncated TesA variant from *E. coli* improved FFA levels eightfold [50], and expression of a recombinant human FAS-system where the thioesterase domain was replaced with one specific towards short- and medium-chain acyl-CoAs enabled the production of 111 mg/L short-chain fatty acids [53]. In yeast, the system offers less flexibility due to the fungal FAB system, where the growing acyl chain is attached to the ACP units of the FAS complex, which is why a soluble thioesterase only has access to the released C_{16}–C_{18} acyl-CoA species.

The opposite reaction, the conversion of FFAs into acyl-CoAs, has also been targeted with metabolic engineering. In *S. cerevisiae*, the genes *FAA1/2/3/4* all encode fatty acid-activating enzymes and are essential for growth if fatty acids are used as the sole carbon source, due to the necessity of acyl-CoA species as substrates for β-oxidation and incorporation into membrane or storage lipids. The protein products of *FAA1* and *FAA4* have been shown to be responsible for the vast majority (≥98%) of the long-chain fatty acid activation, such as C_{16}–C_{18} [61]. Thus, deletion of these two particular genes in yeast enables high levels of FFA production and secretion [50,62]. The combination of *FAA1/FAA4* deletion with expression of a heterologous thioesterase with broad substrate-range led to the production of approximately 490 mg/L extracellular FFA, where a significant proportion (~40%) was unsaturated, compared to the WT, where no unsaturated FFAs were detected [62].

7.2.2.2.3 PREVENTING DEGRADATION OF FATTY ACIDS – BLOCKING β-OXIDATION

In *E. coli*, activation of fatty acids to acyl-CoA constitutes the initiation of β-oxidation. A functional β-oxidation pathway will diminish yields when fatty acids are the aim of production, which is why many studies in *E. coli* involved deletion of the endogenous gene *fadD*, which encodes an acyl-CoA ligase [49,58]. It should however be noted that this is not a preferable choice if a terminal enzyme utilizing acyl-CoA is exploited. Therefore, targeting the second enzyme involved in the oxidation process, *fadE*, is more favourable [56]. Deletion of β-oxidation in yeast, which exclusively takes place in the peroxisomal compartment, has also been targeted, mainly through deletion of *POX1*, the gene encoding the enzyme catalysing the first step in the oxidation process, the conversion of peroxisomal acyl-CoA into trans-2-enoyl-CoA [50,63,64]. A recent study also showed that the effect can be enhanced by combining the *POX1* deletion with deletion of genes encoding Pxa1, which transports acyl-CoA into the peroxisome, and Faa2, an acyl-CoA ligase located in the peroxisome, possibly due to the presence of auxiliary enzymes in the β-oxidation pathway [64].

7.2.2.2.4 MANIPULATING STORAGE LIPID FORMATION IN YEAST FOR INCREASED FREE FATTY ACIDS

Another option to generate FFAs in yeast by reducing by-product formation is to modulate storage lipid and lipid body formation. There are four main genes related to the generation of the two storage lipid species in yeast: *DGA1* and *LRO1* responsible for TAG formation, and *ARE1* and *ARE2* in charge of SE synthesis. Storage lipid formation is not essential in yeast, and deletion of the four genes mentioned above will deplete lipid body formation and reduce fitness in yeast [65]. A study that coupled the deletion of storage lipid formation with improved production of the fatty acid-derived fuel FAEE in *S. cerevisiae* reported a 1.7-fold increase in yield, and if β-oxidation gene *POX1* was additionally knocked out, a 2.9-fold increase in FAEE was seen compared to the control [63].

A less direct approach is to instead increase the flux through the lipid body by overexpressing storage lipid formation genes in concert with lipases that are responsible for degrading TAGs and SEs into FFAs. By overexpressing the TAG synthase Dga1 and the lipase Tgl3 in combination with eliminating the fatty acid-activating enzymes Faa1, Faa4, Fat1, and deleting β-oxidation genes *POX1*, *FAA2*, and *PXA1* in *S. cerevisiae*, an extracellular FFA titre of 2.2 g/L could be achieved – the highest reported so far [64]. A benefit of this strategy in comparison to the deletion of storage lipid formation genes is that it consumes the acyl-CoA molecules that otherwise would inhibit FAB, enabling a faster fatty acid biogenesis.

7.2.2.2.5 INFLUENCING FATTY ACID METABOLISM THROUGH REGULATORY COMPONENTS

Due to the tight regulation of FAB, another strategy has been to focus on global regulators coupled to fatty acid metabolism. For example, a pioneering study made in *E. coli* shows that overexpression of the transcriptional regulator FadR, activating unsaturated FAB and deactivating β-oxidation, could increase the total fatty acid titres to 5.2 ± 0.5 g/L, which is 7.5 times higher than the control strain with normal FadR levels, and corresponds to a yield of 0.26 g FA/g glucose or 74% of the theoretical maximum yield [66]. In yeast, a study was conducted to investigate if the deletion of negative regulators of phospholipid synthesis could increase production of fatty alcohols, which are directly obtained from acyl-CoA [42]. The logic behind this strategy is that it would simultaneously increase the flux through the ACC/FAS system because fatty acids are required as precursors for phospholipid synthesis. Out of six deletions, the one of *RPD3* helped to increase the fatty alcohol titre from 71 to 140 mg/L. A study performed in *Y. lipolytica* also showed that the deletion of the global regulator Snf1 increased the lipid content in the yeast by 2.6 times [67]. However, such a correlation could not be seen between a knockout of Snf1 and fatty alcohol production in *S. cerevisiae*, which was unexpected, since Snf1 is known to negatively regulate Acc1 in this host [42].

7.2.2.2.6 INCREASING STORAGE LIPID FORMATION IN YEAST

Most strategies described above aim to increase the yield of FFAs or activated fatty acids instead of focusing on increasing the storage lipid content. This is due to the fact that fatty acids in the form of TAG/SE cannot directly be converted enzymatically into fuel molecules within the cell. However, the option to produce microbial lipids, which can be converted catalytically to fuels, is an alternative option. Examples of boosting storage lipid formation in *S. cerevisiae* exist. Optimization of TAG formation was investigated through the expression of a native Dga1 with an N-terminal deletion, DgaΔNp, in yeast backgrounds where the native *DGA1* locus and, in particular, its 3′ terminal region had been deleted. These led to a lipid content close to 45% of cell dry weight, corresponding to an oleaginous phenotype. The phenotype is thought to be connected to the disruption and lowered expression of *ESA1*, encoding a histone acetyltransferase, which is chromosomally adjacent to *DGA1* in *S. cerevisiae* [68].

The collective metabolic engineering efforts in the oleaginous yeast *Y. lipolytica* have, on the other hand, primarily been aimed at producing maximal amounts of TAG stored in lipid droplets. In a study from 2014, *Y. lipolytica* engineered to overproduce lipids and cells accumulated lipids to about 90%, reaching titres around 25.3 g/L, which represented a greater than 60-fold improvement over the parental strain [69]. The result was based on deletion of genes connected to β-oxidation and peroxisome biogenesis, overexpression of a gene involved in TAG synthesis, as well as optimization of cultivation conditions. In 2015, a significant achievement was accomplished when Qiao et al. identified a human gene present in cellular obese phenotypes encoding a delta-9-steroyl-CoA desaturase (SCD) generating mono-unsaturated fatty acids. Expression of SCD in *Y. lipolytica* similarly managed to reduce product inhibition (from saturated acyl-CoA) in fatty acid synthesis as well as stimulate TAG-synthesis. Co-expressing SCD with native *ACC1* and *DGA1* genes led to a strain with a product-on-substrate yield of 84.7%, lipid titres close to 55 g/L, and a maximal productivity of about 1 g/L per h [70].

7.2.2.2.7 FAB THROUGH REVERSED β-OXIDATION

Production of bulk chemicals, such as biofuels, requires very high yields in order to be economically viable. The high energy input required in natural FAB means on a practical level that the theoretical yield is fairly low. The energy demand is high due to the use of malonyl-CoA/ACP as an extender unit in FAB, since its biosynthesis requires one ATP. In 2011, a pioneering study was performed where the β-oxidation process in *E. coli* was reversed. As opposed to FAB, β-oxidation is a reversible process, which can function in an opposite direction in the absence of fatty acids. In this mode, the enzymatic steps are identical to the natural FAB except for the initiation, which employs acetyl-CoA as an extending unit. This means that one ATP can be spared per elongation cycle. In the absence of fatty acids, *E. coli* will strongly repress the expression of β-oxidation-related operons in order to prevent the reversed reaction. In the study by Dellomonaco et al., *E. coli* was modified in several regulatory aspects, such as exploiting mutated versions of FadR and AtoC (negative regulators of β-oxidation), allowing for constitutive expression of the β-oxidation genes even in the absence of fatty acids and in the presence of glucose. Furthermore, several fermentation pathways were blocked. In combination with expression of enzymes with terminal activity, high titres and yields could be achieved. For example, when the *n*-butanol-forming enzyme FucO was expressed in the deregulated strain cultivated in high-glucose media, around 14 g/L of *n*-butanol were produced, corresponding to 0.33 g butanol/g glucose consumed and a productivity of 2 g/gCDW/h. In combination with over-expression of a 3-ketoacyl-CoA thiolase with broad chain-length specificity and a thioesterase with long-chain specificity, and a deletion of *fadD*, efficient production of FFAs was enabled with titres and yield of 7 g/L and 0.28 g/g [71]. In a following study, instead of utilizing the regulatory approach, several possible enzyme candidates for the reversed reactions were characterized in vitro. The best candidates were overexpressed with thioesterases of different chain-length specificity, and reasonably high titres for shorter-chain (C_4) products were obtained, while longer-chain products (up to C_{12}) were detected at lower titres [72].

Recently, fatty acid synthesis through reversal of β-oxidation was also implemented in

S. cerevisiae in a synthetic biology approach [73]. In this system, the native Pot1, which catalyses the last step in β-oxidation, or the acetoacetyl synthase Erg10, were used as ketoacyl-CoA synthases (KS). Two mitochondrial β-oxidation enzymes from *Y. lipolytica* normally catalysing hydration and the subsequent dehydrogenation of 2-enoyl-CoA were utilized for their reversed reactions (DH and KR activity, respectively). The last selected gene was the yeast native *ETR1*, whose protein product catalyses the last reductive step in mitochondrial FAB. Instead of targeting regulatory components of the system, the genes were simply overexpressed from a plasmid. By co-expressing genes involved in *n*-butanol, FAEE, and fatty alcohol formation, and detecting the corresponding products, the system proved to be functional up to medium-chain products (4-cycle reversal). However, obtained yields were on the milligram scale, which is far from what was obtained in *E. coli*. The low yields were in part attributed to the limitation of acetyl-CoA concentration in the cytosol. Nevertheless, reversal of β-oxidation remains a promising strategy to produce fatty acid-based products at higher theoretical yield.

7.2.3 Natural Straight-Chain Hydrocarbon-Forming Pathways and Metabolic Engineering Strategies to Produce Alka(e)nes

For many decades it has been known that both microorganisms and higher organisms naturally have capacities to produce alkanes and alkenes (alka(e)nes) [74], but it was not until recently that strong efforts were put into investigating the genetic components of these pathways, most likely as a response to the urgent need of developing renewable biofuels. Here, three mechanistically different routes for natural alkane/alkene formation from fatty acid intermediates are presented, alongside metabolic engineering strategies to improve their production. The pathways include a reductive two-step pathway from activated fatty acid species or FFA, a head-to-head condensation reaction between two fatty acid chains, and a one-step reduction of free fatty acids to terminal alkenes. Of the discussed pathways, the first one is by far the most studied and extensively used in metabolic engineering strategies for microbial alka(e)ne production. To visualize and put the pathways into context, a brief summary of the heterologous pathways and the cellular substrates they would act on in *S. cerevisiae* can be found in Fig. 7.4 (indicated with green (dark grey in print versions) arrows/text), and a detailed overview of the three reaction types is summarized in Fig. 7.5.

7.2.3.1 Two-Step Reduction: Alkane Formation Through the Combined Action of Fatty Acid Reductases and Fatty Aldehyde Deformylating Oxygenases

Schirmer et al. first elucidated the cyanobacterial pathway for straight-chain alkane ($C_{15}-C_{17}$) synthesis in 2010 [75]. By performing subtractive genome analysis between alkane- and nonalkane-producing species, the authors found the pathway to be composed of two enzymes – an acyl-ACP reductase (AAR) and an aldehyde decarbonylase (AD; later suggested to be an aldehyde deformylating oxygenase, ADO, which will be the name used from here onwards), and their activity was verified by heterologous expression of the corresponding genes and product quantification in *E. coli*. The AAR catalyses the reaction from an activated fatty acid to the corresponding aldehyde, while the ADO decarbonylates the aldehyde into an alka(e)ne of chain length C_{n-1}. This is why the alkanes found in organisms that almost exclusively produce even-chain fatty acids consist of uneven carbon chain length. Moreover, in vitro studies of the enzymes also revealed that ADO requires reducing equivalents (NADPH) and an

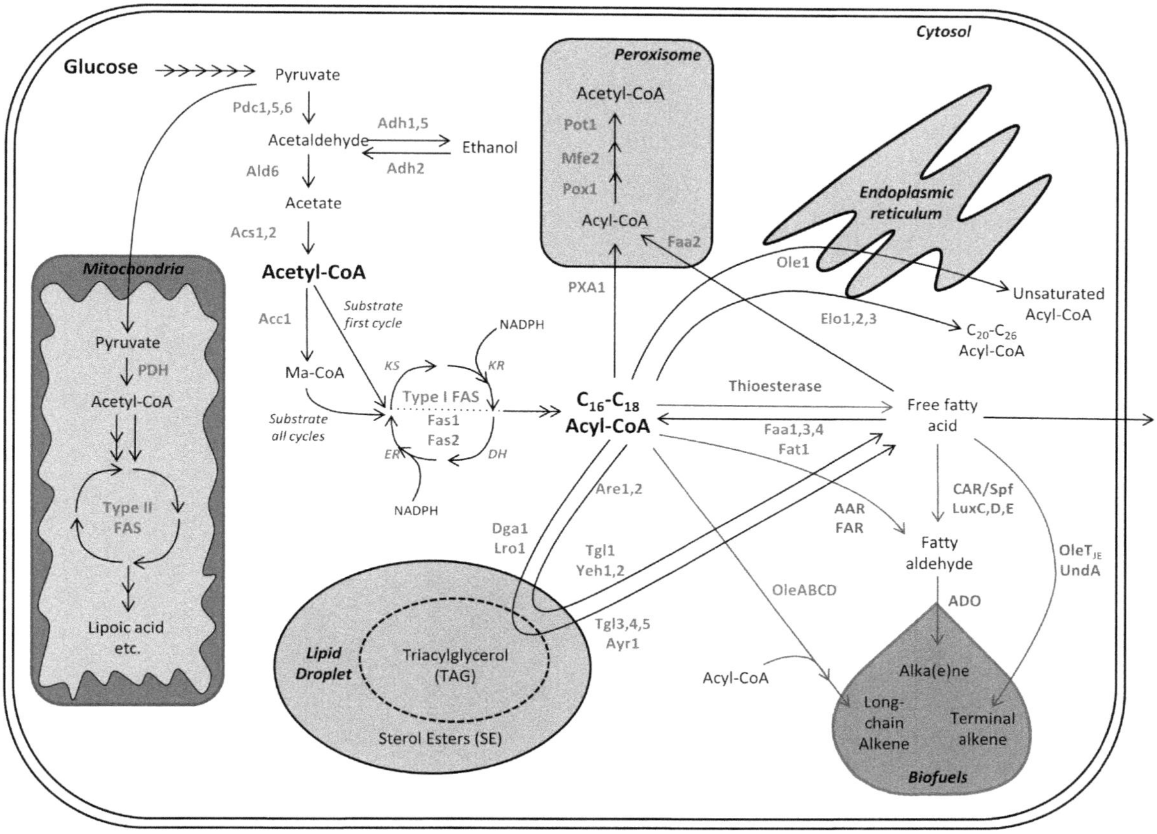

FIGURE 7.4 Overview of fatty acid metabolism in *Saccharomyces cerevisiae* and examples of heterologous enzymes capable of converting fatty acid species into alka(e)ne biofuels. Native genes are shown in orange (light grey in print versions) notation with a blue (black in print versions) arrow while heterologous genes and pathways used in metabolic strategies for alka(e)ne production are marked in green (dark grey in print versions). See main text for more detailed information.

electron donor system, which in *E. coli* can be supplied by ferredoxin (F) and ferredoxin reductase (FNR). When *Synechococcus elongatus* AAR and *Nostoc punctiforme* ADO were expressed in *E. coli*, the total alkane titre was around 300 mg/L where pentadecane was found in majority [75]. The study has generated numerous follow-up reports characterizing the gene products and their reaction components with the main focus on ADO [76–82].

The reduction of activated fatty acids by AAR requires divalent metal ions (eg, Mg^{2+}) and NADPH as co-factor and is activated by K^{+}. The k_{cat} of *S. elongatus* upon stearoyl-CoA was measured to be 0.36 min^{-1}, indicating a rather inefficient enzymatic conversion [82]. However, initial studies suggest that AAR has a specificity for acyl-ACP rather than acyl-CoA, supported by in vitro assays as well as an in vivo experiment [75]; thus, the true activity of AAR might be higher. This is also supported by a subsequent study on *N. punctiforme* AAR [83].

The ADO reaction has been proven to require both oxygen and NADPH while

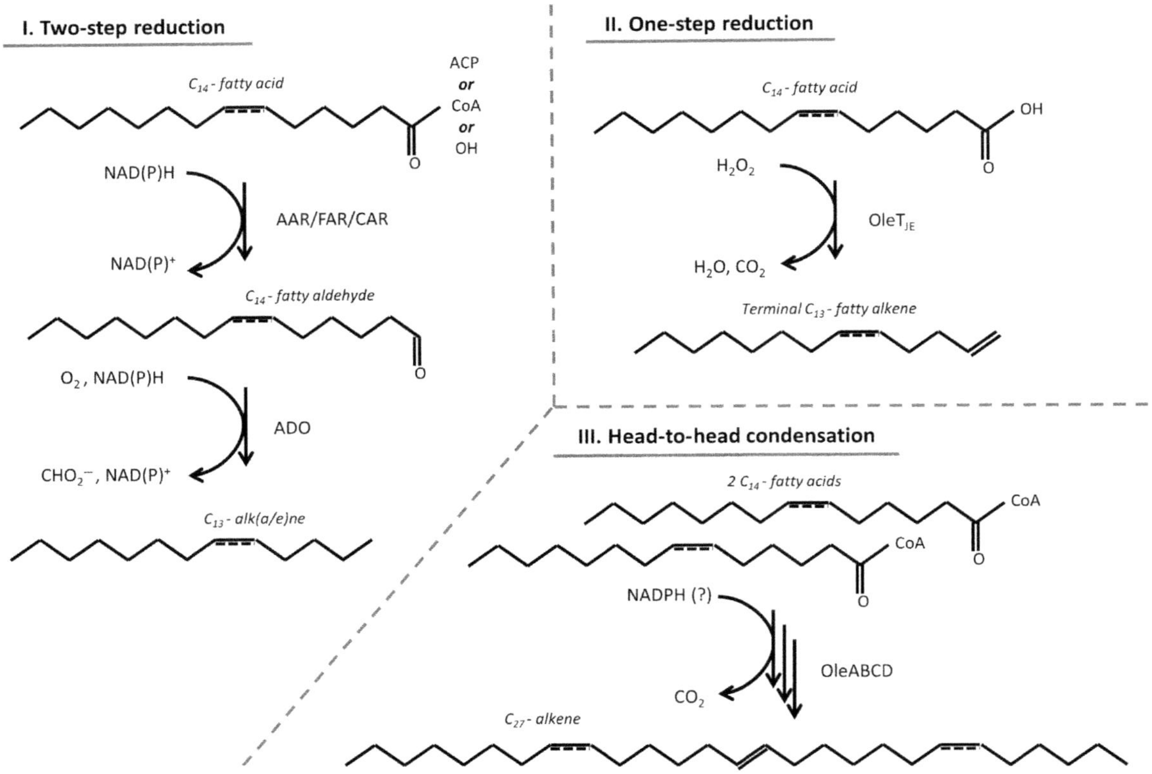

FIGURE 7.5 Overview of alka(e)ne biosynthesis pathways based on fatty acid intermediates. Dashed lines within the molecule structure indicate that the fatty acid intermediate can be unsaturated due to normal cellular metabolism, thus giving rise to an alkene product. AAR, acyl-ACP reductase; FAR, fatty acyl-CoA reductase; CAR, carboxylic acid reductase; ADO, aldehyde deformylating oxygenase; $OleT_{JE}$, P450 fatty acid decarboxylase; OleABCD, head-to-head condensation enzymes from *Micrococcus luteus*.

releasing formate, which is why the enzyme now is referred to as aldehyde-deformylating oxygenase [78,79]. The kinetic properties of ADO are poor, and the enzyme has been seen to only achieve 3–5 catalytic turnovers in in vitro experiments [76,78,79]. ADO is suggested to be the rate-limiting step of the paired enzyme reaction [83]. In vitro studies also revealed that ADO is reversibly inhibited by hydrogen peroxide (H_2O_2), a compound that can form when NADPH and oxygen are in excess. Since both NADPH and oxygen are vital for ADO function, and are likely to form H_2O_2 (also in vivo), ADO activity has been increased fivefold in vivo by fusing ADO to catalase, an enzyme that converts H_2O_2 to the ADO-substrate O_2 [81].

A recent study characterizing both AAR and ADO from the cyanobacterial species *N. punctiforme* showed that the two enzymes form a tight complex, where transfer of the cytosolically insoluble fatty-aldehyde intermediate to ADO is facilitated, which increases the catalytic capacity of ADO as compared to when it is supplied with exogenous fatty aldehydes at high concentration [83]. Whether this interaction is conserved in AAR/ADO enzyme pairs from other species remains to be answered, but the consequence is that utilization of ADOs with alternative, noncognate fatty acid reductases (FARs) might affect the efficiency of the terminal catalytic step. A potential way to reduce such an effect is to bring the pathway enzymes in proximity to each other, either by protein fusion or by using protein scaffolds. Rahman et al. investigated whether alkane production could be increased in this manner, and saw a 4.8-fold increase for an AAR/ADO fusion construct, and an 8.8-fold increase when a scaffold that had space for one AAR and three ADO enzymes (both from *S. elongatus*) compared to the control strain where the enzymes were expressed separately [84]. This indicates a general importance of proximity of the two pathway enzymes for optimal productivity.

The plant *Arabidopsis thaliana* produces very-long chain alkanes as constituents of a protective wax. The gene *CER1* has for many years been presumed to be involved in long-chain alkane biosynthesis [13], and its protein product was recently shown to catalyse the reaction of VLC (very long chain, $\sim C_{30}$) acyl-CoA to VLC alkanes together with CER3 and CYTB5 in the heterologous host *S. cerevisiae* (modified to produce VLC fatty acids) [85]. The proposed enzymatic mechanism of action is that CER1 and CER3 form a heterodimer where CER3 produces an aldehyde intermediate that can be converted to an alkane by CER1, while CYTB5 acts as an electron donor to CER1, which contains a di-iron catalytic core also found in cyanobacterial ADO [85]. Thus, the pathway is presumed to have the same catalytic reaction setup as the cyanobacterial AAR/ADO-system. In an *E. coli* strain engineered to overproduce short- and medium-chain fatty acids (C_8–C_{14}), CER1 from *A. thaliana* was used in combination with a fatty acyl-CoA reductase (Acr) from *Clostridium acetobutylicum* to produce short-chain alkanes in a fed-batch fermentation at a titre of 0.58 g/L – the highest reported titre of microbially produced alkanes until now [86]. Accordingly, substrate specificity of CER1 appears to be relatively broad in vivo, as it is responsible for production of VLC alkane species in its native host. Furthermore, the fact that sole expression of CER1 generated alkanes in *E. coli* when a heterologous acyl-CoA reductase was expressed supports the hypothesis given by Bernhard et al. that CER3 might catalyse the acyl-CoA reductase reaction [85], since individual CER1 expression in yeast failed to produce any alkanes.

The carboxylic acid reductase (CAR) from *Mycobacterium marinum* has been observed to have a broad range activity towards FFAs with chain lengths ranging from C_6 to C_{18}, and converts them into fatty aldehydes [87]. This is in contrast to the pathway previously presented from cyanobacteria, where the AAR relies on an activated form of fatty acids, primarily fatty acyl-ACPs [75]. As previously mentioned, fatty acid synthesis is subjected to product inhibition by activated fatty acid species on multiple levels [34–36]. Thus, utilization of this pathway, in combination with an enzyme that can release FFAs from activated ones, can pose a useful way to circumvent the problem.

A study showed that a CAR-related enzyme from *Nocardia* sp. requires posttranslational activation by a phosphopantetheine transferase (PPTase) for functionality [88]. Akthar et al. expressed *M. marinum* CAR and an activating PPTase from *Bacillus subtilis*, Spf, in *E. coli* to obtain fatty aldehydes. Furthermore, they overexpressed the endogenous thioesterase TesA to obtain high intracellular levels of FFA from acyl-ACP, as well as a fatty aldehyde reductase (AHR) to obtain fatty alcohols from aldehyde intermediates. In a strain overexpressing all the above-mentioned enzymes, they managed to produce around 360 mg/L of fatty alcohols [87]. The CAR/Spf system was also utilized in a study attempting to produce the gaseous alkane propane in *E. coli* [59].

A heterologous butyryl-ACP-specific thioesterase was introduced that was capable of intervening in the native fatty acid synthesis, thus releasing byturic acid (C_4), which subsequently could be converted to C_3 alkane by combining it with the activities of CAR/Spf and a cyanobacterial ADO/FNR-system. By deleting native AHRs responsible for fatty alcohol formation and optimization of O_2 availability, a final titre of 32 mg/L propane was obtained. The low titres were believed to be connected to low affinity of ADO towards short-chain aldehydes.

Proof-of principle production of alkanes in *E. coli* with the ADO-system has been abundant, but was first demonstrated in *S. cerevisiae* in 2015 [89]. The authors expressed the FAR and ADO encoded by *S. elongatus* orf1594 and orf1593, respectively, in yeast, without being able to detect any alkanes. By co-expressing the alkane pathway with the redox-system F/FNR from *E. coli*, still, only trace amounts of heptadecane were produced. By further deleting a native aldehyde dehydrogenase speculated to convert the produced aldehydes to FFAs, thus competing with an already kinetically challenged ADO, the yield was increased to 22 μg/gCDW. The simultaneous production of fatty alcohols at a level of 520 μg/gCDW indicates that ADO is a limiting factor in the system, and that more enzymes (AHRs) need to be potentially eliminated to reduce by-product formation.

7.2.3.2 One-Step Reduction: Terminal Alkene Synthesis

Another biosynthetic route for hydrocarbons is widespread in species of the *Jeotgalicoccus* genus [90]. These organisms produce terminal alkenes through the action of a P450 enzyme that uses FFA as substrates, thus constituting a direct pathway for hydrocarbon synthesis. In 2011, Rude et al. first identified the gene and protein responsible for this synthesis in *Jeotgalicoccus* sp. ATCC 8456, which was named $OleT_{JE}$ and is proposed to perform a decarboxylation of the fatty acid while using H_2O_2 as sole co-factor [90]. The enzyme has been found to accept and convert a range of different fatty acid substrates ($C_{12}-C_{20}$) [91].

Hydrogen peroxide causes oxidative damage in cells and is therefore not an optimal co-factor for efficient biosynthesis of alkenes. To investigate whether $OleT_{JE}$ could function through a monooxygenase mechanism, a research group fused $OleT_{JE}$ to a reductase domain from *Rhodococcus* sp., RhFRED. This fusion construct enabled H_2O_2-independent production of alkenes by *E. coli* in the presence of NADPH and O_2, with a maximum titre of 97.6 mg/L [92].

In a recent report, several homologs to $OleT_{JE}$ were expressed in *S. cerevisiae* to screen for an optimal enzyme candidate. The most active candidate was a codon-optimized version of $OleT_{JE}$, and maximum titres after extensive engineering efforts, including precursor enhancement and co-factor engineering as well as optimization of gene expression and cultivation conditions, reached 3.7 mg/L – below what has been obtained in *E. coli* [93].

In 2014, another one-step decarboxylation pathway for terminal alkene synthesis was discovered. Within the genus *Pseudomonas*, several species produce 1-undecene (C_{11}). Expressing a genomic library from *Pseudomonas fluorescens* in *E. coli* and selecting clones positive for 1-undecene production enabled determination of the gene *undA* to be responsible for alkene formation in the native host. Biochemical characterization of the enzyme in vitro revealed that UndA is a nonheme protein that catalyses an iron- and oxygen-dependent conversion of medium chain ($C_{10}-C_{14}$) FFAs to terminal alkenes, constituting a previously unidentified enzyme family [94]. Expression of several UndA homologs in *E. coli* resulted in titres ranging between 3–6 mg/L.

7.2.3.3 *Head-to-Head Condensation*

In 2010, several orthologous bacterial pathways for long-chain alkene synthesis were also genetically elucidated and characterized. This type of alkene formation relies on a condensation reaction of two molecules of acyl-CoA to bring about alkenes of chain length $C_{23}-C_{33}$ [95–98]. The reactions were found to be catalysed by an enzyme group referred to as OleABCD, where OleA performs the first step through a Claisen condensation analogous to the one executed in FAB by ketoacyl-ACP synthase III, to which OleA shares sequence homology [96]. Subsequent in-depth studies of these biosynthetic enzymes suggested that OleA generates a β-ketoacid that can be converted in the presence of OleC and OleD into the corresponding alkene, while the function of OleB remained elusive [99]. OleD is speculated to be a NADPH-dependent reductase [96]. For aviation fuel purposes, however, long-chain alkenes are not optimal as increased chain length increases the melting point.

7.2.3.4 *Attempts to Diversify Alkane Production*

Since fatty acid species in organisms are even-chained due to the FAB mechanism, the alkane products obtained are odd-chained while fatty alcohols remain even-chained. There have been some efforts to expand this product profile to respond to the wide range of alkane species present in current petroleum-based fuels.

The use of an alternative FAR system from *Photorhabdus luminescens* consisting of three genes (*luxC*, *luxD*, and *luxE*) also enabled reduction of FFAs in *E. coli*. A thioesterase from *Cinnamomum camphora* was used to create a free fatty acid of altered chain-length, enabling both even- and odd-chained alkane production. Yet, in combination with *N. punctiforme* ADO, only modest titres of 2–5 mg/L alka(e)nes were reached. To investigate whether the utilized FAR system was limiting alkane production, fatty aldehyde and fatty alcohol titres were also quantified. The result showed that fatty alcohol titres were high compared to fatty aldehydes, thus the low titres most likely depend on competing enzymes consuming the formed aldehyde intermediate before ADO converts them into alkanes [100].

Another strategy included expression of an alternative enzyme to initiate FAB in *E. coli* [101]. The native enzyme, FabH, has high specificity for acetyl-CoA, which results in a profile of even-chain fatty acids. To alter the profile, an alternative enzyme, FabH2 from *B. subtilis* was utilized, which has a higher activity toward longer-chain acyl-CoAs. When the AAR/ADO pathway from *S. elongatus* was expressed in combination with FabH2, alkane titres were doubled, however, only a low percentage of the alkanes present were even-chained. When propanoate was fed to the cells, the even-chained alkenes increased from 6% to 27% of the total amount, indicating that the precursor propionyl-CoA was limiting.

Most recently, Cao et al. attempted to widen the product scope by utilizing an α-dioxygenase from rice (α-DOX) whose activity forms "C-1" aldehyde intermediates from fatty acids, in contrast to the previously described fatty acid-based reductases. By combining α-DOX with a thioesterase and either an AHR or ADO, odd-chained alcohols, as well as even-chained alkanes, could be formed. By optimizing enzyme expression and cultivation conditions, the highest fatty alcohol titre thus far could be reached: 1.95 g/L [102].

7.2.4 Limitations and Future Directions

There is a huge gap between current microbial alkane yields and titres, currently in the mg/L range, and what is required for commercial production. As described previously, endogenous fatty acid metabolism and cultivation conditions have successfully been targeted to optimize production yields of lipids in prospective cell factories, but the main bottleneck of the alkane pathway is the poor enzymatic activity of AAR/ADO. Natural alkane requirements in hosts such as plants, cyanobacteria, and insects are low, which is why it is plausible that evolutionary pressure to increase activity of these enzymes has remained low. Therefore, to make microbial alkane production more efficient, much work has to be put into engineering FAR, AAR, CAR, and ADO for higher efficacy. To make rational enzyme modifications, considerable knowledge of the structure–function relationship of the enzyme is required. Moreover, each individual mutant/enzyme variant has to be evaluated through product quantification. The overall process of a rational design cycle is therefore labor-intensive.

In order to create a high-throughput screening approach, one alternative is to construct libraries of randomly mutagenized enzymes and combine these with a metabolite sensor [103]. Such sensors can rely on transcription factor-based systems coupled to a reporter gene, for example, yielding a fluorescent output if the metabolite is detected. Thus, enzymes with improved activity could easily be selected by fluorescence-activated cell sorting (FACS). Sensors should optimally have a large dynamic range rather than an on-off signal – meaning that they should produce an increase in output with increasing stimulus – to be able to single out high producers from intermediate ones. Sensors that can detect intermediates in alkane synthesis have been developed; for example, malonyl-CoA sensors functional in several different organisms [104–106], and also a sensor system responsive to fatty acyl-CoA [107]. To increase activity of AAR/ADO-enzymes, metabolic sensors specifically sensing alkanes must be utilized. Both short-chain alkane sensors as well as medium-long-chain alkane sensors have been generated [108,109], and in 2015, a sensor based on transcriptional regulators from the hydrocarbon-utilizing organisms *Acinetobacter baylyi* and *Pseudomonas oleovorans* was also used to detect medium- and long-chain alkanes produced by the AAR/ADO-system [110]. These, and derivatives thereof, will likely become instrumental in the development of more efficient alkane-producing enzymes.

Metabolic sensors also show great promise in the optimization of metabolic pathway expression systems. The most common approach to reach high production yields in cell factories has for a long time been to simply overexpress the essential enzymes in the biosynthetic pathway. However, this drains considerable amounts of energy towards protein synthesis and production of metabolic intermediates that might not be optimally utilized due to low substrate levels and inefficient downstream enzymes, respectively. Thus, the creation of dynamic systems able to respond to environmental stimuli, for example, to overexpress a pathway enzyme only if there is abundant substrate present for it to catalytically convert, is highly sought after. Present examples are a dynamic sensor-regulator system developed to regulate expression of enzymes related to FAEE production based on the presence of the substrate acyl-CoA [107], and the use of a malonyl-CoA sensor to optimize fatty acid production [111]. A sensor activated by fatty acid species could in a similar fashion be applied for the alkane biosynthetic pathway.

If alkane titres can be improved, another issue will inevitably rise – product toxicity. Alkanes are toxic to many microorganisms, and due to their hydrophobic nature, they are likely to cause problems with the membrane integrity of the cells producing them. To find a solution, the toxicity problem can be tackled by different means. By analysing transcriptional data of cells exposed to the stressor [112,113] or performing adaptive laboratory evolution in increasing concentration of stressor [114,115], changes in transcriptome and genome, respectively, can be detected that potentially could be beneficial for the cell when exposed to the stressor. A more direct approach is to take advantage of the fact that cells tend to expel unwanted metabolites from the cell by so-called efflux pumps, and try to identify pumps specific to the product of interest [116]. For example, Chen et al. expressed ABC-transporters from *Y. lipolytica* in *S. cerevisiae*, resulting in increased tolerance (about 80-fold) to exogenously added alkanes mediated by an ability to keep intracellular alkane levels low [117]. A transcriptional data analysis study of *S. cerevisiae* exposed to medium-chain alkanes enabled identification of two endogenous efflux pumps, Snq2 and Pdr5, that were shown to reduce intracellular levels of alkanes and increase tolerance [113]. Transporter engineering in cells can also increase product formation, as a concentration gradient is formed as the products exit the cells, and potential substrate inhibition can be reduced. An exported product is also beneficial for product recovery at the commercial scale.

7.3 ISOPRENOIDS AS AVIATION FUELS

Besides fatty acid-derived alka(e)nes, isoprenoids constitute the second class of compounds derived from cellular metabolism with promising prospects as biofuels. With more than 40,000 identified structures, isoprenoids represent the largest group of chemical products. They occur universally in all classes of organisms, where they fulfil essential functions as, for example, membrane components, vitamins, hormones, pigments, membrane anchors, and electron transport molecules [118]. The major group of isoprenoids, however, function as nonessential secondary metabolites, especially in plants, where they serve, for example, as insect attractants or defense molecules against herbivores or pathogens [119]. Due to their versatile function, many of these secondary metabolites have gained industrial interest as fragrances, flavours, pesticides, or pharmaceuticals [15,119,120].

All isoprenoids are derived from the C_5 isoprene unit, and more precisely, from its two isomeric activated forms, isopentenyl diphosphate (IPP) and dimethylallyl diphosphate (DMAPP). Depending on the number of incorporated isoprene units, they are classified as hemiterpenes (C_5), monoterpenes (C_{10}), sesquiterpenes (C_{15}), diterpenes (C_{20}), etc. Mono- and sesquiterpene hydrocarbons have been proposed as alternative aviation fuels or fuel additives because they have properties (carbon chain length, density, freezing point, and heat of combustion) similar to conventional, petroleum-derived jet fuel [15,121]. Examples include the monoterpenes pinene, sabinene, limonene, phellandrene, and terpinene, as well as the sesquiterpenes farnesene, bisabolene, and valencene (Fig. 7.6).

A major disadvantage of these molecules is, however, that they are unsaturated and therefore contain highly reactive C=C bonds. Reduction of these bonds can be achieved by hydrogenation in the presence of a catalyst, as has been demonstrated for limonene and myrcene, for example, [122]. Another method of chemical modification is the dimerization of cyclic monoterpenes, which can increase the density and heat of combustion to values similar to high-density tactical fuels [123]. Production of naturally occurring mono- [124] and sesquiterpenes by engineered microorganisms can therefore be combined with further chemical conversions to produce suitable aviation fuels with desired properties.

This chapter covers the two major biosynthetic pathways for isoprenoid biosynthesis,

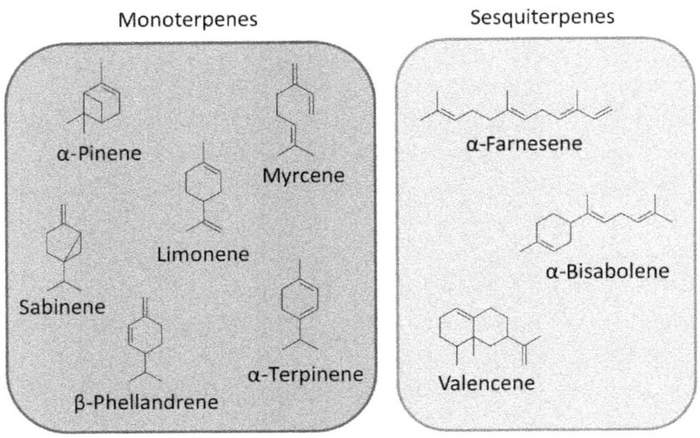

FIGURE 7.6 Examples of monoterpene and sesquiterpene hydrocarbons.

describes general engineering strategies to increase isoprenoid formation in different host organisms (mainly *S. cerevisiae* and *E. coli*), and then gives specific examples on the production of mono- and sesquiterpene hydrocarbons suitable as aviation fuel precursors.

7.3.1 Biosynthesis

The isoprenoid precursors IPP and DMAPP are produced via two different biosynthetic pathways: (1) the mevalonate (MVA) pathway in eukaryotes, in archaea [125], and in some eubacteria [124]; and (2) the nonmevalonate route, also called the methylerythritol (MEP) or deoxyxylulose phosphate (DXP) pathway, found in most eubacteria [126], plastids of plants [127], and in apicomplexa [128].

The MVA pathway uses acetyl-CoA as precursor (Fig. 7.7) [129]. Three molecules of acetyl-CoA are combined in two consecutive condensation reactions to form one molecule of 3-hydroxy-3-methylglutaryl-CoA (HMG-CoA). The first condensation is catalysed by acetyl-CoA acetyltransferase (ACAT), while the second one is catalysed by HMG-CoA synthase (HMGS). In a two-step reduction, which is considered the rate-limiting reaction of the pathway, HMG-CoA is converted to mevalonate by HMG-CoA reductase (HMGR) consuming two NADPH for each mevalonate formed. *S. cerevisiae* contains two HMGR isoenzymes, Hmg1 and Hmg2 [130], which are differentially regulated [131,132] and integrated into the membrane of the endoplasmic reticulum via an N-terminal noncatalytic domain involved in regulating protein degradation [133]. Mevalonate is phosphorylated to mevalonate 5-diphosphate by mevalonate kinase (MK) and phosphomevalonate kinase (PMK). Finally, decarboxylation of mevalonate 5-diphosphate, catalysed by mevalonate diphosphate decarboxylase (MDD), leads to formation of IPP. In some archaea, conversion of mevalonate to IPP follows an alternative route via isopentenyl phosphate [125,134]. IPP can be converted into its isomer, DMAPP, and vice versa by IPP isomerase (IDI).

In contrast to the MVA pathway, the MEP pathway proceeds from glyceraldehyde 3-phosphate and pyruvate as direct precursors (Fig. 7.7) [135,136]. Both molecules are condensed to DXP by DXP synthase (DXS) under release of CO_2. DXP reductoisomerase (DXR) then converts DXP into MEP, a branched aldose. Coupling of MEP to cytidyl monophosphate (CMP) catalysed by MEP

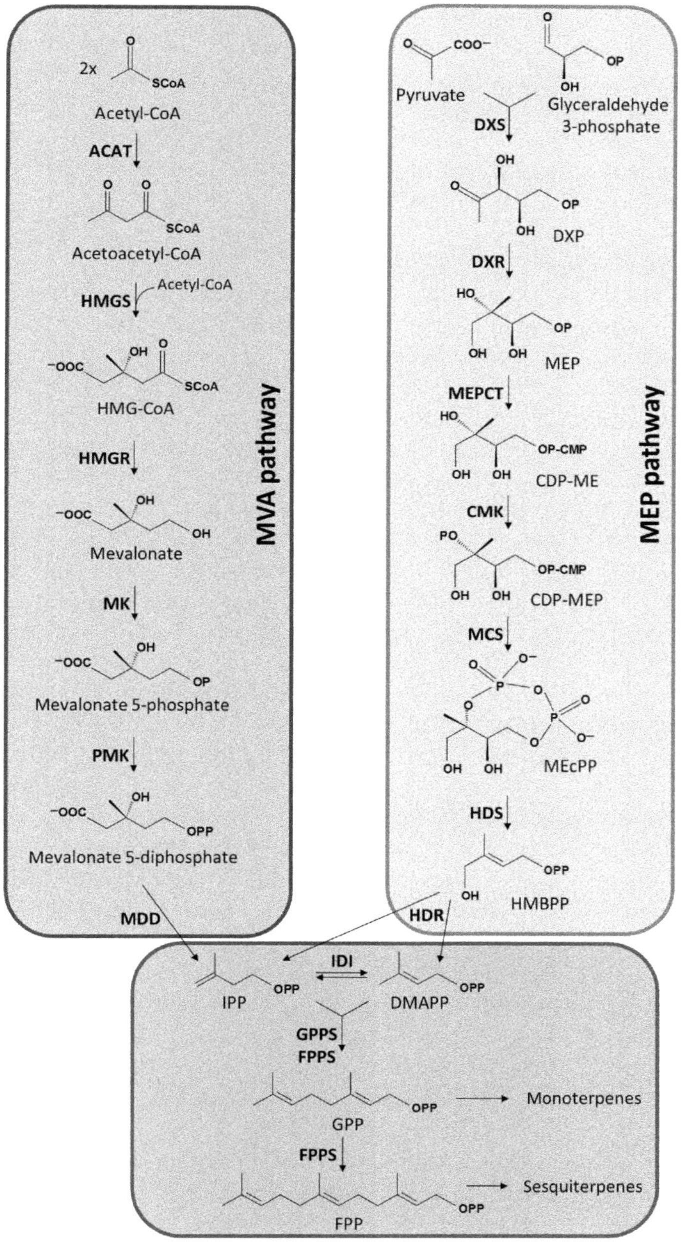

FIGURE 7.7 Biosynthesis of mono- and sesquiterpenes via the MVA and MEP pathway. Metabolites: HMG-CoA, 3-hydroxy-3-methylglutaryl-CoA; DXP, 1-deoxy-D-xylulose 5-phosphate; MEP, 2-C-methyl-D-erythritol 4-phosphate; CDP-ME, 4-diphosphocytidyl-2-C-methyl-D-erythritol; CDP-MEP, 4-diphosphocytidyl-2-C-methyl-D-erythritol 2-phosphate; MEcPP, 2-C-methyl-D-erythritol 2,4-cyclodiphosphate; HMBPP, (*E*)-4-hydroxy-3-methyl-but-2-enyl diphosphate; IPP, isopentenyl diphosphate; DMAPP, dimethylallyl diphosphate; GPP, geranyl diphosphate; FPP, farnesyl diphosphate. Enzymes: ACAT, acetyl-CoA C-acetyltransferase; HMGS, 3-hydroxy-3-methylglutaryl-CoA synthase; HMGR, 3-hydroxy-3-methylglutaryl-CoA reductase; MK, mevalonate kinase; PMK, phosphomevalonate kinase; MDD, mevalonate diphosphate decarboxylase; DXS, DXP synthase; DXR, DXP reductoisomerase; MEPCT, MEP cytidylyltransferase; CMK, CDP-ME kinase; MCS, MEcPP synthase; HDS, HMBPP synthase; HDR, HMBPP reductase; IDI, isopentenyl diphosphate isomerase; GPPS, GPP synthase; FPPS, FPP synthase.

cytidyltransferase (MEPCT) and subsequent phosphorylation of the resulting intermediate 4-diphosphocytidyl-2C-methyl-D-erythritol (CDP-ME) catalysed by CDP-ME kinase (CMK) leads to formation of CDP-ME 2-phosphate (CDP-MEP). In an intramolecular transphosphorylation, 2C-methyl-D-erythritol 2,4-cyclodiphosphate (MEcPP) is formed under release of CMP. The last two enzymes of the pathway – 1-hydroxy-2-methyl-2-(*E*)-butenyl-4-diphosphate (HMBPP) synthase (HMS) and HMBPP reductase (HMR) – both contain an iron-sulphur cluster and are highly sensitive to oxygen [137,138]. HMR produces a mixture of both IPP and DMAPP at a ratio of about 5:1 in *E. coli* [139]. In addition, most organisms that use the MEP pathway for IPP and DMAPP synthesis also possess an IDI to balance the ratio between the two metabolites [140].

When comparing the overall stoichiometry of both pathways for IPP production from glucose, the MVA pathway consumes 1.5 mol of glucose and 3 mol of ATP per mol IPP, while 1 mol of glucose and 2 mol of ATP are required for production of 1 mol of IPP via the MEP pathway [141,142]. This renders the MEP pathway superior in terms of maximum isoprenoid yields that can theoretically be achieved.

After synthesis of IPP and DMAPP, prenyltransferases catalyse the successive head-to-tail condensation of IPP to an allylic diphosphate (DMAPP in the first cycle) to produce the direct precursors of monoterpenes (geranyl diphosphate, GPP), sesquiterpenes (farnesyl diphosphate, FPP), and longer-chain isoprenoids (Fig. 7.7) [143].

7.3.2 Metabolic Engineering to Increase Isoprenoid Production in Microbial Hosts

Numerous studies have investigated possibilities to produce and enhance the production of various mono- and sesquiterpenes in *S. cerevisiae*, *E. coli*, and other microbial hosts. Although not all of these strategies were specifically applied to potential aviation fuel precursors, they can still easily be adapted for this purpose and will also be discussed here.

7.3.2.1 *Saccharomyces cerevisiae*

As *S. cerevisiae* contains the MVA pathway starting from acetyl-CoA for isoprenoid production, approaches similar to the ones described above for increasing the production of fatty acid-derived products have also been employed to increase isoprenoid synthesis in yeast. Engineering of the PDH bypass by expression of a deregulated acetyl-CoA synthetase from *S. enterica* and overexpression of aldehyde dehydrogenase Ald6 – either alone or in combination with overexpression of alcohol dehydrogenase Adh2 and ACAT (Erg10) – increased sesquiterpene production by up to 75% [24,144]. Additional inhibition of acetyl-CoA consumption in the glyoxylate cycle was able to further double sesquiterpene production [144]. Further strategies to increase the cytosolic acetyl-CoA supply in yeast are mentioned in section 7.2.2.1.

Engineering of the MVA pathway in yeast has mainly focused on modulation of key enzymatic reactions, one of them being HMGR (see above). The most common strategy is the overexpression of a truncated version of Hmg – that is, a version only consisting of the C-terminal catalytic domain and devoid of the N-terminal membrane anchor [145–147]. However, overexpression of full-length Hmg1 [148], a stabilized variant of Hmg2, Hmg2 (K6R) [149,150], or expression of a heterologous HMGR from *Staphylococcus aureus* [151], were also proven to be successful in increasing isoprenoid production. Moreover, additional genetic perturbations stabilizing HMGR protein levels and thus leading to improved sesquiterpene production in yeast were identified [152].

Also, balancing the ratio between IPP and DMAPP appears to be a crucial factor for isoprenoid biosynthesis. Overexpression of IDI alone in yeast was able to boost monoterpene production by up to fivefold [150,153].

When producing either mono- or sesquiterpenes, it is important to consider that a single enzyme, FPP synthase Erg20, is responsible in yeast for the two enzymatic steps leading to production of both GPP, the precursor of monoterpenes, and FPP, the sesquiterpene precursor. The wild-type enzyme, however, releases only small amounts of the intermediate GPP [154,155]. Overexpression of Erg20 has therefore mainly been used to increase the production of sesquiterpenes [145,150,156]. To enhance the formation of monoterpenes, on the other hand, mutant variants of the enzyme, such as Erg20(K197E) or Erg20(F96W-N127W), that favour GPP over FPP production, were employed [153,155,157–159]. Alternatively, specific heterologous GPP synthases derived from plants, for instance, can be used [160]. Instead of concentrating on selected enzymatic reactions of the MVA pathway, other studies took a more global approach, such as expression of a mutated version of a transcription factor (Upc2-1) that led to enhanced transcription of several sterol pathway genes – but alone only had a minor effect on sesquiterpene production [145,156,161] – or by simply overexpressing all genes of the MVA pathway, which led to a significant increase in sesquiterpene formation [162].

Apart from increasing the flux through the MVA pathway towards FPP or GPP, it is also crucial to limit the consumption of FPP by other metabolic pathways. Most of these pathways, such as the formation of ergosterol or dolichols, are, however, essential, and can therefore not be entirely disrupted. Since the majority of FPP is converted to squalene by squalene synthase (Erg9), and then further to ergosterol, transcription of the *ERG9* gene has been modulated in order to reduce the flux towards ergosterol and increase the availability of FPP for sesquiterpene production. In several cases, the *ERG9* promoter was exchanged with the *MET3* promoter that is repressed by methionine [163,164]. This, however, requires the constant addition of methionine to the medium because methionine is metabolized by the cells. Other studies therefore used the copper-repressible *CTR3* promoter [165], or a more dynamic approach where *ERG9* was controlled either by the glucose-regulated *HXT1* promoter, which repressed ergosterol formation when glucose was consumed after the exponential growth phase [166], or by promoters repressed by ergosterol itself as feedback inhibition [167].

Another sink for FPP is its dephosphorylation to farnesol. It is believed that two lipid phosphatases, Lpp1 and Dpp1, are involved in this process [168]. In one study, deletion of *DPP1* in an *lpp1*Δ background strain was able to both reduce farnesol levels and increase sesquiterpene production [166], whereas other studies saw no, or even a negative, effect of deletion of *DPP1* either alone or in combination with *LPP1* on sesquiterpene titres [155,169,170].

While FPP represents the major branch point in the isoprenoid biosynthesis, DMAPP is also consumed by other metabolic pathways, namely the prenylation of tRNA. Liu and co-workers therefore overexpressed *MAF1* encoding a negative regulator of RNA polymerase III, which doubled monoterpene production [153].

Deletion or down-regulation of competing pathways represents one way of reducing the drain of precursors or intermediates. Another possibility is spatial control, for example, by exploiting compartmentalization of a eukaryotic cell. Targeting of sesquiterpene synthases to the mitochondria drastically increased product formation, which was further enhanced by targeting a heterologous FPP synthase to the same compartment [170]. However, it remains speculative whether shielding from competing reactions or other possible explanations such as a more favourable environment for enzyme activity are the actual cause for the observed effect.

A more direct method of metabolite channeling by spatial control is the fusion of consecutive pathway enzymes. Fusion of Erg20 to a sesquiterpene synthase as well as fusion of a GPP producing Erg20 mutant to a monoterpene synthase both successfully increased product formation [155,171]. It should be noted, however, that orientation of the enzymes within the fusion construct and the choice of a potential linker peptide may be of importance.

With regard to the fact that the MEP pathway would be the more efficient isoprenoid pathway in terms of carbon preservation and energy consumption, it was also attempted to heterologously express this pathway in yeast, albeit without success [142,172]. It was assumed that this was due to the nonfunctionality of the two iron-sulphur cluster-containing pathway enzymes in the yeast cytosol.

7.3.2.2 *Escherichia coli*

Similar to increasing the flux through the MVA pathway in yeast, engineering of the MEP pathway in *E. coli* mainly focuses on overexpression of several key enzymes, usually in combination. One crucial factor seems to be the early pathway steps because overexpression of DXS and/or DXR often resulted in increased isoprenoid levels (eg, [173–175]). However, in some cases this strategy even had a negative effect on terpenoid production, which suggests that careful fine-tuning of the expression levels and the genetic context are important [176–178]. Similarly, the effect of overexpressing other pathway genes, such as *ispG* and *ispH* encoding HMS and HMR, respectively, on isoprenoid synthesis is dependent on the background strain [179–181].

As in *S. cerevisiae*, overexpression of IDI to balance the ratio between the two precursor isomers is a useful tool to increase isoprenoid accumulation [174,176,182,183]. Also like in yeast, overexpression of FPP synthase (IspA) was used to increase the production of sesquiterpenes and longer terpenes [176,182,184], while either mutated versions of IspA or heterologous GPP synthases were employed to establish production of monoterpenes [185–188].

While heterologous expression of the MEP pathway in yeast seems to be challenging (see above), a functional MVA pathway has been successfully established in *E. coli*, and often, its introduction had a greater effect than engineering the endogenous pathway, despite its theoretically lower yields [178,189–191]. This may be due to the fact that enzymes of the MEP pathway are subject to feedback regulation [192] and expression of a heterologous pathway can therefore benefit from circumventing regulatory mechanisms. However, accumulation of certain MVA pathway intermediates such as HMG-CoA and prenyl diphosphates (IPP, DMAPP, FPP) seems to be toxic to *E. coli* and efficient downstream enzymes and careful pathway balancing are necessary to avoid growth retardation of the engineered strains [189]. Pathway balancing was performed by adding additional gene copies of postulated rate-limiting enzymes [178,193,194], recruitment of enzymes from other sources for improved activity, reduced feedback inhibition or co-factor preference [195–197], or by modulating the strength of ribosome binding sites determining translation efficiency of the different pathway genes [198]. A fourth approach to increase the flux through the MVA pathway in *E. coli* was the assembly of pathway enzymes to a protein scaffold using protein–protein interaction domains, a concept similar to enzyme fusion described above [199]. However, in this case the scaffold allows for modulating the amount of each enzyme per complex, thus compensating for different enzyme activities.

7.3.2.3 Other Organisms

Besides *S. cerevisiae* and *E. coli*, other microorganisms have also been engineered for increased production of isoprenoid precursors, for example, overexpression of endogenous MEP pathway genes or establishment of the heterologous MVA pathway in cyanobacteria [200–202]. Another photosynthetic bacterium that has been engineered through introduction of a heterologous MVA pathway, in this case encoded by an operon derived from *Paracoccus zeaxanthinifaciens*, is *Rhodobacter sphaeroides*, which increased sesquiterpene formation sixfold [203]. Nonphotosynthetic bacteria, in which either endogenous or heterologous isoprenoid pathways have been improved, comprise, among others, *Lactococcus lactis* [204], *Corynebacterium glutamicum* [205,206], *Streptomyces venezuelae* [207], *Pseudomonas putida* [208], and *B. subtilis* [209]. With regard to eukaryotic microorganisms, the endogenous MVA pathway was modulated in several non-*Saccharomyces* species, including *Y. lipolytica* and *Pichia pastoris*, which in most cases focused mainly on overexpression of either full-length or truncated HMGR [210–213].

7.3.3 Production of Monoterpene Hydrocarbons

7.3.3.1 Pinene

Pinene, which occurs as an α- and β-isomer, is a bicyclic monoterpene produced by pine trees and other plants. Pinene synthesis from GPP in a heterologous organism is established by expression of a single enzyme, pinene synthase. Yang et al. [214] expressed a *Pinus taeda* α-pinene synthase in *E. coli* combined with either overexpression of the endogenous FPP/GPP synthase or expression of a GPP synthase derived from *Abies grandis*, whereupon the latter drastically improved pinene production. α-Pinene titres were further enhanced by expression of a hybrid MVA pathway based on genes from *Enterococcus faecalis* and *S. cerevisiae*, respectively, and reached up to 5.4 mg/L in shake flask cultivations and 970 mg/L in a fed-batch fermentation.

To establish a more sustainable process for the production of biofuels, it is important to develop strains that are able to grow on more complex carbon sources such as cellulosic biomass. Bokinsky et al. [215] therefore engineered two *E. coli* strains with the ability to grow on cellulose and xylan, respectively. As described above, a heterologous MVA pathway – in this case entirely from *S. cerevisiae* – GPP synthase from *A. grandis*, and pinene synthase from *P. taeda* were employed to confer pinene synthesis in both strains. Since the two plant enzymes are normally targeted to the chloroplast, their targeting signals were removed and the genes codon optimized for *E. coli*. A co-culture of both strains was able to produce 1.7 mg/L of pinene from switchgrass pre-treated with ionic liquids. While Yang et al. [214] measured pinene in the gaseous phase of the fermentation vessel, Bokinsky et al. [215] used an organic solvent, dodecane, to capture pinene in the liquid phase.

Efficient secretion of the produced biofuel precursor is important for two reasons. On the one hand, it facilitates product recovery, and on the other hand it may alleviate the toxic effect that many terpenes, especially monoterpenes, have on the producing microbes. To address this issue, a gene encoding an inner membrane transporter of *E. coli* was randomly mutagenized and strains expressing this gene library selected for improved tolerance towards (ie, presumably improved secretion of) *n*-octane or 1-octanol [216]. Remarkably, the mutated transporter not only improved the export of *n*-octane, but also of α-pinene, up to fourfold.

Because apart from pinene the precursor GPP may also be toxic to *E. coli*, Sarria et al. [186] tested combinations of three different GPP synthases with three different pinene synthases as well as fused versions of both enzymes in an MVA pathway-expressing strain yielding up to 32 mg/L of pinene. They also showed that the enzyme combination influenced the ratio between α- and β-pinene produced. The best strain derived from this study expressing a fusion protein of GPP synthase and α-/β-pinene synthase (both from *A. grandis*) was also used to produce pinene from macroalgal hydrolysate at up to 21 mg/L [217].

As an alternative to *E. coli* as a host, pinene production was also established in *Corynebacterium glutamicum* using two different GPP synthases and two pinene synthases in connection with overexpression of DXS and IDI of the endogenous MEP pathway, resulting in up to 176 μg/L of pinene [205].

7.3.3.2 Limonene

The monocyclic hydrocarbon limonene was first produced in *E. coli* as an intermediate of the pathway leading to carvone synthesis [218]. Expression of the homodimeric GPP synthase of *A. grandis* together with limonene synthase from *Mentha spicata* omitting their targeting peptides led to a final limonene titre of 5 mg/L. Expression of IDI from *Mentha x piperita* only slightly increased limonene formation. Expression of both genes in a strain containing the yeast MVA pathway together with optimization of certain steps such as replacement of HMGS and HMGR with the respective enzymes from *Staphylococcus aureus* and optimization of gene expression levels drastically enhanced limonene production to 435 mg/L [219]. Optimization of a two-phase fed-batch fermentation process using a similar MVA pathway-containing strain allowed accumulation of up to 2.7 g/L of limonene (in the organic phase) from glycerol with a productivity of 22.6 mg/L per h [187].

In an experimental setup similar to the one described above for transporter optimization, a library of *E. coli* strains expressing 43 different efflux pumps was used to select for enrichments under cultivation in the presence of limonene and other terpenes [116]. A pump conferring higher resistance to limonene derived from *Alcanivorax borkumensis* also increased limonene formation in a producer strain by about 50%.

To enable the production of limonene directly from CO_2, limonene synthases from different sources were introduced into cyanobacterial species, including the two unicellular model organisms *Synechococcus* sp. [220] and *Synechocystis* sp. [202], and the filamentous N_2-fixing *Anabaena* sp. [221]. The produced limonene was either collected in a dodecane overlay [220] or captured from the gas phase. While (over)expression of endogenous or heterologous DXS, IDI, and GPP synthase to improve precursor supply by the endogenous MEP pathway had a positive effect on limonene formation [150,202], inhibiting formation of the storage carbohydrate and by-product glycogen did not [220]. In all cases, production levels have remained rather low, in the μg/L range.

The same is true for limonene production in *S. cerevisiae* so far [159]. In one study, a (+)-limonene synthase from *Citrus limon* and a (−)-limonene synthase from *Perilla frutescens* were expressed in a yeast strain expressing Erg20 (K197G) (see above) both as full-length versions and without their respective plastid-targeting signal, where the truncated versions of the enzymes proved to be superior. When comparing different methods for product recovery, using a dodecane overlay or continuous headspace trapping were most efficient. While many of the above-mentioned engineering strategies will be useful in further increasing limonene production in yeast, the screening of yeast strain libraries producing limonene and other monoterpenes may also be supported by a recently developed colourimetric assay based on 2,2-diphenyl-1-picrylhydrazyl [222].

Apart from low production levels, monoterpene toxicity is a major issue in yeast. Among five different monoterpenes tested, limonene had the strongest inhibitory effect on the yeast growth rate [223] and mainly seems to affect its cell wall [224]. While the use of organic solvents in two-phase fermentation set-ups was able to reduce limonene toxicity [223], the overexpression of endogenous ATP-binding cassette transporters had no effect [225]. Analysis of strains evolved for increased limonene tolerance revealed that a single frame shift mutation in *TCB3* encoding a protein of the ER could confer monoterpene tolerance, potentially by affecting cell wall integrity [226].

7.3.3.3 Sabinene, Phellandrene, and Myrcene

The bicyclic sabinene was produced in both *E. coli* [185] and *S. cerevisiae* [155] employing a sabinene synthase from *Salvia pomifera*. In *E. coli*, expression of the terpene synthase was combined with expression of the *A. grandis* GPP synthase and the hybrid *Enterococcus faecalis/S. cerevisiae* MVA pathway similar to the strategy described for pinene above [214]. Optimization of the fermentation medium and gene induction levels finally resulted in production of 2.65 g/L of sabinene from glycerol in a fed-batch approach. The study in *S. cerevisiae* mainly concentrated on engineering Erg20 to reduce FPP formation and increase GPP formation. The resulting dominant negative mutant enzyme Erg20(F96W-N127W) was then fused to sabinene synthase and expressed in a diploid background strain overexpressing Hmg2(K6R) and carrying heterozygous deletions in *ERG20* and *ERG9* to further reduce FPP formation as well as FPP flux towards ergosterol, leading to production of 17.5 mg/L of sabinene.

Proof-of-principle studies also showed the production of monocyclic phellandrene using a terpene synthase from *Lavandula angustifolia* [227] and acyclic myrcene using myrcene synthase from *Quercus ilex* [228] in *E. coli*. As in the examples mentioned above, engineering strategies included the expression of a truncated plant GPP synthase, the mevalonate pathway, and balancing of gene expression levels. In addition, phellandrene was also produced in *Synechocystis* sp. for evaluating different transcriptional and translational regulatory elements for optimal expression of the terpene synthase gene after chromosomal integration [229].

7.3.4 Production of Sesquiterpene Hydrocarbons

7.3.4.1 Valencene

Like many other mono- and sesquiterpene hydrocarbons, the bicyclic valencene does not only attract interest as a potential fuel, but also as an aroma compound. While valencene represents an aroma substance itself, it also serves as a precursor for the high-value nootkatone. Valencene, among other sesquiterpenes, was first produced in yeast to study the influence of regulating *ERG9* expression with the repressible *MET3* promoter [163]. Here, a valencene synthase gene from *Citrus x paradisi* was expressed under control of the inducible *GAL1* promoter leading production of up to 3 mg/L. Farhi et al. [170] targeted valencene synthesis to the mitochondria with the rational that FPP-consuming enzymes are located in both the cytosol and mitochondria – that is, a precursor pool should be present in different compartments. Their best producer strain was based on an industrial yeast strain expressing both a cytosolic and a mitochondrial version of valencene synthase from *Citrus x sinensis*, a heterologous FPP synthase from *Arabidopsis* also targeted to the mitochondria and cytosolic tHmg1 yielding 1.5 mg/L, which represents a sevenfold improvement over the strain only expressing the terpene synthase in the cytosol.

Because the low valencene titres obtained in this study – also in comparison with other sesquiterpenes – indicated low activity of the synthases, Beekwilder et al. [203] isolated and characterized valencene synthase from *Callitropsis nootkatensis*, which had kinetic properties similar to the one from *Citrus x paradisi*, but showed a better performance when expressed in a nonoptimized yeast strain or in *Rhodobacter sphaeroides*. When combined with expression of the MVA pathway operon from *Paracoccus zeaxanthinifaciens*, valencene production increased to 350 mg/L.

In addition, two other microbial hosts were engineered for production of valencene using the *Callitropsis nootkatensis* enzyme: a *Corynebacterium glutamicum* strain deleted in two prenyltransferase genes and expressing FPP synthase from either *E. coli* or *S. cerevisiae* [206] and, rather unusually, the basidiomycete *Schizophyllum commune* [230]. Maximum titres were 2.4 and 16 mg/L, respectively. In all studies, dodecane was used to capture valencene in situ.

7.3.4.2 Bisabolene

The activity of the terpene synthase often represents a limiting factor in isoprenoid production. To establish bisabolene production in *E. coli* and in *S. cerevisiae*, Peralta-Yahya et al. [231] screened bisabolene synthases from five different sources: two angiosperm and three gymnosperm synthases differing in their domain structure. Most of them were codon-optimized for *E. coli*. In both heterologous hosts, the codon-optimized version of *E*-α-bisabolene synthase from *A. grandis* (AgBIS) exhibited by far the highest activity. Production levels in *E. coli* carrying an optimized MVA pathway reached up to 912 mg/L, whereas a *S. cerevisiae* strain, in which tHMGR, Erg20 and Upc2-1 were overexpressed and *ERG9* was down-regulated, produced 994 mg/L of bisabolene. Later, the protein structure of AgBIS was determined and a catalytic mechanism for bisabolene synthesis was proposed, which opens up opportunities for protein engineering approaches [232]. The fusion of AgBIS to Erg20 increased both enzyme expression levels and bisabolene production in yeast [233].

To identify nonobvious targets for increasing bisabolene production, the authors also screened a yeast deletion library expressing a carotenoid biosynthetic pathway for enhanced colour formation. Gene deletions that were able to increase both carotenoid and bisabolene formation were identified. A strain carrying, among other modifications, deletions in two genes of unknown function was tested in a fed-batch cultivation leading to a production of 5.2 g/L of bisabolene.

Instead of a screening system, Kirby et al. [234] developed a method that allows for selection of strains with enhanced production of bisabolene (and potentially also other terpenes). This method is based on the fact that cells producing bisabolene are more resistant to nonionic surfactants than nonproducers, presumably due to changes in membrane fluidity. This principle was on the one hand used to identify AgBIS gene variants leading to higher bisabolene production, and on the other hand to increase production stability in long-term cultures.

Apart from *E. coli* and *Saccharomyces cerevisiae*, bisabolene production was also introduced in *Synechococcus* sp. [220] and *Streptomyces venezuelae* [207].

7.3.4.3 *Farnesene*

Both farnesene isomers (α- and β-) occur in a wide range of plant and animal species. Both α- and β-farnesene synthases from different plant sources have been heterologously expressed in *Anabaena* sp. [235], *E. coli* [236], and *S. cerevisiae* [237]. While titres in *Anabaena* sp. remained in the μg/L range, engineering an *E. coli* strain through expression of a hybrid MVA pathway and a fusion enzyme of FPP synthase and *Malus x domestica* α-farnesene synthase led to production of up to 380 mg/L of α-farnesene when glycerol was fed as an additional carbon source [238]. Zhu et al. [194] reconstructed the MVA pathway including farnesene synthesis in vitro in order to study its kinetics and derive engineering strategies from these results. By increasing expression of selected MVA pathway genes in *E. coli*, they were able to enhance the α-farnesene titre to 1.1 g/L.

To date, β-farnesene is by far the most advanced example of isoprenoid hydrocarbon production. It is produced on an industrial scale by a heavily engineered *S. cerevisiae* strain developed by Amyris. One of their intermediate strains was described in a study where the CoA precursor pantothenate was used as a metabolic switch to limit farnesene production during the growth phase and thereby increased the genetic stability of the population [239]. This strain contained, among others, the following features: (1) expression of a mutated version of a β-farnesene synthase from *Artemisia annua* with improved catalytic efficiency, the gene being integrated in several copies into the genome; (2) overexpression of the entire endogenous MVA pathway using integration of one or two additional copies per gene as well as expression of an HMGS from *Brassica juncea* and an ACAT from *C. acetobutylicum*; (3) down-regulation of *ERG9* with help of the *MET3* promoter; and (4) overexpression of the PDH bypass using a pyruvate decarboxylase from *Zymomonas mobilis* and endogenous Ald6 and Acs2. Since many of these genes are regulated by the *GAL* promoters, galactose regulation was modified in a way that enabled activity of these promoters in glucose medium.

Crucial for the further improvement of farnesene production strains, however, was the development of an automation platform for high-throughput strain construction and screening. This enabled the construction of the current production strain, which carries about 80,000 new bp and has 40,000 bp deleted (http://www.biofuelsdigest.com/bdigest/2015/08/27/amyris-and-the-velocity-of-renewables/). Production of farnesene from sugarcane is implemented at the Brotas plant in São Paulo, Brazil, which has a capacity of 50 M L per year. The applicability of farnesene-derived fuels as a blending component in jet fuel has been demonstrated in several test flights.

7.3.5 Conclusion and Perspectives

The above examples have shown that although mono- and sesquiterpene yields and titres of primary engineered strains are usually quite low, it is possible to reach industrially relevant production levels. This, however, requires extensive strain engineering efforts, as seen in the farnesene example. One major limitation is limited availability of GPP and FPP in the microbial hosts. Current and future strategies to establish the production of novel compounds will therefore instead rely on platform strains already optimized, amongst other features, for the provision of sufficient terpene precursors using either endogenous, heterologous, or hybrid pathways.

Another challenge is the selection or engineering of efficient terpene synthases. Since terpenes are often only required at low amounts in their native hosts, the evolutionary pressure to select biosynthetic enzymes of high catalytic efficiency has been low – similar to the enzymes of the alka(e)ne pathways discussed above. Although increasing enzyme expression levels through enhancing the gene copy number, selection of efficient promoters and codon optimization may be beneficial, its effect is often limited [229]. As an initial step, it may therefore be useful to screen terpene synthases from different organisms since their activities can vary substantially in the heterologous host [231]. Furthermore, the increasing availability of genome sequences will further enhance the search space for potential gene/enzyme candidates. Rational engineering of terpene synthases for enhanced activity has proven to be difficult, and the screening of enzyme libraries through direct analysis of product formation by GC-MS is only feasible at low throughput.

Therefore, indirect screening or selection methods for high-throughput analysis of large terpene synthase libraries generated by random mutagenesis have been developed. Selection methods can, for example, be based on the toxicity of GPP and FPP that can be relieved by efficient terpene synthases (US Patent 8236512 B1), or on the fact that certain terpenes may reduce the toxic effect of nonionic surfactants [234]. Colour-based screening assays detected, for example, the formation of methanol as a reaction by-product when a synthetic substrate was used [240], or the reduction of colour formation when a competing carotenoid pathway was co-expressed with the terpene synthase library [241]. The implication of such assays is expected to facilitate the identification of improved enzymes in the future.

Apart from low activity, many terpene synthases also show a lack of product specificity. The latter is an even bigger enzyme engineering challenge than the former as no high-throughput screening assays are available. On the other hand, since future biojet fuel blends are likely to consist of more than one compound, mixed product formation may not necessarily be a disadvantage. When screening a mutant library of a bisabolene synthase Kirby et al. [234] identified enzyme variants that were able to produce a mixture of both α-bisabolene and β-farnesene at different ratios.

All in all, it has been shown that the microbial production of aviation fuels or their precursors at industrially relevant levels is possible. The future will show whether extensive efforts in high-throughput strain engineering can boost alka(e)ne production to levels similar to the ones obtained for farnesene. So far, production of farnesene is based on sugarcane. The next challenge will therefore be to turn first-generation jet fuel production into second-generation production based on lignocellulosic biomass – analogous to second-generation bioethanol plants – to achieve a truly sustainable process.

References

[1] J. Nielsen, et al., Metabolic engineering of yeast for production of fuels and chemicals, Curr. Opin. Biotechnol. 24 (3) (2013) 398–404.

[2] G. Ozcengiz, A.L. Demain, Recent advances in the biosynthesis of penicillins, cephalosporins and clavams and its regulation, Biotechnol. Adv. 31 (2) (2013) 287–311.

[3] C.J. Paddon, J.D. Keasling, Semi-synthetic artemisinin: a model for the use of synthetic biology in pharmaceutical development, Nat. Rev. Microbiol. 12 (5) (2014) 355–367.

[4] J. Nielsen, Metabolic engineering, Appl. Microbiol. Biotechnol. 55 (3) (2001) 263–283.

[5] S.A. Angermayr, A.G. Rovira, K.J. Hellingwerf, Metabolic engineering of cyanobacteria for the synthesis of commodity products, Trends Biotechnol. 33 (6) (2015) 352–361.

[6] P. Kallio, et al., Renewable jet fuel, Curr. Opin. Biotechnol. 26 (2014) 50–55.

[7] G.R. Liu, B.B. Yan, G.Y. Chen, Technical review on jet fuel production, Renew. Sust. Energy Rev. 25 (2013) 59–70.

[8] N. Ladygina, E.G. Dedyukhina, M.B. Vainshtein, A review on microbial synthesis of hydrocarbons, Process Biochem. 41 (5) (2006) 1001–1014.

[9] B.F. Pfleger, M. Gossing, J. Nielsen, Metabolic engineering strategies for microbial synthesis of oleochemicals, Metab. Eng. 29 (2015) 1–11.

[10] W. Kricka, J. Fitzpatrick, U. Bond, Metabolic engineering of yeasts by heterologous enzyme production for degradation of cellulose and hemicellulose from biomass: a perspective, Front Microbiol. 5 (2014) 174.

[11] S.Y. Lee, H.M. Kim, S. Cheon, Metabolic engineering for the production of hydrocarbon fuels, Curr. Opin. Biotechnol. 33 (2014) 15–22.

[12] R.C. Coates, et al., Characterization of cyanobacterial hydrocarbon composition and distribution of biosynthetic pathways, PLoS One 9 (1) (2014) e85140.

[13] M.G.M. Aarts, et al., Molecular characterization of the CER1 gene of *arabidopsis* involved in epicuticular wax biosynthesis and pollen fertility, Plant Cell 7 (12) (1995) 2115–2127.

[14] R. Jetter, L. Kunst, Plant surface lipid biosynthetic pathways and their utility for metabolic engineering of waxes and hydrocarbon biofuels, Plant J. 54 (4) (2008) 670–683.

[15] S. Tippmann, et al., From flavors and pharmaceuticals to advanced biofuels: production of isoprenoids in *Saccharomyces cerevisiae*, Biotechnol. J. 8 (12) (2013) 1435–1444.

[16] P. Savakis, K.J. Hellingwerf, Engineering cyanobacteria for direct biofuel production from CO_2, Curr. Opin. Biotechnol. 33 (2015) 8–14.

[17] J.C.F. Ortiz-Marquez, et al., Genetic engineering of multispecies microbial cell factories as an alternative for bioenergy production, Trends Biotechnol. 31 (9) (2013) 521–529.

[18] J. Sheng, X. Feng, Metabolic engineering of yeast to produce fatty acid-derived biofuels: bottlenecks and solutions, Front Microbiol. 6 (2015) 554.

[19] X.Z. Chen, et al., Metabolic engineering of *Escherichia coli*: a sustainable industrial platform for bio-based chemical production, Biotechnol. Adv. 31 (8) (2013) 1200–1223.

[20] A. Beopoulos, et al., *Yarrowia lipolytica* as a model for bio-oil production, Prog. Lipid. Res. 48 (6) (2009) 375–387.

[21] B.J. Sanchez, J. Nielsen, Genome scale models of yeast: towards standardized evaluation and consistent omic integration, Integr. Biol. 7 (8) (2015) 846–858.

[22] H.M. Salis, E.A. Mirsky, C.A. Voigt, Automated design of synthetic ribosome binding sites to control protein expression, Nat. Biotechnol. 27 (10) (2009) 946–950.

[23] B.K.S. Chung, D.Y. Lee, Computational codon optimization of synthetic gene for protein expression, BMC Syst. Biol. 6 (2012) 134.

[24] Y. Shiba, et al., Engineering of the pyruvate dehydrogenase bypass in *Saccharomyces cerevisiae* for high-level production of isoprenoids, Metab. Eng. 9 (2) (2007) 160–168.

[25] J.W. Lee, et al., Systems metabolic engineering of microorganisms for natural and non-natural chemicals, Nat. Chem. Biol. 8 (6) (2012) 536–546.

[26] S.K. Lee, et al., Metabolic engineering of microorganisms for biofuels production: from bugs to synthetic biology to fuels, Curr. Opin. Biotechnol. 19 (6) (2008) 556–563.

[27] O. Tehlivets, K. Scheuringer, S.D. Kohlwein, Fatty acid synthesis and elongation in yeast, Biochim. Biophys. Acta 1771 (3) (2007) 255–270.

[28] K. Magnuson, et al., Regulation of fatty-acid biosynthesis in Escherichia coli, Microbiol. Rev. 57 (3) (1993) 522–542.

[29] J.K. Hiltunen, et al., Mitochondrial fatty acid synthesis and respiration, Biochim. Biophys. Acta Bioenerg. 1797 (6–7) (2010) 1195–1202.

[30] J.Y. Choi, et al., Regulatory elements that control transcription activation and unsaturated fatty acid-mediated repression of the *Saccharomyces cerevisiae OLE1* gene, J. Biol. Chem. 271 (7) (1996) 3581–3589.

[31] D.I. Chan, H.J. Vogel, Current understanding of fatty acid biosynthesis and the acyl carrier protein, Biochem. J. 430 (2010) 1–19.

[32] B. Koch, C. Schmidt, G. Daum, Storage lipids of yeasts: a survey of nonpolar lipid metabolism in *Saccharomyces cerevisiae*, *Pichia pastoris*, and *Yarrowia lipolytica*, FEMS Microbiol. Rev. 38 (5) (2014) 892–915.

[33] S. Shi, et al., Improving production of malonyl coenzyme A-derived metabolites by abolishing Snf1-dependent regulation of Acc1, MBio 5 (3) (2014) e01130–14.

[34] G. Lust, F. Lynen, The inhibition of the fatty acid synthetase multienzyme complex of yeast by long-chain acyl coenzyme A compounds, Eur. J. Biochem. 7 (1) (1968) 68–72.

[35] P. Jiang, J.E. Cronan, Inhibition of fatty-acid synthesis in *Escherichia coli* in the absence of phospholipid-synthesis and release of inhibition by thioesterase action, J. Bacteriol. 176 (10) (1994) 2814–2821.

[36] M.S. Davis, J.E. Cronan Jr., Inhibition of *Escherichia coli* acetyl coenzyme A carboxylase by acyl-acyl carrier protein, J. Bacteriol. 183 (4) (2001) 1499–1503.

[37] T.M. Wasylenko, W.S. Ahn, G. Stephanopoulos, The oxidative pentose phosphate pathway is the primary source of NADPH for lipid overproduction from glucose in *Yarrowia lipolytica*, Metab. Eng. 30 (2015) 27–39.

[38] A. Krivoruchko, et al., Microbial acetyl-CoA metabolism and metabolic engineering, Metab. Eng. 28 (2014) 28–42.

[39] B.W. de Jong, et al., Improved production of fatty acid ethyl esters in *Saccharomyces cerevisiae* through up-regulation of the ethanol degradation pathway and expression of the heterologous phosphoketolase pathway, Microb. Cell Fact. 13 (1) (2014) 39.

[40] X.L. Tang, H.X. Feng, W.N. Chen, Metabolic engineering for enhanced fatty acids synthesis in *Saccharomyces cerevisiae*, Metab. Eng. 16 (2013) 95–102.

[41] J.Z. Lian, et al., Design and construction of acetyl-CoA overproducing *Saccharomyces cerevisiae* strains, Metab. Eng. 24 (2014) 139–149.

[42] X.Y. Feng, J.Z. Lian, H.M. Zhao, Metabolic engineering of *Saccharomyces cerevisiae* to improve 1-hexadecanol production, Metab. Eng. 27 (2015) 10–19.

[43] B.U. Kozak, et al., Replacement of the *Saccharomyces cerevisiae* acetyl-CoA synthetases by alternative pathways for cytosolic acetyl-CoA synthesis, Metab. Eng. 21 (2014) 46–59.

[44] B.U. Kozak, et al., Engineering acetyl coenzyme A supply: functional expression of a bacterial pyruvate dehydrogenase complex in the cytosol of *Saccharomyces cerevisiae*, MBio 5 (5) (2014) e01696–14.

[45] B. Oud, et al., An internal deletion in *MTH1* enables growth on glucose of pyruvate-decarboxylase negative, non-fermentative *Saccharomyces cerevisiae*, Microb. Cell Fact. 11 (2012).

[46] F. Lin, et al., Improving fatty acid availability for bio-hydrocarbon production in *Escherichia coli* by metabolic engineering, PLoS One 8 (10) (2013) e78595.

[47] S. Centeno-Leija, et al., Improving poly-3-hydroxybutyrate production in *Escherichia coli* by combining the increase in the NADPH pool and acetyl-CoA availability, Antonie Van Leeuwenhoek 105 (4) (2014) 687–696.

[48] C.R. Dittrich, et al., Redistribution of metabolic fluxes in the central aerobic metabolic pathway of *E. coli* mutant strains with deletion of the ackA-pta and poxB pathways for the synthesis of isoamyl acetate, Biotechnol. Prog. 21 (2) (2005) 627–631.

[49] Y. Cao, et al., Production of free monounsaturated fatty acids by metabolically engineered *Escherichia coli*, Biotechnol. Biofuels 7 (2014) 59.

[50] W. Runguphan, J.D. Keasling, Metabolic engineering of *Saccharomyces cerevisiae* for production of fatty acid-derived biofuels and chemicals, Metab. Eng. 21 (2014) 103–113.

[51] R.M. Lennen, et al., A process for microbial hydrocarbon synthesis: overproduction of fatty acids in *Escherichia coli* and catalytic conversion to alkanes, Biotechnol. Bioeng. 106 (2) (2010) 193–202.

[52] J.W. Choi, N.A. Da Silva, Improving polyketide and fatty acid synthesis by engineering of the yeast acetyl-CoA carboxylase, J. Biotechnol. 187 (2014) 56–59.

[53] C. Leber, N.A. Da Silva, Engineering of *Saccharomyces cerevisiae* for the synthesis of short chain fatty acids, Biotechnol. Bioeng. 111 (2) (2014) 347–358.

[54] D.T. Eriksen, et al., Orthogonal fatty acid biosynthetic pathway improves fatty acid ethyl ester production in *Saccharomyces cerevisiae*, ACS Synth. Biol. 4 (7) (2015) 808–814.

[55] R.W. Haushalter, et al., Development of an orthogonal fatty acid biosynthesis system in *E. coli* for oleochemical production, Metab. Eng. 30 (2015) 1–6.

[56] E.J. Steen, et al., Microbial production of fatty-acid-derived fuels and chemicals from plant biomass, Nature 463 (7280) (2010) 559–562.

[57] F. Jing, et al., Phylogenetic and experimental characterization of an acyl-ACP thioesterase family reveals significant diversity in enzymatic specificity and activity, BMC Biochem. 12 (2011) 44.

[58] P. Xu, et al., Modular optimization of multi-gene pathways for fatty acids production in *E. coli*, Nat. Commun. 4 (2013) 1409.

[59] P. Kallio, et al., An engineered pathway for the biosynthesis of renewable propane., Nat. Commun. 5 (2014) 4731.

[60] X.J. Zhang, et al., Efficient free fatty acid production in *Escherichia coli* using plant acyl-ACP thioesterases, Metab. Eng. 13 (6) (2011) 713–722.

[61] P.N. Black, C.C. DiRusso, Yeast acyl-CoA synthetases at the crossroads of fatty acid metabolism and regulation, Biochim. Biophys. Acta 1771 (3) (2007) 286–298.

[62] L. Chen, et al., Enhancement of free fatty acid production in *Saccharomyces cerevisiae* by control of fatty acyl-CoA metabolism, Appl. Microbiol. Biotechnol. 98 (15) (2014) 6739–6750.

[63] J.O. Valle-Rodriguez, et al., Metabolic engineering of *Saccharomyces cerevisiae* for production of fatty acid ethyl esters, an advanced biofuel, by eliminating non-essential fatty acid utilization pathways, Appl. Energy 115 (2014) 226–232.

[64] C. Leber, et al., Overproduction and secretion of free fatty acids through disrupted neutral lipid recycle in *Saccharomyces cerevisiae*, Metab. Eng. 28 (2014) 54–62.

[65] L. Sandager, et al., Storage lipid synthesis is non-essential in yeast, J. Biol. Chem. 277 (8) (2002) 6478–6482.

[66] F.Z. Zhang, et al., Enhancing fatty acid production by the expression of the regulatory transcription factor FadR, Metab. Eng. 14 (6) (2012) 653–660.

[67] J. Seip, et al., Snf1 is a regulator of lipid accumulation in *Yarrowia lipolytica*, Appl. Environ. Microbiol. 79 (23) (2013) 7360–7370.

[68] Y. Kamisaka, et al., Overexpression of the active diacylglycerol acyltransferase variant transforms *Saccharomyces cerevisiae* into an oleaginous yeast, Appl. Microbiol. Biotechnol. 97 (16) (2013) 7345–7355.

[69] J. Blazeck, et al., Harnessing *Yarrowia lipolytica* lipogenesis to create a platform for lipid and biofuel production, Nat. Commun. 5 (2014) 3131.
[70] K. Qiao, et al., Engineering lipid overproduction in the oleaginous yeast *Yarrowia lipolytica*, Metab. Eng. 29 (2015) 56–65.
[71] C. Dellomonaco, et al., Engineered reversal of the beta-oxidation cycle for the synthesis of fuels and chemicals, Nature 476 (7360) (2011) 355–359.
[72] J.M. Clomburg, et al., A synthetic biology approach to engineer a functional reversal of the beta-oxidation cycle, ACS Synth. Biol. 1 (11) (2012) 541–554.
[73] J.Z. Lian, H.M. Zhao, Reversal of the beta-oxidation cycle in *Saccharomyces cerevisiae* for production of fuels and chemicals, ACS Synth. Biol. 4 (3) (2015) 332–341.
[74] C.W. Bird, J.M. Lynch, Formation of hydrocarbons by microorganisms, Chem. Soc. Rev. 3 (3) (1974) 309–328.
[75] A. Schirmer, et al., Microbial biosynthesis of alkanes, Science 329 (5991) (2010) 559–562.
[76] D. Das, et al., Oxygen-independent decarbonylation of aldehydes by cyanobacterial aldehyde decarbonylase: a new reaction of diiron enzymes, Angew. Chem. Int. Ed. 50 (31) (2011) 7148–7152.
[77] N. Li, et al., Conversion of fatty aldehydes to alka(e)nes and formate by a cyanobacterial aldehyde decarbonylase: cryptic redox by an unusual dimetal oxygenase, J. Am. Chem. Soc. 133 (16) (2011) 6158–6161.
[78] D.M. Warui, et al., Detection of formate, rather than carbon monoxide, as the stoichiometric coproduct in conversion of fatty aldehydes to alkanes by a cyanobacterial aldehyde decarbonylase, J. Am. Chem. Soc. 133 (10) (2011) 3316–3319.
[79] N. Li, et al., Evidence for only oxygenative cleavage of aldehydes to alk(a/e)nes and formate by cyanobacterial aldehyde decarbonylases, Biochemistry 51 (40) (2012) 7908–7916.
[80] K.G. Aukema, et al., Cyanobacterial aldehyde deformylase oxygenation of aldehydes yields $n-1$ aldehydes and alcohols in addition to alkanes, ACS Catal. 3 (10) (2013) 2228–2238.
[81] C. Andre, et al., Fusing catalase to an alkane-producing enzyme maintains enzymatic activity by converting the inhibitory byproduct H_2O_2 to the cosubstrate O-2, Proc. Natl. Acad. Sci. U. S. A. 110 (8) (2013) 3191–3196.
[82] F.M. Lin, et al., Aldehyde-forming fatty acyl-CoA reductase from cyanobacteria: expression, purification and characterization of the recombinant enzyme, FEBS J. 280 (19) (2013) 4773–4781.
[83] D.M. Warui, et al., Efficient delivery of long-chain fatty aldehydes from the *Nostoc punctiforme* acyl acyl carrier protein reductase to its cognate aldehyde-deformylating oxygenase, Biochemistry 54 (4) (2015) 1006–1015.
[84] Z. Rahman, et al., Enhanced production of n-alkanes in *Escherichia coli* by spatial organization of biosynthetic pathway enzymes, J. Biotechnol. 192 (2014) 187–191.
[85] A. Bernard, et al., Reconstitution of plant alkane biosynthesis in yeast demonstrates that *Arabidopsis* ECERIFERUM1 and ECERIFERUM3 are core components of a very-long-chain alkane synthesis complex, Plant Cell 24 (7) (2012) 3106–3118.
[86] Y.J. Choi, S.Y. Lee, Microbial production of short-chain alkanes, Nature 502 (7472) (2013) 571.
[87] M.K. Akhtar, N.J. Turner, P.R. Jones, Carboxylic acid reductase is a versatile enzyme for the conversion of fatty acids into fuels and chemical commodities, Proc. Natl. Acad. Sci. U. S. A. 110 (1) (2013) 87–92.
[88] P. Venkitasubramanian, L. Daniels, J.P.N. Rosazza, Reduction of carboxylic acids by *Nocardia aldehyde* oxidoreductase requires a phosphopantetheinylated enzyme, J. Biol. Chem. 282 (1) (2007) 478–485.
[89] N.A. Buijs, et al., Long-chain alkane production by the yeast *Saccharomyces cerevisiae*, Biotechnol. Bioeng. 112 (6) (2015) 1275–1279.
[90] M.A. Rude, et al., Terminal olefin (1-Alkene) biosynthesis by a novel P450 fatty acid decarboxylase from *Jeotgalicoccus* Species, Appl. Environ. Microbiol. 77 (5) (2011) 1718–1727.
[91] J. Belcher, et al., Structure and biochemical properties of the alkene producing cytochrome P450 OleTJE (CYP152L1) from the *Jeotgalicoccus* sp. 8456 bacterium, J. Biol. Chem. 289 (10) (2014) 6535–6550.
[92] Y. Liu, et al., Hydrogen peroxide-independent production of alpha-alkenes by OleT(JE) P450 fatty acid decarboxylase, Biotechnol. Biofuels 7 (2014) 28.
[93] B. Chen, D.Y. Lee, M.W. Chang, Combinatorial metabolic engineering of *Saccharomyces cerevisiae* for terminal alkene production, Metab. Eng. 31 (2015) 53–61.
[94] R. Kourist, A new class of enzymes discovered: a non-heme oxidase produces medium-chain 1-alkenes, Angew. Chem. Int. Ed. 54 (14) (2015) 4156–4158.
[95] L. Friedman, B. da Costa, Hydrocarbon-producing genes and methods of their use, Patent WO2008147781.
[96] H.R. Beller, E.B. Goh, J.D. Keasling, Genes involved in long-chain alkene biosynthesis in *Micrococcus luteus*, Appl. Environ. Microbiol. 76 (4) (2010) 1212–1223.
[97] D.J. Sukovich, et al., Structure, function, and insights into the biosynthesis of a head-to-head hydrocarbon in *Shewanella oneidensis* strain MR-1, Appl. Environ. Microbiol. 76 (12) (2010) 3842–3849.
[98] D.J. Sukovich, et al., Widespread head-to-head hydrocarbon biosynthesis in bacteria and role of OleA, Appl. Environ. Microbiol. 76 (12) (2010) 3850–3862.

[99] J.A. Frias, et al., Purification and characterization of OleA from *Xanthomonas campestris* and demonstration of a non-decarboxylative claisen condensation reaction, J. Biol. Chem. 286 (13) (2011) 10930–10938.

[100] T.P. Howard, et al., Synthesis of customized petroleum-replica fuel molecules by targeted modification of free fatty acid pools in *Escherichia coli*, Proc. Natl. Acad. Sci. U. S. A. 110 (19) (2013) 7636–7641.

[101] M. Harger, et al., Expanding the product profile of a microbial alkane biosynthetic pathway, ACS Synth. Biol. 2 (1) (2013) 59–62.

[102] Y.X. Cao, et al., Biosynthesis of odd-chain fatty alcohols in *Escherichia coli*, Metab. Eng. 29 (2015) 113–123.

[103] M. Schallmey, et al., Looking for the pick of the bunch: high-throughput screening of producing microorganisms with biosensors, Curr. Opin. Biotechnol. 26 (2014) 148–154.

[104] J.M. Ellis, M.J. Wolfgang, A genetically encoded metabolite sensor for malonyl-CoA, Chem. Biol. 19 (10) (2012) 1333–1339.

[105] P. Xu, et al., Design and kinetic analysis of a hybrid promoter-regulator system for malonyl-CoA sensing in *Escherichia coli*, ACS Chem. Biol. 9 (2) (2014) 451–458.

[106] S. Li, et al., Development of a synthetic malonyl-CoA sensor in *Saccharomyces cerevisiae* for intracellular metabolite monitoring and genetic screening, ACS Synth. Biol. 4 (12) (2015) 1308–1315.

[107] F.Z. Zhang, J.M. Carothers, J.D. Keasling, Design of a dynamic sensor-regulator system for production of chemicals and fuels derived from fatty acids, Nat. Biotechnol. 30 (4) (2012) 354–359.

[108] D.Y. Zhang, et al., Whole-cell bacterial bioreporter for actively searching and sensing of alkanes and oil spills, Microbial. Biotechnol. 5 (1) (2012) 87–97.

[109] B. Reed, J. Blazeck, H. Alper, Evolution of an alkane-inducible biosensor for increased responsiveness to short-chain alkanes, J. Biotechnol. 158 (3) (2012) 75–79.

[110] W. Wu, et al., Genetically assembled fluorescent biosensor for in situ detection of bio-synthesized alkanes, Sci. Rep. 5 (2015).

[111] P. Xu, et al., Improving fatty acids production by engineering dynamic pathway regulation and metabolic control, Proc. Natl. Acad. Sci. U. S. A. 111 (31) (2014) 11299–11304.

[112] J. Anfelt, et al., Using transcriptomics to improve butanol tolerance of *Synechocystis* sp. strain PCC 6803, Appl. Environ. Microbiol. 79 (23) (2013) 7419–7427.

[113] H. Ling, et al., Transcriptome response to alkane biofuels in *Saccharomyces cerevisiae*: identification of efflux pumps involved in alkane tolerance, Biotechnol. Biofuels 6 (2013) 95.

[114] L. Caspeta, et al., Altered sterol composition renders yeast thermotolerant, Science 346 (6205) (2014) 75–78.

[115] D. Stanley, et al., Generation and characterisation of stable ethanol-tolerant mutants of *Saccharomyces cerevisiae*, J. Ind. Microbiol. Biotechnol. 37 (2) (2010) 139–149.

[116] M.J. Dunlop, et al., Engineering microbial biofuel tolerance and export using efflux pumps, Mol. Syst. Biol. 7 (2011) 487.

[117] B. Chen, H. Ling, M.W. Chang, Transporter engineering for improved tolerance against alkane biofuels in *Saccharomyces cerevisiae*, Biotechnol. Biofuels 6 (1) (2013) 21.

[118] J. Chappell, The biochemistry and molecular biology of isoprenoid metabolism, Plant Physiol. 107 (1) (1995) 1–6.

[119] C.E. Vickers, et al., Metabolic engineering of volatile isoprenoids in plants and microbes, Plant Cell Environ. 37 (8) (2014) 1753–1775.

[120] J. Bohlmann, C.I. Keeling, Terpenoid biomaterials, Plant J. 54 (4) (2008) 656–669.

[121] P.P. Peralta-Yahya, et al., Microbial engineering for the production of advanced biofuels, Nature 488 (7411) (2012) 320–328.

[122] N.I. Tracy, et al., Hydrogenated monoterpenes as diesel fuel additives, Fuel 88 (11) (2009) 2238–2240.

[123] H.A. Meylemans, R.L. Quintana, B.G. Harvey, Efficient conversion of pure and mixed terpene feedstocks to high density fuels, Fuel 97 (2012) 560–568.

[124] T. Kuzuyama, et al., Detection of the mevalonate pathway in *Streptomyces* species using the 3-hydroxy-3-methylglutaryl coenzyme A reductase gene, J. Antibiot (Tokyo) 55 (10) (2002) 919–923.

[125] J.M. Vinokur, et al., Evidence of a novel mevalonate pathway in archaea, Biochemistry 53 (25) (2014) 4161–4168.

[126] M. Rohmer, et al., Isoprenoid biosynthesis in bacteria: a novel pathway for the early steps leading to isopentenyl diphosphate, Biochem. J. 295 (Pt 2) (1993) 517–524.

[127] H.K. Lichtenthaler, The 1-Deoxy-D-Xylulose-5-Phosphate pathway of isoprenoid biosynthesis in plants, Annu. Rev. Plant Physiol. Plant Mol. Biol. 50 (1999) 47–65.

[128] H. Jomaa, et al., Inhibitors of the nonmevalonate pathway of isoprenoid biosynthesis as antimalarial drugs, Science 285 (5433) (1999) 1573–1576.

[129] H.M. Miziorko, Enzymes of the mevalonate pathway of isoprenoid biosynthesis, Arch. Biochem. Biophys. 505 (2) (2011) 131–143.

[130] M.E. Basson, M. Thorsness, J. Rine, *Saccharomyces cerevisiae* contains two functional genes encoding

3-hydroxy-3-methylglutaryl-coenzyme A reductase, Proc. Natl. Acad. Sci. U. S. A. 83 (15) (1986) 5563–5567.

[131] D. Dimster-Denk, M.K. Thorsness, J. Rine, Feedback regulation of 3-hydroxy-3-methylglutaryl coenzyme A reductase in *Saccharomyces cerevisiae*, Mol. Biol. Cell 5 (6) (1994) 655–665.

[132] J.S. Burg, P.J. Espenshade, Regulation of HMG-CoA reductase in mammals and yeast, Prog. Lipid. Res. 50 (4) (2011) 403–410.

[133] R.Y. Hampton, J. Rine, Regulated degradation of HMG-CoA reductase, an integral membrane protein of the endoplasmic reticulum, in yeast, J. Cell Biol. 125 (2) (1994) 299–312.

[134] N. Dellas, et al., Discovery of a metabolic alternative to the classical mevalonate pathway, Elife 2 (2013) e00672.

[135] W.N. Hunter, The non-mevalonate pathway of isoprenoid precursor biosynthesis, J. Biol. Chem. 282 (30) (2007) 21573–21577.

[136] T. Grawert, et al., Biochemistry of the non-mevalonate isoprenoid pathway, Cell Mol. Life Sci. 68 (23) (2011) 3797–3814.

[137] A.K. Kollas, et al., Functional characterization of GcpE, an essential enzyme of the non-mevalonate pathway of isoprenoid biosynthesis, FEBS Lett. 532 (3) (2002) 432–436.

[138] M. Seemann, et al., Isoprenoid biosynthesis via the MEP pathway: in vivo Mossbauer spectroscopy identifies a [4Fe-4S]2 + center with unusual coordination sphere in the LytB protein, J. Am. Chem. Soc. 131 (37) (2009) 13184–13185.

[139] B. Altincicek, et al., LytB protein catalyzes the terminal step of the 2-C-methyl-D-erythritol-4-phosphate pathway of isoprenoid biosynthesis, FEBS Lett. 532 (3) (2002) 437–440.

[140] K. Berthelot, et al., Isopentenyl diphosphate isomerase: a checkpoint to isoprenoid biosynthesis, Biochimie 94 (8) (2012) 1621–1634.

[141] E. Gruchattka, et al., In silico profiling of *Escherichia coli* and *Saccharomyces cerevisiae* as terpenoid factories, Microb. Cell Fact. 12 (2013) 84.

[142] S. Partow, et al., Reconstruction and evaluation of the synthetic bacterial MEP pathway in *Saccharomyces cerevisiae*, PLoS One 7 (12) (2012) e52498.

[143] S. Vandermoten, E. Haubruge, M. Cusson, New insights into short-chain prenyltransferases: structural features, evolutionary history and potential for selective inhibition, Cell Mol. Life Sci. 66 (23) (2009) 3685–3695.

[144] Y. Chen, et al., Establishing a platform cell factory through engineering of yeast acetyl-CoA metabolism, Metab. Eng. 15 (2013) 48–54.

[145] D.K. Ro, et al., Production of the antimalarial drug precursor artemisinic acid in engineered yeast, Nature 440 (7086) (2006) 940–943.

[146] M.A. Asadollahi, et al., Enhancement of farnesyl diphosphate pool as direct precursor of sesquiterpenes through metabolic engineering of the mevalonate pathway in *Saccharomyces cerevisiae*, Biotechnol. Bioeng. 106 (1) (2010) 86–96.

[147] J. Rico, E. Pardo, M. Orejas, Enhanced production of a plant monoterpene by overexpression of the 3-hydroxy-3-methylglutaryl coenzyme A reductase catalytic domain in *Saccharomyces cerevisiae*, Appl. Environ. Microbiol. 76 (19) (2010) 6449–6454.

[148] C. Ohto, et al., Overexpression of the gene encoding HMG-CoA reductase in *Saccharomyces cerevisiae* for production of prenyl alcohols, Appl. Microbiol. Biotechnol. 82 (5) (2009) 837–845.

[149] F. Mantzouridou, M.Z. Tsimidou, Observations on squalene accumulation in *Saccharomyces cerevisiae* due to the manipulation of *HMG2* and *ERG6*, FEMS Yeast Res. 10 (6) (2010) 699–707.

[150] C. Ignea, et al., Improving yeast strains using recyclable integration cassettes, for the production of plant terpenoids, Microb. Cell Fact. 10 (2011) 4.

[151] Q. Li, et al., Enhancing beta-carotene production in *Saccharomyces cerevisiae* by metabolic engineering, FEMS Microbiol. Lett. 345 (2) (2013) 94–101.

[152] C. Ignea, et al., Positive genetic interactors of *HMG2* identify a new set of genetic perturbations for improving sesquiterpene production in *Saccharomyces cerevisiae*, Microb. Cell Fact. 11 (2012) 162.

[153] J. Liu, et al., Overproduction of geraniol by enhanced precursor supply in *Saccharomyces cerevisiae*, J. Biotechnol. 168 (4) (2013) 446–451.

[154] L. Blanchard, F. Karst, Characterization of a lysine-to-glutamic acid mutation in a conservative sequence of farnesyl diphosphate synthase from *Saccharomyces cerevisiae*, Gene 125 (2) (1993) 185–189.

[155] C. Ignea, et al., Engineering monoterpene production in yeast using a synthetic dominant negative geranyl diphosphate synthase, ACS Synth. Biol. 3 (5) (2014) 298–306.

[156] G. Scalcinati, et al., Combined metabolic engineering of precursor and co-factor supply to increase alpha-santalene production by *Saccharomyces cerevisiae*, Microb. Cell Fact. 11 (2012) 117.

[157] M. Oswald, et al., Monoterpenoid biosynthesis in *Saccharomyces cerevisiae*, FEMS Yeast Res. 7 (3) (2007) 413–421.

[158] M.J. Fischer, et al., Metabolic engineering of monoterpene synthesis in yeast, Biotechnol. Bioeng. 108 (8) (2011) 1883–1892.

[159] E. Jongedijk, et al., Capturing of the monoterpene olefin limonene produced in *Saccharomyces cerevisiae*, Yeast 32 (1) (2015) 159–171.

[160] S. Brown, et al., De novo production of the plant-derived alkaloid strictosidine in yeast, Proc. Natl. Acad. Sci. U. S. A. 112 (11) (2015) 3205–3210.

[161] B.E. Jackson, E.A. Hart-Wells, S.P. Matsuda, Metabolic engineering to produce sesquiterpenes in yeast, Org. Lett. 5 (10) (2003) 1629–1632.

[162] P.J. Westfall, et al., Production of amorphadiene in yeast, and its conversion to dihydroartemisinic acid, precursor to the antimalarial agent artemisinin, Proc. Natl. Acad. Sci. U. S. A. 109 (3) (2012) E111–E118.

[163] M.A. Asadollahi, et al., Production of plant sesquiterpenes in *Saccharomyces cerevisiae*: effect of *ERG9* repression on sesquiterpene biosynthesis, Biotechnol. Bioeng. 99 (3) (2008) 666–677.

[164] E.M. Paradise, et al., Redirection of flux through the FPP branch-point in *Saccharomyces cerevisiae* by down-regulating squalene synthase, Biotechnol. Bioeng. 100 (2) (2008) 371–378.

[165] C.J. Paddon, et al., High-level semi-synthetic production of the potent antimalarial artemisinin, Nature 496 (7446) (2013) 528–532.

[166] G. Scalcinati, et al., Dynamic control of gene expression in *Saccharomyces cerevisiae* engineered for the production of plant sesquitepene alpha-santalene in a fed-batch mode, Metab. Eng. 14 (2) (2012) 91–103.

[167] J. Yuan, C.B. Ching, Dynamic control of *ERG9* expression for improved amorpha-4,11-diene production in *Saccharomyces cerevisiae*, Microb. Cell Fact. 14 (2015) 38.

[168] A. Faulkner, et al., The *LPP1* and *DPP1* gene products account for most of the isoprenoid phosphate phosphatase activities in *Saccharomyces cerevisiae*, J. Biol. Chem. 274 (21) (1999) 14831–14837.

[169] S. Takahashi, et al., Metabolic engineering of sesquiterpene metabolism in yeast, Biotechnol. Bioeng. 97 (1) (2007) 170–181.

[170] M. Farhi, et al., Harnessing yeast subcellular compartments for the production of plant terpenoids, Metab. Eng. 13 (5) (2011) 474–481.

[171] L. Albertsen, et al., Diversion of flux toward sesquiterpene production in *Saccharomyces cerevisiae* by fusion of host and heterologous enzymes, Appl. Environ. Microbiol. 77 (3) (2011) 1033–1040.

[172] S. Carlsen, et al., Heterologous expression and characterization of bacterial 2-C-methyl-D-erythritol-4-phosphate pathway in *Saccharomyces cerevisiae*, Appl. Microbiol. Biotechnol. 97 (13) (2013) 5753–5769.

[173] S.W. Kim, J.D. Keasling, Metabolic engineering of the nonmevalonate isopentenyl diphosphate synthesis pathway in *Escherichia coli* enhances lycopene production, Biotechnol. Bioeng. 72 (4) (2001) 408–415.

[174] X. Lv, H. Xu, H. Yu, Significantly enhanced production of isoprene by ordered coexpression of genes *dxs*, *dxr*, and *idi* in *Escherichia coli*, Appl. Microbiol. Biotechnol. 97 (6) (2013) 2357–2365.

[175] M. Albrecht, N. Misawa, G. Sandmann, Metabolic engineering of the terpenoid biosynthetic pathway of *Escherichia coli* for production of the carotenoids beta-carotene and zeaxanthin, Biotechnol. Lett. 21 (9) (1999) 791–795.

[176] P.C. Lee, B.N. Mijts, C. Schmidt-Dannert, Investigation of factors influencing production of the monocyclic carotenoid torulene in metabolically engineered *Escherichia coli*, Appl. Microbiol. Biotechnol. 65 (5) (2004) 538–546.

[177] K.L. Jones, S.W. Kim, J.D. Keasling, Low-copy plasmids can perform as well as or better than high-copy plasmids for metabolic engineering of bacteria, Metab. Eng. 2 (4) (2000) 328–338.

[178] D. Morrone, et al., Increasing diterpene yield with a modular metabolic engineering system in *E. coli*: comparison of MEV and MEP isoprenoid precursor pathway engineering, Appl. Microbiol. Biotechnol. 85 (6) (2010) 1893–1906.

[179] K. Zhou, et al., Metabolite profiling identified methylerythritol cyclodiphosphate efflux as a limiting step in microbial isoprenoid production, PLoS One 7 (11) (2012) e47513.

[180] F.X. Cunningham Jr., T.P. Lafond, E. Gantt, Evidence of a role for LytB in the nonmevalonate pathway of isoprenoid biosynthesis, J. Bacteriol. 182 (20) (2000) 5841–5848.

[181] L.Z. Yuan, et al., Chromosomal promoter replacement of the isoprenoid pathway for enhancing carotenoid production in *E. coli*, Metab. Eng. 8 (1) (2006) 79–90.

[182] C.W. Wang, M.K. Oh, J.C. Liao, Engineered isoprenoid pathway enhances astaxanthin production in *Escherichia coli*, Biotechnol. Bioeng. 62 (2) (1999) 235–241.

[183] S. Kajiwara, et al., Expression of an exogenous isopentenyl diphosphate isomerase gene enhances isoprenoid biosynthesis in *Escherichia coli*, Biochem. J. 324 (Pt 2) (1997) 421–426.

[184] C. Ohto, et al., Prenyl alcohol production by expression of exogenous isopentenyl diphosphate isomerase and farnesyl diphosphate synthase genes in *Escherichia coli*, Biosci. Biotechnol. Biochem. 73 (1) (2009) 186–188.

[185] H. Zhang, et al., Microbial production of sabinene--a new terpene-based precursor of advanced biofuel, Microb. Cell Fact. 13 (2014) 20.

[186] S. Sarria, et al., Microbial synthesis of pinene, ACS Synth. Biol. 3 (7) (2014) 466–475.

[187] C. Willrodt, et al., Engineering the productivity of recombinant *Escherichia coli* for limonene formation from glycerol in minimal media, Biotechnol. J. 9 (8) (2014) 1000–1012.

[188] J. Zhou, et al., Geranyl diphosphate synthase: an important regulation point in balancing a recombinant monoterpene pathway in *Escherichia coli*, Enzyme. Microb. Technol. 68 (2015) 50–55.

[189] V.J. Martin, et al., Engineering a mevalonate pathway in *Escherichia coli* for production of terpenoids, Nat. Biotechnol. 21 (7) (2003) 796–802.

[190] S.H. Yoon, et al., Combinatorial expression of bacterial whole mevalonate pathway for the production of beta-carotene in *E. coli*, J. Biotechnol. 140 (3-4) (2009) 218–226.

[191] C. Wang, et al., Farnesol production from *Escherichia coli* by harnessing the exogenous mevalonate pathway, Biotechnol. Bioeng. 107 (3) (2010) 421–429.

[192] A. Banerjee, et al., Feedback inhibition of deoxy-D-xylulose-5-phosphate synthase regulates the methylerythritol 4-phosphate pathway, J. Biol. Chem. 288 (23) (2013) 16926–16936.

[193] D.J. Pitera, et al., Balancing a heterologous mevalonate pathway for improved isoprenoid production in *Escherichia coli*, Metab. Eng. 9 (2) (2007) 193–207.

[194] F. Zhu, et al., In vitro reconstitution of mevalonate pathway and targeted engineering of farnesene overproduction in *Escherichia coli*, Biotechnol. Bioeng. 111 (7) (2014) 1396–1405.

[195] H. Tsuruta, et al., High-level production of amorpha-4,11-diene, a precursor of the antimalarial agent artemisinin, in *Escherichia coli*, PLoS One 4 (2) (2009) e4489.

[196] S.M. Ma, et al., Optimization of a heterologous mevalonate pathway through the use of variant HMG-CoA reductases, Metab. Eng. 13 (5) (2011) 588–597.

[197] L.J. Weaver, et al., A kinetic-based approach to understanding heterologous mevalonate pathway function in *E. coli*, Biotechnol. Bioeng. 112 (1) (2015) 111–119.

[198] F.F. Nowroozi, et al., Metabolic pathway optimization using ribosome binding site variants and combinatorial gene assembly, Appl. Microbiol. Biotechnol. 98 (4) (2014) 1567–1581.

[199] J.E. Dueber, et al., Synthetic protein scaffolds provide modular control over metabolic flux, Nat. Biotechnol. 27 (8) (2009) 753–759.

[200] F.K. Bentley, A. Zurbriggen, A. Melis, Heterologous expression of the mevalonic acid pathway in cyanobacteria enhances endogenous carbon partitioning to isoprene, Mol. Plant 7 (1) (2014) 71–86.

[201] K. Kudoh, et al., Prerequisite for highly efficient isoprenoid production by cyanobacteria discovered through the over-expression of 1-deoxy-d-xylulose 5-phosphate synthase and carbon allocation analysis, J. Biosci. Bioeng. 118 (1) (2014) 20–28.

[202] H. Kiyota, et al., Engineering of cyanobacteria for the photosynthetic production of limonene from CO_2, J. Biotechnol. 185 (2014) 1–7.

[203] J. Beekwilder, et al., Valencene synthase from the heartwood of Nootka cypress (*Callitropsis nootkatensis*) for biotechnological production of valencene, Plant Biotechnol. J. 12 (2) (2014) 174–182.

[204] A.A. Song, et al., Engineering the lactococcal mevalonate pathway for increased sesquiterpene production, FEMS Microbiol. Lett. 355 (2) (2014) 177–184.

[205] M.K. Kang, et al., Biosynthesis of pinene from glucose using metabolically-engineered *Corynebacterium glutamicum*, Biotechnol. Lett. 36 (10) (2014) 2069–2077.

[206] J. Frohwitter, et al., Production of the sesquiterpene (+)-valencene by metabolically engineered *Corynebacterium glutamicum*, J. Biotechnol. 191 (2014) 205–213.

[207] R.M. Phelan, et al., Engineering terpene biosynthesis in *Streptomyces* for production of the advanced biofuel precursor bisabolene, ACS Synth. Biol. 4 (4) (2015) 393–399.

[208] H. Beuttler, et al., Biosynthesis of zeaxanthin in recombinant *Pseudomonas putida*, Appl. Microbiol. Biotechnol. 89 (4) (2011) 1137–1147.

[209] D. Xue, et al., Enhanced C30 carotenoid production in *Bacillus subtilis* by systematic overexpression of MEP pathway genes, Appl. Microbiol. Biotechnol. 99 (14) (2015) 5907–5915.

[210] F. Matthaus, et al., Production of lycopene in the non-carotenoid-producing yeast *Yarrowia lipolytica*, Appl. Environ. Microbiol. 80 (5) (2014) 1660–1669.

[211] T. Wriessnegger, et al., Production of the sesquiterpenoid (+)-nootkatone by metabolic engineering of *Pichia pastoris*, Metab. Eng. 24 (2014) 18–29.

[212] H. Shimada, et al., Increased carotenoid production by the food yeast *Candida utilis* through metabolic engineering of the isoprenoid pathway, Appl. Environ. Microbiol. 64 (7) (1998) 2676–2680.

[213] K.Y. Hara, et al., Development of a multi-gene expression system in *Xanthophyllomyces dendrorhous*, Microb. Cell Fact. 13 (2014) 175.

[214] J. Yang, et al., Metabolic engineering of *Escherichia coli* for the biosynthesis of alpha-pinene, Biotechnol. Biofuels 6 (1) (2013) 60.

[215] G. Bokinsky, et al., Synthesis of three advanced biofuels from ionic liquid-pretreated switchgrass using engineered *Escherichia coli*, Proc. Natl. Acad. Sci. U. S. A. 108 (50) (2011) 19949–19954.

[216] J.L. Foo, S.S. Leong, Directed evolution of an *E. coli* inner membrane transporter for improved efflux of biofuel molecules, Biotechnol. Biofuels 6 (1) (2013) 81.

[217] C. Scullin, et al., Optimization of renewable pinene production from the conversion of macroalgae *Saccharina latissima*, Bioresour. Technol. 184 (2015) 415–420.

[218] O.A. Carter, R.J. Peters, R. Croteau, Monoterpene biosynthesis pathway construction in *Escherichia coli*, Phytochemistry 64 (2) (2003) 425–433.

[219] J. Alonso-Gutierrez, et al., Metabolic engineering of *Escherichia coli* for limonene and perillyl alcohol production, Metab. Eng. 19 (2013) 33–41.

[220] F.K. Davies, et al., Engineering limonene and bisabolene production in wild type and a glycogen-deficient mutant of *Synechococcus* sp. PCC 7002, Front Bioeng. Biotechnol. 2 (2014) 21.

[221] C. Halfmann, L.P. Gu, R.B. Zhou, Engineering cyanobacteria for the production of a cyclic hydrocarbon fuel from CO_2 and H_2O, Green Chem. 16 (6) (2014) 3175–3185.

[222] J.B. Behrendorff, et al., 2,2-Diphenyl-1-picrylhydrazyl as a screening tool for recombinant monoterpene biosynthesis, Microb. Cell Fact. 12 (2013) 76.

[223] T.C. Brennan, et al., Alleviating monoterpene toxicity using a two-phase extractive fermentation for the bioproduction of jet fuel mixtures in *Saccharomyces cerevisiae*, Biotechnol. Bioeng. 109 (10) (2012) 2513–2522.

[224] T.C. Brennan, J.O. Kromer, L.K. Nielsen, Physiological and transcriptional responses of *Saccharomyces cerevisiae* to d-limonene show changes to the cell wall but not to the plasma membrane, Appl. Environ. Microbiol. 79 (12) (2013) 3590–3600.

[225] F.F. Hu, et al., Key cytomembrane ABC transporters of *Saccharomyces cerevisiae* fail to improve the tolerance to D-limonene, Biotechnol. Lett. 34 (8) (2012) 1505–1509.

[226] T.C. Brennan, et al., Evolutionary engineering improves tolerance for replacement jet fuels in *Saccharomyces cerevisiae*, Appl. Environ. Microbiol. 81 (10) (2015) 3316–3325.

[227] C. Formighieri, A. Melis, Carbon partitioning to the terpenoid biosynthetic pathway enables heterologous beta-phellandrene production in *Escherichia coli* cultures, Arch. Microbiol. 196 (12) (2014) 853–861.

[228] E.M. Kim, et al., Microbial synthesis of myrcene by metabolically engineered *Escherichia coli*, J. Agric. Food Chem. 63 (18) (2015) 4606–4612.

[229] C. Formighieri, A. Melis, Regulation of beta-phellandrene synthase gene expression, recombinant protein accumulation, and monoterpene hydrocarbons production in *Synechocystis* transformants, Planta 240 (2) (2014) 309–324.

[230] K. Scholtmeijer, et al., Production of (+)-valencene in the mushroom-forming fungus *S. commune*, Appl. Microbiol. Biotechnol. 98 (11) (2014) 5059–5068.

[231] P.P. Peralta-Yahya, et al., Identification and microbial production of a terpene-based advanced biofuel, Nat. Commun. 2 (2011) 483.

[232] R.P. McAndrew, et al., Structure of a three-domain sesquiterpene synthase: a prospective target for advanced biofuels production, Structure 19 (12) (2011) 1876–1884.

[233] B. Ozaydin, et al., Carotenoid-based phenotypic screen of the yeast deletion collection reveals new genes with roles in isoprenoid production, Metab. Eng. 15 (2013) 174–183.

[234] J. Kirby, et al., Use of nonionic surfactants for improvement of terpene production in *Saccharomyces cerevisiae*, Appl. Environ. Microbiol. 80 (21) (2014) 6685–6693.

[235] C. Halfmann, et al., Genetically engineering cyanobacteria to convert CO(2), water, and light into the long-chain hydrocarbon farnesene, Appl. Microbiol. Biotechnol. 98 (23) (2014) 9869–9877.

[236] J. Crock, M. Wildung, R. Croteau, Isolation and bacterial expression of a sesquiterpene synthase cDNA clone from peppermint (*Mentha* x *piperita*, L.) that produces the aphid alarm pheromone (*E*)-beta-farnesene, Proc. Natl. Acad. Sci. U. S. A. 94 (24) (1997) 12833–12838.

[237] S. Tippmann, et al., Production of farnesene and santalene by *Saccharomyces cerevisiae* using fed-batch cultivations with RQ-controlled feed, Biotechnol. Bioeng. 113 (1) (2016) 72–81.

[238] C. Wang, et al., Metabolic engineering of *Escherichia coli* for alpha-farnesene production, Metab. Eng. 13 (6) (2011) 648–655.

[239] C.M. Sandoval, et al., Use of pantothenate as a metabolic switch increases the genetic stability of farnesene producing *Saccharomyces cerevisiae*, Metab. Eng. 25 (2014) 215–226.

[240] R. Lauchli, et al., High-throughput screening for terpene-synthase-cyclization activity and directed evolution of a terpene synthase, Angew. Chem. Int. Ed. Engl. 52 (21) (2013) 5571–5574.

[241] M. Furubayashi, et al., A high-throughput colorimetric screening assay for terpene synthase activity based on substrate consumption, PLoS One 9 (3) (2014) e93317.

CHAPTER

8

Pyrolysis of Biomass for Aviation Fuel

R.W. Jenkins[1], *A.D. Sutton*[1] *and D.J. Robichaud*[2]

[1]Chemistry Division, Los Alamos National Laboratory, Los Alamos, NM, United States, [2]National Bioenergy Center, National Renewable Energy Laboratory, Golden, CO, United States

8.1 INTRODUCTION

The production of sustainable liquid fuel from biomass is one of the major scientific challenges of the 21st century. While suitable alternative technologies to combustion engines are being developed and commercialized for road transport vehicles (electric cars and hydrogen fuel cells), the energy density required for aviation transport means a liquid fuel-powered internal combustion engine is the only viable technology for the foreseeable future. There are numerous technologies currently being investigated to convert renewable biomass feedstocks into liquid mixtures suitable for aviation [1], however, the stringent international standards for jet fuel lead to a relatively narrow range of molecules [2,3]. The minimum specific energy density of 42.80 MJ/kg limits the presence of oxygen in a suitable jet fuel, the maximum freezing point of $-47°C$ eliminates long-chain linear alkanes ($C_{10}+$) and further limits oxygen due to its contribution to intermolecular forces, and minimum flash point of 38°C excludes short-chain linear alkanes (up to C_9) and other volatile hydrocarbons. While these issues can be ameliorated by the blending of numerous components or the use of additives, the only unimolecular species that are suitable for jet fuel applications are unsaturated, branched, or cyclic hydrocarbons in the $C_{10}-C_{16}$ region (eg, kerosene).

A significant area of research that concentrates on producing liquid fuels from biomass is thermochemical conversion – that is, the breaking down of the large biological polymers present in biomass via treatment with high pressures and temperatures. This is an attractive technology, as it has the potential to produce liquid fractions directly from biomass in high yields. The term given to the thermochemical treatment, as well as the product yields and distribution, depends on the operating conditions (ie, temperature, pressure, residence time, presence of water), and while these are numerous (each of which will be discussed in later chapters), this section will concentration on the subject of pyrolysis – that is, the thermal cracking of biomass in the absence of oxygen. It is similar to a method that has been used for hundreds of years to produce charcoal [4], but also produces gaseous and

Biofuels for Aviation.
DOI: http://dx.doi.org/10.1016/B978-0-12-804568-8.00008-1

TABLE 8.1 Summary of Different Pyrolysis Technologies, Outlining Operating Conditions and Approximate Product Yields

Conversion Technology	Operating Conditions	Product Yields
Slow pyrolysis	Temperature: 300–700°C	Bio-oil: ~30 wt.%
	Vapour residence time: 10–100 min	Biochar: ~35 wt.%
	Heating rate: 0.1–1°C/s	Gases: ~35 wt.%
	Feedstock size: 5–50 mm	
Fast pyrolysis	Temperature: 400–800°C	Bio-oil: ~50 wt.%
	Vapour residence time: 0.5–5 s	Biochar: ~20 wt.%
	Heating rate: 10–200°C/s	Gases: ~30 wt.%
	Feedstock size: <3 mm	
Flash pyrolysis	Temperature: 800–1000°C	Bio-oil: ~75 wt.%
	Vapour residence time: <0.5 s	Biochar: ~12 wt.%
	Heating rate: >1000°C/s	Gases: ~13 wt.%
	Feedstock size: <0.2 mm	

Adapted from Nanda, S., et al., Pathways of lignocellulosic biomass conversion to renewable fuels. Biomass Convers. Biorefin. 4 (2) (2014) 157–191.

liquid fractions. Depending on the reaction conditions, the ratio of these fractions alters drastically (Table 8.1). Slower, low-temperature pyrolysis (400°C, >10 min residence time) promotes a larger solid fraction – termed 'biochar' – whereas faster, high-temperature pyrolysis ('flash pyrolysis') (800°C, <0.5 s) gives liquid fractions of up to 75% [5]. This chapter focuses mostly on the fast pyrolysis regime, which can produce liquid yields up to 65%. Fast pyrolysis is generally the preferred method for a liquid product in part due to scalability of the method and economics of particle grinding. The liquid fraction is termed 'bio-oil' and, though not suitable as a fuel itself, can be upgraded and refined to high-quality fuels and chemicals.

8.2 PYROLYSIS FEEDSTOCKS

As well as the processing conditions, the resultant range of products will depend on the original biomass composition. The vast majority (approximately 90 wt.%) of dry biomass can generally be classified into three main groups – cellulose, hemicellulose, and lignin (collectively termed 'lignocellulose') – though the ratios between these three groups is highly dependent on the feedstock source (Table 8.2). Typically, biomass contains between 30–60% cellulose, 20–40% hemicellulose, and 15–25% lignin. The remaining ca. 10 wt.% is comprised of inorganic minerals (ie, ash content) and organic extractives such as lipids (which include terpenes, triglycerides, and steroids), simple sugars, and starches [6]. As lignocellulose is the cheapest and most abundant form of terrestrial biomass, the majority of pyrolysis literature is concerned with its conversion [7], and is the feedstock that will be considered and discussed in this chapter. However, other feedstock sources for pyrolysis have been studied, including triglycerides [8], algal biomass [9], and chitin/chitosan [10].

Cellulose and hemicellulose are both large polymers of sugar monomers (ie, polysaccharides) of differing structure. Cellulose is strictly a linear polymer of C_6 D-glucose units linked

TABLE 8.2 Biopolymer, Elemental, and Physical Composition of Common Lignocellulosic Feedstocks for Pyrolysis Applications [13–21]

Biomass Source	Biopolymer Composition (wt.%)			Elemental Composition (wt.%)					Physical Composition (wt.%)			
	Cellulose	Hemicellulose	Lignin	Carbon	Hydrogen	Oxygen	Nitrogen	Sulphur	Moisture Content	Ash	Volatile Matter	Fixed Carbon
Wood	35–50	20–30	25–30	51.6	6.3	41.5	0.1	1	20	0.4–1	82	17
Pine sawdust	35	29	28	51.6	5.2	43.2			10.2	0.3		
Switch grass	30–50	10–40	5–20	44.8	5.8	49.1	0.3		13–15	4.5–5.8		
Wheat straw	33–40	20–25	15–20	48.5	5.5	3.9	0.3		16	4	59	21
Wheat shell	10–15	30	4–8	46.8	6.0	43.8	3.4	<0.1	18.8	5.4		
Sugarcane bagasse	19–24	32–48	23–32	49.6	6.0	43.8	0.5			1.6	82.1	16.3
Miscanthus	24	44	17						11.5	1.5–4.5	66.8	15.9
Corncob	50.5	31	15	49.3	5.4	44.7	0.6		7.4	1.5	79.6	11.6
Pistachio shell	60.6		12.8	50.0	6.1	42.3	1.6		7.4	1.3	76.9	14.3
Rice straw	32.1	24	18						6	4.3	79	10.7
Rice husk	36.8	20.6	14.9	10.9	4.8	53.8	0.4		8–10	13.6	62.4	14.6
Barley straw	31–34	24–29	14–15						30	6	46	18
Oregenum stalk	33.8	9.3	10.9	42.5	6.0		0.7	0.3	6.3	2.1		
Cottonseed			14.29	55.4	8.8	31.4	4.4		4.6	3.3	79.5	12.6
Tobacco residue		44.3	9.2	52.7	6.2	40.1	2.1		7.7	11.2	67.7	13.4
Sweet sorghum bagasse	34.2	24.3	6.5	45.7	5.8	48.2	0.3		4.2	3.5	76.3	16.0
Paper	85–99	0–15	0									
Industrial waste paper	60–70	10–20	5–10									

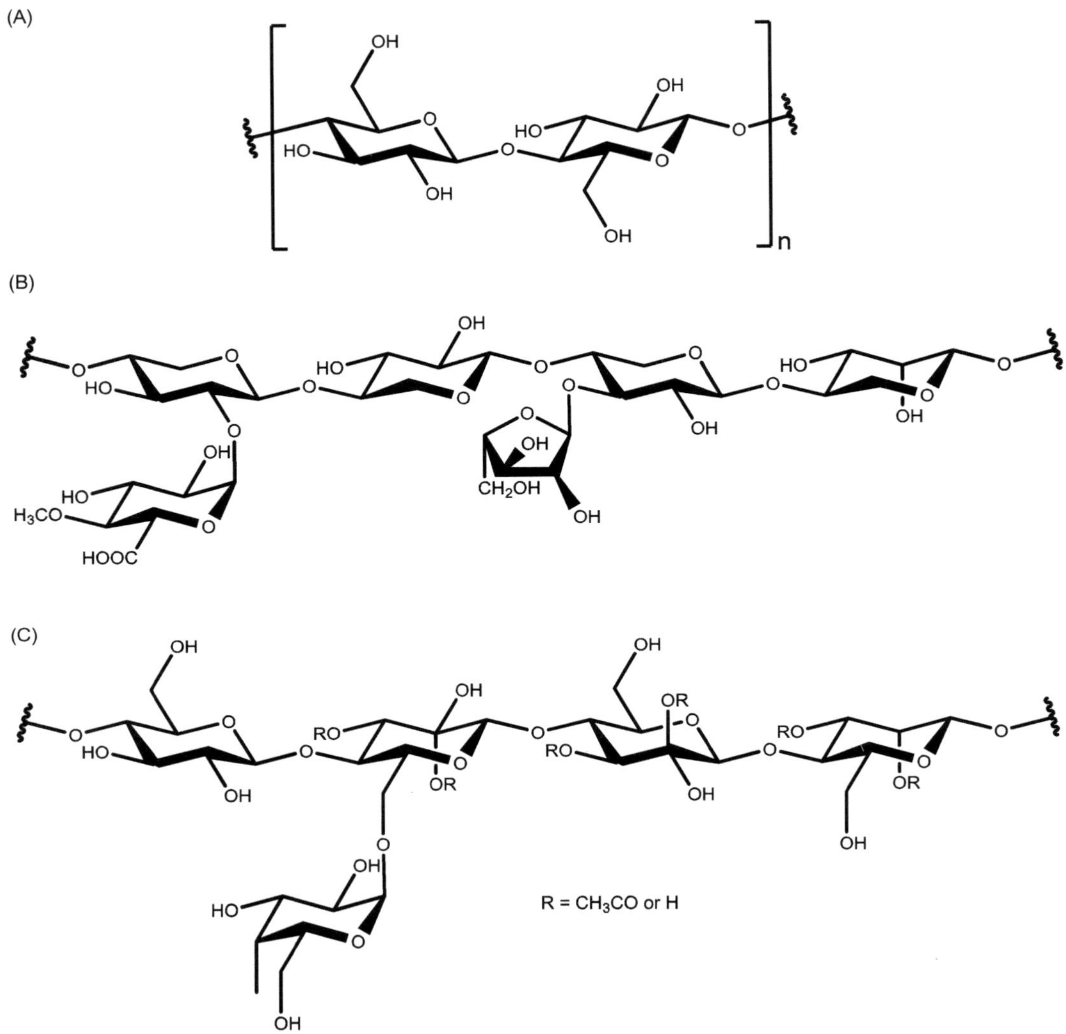

FIGURE 8.1 The structure of polysaccharides: (A) Cellulose; (B) Xylan (example of hemicellulose); (C) Glucomannan (example of hemicellulose).

via glyosidic bonds (β [1→4]) of a molecular weight of approximately 100,000 daltons, which provides structure via intermolecular H-bonds and van der Waals forces between the chains. Due to its strong structure it is insoluble in the majority of solvents. Hemicellulose, however, is a branched chain of an approximate molecular weight of 30,000 daltons, and is comprised of many different sugar monomers, including pentoses (C_5, ie, xylose) and hexoses (C_6, ie, glucose), as well as sugar acids such as methylgluconorinic and galacturonic acids. As such, it has an amorphous structure (Fig. 8.1) and, unlike cellulose, is soluble in dilute alkali solutions [11,12].

Lignin is a large aromatic polymer of an approximate molecular weight of 20,000 daltons that is varied in its molecular structure

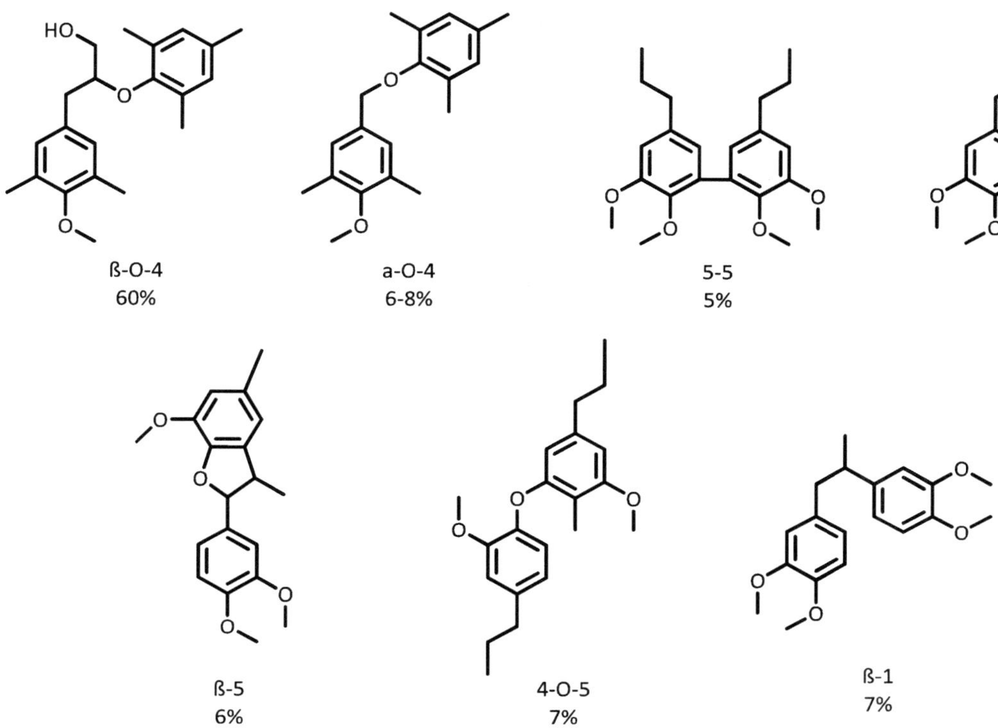

FIGURE 8.2 Common linkages found in lignin structure, with typical composition found in hardwood [25].

and connectivity. Lignin provides structure to plants, and a physical barrier that protects degradation of the cellulosic material present in biomass. As such, it is a robust biopolymer that resists depolymerization and defunctionalization. Its variance of structure also leads to difficult selective transformation. However, lignin primarily consists of interlinked *p*-hydroxyphenylpropane monomeric units, or 'monolignols'. The three main monolignols are *p*-coumaryl alcohol, coniferyl alcohol, and sinapyl alcohol, and the proportion of these monolignols changes depending on the source – for example, hardwoods contain a higher content of coniferyl and sinapyl alcohol units and thus they have a higher methoxy content than grasses, which are almost completely comprised of coniferyl alcohol units, or 'G'-type lignin [22]. The linkages between monolignol units include C—C (β-5, 5-5, β-1, β-β) and C—O bonds (β-O-4, 4-O-5, α-O-4) (Fig. 8.2), however, the dominant linkages are β-O-4 aryl ethers (approximately 46% in softwood and 60% in hardwood) [23], which are the most susceptible to cleavage. As well as these linkages, the alkyl chains between the aromatic rings also contain other oxygenated groups such as alcohols, carbonyls, and carboxylic acid moieties [24].

8.3 BIO-OIL COMPOSITION AND PROPERTIES

The liquid fraction obtained from biomass fast pyrolysis is produced via short residence times at moderate temperatures (400–600°C). The resulting vapour from this reaction is rapidly quenched and condensed to yield the

TABLE 8.3 Typical Properties and Composition of Pyrolysis Bio-oil and Crude Oil

Property		Bio-oil[a]	Crude Oil
Energy density (MJ/kg)		13–18	44
Water (wt.%)		15–30	0.1
pH		2.0–3.7	–
Solids (wt.%)		0.01–1	1
Density (kg/m)		1.1–1.3	0.86–0.94
Kinematic viscosity at 50°C (mm^2/s)		13–80	180
Ash content (wt.%)		0.004–0.3	0.1
	Carbon (wt.%)	55–65	83–86
	Hydrogen (wt.%)	5–7	11–14
	Nitrogen (wt.%)	<0.4	<1
Elemental Composition	Sulphur (wt.%)	<0.05	<4
	Oxygen (wt.%)	28–40	<1
	Na + K (ppm)	5–500	–
	Ca (ppm)	4–600	–
H/C		0.9–1.5	1.5–2.0
C/O		2–3.3	–

[a]*Includes properties of approximately 150 different bio-oil samples generated from different pyrolysis systems and different feedstocks [7, 27, 30].*

pyrolysis bio-oil. The composition of bio-oil is a complex mixture of over 400 identified species [26], 99.7% of which is composed of carbon, hydrogen, and oxygen, and includes acids, alcohols, aldehydes, esters, ketones, sugars, phenols, guaiacols, syringols, furans, phenols, and extractible terpenes with multifunctional groups [27,28]. Pure hydrocarbons are rarely observed. The high proportion of oxygen present (28–40%) in the form of oxygenated hydrocarbons and the presence of alkali components (eg, Na or K) leads to a variety of undesirable properties. In comparison to crude oil (Table 8.3), bio-oil possesses a low energy density, is corrosive due to the presence of organic acids, and is viscous due to the increased intermolecular forces. Bio-oil has also been shown to have poor stability, as the rapid quenching of the pyrolysis vapour produces a liquid that is not in thermodynamic equilibrium. Studies have shown that bio-oil tends towards thermodynamic equilibrium (via degradation and repolymerization) in storage conditions, with increases in viscosity, molecular weight, and even complete phase separation observed after 6 months of ambient storage [29]. Bio-oil also contains a significant amount of water when compared to crude oil (15–30% compared to 0.1%), derived both from the dehydration reactions that occur during biomass pyrolysis (and subsequent storage), and from the original moisture present in the lignocellulosic feedstock. The presence of water in bio-oil, and the difficulty in removing it, presents a number of problems, including: corrosiveness, cold filter plugging point (the temperature at which flow is impaired to the point where an engine is rendered inoperable), and catalyst deactivation.

Bio-oil, therefore, cannot be used as a fuel directly, and the production of stable fuel-like molecules of desirable physical properties is imperative. While a number of physical methods have been researched to improve the physical properties and stability of bio-oil (ie, hot gas filtering, solvent addition, emulsification [31]), the reactive oxygenated groups present in bio-oil remain. The chemical upgrading of bio-oil via deoxygenation, therefore, is required, though unfortunately has proved to be challenging. While there is sparse literature in the specific area of pyrolysis for jet fuel production, the modern literature in this area focuses on the increased yield, control, and selectivity of conversion, the ultimate aim of which is the ability to produce specific fuel formulations [32]. The cost of upgrading, however, must be taken into account. While the presence of oxygen leads to undesirable properties, complete or near-complete removal of oxygen (<0.5%) would increase the cost of production above what is viable [33,34]. Less extreme hydrotreatment has been proposed (to approximately 7 wt.% oxygen), which could provide a feasible balance between the cost of production and desirable fuel properties [35].

8.4 FAST PYROLYSIS OF BIOMASS

To produce fuel molecules via the upgrading of pyrolysis bio-oil, one must first understand the fundamental chemistry of the reactions taking place during pyrolysis, and have knowledge of the final products. The decomposition of lignocellulose (and its three main components) via thermogravimetric analysis has been considered and studied, and it is generally accepted that the three main components degrade separately, which reflects their relative stability. In a study by Yang et al. [36], hemicellulose and cellulose were found to decompose between 220°C and 315°C and between 315°C and 400°C, respectively, while lignin did not exhibit a particular decomposition event. Rather, its decomposition occurred over a wide temperature range (160–900°C), and generated significant solid residue. The high solid residue production by the degradation of lignin, however, is unsurprising due to its high aromatic content. Under pyrolysis conditions, aromatic species will fuse and oligomerize, eventually adding to the 'char' yield.

8.4.1 Mechanisms of Action

During pyrolysis, a wide range of reactions occur [37]. The initial reactions (called primary reactions [38]), concern the initial breaking of the bonds within the biopolymers present in biomass and the subsequent stabilizing rearrangements that take place. These include depolymerization, the cleavage of bonds between the monomer units of the biopolymers eventually leading to the formation of condensable oligomers; [39] fragmentation, where the covalent bonds present in the polymers (and in some cases the monomers) are broken to produce a range of condensable low-molecular-weight organic compounds and incondensable gas [40]; and char formation, caused by the formation and fusing of aromatic rings to form graphitic species [41]. The following reactions (called secondary reactions [42]) are due to the unstable volatile species produced in the primary reactions, which include cracking, whereby the bonds of the volatile species are broken and produce low-molecular-weight organic compounds (similar to fragmentation); or by recondensation, where the volatile species combine to form a larger, nonvolatile, stable molecule. It should be noted that recondensation can lead to the formation of secondary char [42]. The occurrence of these reactions is dependent on the conditions of the pyrolysis. While depolymerization can occur between temperatures of between 250°C and500°C, few fragmentation reactions occur within this range. Therefore, maximum bio-oil yields are typically achieved between 450°C and 550°C [43]. Above this temperature, more fragmentation and volatile species cracking reactions occur, producing more incondensable gases [38]. For all biomass constituents, at higher temperatures and long residence times, more recondensation reactions occur leading to the formation of more polyaromatic species and charring [42].

The pyrolytic degradation of cellulose starts with its depolymerization, whereby the glycosidic bond becomes unstable and breaks. However, with no water to carry out the hydrolysis, anhydro-oligosaccharides or anhydro-saccharides (depending on the number of monomer units) are formed [44]. The most common anhydro-saccharide – levoglucosan – is formed when the glycosidic bond breaks, and the C_1 carbon of the sugar monomer forms a bond with the hydroxyl group attached to C_6, making a new ring which contains an ether bond. Levoglucosan can actually reach yields of up to 60% [45], and is relatively stable – it has been found to not break down when pyrolysed alone [46]. Furan species also form via the contraction of the cyclic saccharides, specifically the breaking of the C_1–O bond and recyclization at the C_2, followed by elimination of oxygenated groups to form the carbon-carbon double bonds [47]. These contractions can be accompanied by glycosidic bond cleavage, leading ultimately to the production of furans such as 5-hydroxymethylfuran (5-HMF). This can further degrade, yielding other furans such as 5-methylfuran (5-MF), furfuryl alcohol, and furfural [48]. Depolymerization reactions produce unstable species which include carbonyl and carboxylic acid moieties, and can undergo dehydration and/or fragmentation reactions to produce volatile species such as CO, CO_2, hydroxyacetaldehyde, formic acid, acetic acid, and aldehyde compounds [49].

Due to the varied structure of the hemicellulose, certain moieties of lower stability within its structure are removed and/or broken prior to depolymerization. Acetyl moieties (found in up to 10 wt.% in hemicelluloses) fragment to produce acetic acid, carboxylic acid groups break off to form formic acid, and methoxy groups from monomers such as 4-*O*-methyl-α-D-glucuronic acid cleave to form methanol [50]. Upon depolymerization, a wide range of anhydro-saccharides are produced, reflecting the variety of saccharide monomers present in hemicellulose. As with cellulose, the pyran rings can rearrange to form furan rings and ultimately form 5-HMF, 5-MF, and furfural [51]. The depolymerization of hemicellulose occurs much faster than cellulose, and thus a significant amount of CO, CO_2, and H_2 is produced. As with cellulose, similar small organic volatiles are produced from the various depolymerization reactions, with the inclusion of 1-hydroxy-butanone (from xylan) [37]. Hemicellulose produces significantly more char than cellulose. This is likely due to its lower relative stability due to the lack of crystallinity, and of the lower stability of the monomer units, which produces more reactive volatile species and a greater proportion of secondary reactions [52].

Lignin is also varied in structure and so produces a number of different products during the primary pyrolysis reactions. Following the dehydration of the hydroxyl groups present, the ether linkages (α-O-4, β-O-4) break down, releasing phenolic compounds (such as guaiacol, eugenol, or dimethoxyphenol), as well as small volatiles such as CO, CO_2, and H_2O from the stabilizing rearrangement reactions [53,54]. The alkyl chains between the monomers can become unstable and react, producing short-chain organic compounds such as methane, acetic acid, or acetaldehyde [55]. The methoxy groups present throughout lignin undergo homolytic cleavage, generating either methane or methanol, and leaving either an –OH or –H behind [24]. At pyrolysis conditions (ie, above 450°C), most of the lignin linkages break, though the more stable (ie, 5-5, 4-O-5) remain present and require higher temperatures to be broken [56]. Being an aromatic polymer, lignin produces more char than the pyrolysis of hemicellulose or cellulose, and has been observed during reactions above 500°C, via the significant evolution of H_2 at these conditions [36]. If a process by which the aromatic production could be reduced, the resultant solid char formation would decrease, which would therefore increase the ultimate bio-oil yield. While this is a significant challenge (as typical biomass contains up to 30 wt.% aromatic species in the form of lignin), research into pyrolysis optimization, as well as the catalysis and process of upgrading pyrolysis bio-oil to maximize yield, has improved properties and reduced the molecular range.

8.4.2 Fast Pyrolysis Review

The pyrolysis of biomass to produce bio-oil has been extensively reviewed in recent years [8,57,58]. Below is an updated, extensive review of the research that has been carried out to produce bio-oil, a summary of which can be found in Table 8.4.

The perennial grass, miscanthus, was pyrolysed using a fluidized bed pyrolysis reactor by Kim et al. [59]. Investigating a range of temperatures (350–500°C) and residence times (1.2–3.8 s), the maximum bio-oil yield was found to be 58.9% at 350°C and a residence time of 1.9 s. Analysing the physical properties and chemical composition of the bio-oil extensively, it was found that the major constituents of the bio-oil were 1-(acetyloxy)-2-butanone, furfural, dihydromethylfuranone, and levoglucosan (from cellulose and hemicellulose); and guaiacol, 4-vinylphenol, and syringol (from lignin).

The pyrolysis of stem wood using a pilot-scale cycle reactor was assessed by Wiinikka et al. at 750°C and 20 kg/h biomass feed, reaching a bio-oil yield of 54.6% [60]. The physicochemical properties of the oil were measured. A different wood source (rubber wood) sawdust was pyrolysed using a fixed-bed reactor by Mazlan et al. with a maximum oil yield of 33 wt.% reached at 550°C. Eucalyptus waste was pyrolysed by Amutio et al. using a conical spouted reactor. The maximum bio-oil yield of 75.4% was reached at 500°C, with a biomass feed of 0.12 kg/h. Phenols and ketones were found to be the main organic compounds in the resulting bio-oil, with a concentration of 26 and 10 wt.%, respectively.

Maize stalk was pyrolysed and studied by Zheng et al. using a fluidized bed reactor [61]. Studying a range of temperatures (420–580°C), the highest bio-oil yield (66%) was reached at 500°C. The bio-oil contained mostly water, ketones, acids, phenolics, and furans. A different kind of stalk – cotton stalk – was pyrolysed using a fluidized bed reactor by Ali et al. [62]. The effects of temperature and particle size on the bio-oil yield were investigated. At optimum conditions – 490°C and a feed size of 1 mm – the bio-oil yield reached 36%. The major components of the bio-oil were found to be phenols, carboxylic acids, ketones, aldehydes, furans, and sugars. Cotton stalk pyrolysis was also carried out by Zheng et al. using a fluidized bed reactor at a temperature range of 480–530°C [63]. The maximum bio-oil yield of 55% was reached at 510°C, with the major constituents of the bio-oil being acids, aromatics, and water. Cotton seed pyrolysis was investigated by Apaydin-Varol et al. at a range of temperatures (400–700°C), heating rates (5–700°C/min), and carrier gas flow rates (100–800 mL/min) [15]. The optimum

TABLE 8.4 Various Fast Pyrolysis Experiments (Without Catalyst), Showing Experimental Conditions and Resulting Bio-oil Yield, Properties, and Composition

	Experimental Conditions				Bio-oil Properties								
									Elemental Composition (%)				
Biomass Feedstock	Reactor Type	Temperature (°C)	Carrier Gas Rate (L/min)	Bio-oil Yield (%)	Density (kg/m)	Kinematic Viscosity (Temperature) (mm^2/s)	Water Content (%)	Energy Density (MJ/kg)	Carbon	Hydrogen	Nitrogen	Oxygen	References
Chestnut capulae	Fixed-bed	600	0.1	52.5				26.16	66.4	6	1.2	30.2	[73]
Corncob	Fluidized bed	550	1.5	54.6	1.13			19.92	52.4	6.8	0.74	40.1	[64]
Corncob	Pilot-scale fluidized bed	500		61	>1.0			26.2	55.1	7.6	0.6	37	[65]
Cotton seed	Tubular	500	0.2	49.5				39.49	72.9	11.5	9.7	5.9	[15]
Cotton stalks	Fluidized bed	510		55	1.16	125 (40°C)	24.4	17.8	42.3	7.9	0.3	49.4	[63]
Eucalyptus waste	Conical spouted	500	10	75.4			35.7						[76]
Jute stick	Fluidized bed	500	0.2	66.7	1.11	2.34 (30°C)	12–17	18.25					[70]
Maize stalk	Fluidized bed	500		66	1.22	138 (40°C)		19.6	44.3	6.3	0.6	47.5	[61]
Miscanthus	Fluidized bed	350		58.9		16.5	18	16.6	40.8	7	0.6	51.7	[59]
Palm kernel shell	Fluidized bed	550	2	56	1.04								[69]
Rapeseed	Tubular	550	0.1	73				39.4	73.1	11.5	5.9	5.6	[77]
Rice husk	Fluidized bed	465		56	1.19	128 (40°C)	25.2	17.42	41.7	7.7	0.3	50.3	[66]
Rice husk	Fixed-bed	550	0.15	49.9	1.18		15.3	24.7	49.2	11.7	2.4	35.6	[67]
Rice husk	Conical Spouted bed	450	1	70			22.8						[68]
Spent coffee	Micro-tubular	550	0.5	63.4									[74]
Stem wood	Pilot-scale cyclone	750	750	54.6				17.5	46.6	7.3	0.1	45.8	[60]
Sugarcane bagasse	Fixed-bed	500	0.2	66.1	1.13	3.9 (20°C)	15	17.25					[71]
Sugarcane bagasse	Fluidized bed	500	50	72.9									[72]
Sweet sorghum	Fluidized bed	500	69	43.5			31.4	18.57	47.8	6.38	0.69	45.08	[18]

conditions for bio-oil production were found to be 500°C, 200 mL/min carrier gas flow and a heating rate of 300°C/min, where the bio-oil yield reached 49.5%. The bio-oil possessed a high energy density (39.5 MJ/kg) compared to what is usually found to pyrolyse bio-oils (<25 MJ/kg), likely due to the major presence of fatty acids in the bio-oil from the lipids in cotton seeds (23.8% hexandecanoic acids, 39.5% octadecadienoic acid), which have a high C/O ratio and, thus, a high energy density themselves.

The effects of the pyrolysis product yields and chemical composition of the bio-oil obtained from corncob using a fluidized bed reactor were investigated by Zhang et al. [64]. While the investigation does use cracking catalysts, experiments were carried out without the catalysts for comparison and reached a 54.6% yield at 550°C, 1.5 L/min carrier gas flow, and 1.0–2.0 mm particle size. The major products formed were found to be acids, phenols, and furans. Mullen et al. also investigated the pyrolysis of corncob using a pilot-scale fluidized-bed reactor, achieving bio-oil yields of 61% at 500°C [65].

The fast pyrolysis of rice husk has been carried out by Ji-Lu [66], Zhou et al. [67], and Alvarez et al. [68]. Using a fluidized bed reactor, Ji-Lu investigated a range of temperatures (420–540°C), finding the maximum bio-oil yield of 56% to be reached at 465°C. The major products found in the bio-oil were acids, aromatics, and water. Zhou used a fixed-bed reactor at 550°C, reaching a maximum bio-oil yield of 49.9°C with a heating rate of 25°C/min and a nitrogen flow rate of 150 mL/min. The main products in the bio-oil were found to be phenolic species and long-chain aliphatics such as 1-octadecene and docosane. Alvarez et al. carried out investigations into the fast pyrolysis of the rice husk using a conical spouted bed reactor, with continuous removal of char [68]. Considering a range of temperatures, the maximum bio-oil yield of 70 wt.% was reached at 450°C. While the physical properties of the bio-oil were not assessed, the chemical compositions of the products were assessed. Increasing temperature was found to increase the amount of phenolics and furans, while reducing the concentration of acid and ketonic species.

Asadullah et al. investigated the pyrolysis of palm kernel shells using a fluidized bed reactor, reacting 57% bio-oil yield at 550°C with a carrier gas flow of 2 L/min [69]. The energy density of the bio-oil achieved was 23.48 MJ/kg, with phenol being the major product formed. Asadullah et al. also studied the pyrolysis of jute stick in a fluidized bed reactor [70], investigating temperatures between 300–600°C, reaching the maximum bio-oil yield of 66.7% at 500°C, and the pyrolysis of sugar bagasse in a fixed-bed reactor between 300–600°C, again reaching the maximum yield of 66.1% at 500°C [71]. Montoya et al. also pyrolysed sugarcane bagasse in a fluidized bed reactor, investigating the effect of temperature, biomass feed rate, gas flow, and particle on the global product distribution [72]. Maximum bio-oil yield of 72.9% was reached at 500°C, with a carrier gas flow of 50 L/min, and a biomass feed rate of 2 kg/h. Heating rate was found to have the largest effect on yield, where increasing the heating rate increased the yield. The composition and physical properties were not assessed. The pyrolysis of another form of bagasse – sweet sorghum – was carried out by Yin et al. using a fluidized bed reactor at 500°C and 60 L/min carrier gas flow [18]. Using these conditions, a total bio-oil yield of 43.5% was reached, which was the combined yield of four fractional condensers at different temperatures and an electrostatic precipitator. The major products were found to be mostly phenolic compounds.

Rapeseed pyrolysis was assessed by Onay et al. using a tubular reactor. Investigating the effects of temperature (400–700°C), particle size, and carrier gas flow, the maximum bio-oil yield of 73% was reached at 550°C, 0.6–1.25 mm particle size, and 100 mL/min. It was found by ^{1}H NMR that 85% of the hydrogens present in the bio-oil were saturated aliphatic hydrogens. This is supported by the high energy density of the bio-oil obtained (39.45 MJ/kg), due to the presence of lipids in the rapeseed which will likely be present in the bio-oil as fatty acids.

Kar et al. studied the pyrolysis of chestnut cupulae within a fixed-bed reactor, and investigated the effect of temperature, particle size, and sweeping gas (nitrogen) flow rate on the product yields [73]. The maximum bio-oil yield of 52.50 wt.% was obtained at 600°C and 100 mL/min sweeping gas rate. The ideal particle size for maximum liquid yield was found to be below 0.25 mm, while changing sweeping gas rate had little effect. Spent coffee grounds were pyrolysed by Ngo et al. [74] using a micro-tubular reactor. Investigating the effects of temperature and sweeping gas (N_2) rate, the maximum bio-oil yield was found to be 63.4% at 550°C and 500 mL/min. The bio-oil was found to contain mostly phenol derivatives, aldehydes, ketones, acids, and alcohols. An interesting component of the bio-oil was a significant amount of fatty acids (due to the high lipid content of waste coffee [75]), which would contribute to the stability of the bio-oil.

There are a number of technologies and feedstocks (of varied composition) that can produce bio-oil successfully in significant yields. The variety, however, leads to a number of different products and a difference in concentration of each product. Selectivity, therefore, is difficult. While some feedstocks produce bio-oils of high energy density, those feedstocks are usually high in triglycerides (such as seeds) and therefore higher in C/O ratio than lignocellulose on average. These also contain significant amounts of nitrogen, due to the higher presence of nitrogen-containing compounds such as protein, which would need to be removed from a final fuel formulation due to the negative effects of NO_x produced from nitrogen-containing fuels [75]. Regardless of the ultimate composition of a bio-oil, however, the quality is not conducive to fuel applications without significant upgrading.

8.4.3 Fast Pyrolysis Bio-oil Upgrading

As has been established, the bio-oil obtained from fast/flash pyrolysis of biomass must be upgraded to improve its properties to what is acceptable for fuel applications, reduce the number of molecules obtained, and allow for directed and selective fuel formulation production. In many biorefinery processes, the bio-oil is condensed to its liquid form prior to upgrading (Fig. 8.3). Raw bio-oil suffers from a variety

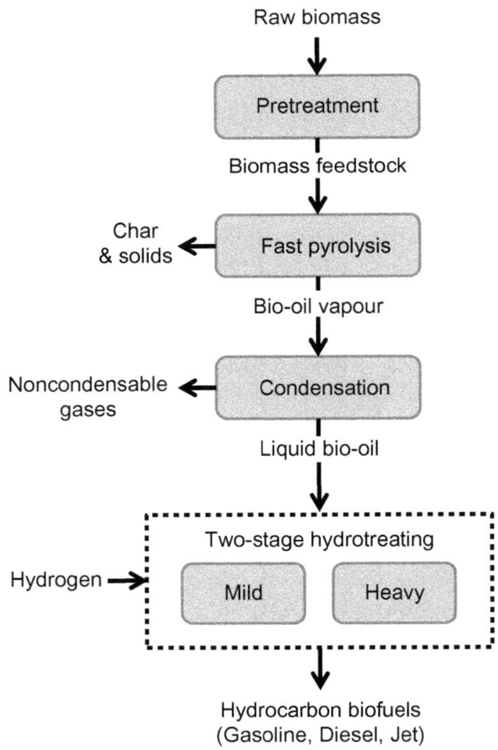

FIGURE 8.3 Process flow diagram for biomass fast pyrolysis and subsequent upgrading.

of issues. Organic acids derived from degradation of hemicelluloses lead to high acidity [78], carbonyl compounds result in instability and cause polymerization [79,80], and the high polarity makes bio-oil insoluble in hydrocarbons (eg, petroleum products) [81]. Therefore, these bio-oils must be upgraded (traditionally through hydrotreating) to be suitable for use as a fuel [82–84]. However, catalytic upgrading of bio-oils can suffer from tar plugging and catalyst fouling due to significant coking. As a result of this thermal instability, hydrotreating of bio-oils is usually performed in two-stages in order to deal with the complexity of the oil (hundreds of compounds) and thermal reactive nature of the oil [82].

8.5 CATALYTIC FAST PYROLYSIS OF BIOMASS

One promising process for the production of more stable bio-oils is catalytic fast pyrolysis, whereby the pyrolysis oil is upgraded in the vapour phase before being condensed into liquid, using a catalyst. This comes in two forms: in situ, whereby the deoxygenation catalyst is within the pyrolysis reactor itself and the pyrolysis vapour is deoxygenated as it is produced; and ex situ, whereby the catalytic upgrading is performed in a separate reactor and the pyrolysis vapour produced is separated from the char, ash, and other solids produced before being passed through the catalyst (Fig. 8.4) [85]. In both processes, upgrading can be performed in the presence of hydrogen or an inert carrier gas. Depending on the conditions and the catalyst, the mechanisms of action and therefore products of catalytic fast pyrolysis are different. However, the aim is always to reduce the oxygen, produce hydrocarbons to improve resultant fuel properties, and reduce coking to increase the bio-oil yield. The chemistry of catalytic fast pyrolysis includes hydrotreatment, cracking, ketonization, aldol condensation, and aromatization.

In hydrotreatment, the pyrolysis vapour is exposed to hydrogen (and catalyst) and can undergo a range of reactions, including hydrogenation (whereby hydrogen is incorporated into a molecule's structure without breaking a bond, that is, C–C double bond to C–C single bond, ketone to alcohol, or imine to amine), hydrogenolysis (whereby a bond is cleaved by hydrogen), hydrodeoxygenation (whereby oxygen is removed in the form of water), decarbonylation (whereby oxygen is removed in the form of carbon monoxide [and water if originated from a carboxylic acid]), and decarboxylation (whereby the oxygen is removed in the form of carbon dioxide). While increasing the hydrogen content of the bio-oil improves the majority of the physical properties, complete saturation is not required – an oxygen content of below 5 wt.% will likely be necessary for aviation biofuel applications [86]. Hydrodeoxygenation can minimize the hydrogen consumption, as phenol and its derivatives are the least active for this, and do not reach full saturation in this process [87]. Conventional hydrodeoxygenation of bio-oil is carried out in liquid phase between 250°C and 450°C at 75–300 bar, with sulphided CoMo or NiMo catalysts, due to their use as hydrodesulphurization catalysts in the fossil-fuel industry [88]. However, these catalysts can leach sulphur into the bio-oil, which is not ideal due to the need for low-sulphur fuels (as sulphur in fuels can generate SO_x upon combustion, leading to acid rain formation), and as sulphur is needed to maintain the activity of the catalyst.

Cracking produces aromatics and olefins by a variety of pathways, including C–C bond cleavage, hydrogen transfer, isomerization, and deoxygenation reactions such as decarboxylation, decarbonylation, and dehydration [86]. Olefins are produced by repeated dehydration steps of oxygenated compounds formed

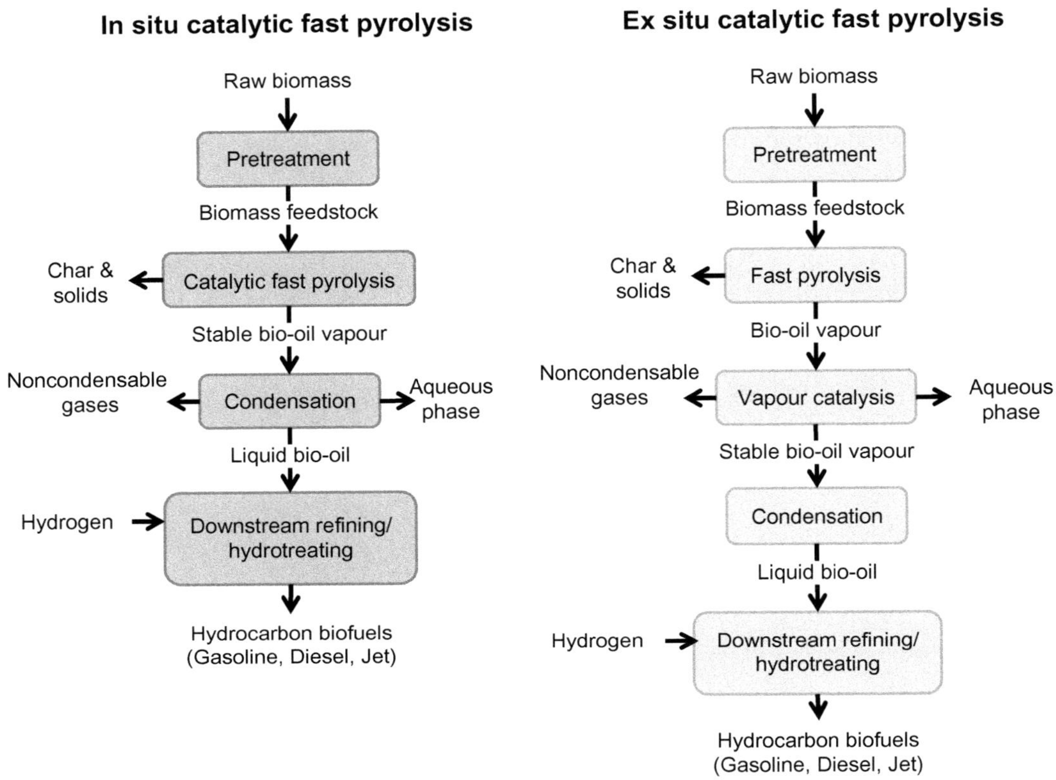

FIGURE 8.4 Process flow diagrams for catalytic fast pyrolysis pathways [85].

during pyrolysis, which can oligomerize via condensation and Diels-Alder reactions to produce aromatics [89]. This aromatic production leads, unfortunately, to the increased production of charring. This, along with the production of short-chain uncondensable hydrocarbons (C_1-C_4), can reduce the overall bio-oil yield. Experimental methods for catalytic cracking of bio-oil (usually carried out in the presence of zeolites, to be discussed later) typically results in bio-oil yields of 14–23%.

As well as cracking and hydrotreatment, catalytic fast pyrolysis can also include the species within the pyrolysis vapour reacting with each other. As has been discussed, bio-oil typically contains a large portion of carboxylic acids, esters, ketones, and aldehydes [27,28]. The carbonyl species within these molecules are active for C–C coupling reactions such as aldol condensation, Michael addition, and ketonization, which produces longer-chain molecules that can be converted into fuel-like molecules via their hydrodeoxygenation [90]. It is known, however, that the pyrolysis of carbonyl-containing species leads to significant aromatic production (via cyclization of the condensation products) and subsequent coking [91], thus reducing the yield of bio-oil.

8.5.1 Catalysts

The catalysts used for catalytic fast pyrolysis have a significant effect on the resulting composition of the bio-oil, as they will have different activities for the mechanisms described above. Below is a discussion of the catalysts used for catalytic fast pyrolysis, along with research carried out with them. A summary is included in Table 8.5.

8.5.1.1 Zeolites

Some of the most promising catalysts for bio-oil upgrading are zeolite-based systems. Zeolites occur naturally in geological formation, with over 40 naturally occurring forms known such as mordenite, natrolite, and faujasite.

TABLE 8.5 Various Catalytic Fast Pyrolysis Studies Showing Yields and (Where Available) Properties and Composition of the Liquid Produced

	Experimental Conditions							
Biomass Feedstock	**Catalyst**	**Reactor type**	**Temperature (°C)**	**Bio-oil Yield (%)**	**Hydrocarbon Yield (%)**	**Major Products**	**Bio-oil Properties and Composition**	**References**
Palm oil	HZSM-5	Fixed-bed	350	53	N/A	Aromatics		[96]
Pine sawdust	HZSM-5	Fixed-bed	300	47.8	N/A	Phenolics		[98]
Poplar wood	HZSM-5	Fluidized bed	500	33.60	N/A	Aromatics, ketones	$\nu = 147\ mm^2/s$ (@ 25°C), C = 71.2%, H = 6.8%, O = 21.9%	[22]
Corn stover	HZSM-5	Microwave semi-continuous	500	~20	N/A	Oxygenated aliphatics and aromatics		[107]
Waste pepper stems	HZSM-5	Py-GC/MS	500	N/A	N/A	Aromatics		[106]
Wood sawdust	HZSM-5	Fluidized bed	600	N/A	N/A	14% aromatics, 5% olefins		[97]
Pine wood	HZSM-5	Fluidized bed	600	N/A	23.2	Aromatics, C_2–C_4 olefins		[100]
Rice husks	HZSM-5	Fluidized bed	400	7.2	N/A	Phenols, cresols and aromatics	C = 77.6%, H = 7.7%, O = 14.5%	[103]
Corncob	HZSM-5	Fixed-bed	550	13.7	N/A	Aromatics	$\rho = 0.95\ kg/m^3$, H_2O = 22.9%, C = 74.1%, H = 9.3%, O = 14.7%	[101]

(Continued)

TABLE 8.5 (Continued)

Biomass Feedstock	Experimental Conditions: Catalyst	Reactor type	Temperature (°C)	Bio-oil Yield (%)	Hydrocarbon Yield (%)	Major Products	Bio-oil Properties and Composition	References
Food waste	HZSM-5	Py-GC/MS	600	N/A	N/A	Aromatics		[105]
Rice stalk	HZSM-5	Fluidized bed	550	N/A	23.3	Aromatics, C_2-C_4 olefins		[102]
Corn stover	HZSM-5	Py-GC/MS	500	N/A	N/A	Phenolics		[104]
Corn stover	HZSM-5 (used)	Py-GC/MS	500	N/A	N/A	Phenolics		[104]
Corn stover	HZSM-5 (regenerated)	Py-GC/MS	500	N/A	N/A	Phenolics		[104]
Palm oil	H-β	Fixed-bed	400	53.7	N/A			[96]
Pinewood sawdust	Ga/ZSM-5	Fluidized bed	550	N/A	34.3	23.2% aromatics		[111]
Pine wood	Mo/ZSM-5	Py-GC/MS	650	N/A	N/A	Aromatics		[113]
Pine wood	Ni/ZSM-5	Quartz tube	600	N/A	16	Aromatics		[112]
Pine sawdust	Meso-HZSM-5	Fixed-bed	300	53.6	N/A	Phenolics		[98]
Pine sawdust	Meso-MFI	Fixed-bed	300	47.5	N/A	Phenolics, PAHs		[98]
Lignocel HBS	Al2O3	Fixed-bed	500	37.4	5.5	Phenolics, aromatics	$H_2O = 32.5\%$, $C = 68.4$, $H = 7.6\%$. $O = 24\%$	[118]
Lignocel HBS	MgO	Fixed-bed	500	51	26.8	Unsaturates	$H_2O = 29.6\%$, $C = 70.1$, $H = 8\%$, $O = 21.9\%$	[118]
Guaiacol	Pt/C	Fixed-bed	250	N/A	30.8	Phenol		[114]
Guaiacol	Pd/C	Fixed-bed	250	N/A	50.4	Phenol		[114]
Furan	Ni/SiO_2	Tubular	250	N/A	85	THF, furfural alcohol		[115]
Furan	Pt/Al_2O_3	Packed bed	300	N/A	N/A	C_3-C_4 aliphatics		[116]

Specifically, a zeolite is a crystalline, microporous aluminosilicate material consisting of interconnecting SiO_4 and AlO_4 tetrahedra. In these forms, silicon has a +4 charge, while alumina has a +3 charge, and so the AlO_4 tetrahedron carries a negative charge. The interconnecting tetrahedra form a three-dimensional framework, with uniform pores of roughly molecular dimensions that run through the structure. As such, monopositive cations can be accommodated in these pores, cancelling out the AlO_4 negative charge and allowing the overall structure to remain electroneutral [92]. Due to their unique combination of chemical nature and pore structure, zeolites have found many uses, including gas physisorption, liquid adsorption, molecular sieves, and catalytic applications.

The most prevalent and widely researched zeolite is ZSM-5. It is such called due to its original synthesis by Mobil scientists (Zeolite Socony Mobil-5). Its structure has been well defined in the literature [93], containing straight and sinusoidal channels of ca. 5.5 Å diameter, and pores of ca. 8 Å. Commercially, ZSM-5 is generally obtained in its ammonium form. Upon calcination, however, the ammonium ion breaks down, ammonia is given off leaving behind a proton, and HZSM-5 is produced. HZSM-5 can exhibit both Brønsted acidity and – when a Brønsted acid site is dehydrated – Lewis acidity, and can be characterized by its silica-to-alumina ratio (SAR, SiO_2/Al_2O_3). A higher SAR signifies less negatively charged AlO_4 tetrahedrons which need to be neutralized by cations, and therefore has less acidic sites overall, leading to a lower reactivity. HZSM-5 is an effective dehydration, isomerization, and oligomerization catalyst. It is most significantly used to produce synthetic gasoline via the methanol-to-gasoline process, whereby two moles of methanol are dehydrated to form dimethyl ether, before being further dehydrated to C_2-C_5 olefins and eventually oligomerized to a mixture of alkanes, cycloalkanes, alkenes, and aromatics [94]. The protonated form of ZSM-5, HZSM-5, is shape selective due to its pore size. Molecules no bigger than trimethylbenzene can be produced in the pores [7]. Other zeolites of differing structure are also considered for catalytic fast pyrolysis, such as H-Y and H-β [95].

The catalytic fast pyrolysis of palm oil to hydrocarbons with HZSM-5 and H-β was investigated by Twaiq et al. [96]. Using a fixed-bed reactor at temperatures between 350–450°C, weight hourly space velocities of $1-4\ h^{-1}$ and atmospheric pressure, HZSM-5 reached a maximum bio-oil yield of 53% at 350°C and a weight hourly space velocity of $2.5\ h^{-1}$.

While aromatic production would preferentially be avoided during pyrolysis reactions, if aromatics can be produced in high yields without the subsequent oligomerization and eventual coke formation, they could be a valuable fuel addition due to their requirement for lubricity and material compatibility purposes. Carlson et al. carried out the catalytic fast pyrolysis of wood sawdust using HZSM-5-based catalysts in an attempt to produce aromatics (the production of which HZSM-5 is ideal for) and olefins [97]. Using a bubbling fluidized bed reactor, the highest aromatic yield from the sawdust reached 14% at 600°C and a low biomass weight hourly space velocity ($<0.5\ h^{-1}$). There was, however, significant loss of carbon-to-coke formation at these conditions (>30%).

Pinewood is a popular feedstock for pyrolysis due to its abundance. Park et al. carried out catalytic fast pyrolysis of pinewood using zeolite catalysts, including HZSM-5, a mesoporous form of HZSM-5 (Meso-HZSM-5), and a mesoporous MFI catalyst (Meso-MFI) [98]. The maximum bio-oil yields reached for HZSM-5, MMZ, and Meso-MFI were 47.8%, 53.6%, and 47.5%, respectively, all at 300°C. Meso-MFI did, however, produce a larger proportion of gasoline range (C_8-C_9) aromatic species. The catalytic fast pyrolysis of poplar wood with

HZSM-5 was carried out by Agblevor et al. using a fluidized bed reactor [99]. At 500°C and a biomass feed rate of 100 g/h, a bio-oil yield of 33.6% was reached, mostly comprised of aromatic species and ketones, though the viscosity was much higher than would be allowed for fuel molecules.

Zhang et al. carried out the catalytic fast pyrolysis of pine wood in a fluidized bed reactor using HZSM-5 as the catalyst [100]. Assessing the effect of temperature and weight hourly space velocity on yield and selectivity, the maximum carbon yield was 23.7% at a temperature of 600°C and a weight hourly space velocity of 0.35 h^{-1}. Zhang et al. also carried out the catalytic fast pyrolysis of corncob using HZSM-5 and a fluidized bed reactor [101]. The oil yield achieved was 13.7%, 74.2% of which was comprised of aromatic hydrocarbons. While the oil yield was significantly lower than what was reached via non-catalytic methods (33.9%), the quality of the bio-oil was much increased as the C/O ratio was significantly higher. Further work by Zhang et al. includes the catalytic fast pyrolysis of rice straw biomass in a fluidized bed reactor using HZSM-5 at 550°C and a biomass feeding rate of 44 g/h [102], reaching a hydrocarbon yield of 23.3%, most of which consisted of aromatic species and C_2–C_4 olefins.

Rice husks are another common feedstock used in pyrolysis, as a waste product of rice cultivation. Williams and Nugranad studied the pyrolysis catalytic fast pyrolysis of rice husks using a fluidized bed reactor using HZSM-5 at various temperatures (400–600°C) [103]. The yield of oil from catalytic fast pyrolysis was drastically reduced at 550°C when compared to noncatalytic pyrolysis (7.2% maximum yield compared to 28.5%), though the quality of the bio-oil was much higher, as was evidenced by the higher C/O ratio.

Recently, Bo Zhang et al. have carried out interesting research into using zeolites for catalytic fast pyrolysis, including looking into the effect of fresh, spent, and regenerated HZSM-5 on the catalytic fast pyrolysis of corn stover using a pyroprobe GC-MS [104]. Carrying out reactions at 550°C it was found that there was little difference between the fresh and regenerated HZSM-5, while the spent HZSM-5 (pre-coked) formed less coke and less phenolics in the reaction while producing more of the target products, with higher hydrocarbons and lower oxygen contents. Again using a pyroprobe GC-MS, Bo Zhang et al. studied the catalytic fast pyrolysis of food waste, a waste product gaining popularity in research in recent years [105]. Carrying out the reactions at a range of temperatures (500–700°C), it was found that 600°C was the optimum temperature for production of aromatics and other organic pyrolysis products. Also looking into the catalytic fast pyrolysis of food waste, Park et al. studied its effect on waste pepper stems using HZSM-5 with a pyroprobe GC/MS, and found that HZSM-5 produced significantly higher proportions of aromatics (benzene, toluene, xylene) than noncatalytic pyrolysis [106]. In another experimental set-up, Bo Zhang et al. carried out catalytic fast pyrolysis on a modified HZSM-5 catalyst (with a deposition of SiO_2) using a microwave-assisted method [107]. Carrying out reactions at 500°C, with varying levels of deposited SiO_2 (1.3–5.9%), the bio-oil yield achieved was 15.6–22.6% with no significant trend observed. The amount of SiO_2 also did not affect the chemical composition of the bio-oil achieved, with the majority being oxygenated aliphatic and aromatics. However, increasing the proportion of SiO_2 reduced the amount of coke produced.

In situ and ex situ catalytic fast pyrolysis of miscanthus with HZSM-5 was carried out by Gamliel et al. using a pyroprobe GC-MS [108]. While ex situ catalytic fast pyrolysis led to higher selectivity or aromatic species, in situ catalytic fast pyrolysis yielded more bio-oil, while the solid material for both was similar.

Recently, Hammer et al. developed a two-step process to produce high quality bio-oils from biomass [109]. In this process, the biomass is initially pyrolysed between 300–350°C to decompose cellulose and hemicellulose either noncatalytically or with HZSM-5, and in the second step the remaining lignin-rich biomass was pyrolysed in the presence of HZSM-5 at 500°C (ie, in situ CFP arrangement), ultimately producing two separate bio-oils. This method was found to increase the aromatic portion of the bio-oil while reducing the amount of oxygenates when compared to conventional one-step pyrolysis at 500°C.

8.5.1.2 Metal-impregnated Zeolites

By impregnating catalytic metal centres within the structure of zeolites, multifunctional catalysts can be produced, whereby the Brønsted acid zeolite can carry out cracking reactions (as discussed in the previous section), and the metal centre is active for additional reaction types, such as hydrogenation reactions (eg, hydrodeoxygenation). By creating a bifunctional catalyst, there is potential for an improved product suite, closer resembling current fuels. However, introducing certain metals, such as Fe or Cu does not significantly change the resulting products [110]. Furthermore, how the new metal-doped catalyst is synthesized can have a significant impact on its activity.

Cheng et al. produced significant amounts of aromatic compounds from furan using a bifunctional Ga/ZSM-5 catalyst, reaching a maximum aromatic yield of 23.7% at 600°C with a weight hourly space velocity of $0.3\ h^{-1}$ [111].

French and Czernik carried out fast pyrolysis of wood, lignin, and cellulose in a quartz tube reactor with a range of metal-impregnated ZSM-5 catalysts, the products were analysed by a molecular beam mass spectrometer [112]. While a vast range of metal-impregnated catalysts were tested, including gallium, cobalt, and iron-impregnated ZSM-5, Ni/ZSM-5 achieved the highest hydrocarbon yield of 16% with a toluene yield of 3.5%. Thangalazhy-Gopakumar et al. also carried out catalytic fast pyrolysis of pine wood with metal-impregnated ZSM-5 – specifically Ni, Co, Mo, and Pt – and carried out the reactions at a range of hydrogen pressures (7–27.5 bar) [113]. While the Mo/ZSM-5 was not as active as HZSM-5 at lower pressures (7–21 bar), upon carrying out the reactions at 27.5 bar the metal-impregnated zeolites reached a higher aromatic yield (~41.5 wt.% carbon yield) than HZSM-5 (36.2 wt.% carbon yield).

8.5.1.3 Supported Metal Catalysts

While some supported metal catalysts have been discussed previously, those were coupled with active porous catalysts for a bifunctional catalyst system. By supporting active metal catalyst on a less active surface, hydrodeoxygenation reactions dominate. The literature surrounding the use of metal-supported catalysts on biomass pyrolysis vapours is sparse; however, model compounds such as guaiacol and furan are commonly used.

Sun et al. investigated a number of metal and precious metal catalysts supported on carbon, including Cu/C, Fe/C, Pd/C, Pt/C, and Ru/C on the hydrodeoxygenation of guaiacol [114]. Ru/C was found to be the best conversion catalyst, with complete conversion at 350°C; however, this resulted in solely volatile gas-phase products. For Pd/C and Pt/C, significantly high conversions were reached at 250°C (30.8% and 50.4%, respectively), with the major product being phenol.

The hydrodeoxygenation of furfural was investigated by Sitthisa et al. using Cu, Ni, and Pd supported on SiO_2 [115]. Investigating a range of temperatures (230–290°C), and an H_2 flow of 60 mL/min, maximum conversion of 77% was achieved by 10% Cu/SiO_2 at 270°C, with the major product being furfuryl alcohol, 100% conversion was reached by 5% Ni/SiO_2 at 230°C, with the major products being furan and furfuryl alcohol, and 85% conversion was reached by 1% Pd/SiO_2 at 250°C, with the major products being THF and furfuryl alcohol. The hydrodeoxygenation of furan was also investigated by Runnebaum et al. using 1% Pt/Al_2O_3 at 300°C and 1.4 bar in a packed-bed reactor, leading to the formation of C_3-C_4 aliphatics and a small amounts of C_7 aliphatics [116].

8.5.1.4 Metal Oxides

Metal oxides have been utilized for a number of catalytic processes [117], and have been considered for catalytic fast pyrolysis. While acidic metal oxides such as Al_2O_3 have been reported to reduce the bio-oil yield while increasing water and charring when compared to noncatalytic conditions in a fixed-bed reactor, basic metal oxides such as MgO have been found to improve the bio-oil quality in terms of the proportion of carbon and reduction of oxygen in the resulting bio-oil. It was, however, reported to reduce the overall bio-oil yield [118]. Transitional metal oxides, such as TiO_2, Fe_2O_3, NiO, and ZnO, generally reduce the bio-oil and hydrocarbon yields while increasing the water, gas, and charring [119].

8.6 JET FUEL-SPECIFIC PYROLYSIS

The studies carrying out pyrolysis for the specific aim of producing jet fuel are sparse. During the pyrolysis of palm oil by Twaiq et al. [96], discussed previously using a fixed-bed reactor, the products achieved from HZSM-5 and H-β were analysed from the fractions of the liquid fractions produced, assigning distillation temperature ranges for gasoline (boiling point 60–135°C), kerosene (boiling point 135–160°C), and diesel (boiling point 160–200°C). Using HZSM-5, the bio-oil procured was determined to contain 26.6% gasoline, 15.8% kerosene, and 4.2% diesel, while the bio-oil produced using H-β was found to contain 22.1% gasoline, 12.1% kerosene, and 10.7% diesel. While this is one of the few publications to determine any such fractions, boiling point alone is a poor measure for the suitability for jet fuel applications. Furthermore, the large-scale production of jet fuel from a triglyceride source is unsustainable, as it requires the production of a significant portion of biomass that goes unused, and can therefore be considered waste.

The only research to produce jet fuel from lignocellulosic biomass pyrolysis is the recent work by Wang et al. [120], though the entire process contains several steps. In the investigation, fast pyrolysis of straw stalks was carried out using a circulating fluidized bed with a capacity of 120 kg/h. This bio-oil was then catalytically cracked with HZSM-5 to produce aromatics and light olefins over a range of temperatures (450–650°C). Both aromatic yield and light olefin production peaked at 550°C, reaching 29.2°C-mol.% and 17.4°C-mol.%. In the next step, alkylation of the aromatics, using the light olefins produced as the alkylating agents, was carried out in 1-butyl-3-methylimidazolium chloroaluminate ionic liquid ([bmim]Cl-2$AlCl_3$), which acts as both a solvent and a catalyst for the reactions. At facile conditions – that is, atmospheric pressure and 20°C – a high selectivity of 88.4% of C_8-C_{15} aromatic hydrocarbons was produced. Lastly, hydrogenation of these alkylated products was carried out between temperatures of 120°C and 210°C for 6 h in the presence of 5% Pd/C and 5 MPa of hydrogen. After this reaction period, 72.5 wt.% of the resulting liquid was comprised of C_8-C_{15} cyclic alkanes. The jet fuel produced was then analysed for its physical properties and was found to have a higher heat of combustion and a lower freezing temperature than conventional jet fuels (Jet A, Jet A-1, and JP-8 [military grade]), while possessing similar density, viscosity, and H/C ratio than its crude oil-derived counterparts. It should be noted, however, that one aspect of the fuel that was not determined was flash point. While this is set at a minimum of 38°C, is it unclear whether this specification is for the safety of handling or whether it has any effect on the resulting combustion of the fuel in jet engines. The fuel tanks of a commercial airliner, however, act as heat sinks during flight and so flash point should be taken into account and determined before a fuel is deemed suitable. While there are many steps to the Wang process, and each step leads to a loss of carbon and a lower ultimate yield of jet fuel, it shows how biomass can be catalytically broken apart and subsequently recombined to produce jet fuel hydrocarbons.

8.7 CONCLUSIONS AND FINAL REMARKS

Pyrolysis, specifically fast pyrolysis, is a technology that produces liquid fractions (termed 'bio-oil') in high yield (up to 65 wt.%) from biomass by subjecting it to moderate temperatures (400–600°C), short residence times (0.5–5 s), and high heating rates (10–200°C/s). The bio-oil produced, however, is unstable, viscous, and corrosive due to the high presence of oxygen, and requires significant upgrading before being able to meet the stringent specifications outlined by jet fuel specifications. For any liquid produced to be considered as a jet fuel, its low-temperature properties (melting point and viscosity), energy density, and flash point require determination.

Depending on the feedstock, and the feedstock's ratio of cellulose, hemicellulose, and lignocellulose within its composition, the resultant bio-oil will have a very different composition. These products can also react with each other, providing an even more varied bio-oil composition. Upgrading of bio-oil via deoxygenation, hydrotreatment, and cracking via the use of heterogeneous catalysts (largely zeolites, but also metal oxides and supported metal catalysts) has been considered and researched extensively in the literature. However, the liquid products are usually relatively low in yield (10–50%), the majority of which is comprised of aromatic species or oxygenated aliphatics. The fuel properties of most liquid products are also not determined, either due to the small amount being produced or the fundamental nature of the investigations. While aromatics are desirable in small quantities in jet fuel due to their seal swell characteristics, to meet specifications the majority of the fuel's formulations should consist of alkanes, cycloalkanes, and potentially, alkenes.

It is clear that, despite the ability of pyrolysis to produce significant liquid yields, it is unsuitable as a fuel production technology without significant further upgrading. One study, where jet fuel range cycloalkanes (C_8–C_{15}) were produced, showed that pyrolysis can have a place in the production of jet fuel when coupled with additional reactions, such as alkylation of the aromatics produced using the olefins produced from pyrolysis, and subsequent hydrogenation of those aromatics. While this process, and others like it, warrant further research for optimization, the economics of such a process are likely to be significantly limiting.

Lastly, the technology of pyrolysis has many different variables (temperature, gas flow, residence time, feedstock composition, feedstock particulate size, catalyst activity), each of which has a drastic effect on the resultant composition of the bio-oil. Any technology that can reduce this variety, increase selectivity, and therefore allow for directed production of fuel formations, should be extensively researched.

References

[1] J.Q. Bond, et al., Production of renewable jet fuel range alkanes and commodity chemicals from integrated catalytic processing of biomass, Energy Environ. Sci. 7 (4) (2014) 1500–1523.

[2] British Ministry of Defence, DEF STAN 91-91 Turbine Fuel, Kerosene Type, Jet A-1, MOD, Glasgow, 2011.

[3] G. Liu, B. Yan, G. Chen, Technical review on jet fuel production, Renew. Sustain. Energy Rev. 25 (2013) 59–70.

[4] C. Stevens, R.C. Brown, *Thermochemical Processing of Biomass: Conversion into Fuels, Chemicals and Power*, John Wiley & Sons, Chichester, United Kingdom, 2011.
[5] S. Nanda, et al., Pathways of lignocellulosic biomass conversion to renewable fuels, Biomass Convers. Biorefin. 4 (2) (2014) 157–191.
[6] D. Mohan, C.U. Pittman, P.H. Steele, Pyrolysis of wood/biomass for bio-oil: a critical review, Energy Fuels 20 (3) (2006) 848–889.
[7] C. Liu, et al., Catalytic fast pyrolysis of lignocellulosic biomass, Chem. Soc. Rev. 43 (22) (2014) 7594–7623.
[8] T. Stedile, et al., Comparison between physical properties and chemical composition of bio-oils derived from lignocellulose and triglyceride sources, Renew. Sustain. Energy Rev. 50 (2015) 92–108.
[9] J. Yanik, et al., Pyrolysis of algal biomass, J. Anal. Appl. Pyrolysis 103 (2013) 134–141.
[10] T. Wanjun, W. Cunxin, C. Donghua, Kinetic studies on the pyrolysis of chitin and chitosan, Polym. Degrad. Stab. 87 (3) (2005) 389–394.
[11] M. Jermyn, Cellulose and hemicelluloses, Modern Methods of Plant Analysis/Moderne Methoden der Pflanzenanalyse, Springer, Berlin, Germany, 1955, pp. 197–225.
[12] S. Dutta, et al., Advances in conversion of hemicellulosic biomass to furfural and upgrading to biofuels, Catal. Sci. Technol. 2 (10) (2012) 2025–2036.
[13] T. Chen, et al., Effect of hot vapor filtration on the characterization of bio-oil from rice husks with fast pyrolysis in a fluidized-bed reactor, Bioresour. Technol. 102 (10) (2011) 6178–6185.
[14] E. Apaydin-Varol, E. Pütün, A.E. Pütün, Slow pyrolysis of pistachio shell, Fuel 86 (12) (2007) 1892–1899.
[15] E. Apaydin-Varol, et al., Synthetic fuel production from cottonseed: Fast pyrolysis and a TGA/FT-IR/MS study, J. Anal. Appl. Pyrolysis 105 (2014) 83–90.
[16] A.E. Pütün, et al., Comparison between the "slow" and "fast" pyrolysis of tobacco residue, Ind. Crops Prod. 26 (3) (2007) 307–314.
[17] M. Garcìa-Pèrez, A. Chaala, C. Roy, Vacuum pyrolysis of sugarcane bagasse, J. Anal. Appl. Pyrolysis 65 (2) (2002) 11–136.
[18] R. Yin, et al., Characterization of bio-oil and biochar obtained from sweet sorghum bagasse fast pyrolysis with fractional condensers, Fuel 112 (2013) 96–104.
[19] J. Yanik, et al., Fast pyrolysis of agricultural wastes: characterization of pyrolysis products, Fuel Process. Technol. 88 (10) (2007) 942–947.
[20] M.I. Jahirul, et al., Biofuels production through biomass pyrolysis – a technological review, Energies 5 (12) (2012) 4952–5001.
[21] İ. Demiral, A. Eryazıcı, S. Şensöz, Bio-oil production from pyrolysis of corncob (*Zea mays L.*), Biomass Bioenergy 36 (2012) 43–49.
[22] C.J. Chuck, et al., Renewable biofuel additives from the ozonolysis of lignin, Bioresour. Technol. 143 (2013) 549–554.
[23] M.P. Pandey, C.S. Kim, Lignin depolymerization and conversion: a review of thermochemical methods, Chem. Eng. Technol. 34 (1) (2011) 29–41.
[24] D. Shen, et al., The pyrolytic degradation of wood-derived lignin from pulping process, Bioresour. Technol. 101 (15) (2010) 6136–6146.
[25] F.S. Chakar, A.J. Ragauskas, Review of current and future softwood kraft lignin process chemistry, Ind. Crops Prod. 20 (2) (2004) 131–141.
[26] N.E. Carpenter, Chemistry of Sustainable Energy, Taylor & Francis, Boca Raton, FL, 2014.
[27] Q. Zhang, et al., Review of biomass pyrolysis oil properties and upgrading research, Energy Convers. Manag. 48 (1) (2007) 87–92.
[28] M.S. Talmadge, et al., A perspective on oxygenated species in the refinery integration of pyrolysis oil, Green Chem. 16 (2) (2014) 407–453.
[29] J. Meng, et al., Thermal and storage stability of bio-oil from pyrolysis of torrefied wood, Energy Fuels 29 (8) (2015) 5117–5126.
[30] A. Oasmaa, S. Czernik, Fuel oil quality of biomass pyrolysis oils state of the art for the end users, Energy Fuels 13 (4) (1999) 914–921.
[31] R.M. Baldwin, C.J. Feik, Bio-oil stabilization and upgrading by hot gas filtration, Energy Fuels 27 (6) (2013) 3224–3238.
[32] T. vom Stein, J. Klankermayer, W. Leitner, Tailor-Made Fuels and Chemicals from Biomass, Catalysis for the Conversion of Biomass and its Derivatives, Epubli, Berlin, Germany, 2013, pp. 411–434.
[33] S. Arbogast, et al., Advanced bio-fuels from pyrolysis oil: the impact of economies of scale and use of existing logistic and processing capabilities, Fuel Process. Technol. 104 (2012) 121–127.
[34] R.L. McCormick, et al., Properties of oxygenates found in upgraded biomass pyrolysis oil as components of spark and compression ignition engine fuels, Energy Fuels 29 (4) (2015) 2453–2461.
[35] R.J. French, J. Stunkel, R.M. Baldwin, Mild hydrotreating of bio-oil: effect of reaction severity and fate of oxygenated species, Energy Fuels 25 (7) (2011) 3266–3274.
[36] H. Yang, et al., Characteristics of hemicellulose, cellulose and lignin pyrolysis, Fuel 86 (12) (2007) 1781–1788.
[37] F.-X. Collard, J. Blin, A review on pyrolysis of biomass constituents: mechanisms and composition of the products obtained from the conversion of cellulose,

hemicelluloses and lignin, Renew. Sustain. Energy Rev. 38 (2014) 594–608.
[38] M. Van de Velden, et al., Fundamentals, kinetics and endothermicity of the biomass pyrolysis reaction, Renew. Energy 35 (1) (2010) 232–242.
[39] J. Banyasz, et al., Gas evolution and the mechanism of cellulose pyrolysis, Fuel 80 (12) (2001) 1757–1763.
[40] Q. Lu, et al., Influence of pyrolysis temperature and time on the cellulose fast pyrolysis products: analytical Py-GC/MS study, J. Anal. Appl. Pyrolysis 92 (2) (2011) 430–438.
[41] I. Pastorova, et al., Cellulose char structure: a combined analytical Py-GC-MS, FTIR, and NMR study, Carbohydr. Res. 262 (1) (1994) 27–47.
[42] P. Morf, P. Hasler, T. Nussbaumer, Mechanisms and kinetics of homogeneous secondary reactions of tar from continuous pyrolysis of wood chips, Fuel 81 (7) (2002) 843–853.
[43] D. Neves, et al., Characterization and prediction of biomass pyrolysis products, Prog. Energy Combust. Sci. 37 (5) (2011) 611–630.
[44] D. Liu, Y. Yu, H. Wu, Differences in water-soluble intermediates from slow pyrolysis of amorphous and crystalline cellulose, Energy Fuels 27 (3) (2013) 1371–1380.
[45] P.R. Patwardhan, et al., Distinguishing primary and secondary reactions of cellulose pyrolysis, Bioresour. Technol. 102 (8) (2011) 5265–5269.
[46] M.S. Mettler, et al., Pyrolytic conversion of cellulose to fuels: levoglucosan deoxygenation via elimination and cyclization within molten biomass, Energy Environ. Sci. 5 (7) (2012) 7864–7868.
[47] D. Shen, S. Gu, The mechanism for thermal decomposition of cellulose and its main products, Bioresour. Technol. 100 (24) (2009) 6496–6504.
[48] Z. Wang, et al., Effect of cellulose crystallinity on the formation of a liquid intermediate and on product distribution during pyrolysis, J. Anal. Appl. Pyrolysis 100 (2013) 56–66.
[49] W.N.R.W. Isahak, et al., A review on bio-oil production from biomass by using pyrolysis method, Renew. Sustain. Energy Rev. 16 (8) (2012) 5910–5923.
[50] D. Shen, S. Gu, A.V. Bridgwater, Study on the pyrolytic behaviour of xylan-based hemicellulose using TG–FTIR and Py–GC–FTIR, J. Anal. Appl. Pyrolysis 87 (2) (2010) 199–206.
[51] T. Hosoya, H. Kawamoto, S. Saka, Pyrolysis behaviors of wood and its constituent polymers at gasification temperature, J. Anal. Appl. Pyrolysis 78 (2) (2007) 328–336.
[52] R. Alén, E. Kuoppala, P. Oesch, Formation of the main degradation compound groups from wood and its components during pyrolysis, J. Anal. Appl. Pyrolysis 36 (2) (1996) 137–148.
[53] S. Wang, et al., Comparison of the pyrolysis behavior of lignins from different tree species, Biotechnol. Adv. 27 (5) (2009) 562–567.
[54] W. Mu, et al., Lignin pyrolysis components and upgrading – technology review, Bioenergy Res. 6 (4) (2013) 1183–1204.
[55] K. Candelier, et al., Utilization of thermodesorption coupled to GC–MS to study stability of different wood species to thermodegradation, J. Anal. Appl. Pyrolysis 92 (2) (2011) 376–383.
[56] T. Nakamura, H. Kawamoto, S. Saka, Pyrolysis behavior of Japanese cedar wood lignin studied with various model dimers, J. Anal. Appl. Pyrolysis 81 (2) (2008) 173–182.
[57] S. Xiu, A. Shahbazi, Bio-oil production and upgrading research: a review, Renew. Sustain. Energy Rev. 16 (7) (2012) 4406–4414.
[58] A.V. Bridgwater, Review of fast pyrolysis of biomass and product upgrading, Biomass Bioenergy 38 (2012) 68–94.
[59] J.-Y. Kim, et al., Assessment of miscanthus biomass (*Miscanthus sacchariflorus*) for conversion and utilization of bio-oil by fluidized bed type fast pyrolysis, Energy 76 (2014) 284–291.
[60] H. Wiinikka, et al., Fast pyrolysis of stem wood in a pilot-scale cyclone reactor, Energy Fuels 29 (5) (2015) 3158–3167.
[61] J.-L. Zheng, Pyrolysis oil from fast pyrolysis of maize stalk, J. Anal. Appl. Pyrolysis 83 (2) (2008) 205–212.
[62] N. Ali, et al., Bio-oil production from fast pyrolysis of cotton stalk in fluidized bed reactor, Arab. J. Sci. Eng. 40 (11) (2015) 3019–3027.
[63] J.-L. Zheng, W.-M. Yi, N.-N. Wang, Bio-oil production from cotton stalk, Energy Convers. Manag. 49 (6) (2008) 1724–1730.
[64] H. Zhang, et al., Catalytic fast pyrolysis of biomass in a fluidized bed with fresh and spent fluidized catalytic cracking (FCC) catalysts, Energy Fuels 23 (12) (2009) 6199–6206.
[65] C.A. Mullen, et al., Bio-oil and bio-char production from corn cobs and stover by fast pyrolysis, Biomass Bioenergy 34 (1) (2010) 67–74.
[66] Z. Ji-Lu, Bio-oil from fast pyrolysis of rice husk: Yields and related properties and improvement of the pyrolysis system, J. Anal. Appl. Pyrolysis 80 (1) (2007) 30–35.
[67] L. Zhou, et al., Catalytic pyrolysis of rice husk by mixing with zinc oxide: characterization of bio-oil and its rheological behavior, Fuel Process. Technol. 106 (2013) 385–391.
[68] J. Alvarez, et al., Bio-oil production from rice husk fast pyrolysis in a conical spouted bed reactor, Fuel 128 (2014) 162–169.

[69] M. Asadullah, et al., Production and detailed characterization of bio-oil from fast pyrolysis of palm kernel shell, Biomass Bioenergy 59 (2013) 316–324.
[70] M. Asadullah, et al., Jute stick pyrolysis for bio-oil production in fluidized bed reactor, Bioresour. Technol. 99 (1) (2008) 44–50.
[71] M. Asadullah, et al., Production of bio-oil from fixed bed pyrolysis of bagasse, Fuel 86 (16) (2007) 2514–2520.
[72] J. Montoya, et al., Bio-oil production from Colombian bagasse by fast pyrolysis in a fluidized bed: an experimental study, J. Anal. Appl. Pyrolysis 112 (2015) 379–387.
[73] T. Kar, S. Keles, Fast pyrolysis of chestnut cupulae: yields and characterization of the bio-oil, Energy Explor. Exploit. 31 (6) (2013) 847–858.
[74] T.-A. Ngo, J. Kim, S.-S. Kim, Fast pyrolysis of spent coffee waste and oak wood chips in a micro-tubular reactor, Energy Sources, Part A 37 (11) (2015) 1186–1194.
[75] R.W. Jenkins, et al., Effect of the type of bean, processing, and geographical location on the biodiesel produced from waste coffee grounds, Energy Fuels 28 (2) (2014) 1166–1174.
[76] M. Amutio, et al., Fast pyrolysis of eucalyptus waste in a conical spouted bed reactor, Bioresour. Technol. 194 (2015) 225–232.
[77] O. Onay, O.M. Kockar, Slow, fast and flash pyrolysis of rapeseed, Renew. Energy 28 (15) (2003) 2417–2433.
[78] A. Oasmaa, D.C. Elliott, J. Korhonen, Acidity of biomass fast pyrolysis bio-oils, Energy Fuels 24 (12) (2010) 6548–6554.
[79] Czernik, S., Storage of biomass pyrolysis oils, in: Proceedings of Specialist Workshop on Biomass Pyrolysis Oil Properties and Combustion, 1994, 67–76.
[80] S. Czernik, D.K. Johnson, S. Black, Stability of wood fast pyrolysis oil, Biomass Bioenergy 7 (1–6) (1994) 187–192.
[81] S. Czernik, A.V. Bridgwater, Overview of applications of biomass fast pyrolysis oil, Energy Fuels 18 (2) (2004) 590–598.
[82] D.C. Elliott, Historical developments in hydroprocessing bio-oils, Energy Fuels 21 (3) (2007) 1792–1815.
[83] L. Moens, et al., Study of the neutralization and stabilization of a mixed hardwood bio-oil, Energy Fuels 23 (5) (2009) 2695–2699.
[84] R.H. Venderbosch, et al., Stabilization of biomass-derived pyrolysis oils, J. Chem. Technol. Biotechnol. 85 (5) (2010) 674–686.
[85] D.A. Ruddy, et al., Recent advances in heterogeneous catalysts for bio-oil upgrading via "ex situ catalytic fast pyrolysis": catalyst development through the study of model compounds, Green Chem. 16 (2) (2014) 454–490.
[86] P.M. Mortensen, et al., A review of catalytic upgrading of bio-oil to engine fuels, Appl. Catal. A 407 (1) (2011) 1–19.
[87] Z. Ma, et al., Overview of catalyst application in petroleum refinery for biomass catalytic pyrolysis and bio-oil upgrading, RSC Adv. 5 (107) (2015) 88287–88297.
[88] M. Crocker, Thermochemical Conversion of Biomass to Liquid Fuels and Chemicals, RSC Publishing, London, UK, 2010.
[89] A. Corma, et al., Processing biomass-derived oxygenates in the oil refinery: catalytic cracking (FCC) reaction pathways and role of catalyst, J. Catal. 247 (2) (2007) 307–327.
[90] S. De, B. Saha, R. Luque, Hydrodeoxygenation processes: advances on catalytic transformations of biomass-derived platform chemicals into hydrocarbon fuels, Bioresour. Technol. 178 (2015) 108–118.
[91] A.G. Gayubo, et al., Transformation of oxygenate components of biomass pyrolysis oil on a HZSM-5 zeolite. II. Aldehydes, ketones, and acids, Ind. Eng. Chem. Res. 43 (11) (2004) 2619–2626.
[92] E.M. Flanigen, J.C. Jansen, H. van Bekkum, Introduction to Zeolite Science and Practice, Elsevier Science, Amsterdam, The Netherlands, 1991.
[93] D. Olson, et al., Crystal structure and structure-related properties of ZSM-5, J. Phys. Chem. 85 (15) (1981) 2238–2243.
[94] C.D. Chang, A.J. Silvestri, The conversion of methanol and other O-compounds to hydrocarbons over zeolite catalysts, J. Catal. 47 (2) (1977) 249–259.
[95] C. Mukarakate, et al., Upgrading biomass pyrolysis vapors over [small beta]-zeolites: role of silica-to-alumina ratio, Green Chem. 16 (12) (2014) 4891–4905.
[96] F.A. Twaiq, N.A. Zabidi, S. Bhatia, Catalytic conversion of palm oil to hydrocarbons: performance of various zeolite catalysts, Ind. Eng. Chem. Res. 38 (9) (1999) 3230–3237.
[97] T.R. Carlson, et al., Production of green aromatics and olefins by catalytic fast pyrolysis of wood sawdust, Energy Environ. Sci. 4 (1) (2011) 145–161.
[98] H.J. Park, et al., Highly valuable chemicals production from catalytic upgrading of radiata pine sawdust-derived pyrolytic vapors over mesoporous MFI zeolites, Appl. Catal. B 95 (3) (2010) 365–373.
[99] F.A. Agblevor, et al., Fractional catalytic pyrolysis of hybrid poplar wood, Ind. Eng. Chem. Res. 49 (8) (2010) 3533–3538.
[100] H. Zhang, et al., Catalytic fast pyrolysis of wood and alcohol mixtures in a fluidized bed reactor, Green Chem. 14 (1) (2012) 98–110.
[101] H. Zhang, et al., Comparison of non-catalytic and catalytic fast pyrolysis of corncob in a fluidized

bed reactor, Bioresour. Technol. 100 (3) (2009) 1428–1434.

[102] H. Zhang, et al., Catalytic fast pyrolysis of straw biomass in an internally interconnected fluidized bed to produce aromatics and olefins: effect of different catalysts, Bioresour. Technol. 137 (2013) 82–87.

[103] P.T. Williams, N. Nugranad, Comparison of products from the pyrolysis and catalytic pyrolysis of rice husks, Energy 25 (6) (2000) 493–513.

[104] B. Zhang, et al., Catalytic upgrading of fast pyrolysis biomass vapors over fresh, spent and regenerated ZSM-5 zeolites, Fuel Process. Technol. 138 (2015) 430–434.

[105] B. Zhang, et al., Catalytic fast co-pyrolysis of biomass and food waste to produce aromatics: analytical Py–GC/MS study, Bioresour. Technol. 189 (2015) 30–35.

[106] Y.-K. Park, et al., Catalytic fast pyrolysis of waste pepper stems over HZSM-5, Renew. Energy 79 (2015) 20–27.

[107] B. Zhang, et al., Microwave-assisted catalytic fast pyrolysis of biomass for bio-oil production using chemical vapor deposition modified HZSM-5 catalyst, Bioresour. Technol 197 (2015) 79–84.

[108] D.P. Gamliel, et al., Investigation of in situ and ex situ catalytic pyrolysis of miscanthus × giganteus using a PyGC–MS microsystem and comparison with a bench-scale spouted-bed reactor, Bioresour. Technol. 191 (2015) 187–196.

[109] N.L. Hammer, et al., Two-step pyrolysis process for producing high quality bio-oils, Ind. Eng. Chem. Res. 54 (43) (2015) 10629–10637.

[110] A. Campanella, M.P. Harold, Fast pyrolysis of microalgae in a falling solids reactor: effects of process variables and zeolite catalysts, Biomass Bioenergy 46 (2012) 218–232.

[111] Y.T. Cheng, et al., Production of renewable aromatic compounds by catalytic fast pyrolysis of lignocellulosic biomass with bifunctional Ga/ZSM-5 catalysts, Angew. Chem. 124 (6) (2012) 1416–1419.

[112] R. French, S. Czernik, Catalytic pyrolysis of biomass for biofuels production, Fuel Process. Technol. 91 (1) (2010) 25–32.

[113] S. Thangalazhy-Gopakumar, S. Adhikari, R.B. Gupta, Catalytic pyrolysis of biomass over H + ZSM-5 under hydrogen pressure, Energy Fuels 26 (8) (2012) 5300–5306.

[114] J. Sun, et al., Carbon-supported *bimetallic Pd–Fe catalysts for vapor-phase hydrodeoxygenation of guaiacol,* J. Catal. 306 (2013) 47–57.

[115] S. Sitthisa, D.E. Resasco, Hydrodeoxygenation of furfural over supported metal catalysts: a comparative study of Cu, Pd and Ni, Catal. Lett. 141 (6) (2011) 784–791.

[116] R.C. Runnebaum, et al., Catalytic conversion of furan to gasoline-range aliphatic hydrocarbons via ring opening and decarbonylation reactions catalyzed by Pt/γ-Al2O3, Catal. Lett. 142 (6) (2012) 664–666.

[117] H.H. Kung, Transition Metal Oxides: Surface Chemistry and Catalysis, Elsevier Science, Amsterdam, The Netherlands, 1989.

[118] S. Stefanidis, et al., In-situ upgrading of biomass pyrolysis vapors: catalyst screening on a fixed bed reactor, Bioresour. Technol. 102 (17) (2011) 8261–8267.

[119] Q. Lu, et al., Catalytic upgrading of biomass fast pyrolysis vapors with nano metal oxides: an analytical Py-GC/MS study, Energies 3 (11) (2010) 1805–1820.

[120] J. Wang, et al., Preparation of jet fuel range hydrocarbons by catalytic transformation of bio-oil derived from fast pyrolysis of straw stalk, Energy 86 (2015) 488–499.

CHAPTER

9

Towards an Aviation Fuel Through the Hydrothermal Liquefaction of Algae

S. Raikova[1], C.D. Le[2], J.L. Wagner[1], V.P. Ting[3] and C.J. Chuck[3]

[1]Centre for Doctoral Training, Centre for Sustainable Chemical Technologies, Department of Chemical Engineering, University of Bath, Bath, United Kingdom, [2]Department of Oil Refining and Petrochemistry, Hanoi University of Mining and Geology, Hanoi, Vietnam, [3]Department of Chemical Engineering, University of Bath, Bath, United Kingdom

9.1 INTRODUCTION

There are numerous potential benefits of using microalgae as a biofuel feedstock, which address many of the issues associated with both first- and second-generation fuels [1]. Algae have faster growth rates, benefiting from higher photosynthetic efficiencies [2] (~3–8% compared to 0.5% for terrestrial plants), shorter growth cycles, and lower water demands compared to terrestrial plants [3,4], and as algae cultivation does not require arable land, algae can be cultivated without competing with food production. Additionally, the high CO_2 fixation capacity of algae means that widespread algal production could potentially be used as a means of sequestering industrial CO_2 emissions. These environmental benefits have clear links to potential economic and social benefits, particularly for developing nations [5]. The potential of microalgae and macroalgae as feedstocks for aviation fuels has already been covered in-depth in chapter 'Feedstocks for Aviation Biofuels'. However, the large-scale implementation of algal technologies as a replacement for fossil fuel resources hinges on improvements in the efficiency of algal production and conversion.

9.2 HYDROTHERMAL PROCESSING OF ALGAL FEEDSTOCKS

Most early investigations into algal biofuels focused on adapting existing biodiesel production technologies such as the use of microalgal lipids as a feedstock. An alternative to biodiesel production from algal lipids is thermochemical processing, which refers to processes such as gasification, pyrolysis, or hydrothermal liquefaction (HTL). These techniques utilize the

Biofuels for Aviation.
DOI: http://dx.doi.org/10.1016/B978-0-12-804568-8.00009-3

entire algal biomass, including proteins and carbohydrates, to generate algal oils. However, although both gasification and pyrolysis are effective for processing of dry feedstocks such as lignocellulose, they are unappealing from an economic standpoint due to the high mass fraction of water in algal feedstock, and the need for energy-intensive drying [6].

In contrast, thermochemical processes conducted in the presence of water, such as hydrothermal gasification, hydrothermal carbonization, and HTL, are more suitable for wet feedstocks. HTL, in particular, has recently been examined for the processing of whole microalgal feedstocks, and can offer significant energetic cost savings, as lipid content and overall biomass productivity are often inversely related [6]. It utilizes water at sub-/near-critical conditions (200–374°C, 5–28 MPa) as both the reaction medium and solvent for a host of simultaneous reactions, converting algal biomass into a bio-crude oil, alongside a nutrient-rich aqueous phase, a solid char, and a number of gaseous products (Fig. 9.1). The bio-crude oil produced is acidic, highly viscous, and contains high proportions of N and O. Much like pyrolysis oil, this bio-crude can be catalytically upgraded to produce various fractions of hydrocarbon fuels, including aviation kerosene. However, the nitrogen content is substantially higher in algal bio-oils than alternative terrestrial oils, and as such, removal of the N-compounds in the bio-crude is one of the most significant challenges that must be overcome in order to advance HTL technology for biofuel production from microalgae.

HTL can be used to process biomass at a concentration of ca. 10–25 wt% in water, reducing the energy consumption of biomass preparation by 88% (with respect to using dry biomass) if a 16% slurry is generated using centrifugation and processed without further drying steps [7]. The mild temperatures used in HTL are well within the range of temperatures encountered in many conventional oil refinery operations [8], and as such, HTL processing for algal biomass can be made energy-efficient and easily scalable. Liu et al. evaluated the life cycle performance of lab-, pilot-, and full-scale scenarios, finding significant improvements in Greenhouse Gas emissions with respect to gasoline and corn ethanol, and a potential Energy Return on Energy Invested of ~2.5 for the full- scale scenario [9], subject to the optimization of a closed-loop system incorporating energy and nutrient recycling.

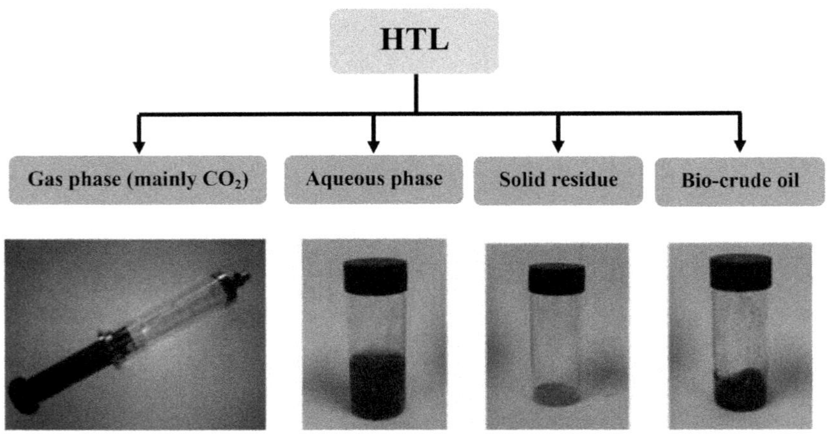

FIGURE 9.1 Products of hydrothermal liquefaction.

9.2.1 Reaction Mechanism

HTL employs water at near-critical conditions as a reaction medium and solvent. At near-critical conditions, the solvent properties of liquid water change substantially (including changes in dielectric constant, density, diffusivity, polarity, viscosity, H-bonding, and H^+ donor capabilities), transforming it from a polar, highly H-bonded solvent to exhibiting behaviour more typical of a nonpolar organic solvent. In this form, it can act as a solvent for a range of organic reactions and facilitate the breakdown of biomass to bio-crude [10] (Fig. 9.2).

The prevention of vapour formation at high pressure means the enthalpy associated with the phase change of water is largely reduced, giving vastly increased energy efficiencies for HTL processing over pyrolysis [12].

HTL is comprised of hundreds of simultaneous reactions, which are not well-characterized in the literature. In hot compressed liquid water (near the critical point of 374°C and 22.1 MPa), there are two competitive reactions: hydrolysis and repolymerization [13,14]. The former is more important and dominant at the early stages of the process, when the microalgae is decomposed and depolymerized into small compounds. These are highly reactive, and rapidly polymerize and form liquid bio-crude, gaseous, and solid products [15–17]. At higher reaction temperatures and longer reaction times, repolymerization, condensation, and decomposition of substances from different phases may occur. This results in an increase of the solid and gas yields and reduction of the bio-crude yield [18–20]. Dehydration reactions, leading to oxygen content reduction, also play a key role in improving the quality of the bio-crude.

9.3 HYDROTHERMAL LIQUEFACTION OF MICROALGAE

Bio-crude yields and properties are dependent on the algal species and its composition. However, operational parameters, such as reaction temperature, heating rate, retention time, the initial biomass loading, and the presence of catalysts also have a significant effect on yields [14]. A wide range of operating conditions have been explored, with liquefaction typically carried out in high pressure batch reactors. Oil yields are generally calculated as a weight percentage of the feedstock weight, using either the dry (d) or the dry ash-free (daf) weight as a basis.

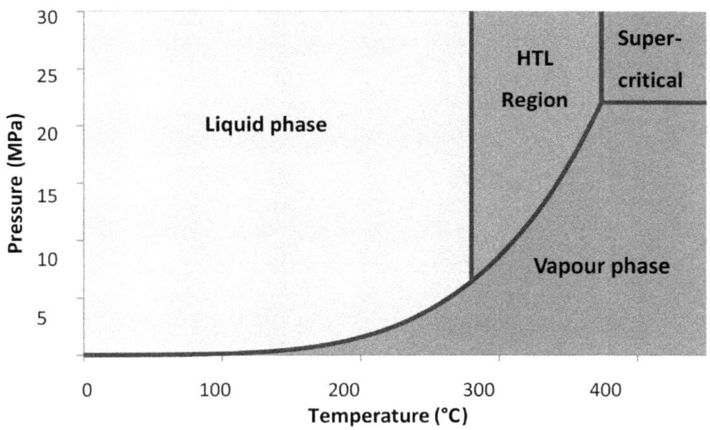

FIGURE 9.2 Hydrothermal liquefaction phase diagram [11].

9.3.1 Effect of Biomass Composition on Bio-Crude Oil Production

It has been previously shown that bio-crude yields from HTL of microalgae can be strongly correlated to the biomass composition. The lipid fraction of the biomass is much more readily converted into bio-crude than other biomass components, with a study by Biller and Ross reporting that conversion efficiencies of different algal components to bio-crude oil were in the order lipids > proteins > carbohydrates [15]. This explains the particularly high yields obtained from lipid-rich microalgae. However, as HTL utilizes all biomass components, a high lipid content is not essential for obtaining good bio-crude yields, in contrast to algal biodiesel, which relies entirely on lipids. This effect has been demonstrated by Yu et al., who reported a yield of 39.0% bio-crude oil from an algae strain containing only 0.1% lipids [21]. Utilizing the entire algal biomass for processing lifts the constraints on high-lipid algal strain selection, and can significantly reduce cost and energy requirements of cultivation by using faster-growing strains or algal communities.

Although the specifics of the effects of biomass composition remain unclear, a number of studies have examined the role of the different component fractions (eg, lipids, proteins, and carbohydrates) in bio-crude production via HTL. Most liquefaction processes under optimized conditions have resulted in bio-crude yields around 30–45% [15,22–25], regardless of algae strain [12], although, notably, Li et al. obtained yields of 55% for *Nannochloropsis* sp. under HTL at 260°C for 60 min and at 25% algal loading, and 83% for *Chlorella* sp. (220°C, 90 min, 25% algal load) [26]. The two algal species have very different compositions (*Nannochloropsis* contains protein, lipids, and carbohydrates in an approximate ratio of 52:14:22, whereas the same components in *Chlorella* have a ratio of 9:60:13), but in both cases, bio-crude yield was enhanced with respect to initial lipid content, further reinforcing the advantages of hydrothermal processing over lipid extraction for algal biodiesel [22].

Higher heating values (HHV) of 25–35 MJ/kg are typical for bio-crude, and higher lipid levels in the biomass appear to correspond to higher HHV, although the two properties are not linearly related. Li et al. found, under optimal conditions, *Nannochloropsis* and *Chlorella* bio-crude HHVs were 31.5 and 34.2 MJ/kg, respectively, with a maximum HHV of 37 MJ/kg. Although this constitutes a significant increase with respect to the starting biomass (22.4 and 32.3 MJ/kg initially), it still falls short of the energy content of mineral crude (41–48 MJ/kg) [27]. Interestingly, higher bio-crude yields do not necessarily correspond to better oil properties; *Nannochloropsis* processed at 350°C yielded 34% oil with an HHV of 38.1 MJ/kg, while 46% oil was obtained at a processing temperature of 310°C, with a much lower HHV of 27.7 MJ/kg [25] (Table 9.1).

The numerous simultaneous reactions occurring under HTL conditions lead to a bio-crude containing a diverse range of chemical compounds, the main constituents of which have been found to be C5–C16 cyclic nitrogen compounds, C15–C33 branched and unbranched hydrocarbons, branched oxygenates, aromatic compounds, and heterocycles [26]. Brown et al. carried out a detailed analysis of bio-crude produced by liquefaction of *Nannochloropsis* sp. Over 90 compounds were detected using GC–MS, although some lighter organics are likely to have been lost during sample extraction, and heavier compounds may not have been detected by GC. A brief overview of the major compounds is presented in Table 9.2.

Elevated heteroatom (O and N) content with respect to mineral crude oil is typical of bio-crude oils [9,10,29]. Higher O and N levels give rise to undesirable fuel properties, such as high acidity and viscosity [24], and the increased diversity of chemical composition can negatively affect combustion performance, storage

TABLE 9.1 Comparison of Documented Bio-Crude Yields, Compositions, and Energy Values

Feedstock	Processing Conditions	Oil Yield (%)	C (%)	H (%)	N (%)	O (%)	S (%)	HHV	References
Spirulina	Biomass only	–	42.26	5.86	3.47	47.26	1.15	20.4	[25]
Spirulina	300°C, 10–12 min, 20% TS, DCM extraction	32.6	68.9	8.9	6.5	14.9	0.86	33.2	[24]
Chlorella	350°C, 60 min, 10% TS	35.8	70.7	6.8	5.9	14.8	0.0	35.1	[23]
Nannochloropsis	350°C, 60 min, 10% TS	34.3	68.1	8.8	4.1	19.0	0.0	34.5	[23]
Spirulina	350°C, 60 min, 10% TS	29	73.3	9.2	7	10.4	0.0	36.8	[15]
Spirulina	220°C, 20 bar, 30 min, 25% TS	38	59.15	5.50	10.47	18.19	1.22	28.7	[25]
Spirulina	310°C, 115 bar, 30 min, 25% TS	38	71.29	8.01	7.66	16.82	0.81	35.2	[25]
Spirulina	350°C, 195 bar, 30 min, 25% TS	30	70.69	8.05	7.22	10.06	0.77	34.3	[25]

TS, Total Solid Fraction.

TABLE 9.2 List of Major Bio-Crude Components Detected by GC–MS [28]

Major Compound	Structure
Indole	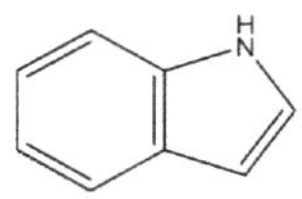
Methylindole	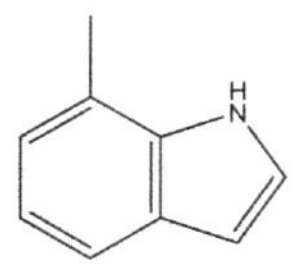
1-Pentadecene	
Heptadecane	
Isomers of 2-phytene	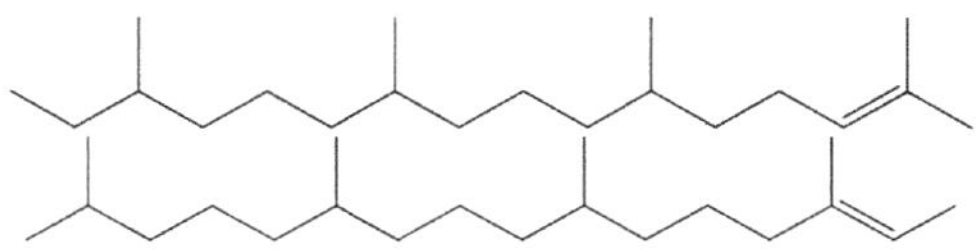
Myristic acid	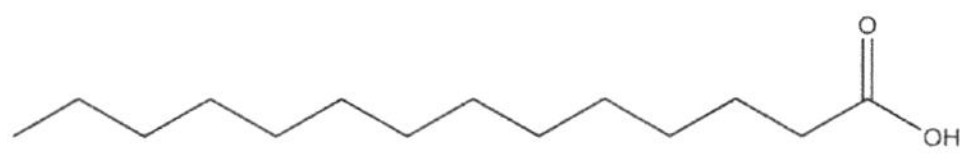
Phytane	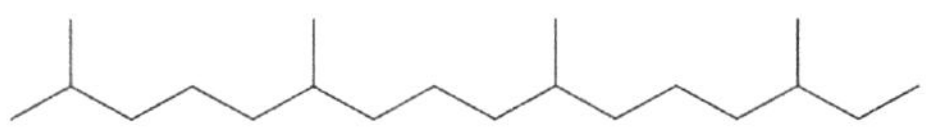

(*Continued*)

TABLE 9.2 (Continued)

Major Compound	Structure
Palmitoleic acid	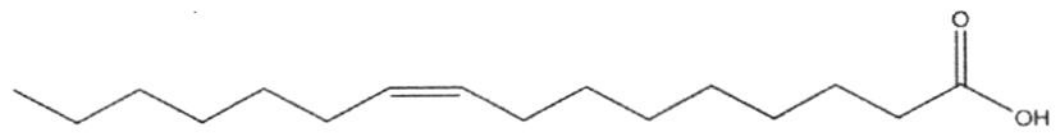
Palmitic acid	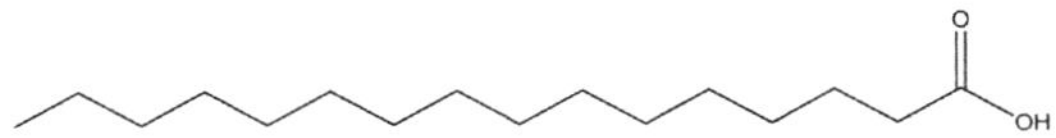
Oleic acid	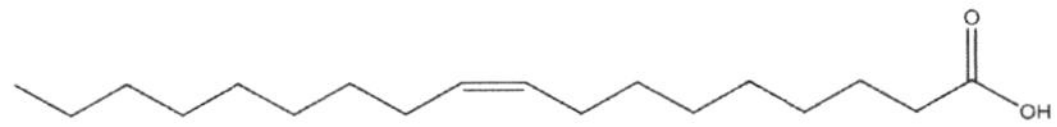
Stearic acid	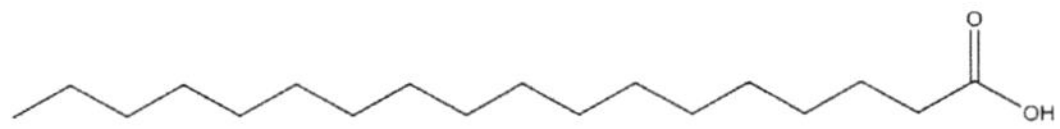
Cholest-4-ene	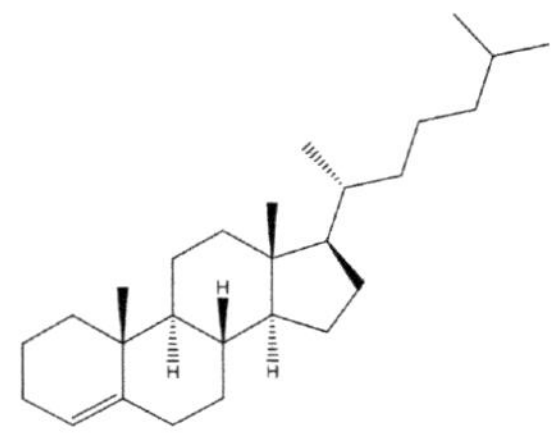
Cholest-5-ene	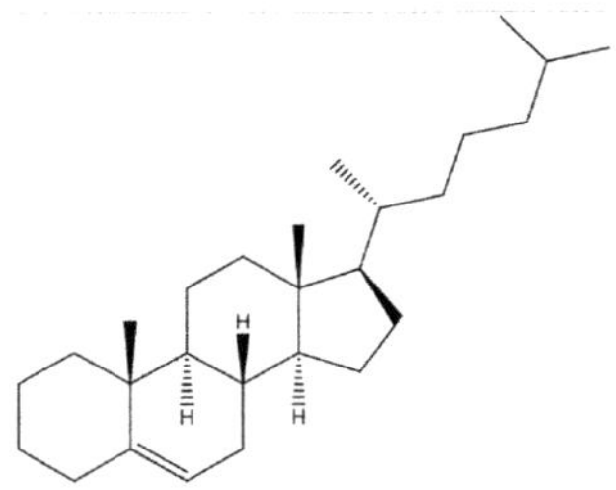
Cholesta-3,5,-diene	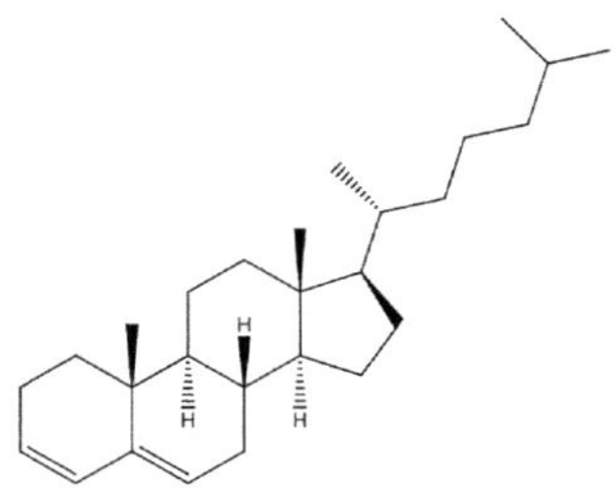
Cholesterol	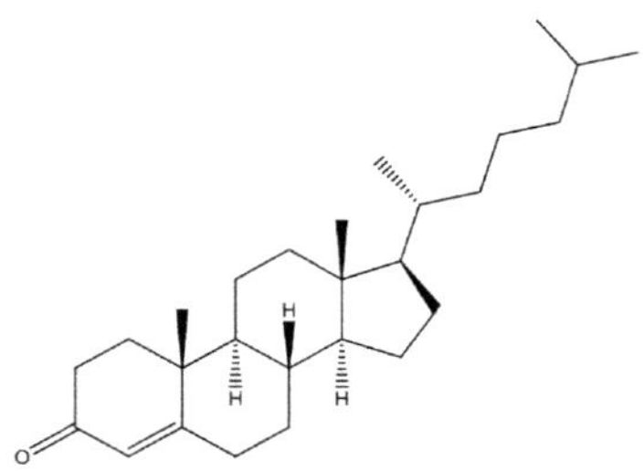

(Continued)

TABLE 9.2 (Continued)

Major Compound	Structure
Cholest-4-en-3-one	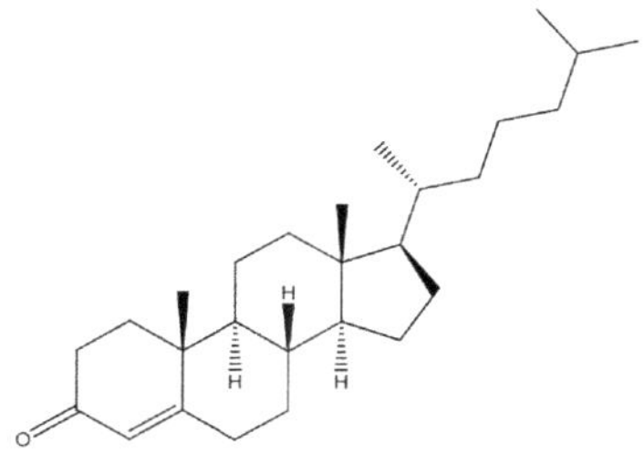
Cholest-4,6-diene-3-one	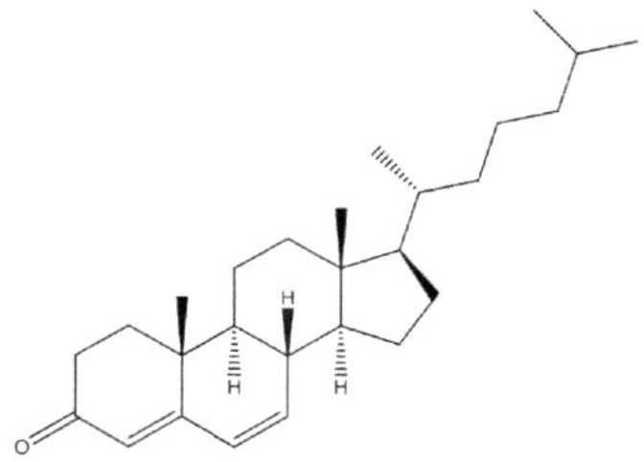

stability, and economic value [24,29]. The heteroatom content of the bio-crude can generally be attributed to the high O and N content of the starting biomass, although there is not always a direct correlation: *Chlorella* biomass containing 1.9% N was processed to bio-crude with a 0.3% N content, while *Nannochloropsis* (7.5% N initially) resulted in a bio-crude with 5.4% N. *Chlorella* bio-crude was also found to have a higher O content (11.5% compared to 9.5% for *Nannochloropsis*), although the O content of the biomass was 21.8%, compared to 29.1% for *Nannochloropsis* [26].

Additionally, the elevated N content of bio-crude of up to 11%, arising from the proteins in the starting biomass (compared to around 0.1, and seldom above 1%, for mineral crude) [30,31], limits its direct usability as a fuel, as it can lead to increased NO*x* emissions. Significantly, higher nitrogen levels may contribute to catalyst poisoning, making these bio-crude oils unsuitable for co-refining in existing refineries [12,31]. Hence, although bio-crude produced via HTL presents numerous advantages compared to lipid extraction, the product needs significant upgrading to obtain a suitable aviation fuel [24]. An alternative solution posed by Garcia Alba et al. suggests that proteins could be extracted prior to HTL processing and sold as a high-value co-product in a biorefinery paradigm [13]. However, this clearly has implications for bio-crude volumes, particularly when low-lipid algae such as *Spirulina* are used.

9.3.2 Effect of Microalgal Loading on Product Formation

One of the key benefits of wet biomass processing in comparison to lignocellulosic biomass or dry processing is that microalgae forms a slurry with water, which can be easily pumped and flowed through the reactor. Particularly for continuous HTL operation, determination of the optimal loading of microalgae in the slurry is crucial.

From the literature, many factors such as the scale of the reactor and economics need to be considered with the selection of the microalgal load used in HTL. Hence, there is no clear connection between the microalgal load (total solid fraction, TS) in the feed and the bio-crude yield. Studies of HTL of algae performed over a large range of microalgal loads (ranging from 4% to 50%) have apparently yielded contradictory results [6]. For example, Jena et al. concluded that an increase of the biomass fraction in the feed by 10% resulted in a jump of more than 20% in the bio-crude yield [32], though the opposite findings were reported by Garcia Alba et al. [13].

The optimum ratio of microalgae and water in the feed is presumably dependent on species and reactor design. A high microalgal loading would be economically beneficial, due to reduction of the water content of the feed, leading to a reduction in the energy required for heating the water, and a correspondingly higher volume efficiency and productivity of the HTL reactor [12]. However, there is expected to be an upper limit due to mass transfer limitations and the formation of blockages in continuous thermochemical conversion [33]. Continuous liquefaction studies have shown that while some practical operational issues of reactor blockage may occur at microalgal loads above 5 wt% in the feed [31], processing of microalgae slurries with microalgal loads of up to 35% (obtaining a bio-crude yield of up to 63.6%) are possible [34]. Another key consideration on the loading of microalgae is the cost of drying the original feedstock, which plays a significant role in dictating the most suitable loading achievable.

9.3.3 Effect of Holding Temperature on Microalgal Processing

The holding temperature has a remarkable influence on the yield and properties of HTL products. In general, an initial rise in the holding temperature triggers bio-crude production. After reaching a maximum value for the bio-crude yield, further increase in the holding temperature inhibits biomass liquefaction.

The optimization of holding temperature varies depending on microalgal species; however, in the majority of cases, compared to a bio-crude yield obtained at ca. 350°C, lower holding temperatures (~310–320°C) generated higher bio-crude yields. This effect was observed by Dote et al. [35], and subsequently by Minowa et al. [36], who observed that yields of bio-crude oil from the liquefaction of *Dunaliella tertiolecta* increased from 30.9% to 43.8% when increasing the holding temperature from 250°C to 300°C, but dropped to 42.6% when processing temperature was increased further to 340°C [36].

These findings are mirrored in the work of Gai et al. for each retention time examined, oil yields from *Chlorella pyrenoidosa* increased when holding temperature was increased from 260°C to 280°C, and proceeded to decline when reaction temperatures were brought to 320°C [37]. The authors suggested that with increasing temperatures, more secondary decomposition reactions may be triggered, resulting in the production of gases and char rather than oils. Regrettably, they were not able to satisfactorily confirm this using the gas yields obtained in their investigation, owing to the high experimental error.

In addition, the holding temperature has a strong influence on the bio-crude properties: increasing the temperature results in a decline in oxygen content and consequently increases the HHV of the oil [36,38]. The nitrogen content in turn appears to increase with holding temperature, likely due to the promotion of protein decomposition at high temperatures, although the opposite effect has been reported in other studies [21].

The length of time the reaction is held at temperature also has a significant effect on product formation. It is expected that short retention times should be sufficient to degrade the biomass effectively, since hydrolysis and decomposition reactions are expected to be rapid under supercritical conditions [39]. However, there is still a relationship between holding temperature and holding time when bio-crude yield is considered. In many cases, to attain a high bio-crude yield, higher maximum temperatures require shorter holding times. For example, Karagoz et al. determined that at low temperatures (150°C) long holding times favoured the conversion of sawdust to oil, with the opposite trend observed at higher temperatures (250–280°C) [40]. Yu et al. also found that while the overall conversion of biomass and gas yield increased at higher temperature [41], a 120-min hold time was needed at 280°C to produce a bio-crude yield of 39.4%, whereas at 370°C, only 5 min was needed to attain a bio-crude yield of 49.4% [13].

While the holding time has an effect on the bio-crude yield, due to the influence of other operational parameters such as temperature, algae strain, and algal loading, it is difficult to quantify differences across all published studies [42]. There is no consensus on the optimum holding time for continuous operation in the literature. However, as hold time has such a significant effect on the oil yield and composition, this needs to be carefully optimized for continuous HTL reactions, requiring more detailed research and analysis of kinetic data in this area.

9.3.4 Effect of Heating Rate on the Crude Bio-Oil

While the effect of heating rate on the bio-crude yield is also dependent on the algal strain, maximum temperature, and algal load, it has recently been suggested that rapid heating rates and shorter reaction times may result in comparable, if not better, bio-oil yields than slow temperature ramps with longer retention times. For example, bio-crude yields of up to 66% were obtained during the conversion of *Nannochloropsis* using a holding time of only one minute. This was achieved by placing batch reactors into a sand bath with a temperature significantly above the required set-point temperature, enabling very fast heating rates [43]. This would support conversion by continuous HTL reaction. However, other studies report that while higher heating rates are beneficial for bulk fragmentation of biomass and inhibit char formation, they do not have a great effect on the bio-crude yield [42].

9.3.5 Effect of Catalysts on the HTL of Microalgae

The effect of both homogeneous catalysts, such as organic acids and organic bases, and heterogeneous catalysts, such as metallic oxides, on the HTL of microalgae has been investigated. Homogeneous organic catalysts are desirable since they are inexpensive, do not require separation prior to further upgrading, and reduce the degree of coking [44]. In many cases, HTL experiments have been conducted in the presence of inorganic salts such as sodium carbonate and other water-soluble species [15,44,45]. However, while these catalysts appear to selectively promote the decarboxylation of carbohydrates, they have been found to have a detrimental effect on oil formation from lipids and proteins [15]. One study on the conversion of albumin suggests that the nitrogen partitioning to the oil may be reduced with the presence of sodium carbonate [35], but this observation has not been reported in other similar studies [44].

While homogeneous catalysts are difficult to recycle, heterogeneous catalysts allow post-HTL separation and high reaction selectivity. Liquefaction in the presence of heterogeneous metal catalysts has been reported to give increased liquefaction yields compared to catalyst-free conditions. Furthermore, the presence of the catalyst had a significant influence on the physical properties (eg, the colour and apparent viscosity) of the product oils [45]. The highest oil yields (57%) were achieved with supported Pd catalysts; however, Ni- and Ru-based catalysts produced bio-crudes with the lowest N content [45]. Pt, Ni, and CoMo catalysts were also reported to enhance the degree of deoxygenation, but with reduced bio-crude yields [23].

9.3.6 Effect of Initial Pressure and Additional Reductive Gas

For batch HTL operation, high pressure is required to ensure the water remains in the liquid phase, reduces the enthalpy of phase change of water, and improves the efficiency of the HTL process. Several investigations have introduced additional pressure by charging the reactor with a gas prior to reaction. For batch HTL operation, bio-crude yield was reportedly not affected by the initial pressure [41]. The gases most commonly used are either inert gases (eg, nitrogen, argon) or reducing gases such as CO and H_2. While reducing gases are added in order to reduce the oxygen content in the bio-crude, and stabilize the fragmented products of liquefaction, their hazardous nature, gas channelling, and maldistribution need to be carefully considered during the operation [42]. These problems can be overcome by using hydrogen donor solvents, such as tetralin and tetrahydrophenanthrene [42], which have been found to be more favourable than the parent gases in the hydrogenation of biomass fragments. Hydrogen donors (eg, tetralin) were reported by Wang et al. to be more effective than H_2 for increasing the bio-crude yield and reducing the level of oxygenates in the bio-crude, which typically corresponds to an increase in HHV and overall improvements in bio-crude quality [46].

9.4 MACROALGAE AS A FEEDSTOCK: PROSPECTS AND CHALLENGES

As an alternative to microalgae, seaweed (macroalgae) can be processed through HTL using a similar process to that described above for microalgal feedstocks [47]. While currently underdeveloped as a feedstock, macroalgae has enormous potential as a future feedstock for fuel production (see chapter: Feedstocks for Aviation Biofuels) [48]. Currently, the global consumption of seaweeds for fertilizers, cosmetics ingredients, foodstuffs, and phycocolloids is a multi-billion dollar industry [47]. The majority of macroalgae comes from a relatively small number of species with five genera, *Laminaria* (also called Saccharina for some species), *Undaria*, *Porphyra*, *Eucheuma*, and *Gracilaria*, representing 76% of the total tonnage of cultured macroalage [47].

Macroalgae tend to contain a high carbohydrate content and represent a potential source of biomass for bioenergy and chemicals [49]. According to Anastasakis and Ross, wet biomass with an assumed moisture content of 80% requires 10.4 MJ to evaporate the water, equivalent to 87% of the total calorific value of the seaweed (the HHV of the dry biomass of *Laminaria saccharina* was calculated to be 12 MJ/kg) [47]. Therefore, it has been suggested that only wet processes such as HTL will be energy efficient because of the high energy required to dry the algae [50]. Moreover, macroalgae from marine environments contain high alkali contents, which can generate issues such as slagging and fouling during combustion and potentially during pyrolysis as well [51].

In addition, macroalgae tend to contain higher carbohydrate and much lower lipid contents than microalgae, and, if from a marine origin, can also possess an ash content of up to 25%. This will substantially affect HTL reaction pathways; further research is needed in this area.

All studies to date have focused on batch systems [49], where macroalgal samples were harvested, treated by washing with fresh water, dried, and pulverized prior to the HTL stage [51]. Bio-crude yields up to 23% have been commonly reported from macroalgal HTL [19,50]. However, Bach et al. reported an oil yield of 79% from the HTL of *L. saccharina* at 350°C with a high heating rate (585°C per minutes) [52]. Neveux et al. found that HTL of *Oedogonium* spp. and *Cladophora* spp. (freshwater macroalgae) gave bio-crude yields of 26% and 20%, respectively, on a dry mass basis – higher than marine macroalgae *Derbesia* spp., which only yielded 20%, though this generated less solid residue than the freshwater macroalgae species [53].

Jin et al. carried out co-liquefaction experiments of microalgae (*Spirulina platensis*) and macroalgae (*Entermorpha prolifera*) at 340°C, and discovered optimal synergistic conversion effects for a 50/50 mixture of the two algae. While the fatty acid content of *Spirulina* strongly improved the conversion of *Entermorpha*, inorganic salts present in the latter feed did not considerably enhance bio-oil production [54]. In general, a greater amount of energy was recovered in the bio-crude produced, though it contained more nitrogen than the separately produced micro- and macroalgae bio-oils.

Macroalgal biomass has high potential for biofuel production, although there are still significant technological challenges to be overcome. These challenges include cost-effective cultivation and economic harvesting, and more efficient preprocessing of the feedstock. In addition to cleaning, dewatering, and milling steps [55], marine biomass requires desalination in order to prevent equipment corrosion, which can carry a particularly heavy environmental and energetic cost [56]. Additional challenges include developing a continuous HTL reaction system that can readily tolerate high ash content, and designing a biorefinery concept where the supply of aquatic biomass is substantial and secure, and where the aqueous phase can be used as an effective fertilizer. Lenstra et al. reasoned that as the current nearshore cultivation, harvest, and preprocessing costs for macroalgae range between €336 and €669 per tonne, substantial technological development is needed if a sustainable biofuel is to be produced from this feedstock [50,57]. Moreover, the seasonal nature of seaweed growth and culture must be considered for commercial-scale biofuel production, and the technologies of storage or preservation of macroalgae after harvesting need to be developed to enable year-round fuel production.

Similarly to microalgae, the batch experiments only give limited insights into a fully functioning industrial continuous process. Advancement and commercialization of HTL processing of macroalgae require the use of a continuous process with incorporated heat recovery, increasing the energy efficiency and allowing for high throughput. Particular engineering challenges on the processing side include the high salt content of macroalgae and the associated corrosion risk.

The cultivation of macroalgae as a sole biofuel feedstock may not be currently profitable [47]. However, macroalgae contains a diverse range of extractable biochemicals, which could potentially be used to add value to macroalgal fuels in a biorefinery paradigm. The vast majority of farmed macroalgae is currently utilized for human consumption (accounting for 83–90% of the global seaweed industry [55]), but other macroalgae-derived materials also have a significant market value. In particular, seaweed hydrocolloids (predominantly comprising agar, alginate, and carrageenan) account for a significant portion of the remaining value (estimated around $545 million annually) [58,59]. Other macroalgae derivatives are also utilized in the agricultural, chemical, and pharmaceutical industries [60]. Lammens et al. have also evaluated the potential of algal proteins as a source of high-value bio-based chemicals, such as *N*-methylpyrrolidinone, *N*-vinylpyrrolidinone, and acrylonitrile [61]. Overall, there is significant scope for extraction of higher-value materials from macroalgae prior to conversion to fuels, and the application of a biorefinery concept could increase the value of the seaweed biomass and improve the economics of production of bioenergy from macroalgae, thereby leading to a more rapid commercial realization of algal-derived biofuels.

9.5 CHARACTERISTICS OF BIO-CRUDE OIL FROM HYDROTHERMAL LIQUEFACTION OF ALGAE

The bio-crude oil produced by the HTL of microalgae requires significant upgrading before it can enter the conventional fuel stream and be used to produce aviation kerosene. The high levels of nitrogen (typically between 4% and 8%) and sulphur (up to 1%), in particular, must be almost completely removed if the resulting fuels are to comply with emission standards. The crude bio-oils from HTL of algae can be blended with fossil oil and upgraded to conventional kerosene and other hydrocarbon fuels in a conventional refinery. While it is true that these facilities are capable of processing oils with high levels of sulphur, common industrial catalysts are not able to tolerate the high levels of oxygen and nitrogen seen in the bio-oils. Therefore, at least partial upgrading of the oils is required to reduce the oxygen and nitrogen concentrations to a level comparable to fossil crude oil before they can be sent to a refinery for fractionation.

Detailed chemical analysis of the oils is difficult, as they can contain up to several hundred compounds, including phenols, alkanes, alkenes, fatty acids, ketones, aldehydes, nitriles, amides, and nitrogen heterocycles such as indoles and pyridines [62,63]. In addition, due to the high boiling point range of the oils, most reports have only achieved partial characterization [6,64]. Elemental analysis is therefore one of the most reliable methods to achieve a like-for-like comparison of different crude oils and of the performance of different upgrading studies.

9.5.1 Upgrading of Bio-Crude Oil Produced by the HTL of Microalgae

Only a few studies have investigated the upgrading of HTL oils from microalgae (Table 9.3), with the majority of these studies conducted in batch and achieving only partial nitrogen removal. While the choice of catalyst was found to impact the physical properties of the reaction products, it had little impact on the denitrogenation performance itself [66]. Instead, the degree of denitrogenation appears to be mainly dependent on the reaction temperature, which suggests that these reactions are thermally controlled [66]. Unfortunately, the high temperatures also result in a significant reduction in the hydrogen content of the oils, presumably due to an increase in the aromatic content of the reduction product above desirable levels. Potentially, the poor performance of the selected catalysts under these conditions could lead to insufficient hydrogen being provided for complete hydrogenation. It has also been suggested that the accumulation of ammonia may be limiting the complete reduction of nitrogen compounds [69]. Nonetheless, more recent studies have demonstrated the full catalytic upgrading of bio-crude directly to suitable hydrocarbon fuels. Research has focused on both in situ heterogeneous catalysis and hydrotreatment of the bio-crude after separation [70]. Duan and Savage reported on the use of a 5% Pd/C catalyst for

TABLE 9.3 Summary of Bio-Crude Upgrading Studies Conducted to Date

Feedstock (Microalgae)	Bio-Crude Quality (wt%)			Catalyst	Reaction Temperature	Pressure	Reaction Mode	Reaction Time	Product Quality (wt%)			References	Comments
	N	S	O						N	S	O		
Bio-crude from HTL of *Nannochloropsis* at 350°C for 1 h	5.32	0.56	8.35	HZSM-5 (0–50 wt% loading)	400–500°C	4.35 MPa (hydrogen)	Batch	0.5–4 h	1.6–2.71	Bdl	0.39–2.81	[65]	
Bio-crude from HTL of *Nannochloropsis* at 340°C for 4 h	4.80	0.48	8.07	Pt/C, Mo_2C, HZSM-5 (5–20 wt%)	430–530°C	3.5 MPa (hydrogen)	Batch in supercritical water (water/oil mass ratio of 4:5)	2–6 h	1.50–3.61	Bdl	0.13–5.31	[66]	Temperature has biggest impact on oil properties
Bio-crude from HTL of *Nannochloropsis* at 320°C for 4 h	4.89	0.68	6.52	Pt/C (25 wt %), HCl, NaOH	400°C	3.4 MPa (hydrogen)	Batch in supercritical water (water/oil mass ratio of 1:1)	4 h	2.17–2.79	Bdl	4.31–4.71	[67]	
Bio-crude from HTL of *Chlorella* sp. at 350°C for 1 h	7.3	Nd	7.8	Pt/γ-Al_2O_3 (0–40 wt%) HCOOH (0–88 wt%)	400°C	6 MPa (hydrogen)	Batch in supercritical water (0–43 wt%)	1 h	2.4–5.8	N/A	4.7–17.9	[8]	With formic acid, reacted with the crude algal oil resulting in increased yield
Bio-crude from HTL of *Desmodesmus* sp. at 200–375°C for 5–60 min	0.2–6.3	Nd	10–39	HZSM-5 (SiO_2/Al_2O_3 = 280), zeolite: sample mass ratio of 20:1	600°C		Pyrolysis probe (heating rate: 20°C per second, pyrolysis time: 10 s)		0.02–0.14	N/A	0.08–0.30	[68]	Elemental calculated from structure of compounds identified by GC
Bio-crudes from HTL of *Nannochloropsis* sp. at 344–362°C, in continuous reactor	4.0–4.7	0.3–0.5	5.3–8.0	Co-promoted MoS_2 on fluorinated alumina support	125–170°C (1st quarter of reactor); 405°C (Main reactor)	13.6 MPa in hydrogen flow	Bench-scale hydroprocessing system	LHSV of 0.14–0.20 h^{-1}	<0.05–0.16	<50 ppm	0.8–1.2	[34]	

hydrotreatment of algal bio-crude oil, finding that the bio-crude HHV from *Nannochloropsis* was significantly improved from approximately 37 MJ/kg to 44 MJ/kg, with corresponding improvements in the kinematic viscosity, and significant reductions in N, O, and S content. However, these results were achieved over long retention times of up to 4 h and 80 wt% catalyst loadings, and the yield was substantially decreased [71]. Other studies have centred around supported Pd catalysts, though more recently, Li and Savage reported the upgrading of algal crude oil produced by HTL using HZSM-5, generating a paraffin-like oil composed of >95 wt% C and H, which retained 80% of the energy content of the initial bio-crude and was suitable for use as a liquid fuel [65].

A number of studies have attempted to upgrade crude bio-oils produced by the liquefaction of *Nannochloropsis* sp. [66,67] or *Chlorella* sp. [8] in the presence of supercritical water. It was hoped that the reaction of water with hydrocarbons would generate additional hydrogen which, in turn, would promote the removal of heteroatoms from the bio-crude. However, while the presence of water did not appear to have a beneficial effect on the denitrogenation performance, it was found to increase the oxygen content of the reaction product [8]. The best results in the presence of supercritical water were obtained over Pt/C catalyst at 530°C with a reaction time of 6 h, with almost complete sulphur removal and a reduction in nitrogen content from 4.0% to 1.5% [66]. In contrast, upgrading of a liquefaction oil obtained from *Nannochloropsis* in the absence of water over an HZSM-5 catalyst, at 500°C, and with a reaction time of 4 h, resulted in a reduction in the nitrogen content from 5.3% to 1.6% [65].

Alternatively, Roussis et al. performed experiments to thermally treat HTL bio-crude oil over a temperature range of 350–450°C, with a residence time of 60 min [72]. While the total nitrogen content was not affected by the thermal treatment, the total oxygen content was significantly reduced from 5.7% in the bio-crude to 0.2% in the product oil treated at 400°C. Thermal treatment was also found to reduce the total acid content, as well as the boiling point range of the bio-crude oils, making them more volatile and less viscous. In addition, trace metals were partly removed from the bio-crude. Similar findings were also reported by Bai et al., who used a two-step process to hydrothermally treat the bio-crude oil [73].

These findings demonstrate that it is possible to process the HTL bio-crude in a similar way to heavy crude oil, continuously treating in a number of subsequent processes, including thermal treatment, hydrotreatment, and hydrocracking, to make fractions suitable for further processing towards production of different commercial products, such as gasoline, diesel, and jet fuel.

In a similar effort, promising results were achieved during the continuous hydrotreating of bio-oils produced by the HTL of various strains of *Nannochloropsis* over sulphided CoMo catalyst supported on fluorinated alumina [34]. Single-stage upgrading at 405°C with a space velocity of $0.20\ h^{-1}$ and a pressure of 14 MPa in excess hydrogen produced oils with a nitrogen content ranging from 0.07 wt% to 0.25 wt%. An even lower nitrogen content could be achieved by pretreating the oil at temperatures between 125°C and 170°C in the first quarter of the reactor, before conversion at 405°C. While this study demonstrated that almost complete denitrogenation of microalgal crude oils can be achieved under continuous conditions, further studies are required to verify these findings and optimize the reaction conditions.

9.6 CONTINUOUS HTL SYSTEMS AND CHALLENGES IN ADVANCING THE TECHNOLOGY

The vast majority of the work to date has focused on batch systems for hydrothermal processing, which, while providing a wealth of information regarding optimal system conditions and chemical mechanisms, is of limited use in developing processing systems to deliver sufficient volumes of algal bio-crude for widespread use as an aviation fuel feedstock. Moreover, the use in most cases of small batch reactors has necessitated use of solvents for the recovery of the bio-crude oil product, leading to complications in determination of the oil yield and distortion of the compositional analysis and properties by the partitioning of solvent-extractable and water-soluble components.

While feeding wet biomass slurry at pressure is highly challenging, in order to advance HTL technology, it is necessary to move to continuous bench/pilot-scale HTL reaction systems. This would help to avoid the excessively long warm-up and cool-down periods that are problematic with batch reactors. In addition, heat exchange/recycling can be integrated into continuous systems, making the process simultaneously more economically feasible and more controllable. To date, only a small number of studies have examined the continuous HTL of biomass [31,34,74], with the majority of these systems not yet approaching a demonstration scale of operation.

Ocfemia et al. developed a small-scale continuous HTL reactor system, including a high-pressure slurry feeder, a process gas (eg, CO) feeder, a continuous stirred tank reactor (CSTR), and a vapour-liquid separation vessel, to process swine manure [74]. A CSTR was selected over a plug-flow reactor for ease of handling, to reduce plugging, and to allow good temperature control throughout the reaction vessel. At a processing temperature of 305°C and a residence time of 80 min, bio-crude oil was produced with an HHV of 31 MJ/kg and bio-crude yields ranging from 60% to 70.4%. The reactor was successfully operated for 16 h continuously with a capacity of up to 48 kg of manure slurry (eg, a solid content of 20%). Although no clogging or accumulation was observed using the CSTR, a pump durability issue was encountered after several experiments due to rapid wearing of the piston and sleeve.

Jazrawi et al. described a continuous system for biomass processing to bio-crude using *Chlorella* and *Spirulina* at relatively low algal loading (1–10%) [31]. The design of the HTL reactor differed from conventional continuous reactors in that the slurry was not stirred and the reactants were flowed through coiled stainless steel tubes submerged in a fluidized sand bath. Due to the low biomass loading of some samples, quantitative gravity separation of the bio-crude from the aqueous phase was challenging, and was achieved instead using DCM extraction of the oil phase from the reaction mixture. The group struggled to operate using algal loadings >10%, but nonetheless obtained maximum yields of 41.7% from *Chlorella* processed at 350°C, comparable to yields obtained under batch conditions. They also found that more severe operating conditions resulted in a decrease in bio-crude O content and a lower-molecular-weight bio-crude. Jazrawi et al. also encountered problems with pumping wet biomass slurry at high pressure [31]. However, they argued that it is more difficult to control the pressure in a smaller slurry system to maintain stable flows, as lower flow rates are accompanied by an increased risk of flow disruption and blockage due to greater incidence of particles damaging the valve seals of smaller pumps. The authors believed that this problem could be reduced if the HTL process were to be implemented at larger scale.

Elliott et al. developed a bench-scale continuous system for HTL of algal biomass in 2013 [34]. The system included the combination of a CSTR and a plug-flow reactor, with integrated modules for catalytic upgrading and sulphur stripping. This more complex hybrid HTL reactor configuration was developed as a direct result of plugging problems experienced previously with a plug-flow reactor system. The solids and bio-crude product were separated without the use of solvents through the use of an in-line filtration unit. The group was able to operate at far higher loadings than Jazrawi et al. (up to 35%), obtaining maximum bio-crude yields of 63.6% using a specially cultured *Nannochloropsis* strain. The group presented a proposal for a biorefinery, with water and nutrients in the post-HTL water recycled into the algal growth stage, although the model did not explore the potential of using the CO_2-rich gaseous phase as the CO_2 source to supplement algal cultivation [34].

9.7 PROCESS INTEGRATION FOR AN ADVANCED BIOREFINERY

Despite the many potential benefits of algal biofuel technology, for costs and sustainability benefits to be optimized, secondary value streams must be considered. Conventional crude oil refineries generate a wide range of products, including paraffin, lubricants, gases, sulphuric acid, petrochemicals, and feedstocks for plastic manufacture, alongside fuels. Similarly, algal fuel production has the potential to co-produce high-value chemicals, such as proteins, vitamins, and trace minerals in addition to liquid fuels [5]. However, few industries have the economic capacity to accommodate high volumes of co-products. The only industries on a similar scale to fuels are mining, agriculture, plastics, and environmental remediation.

9.7.1 Algae Cultivation and Waste Water Treatment

Significant value could be added by combining the microalgal production process with a secondary objective, such as waste water treatment, the recovery of metals from mining waste, or carbon sequestration from power plant effluents. Waste water treatment in particular could provide substantial quantities of algal biomass, without requiring the addition of costly nutrients. The environmental remediation industry has vastly expanded in recent years, with the remediation market in the United States alone generating an estimated $12.8 billion in 2010, of which waste water treatment represented just under 50% [75]. Wastewater treatment via algal remediation lends itself well to synergistic combination with biomass production, and numerous investigations into the cultivation of algae on industrial, municipal, and agricultural waste waters have already been carried out [76,77].

Global municipal waste water production amounts to approximately 300 billion m^3, of which just over 50% is currently treated [78]. Assuming an average biomass yield of 1 g/L, and a liquefaction yield of 30%, complete conversion of the existing waste water treatment facilities could result in an annual bio-crude production of up to 90 billion L.

In addition, waste water treatment using microalgae has been proven to be highly effective in reducing the concentration of N and P pollutants, which are used as nutrients, in the effluent [79]. Compared to conventional methods such as chemical precipitation or the production of an activated sludge, forming waste products which are often disposed of by landfill, microalgal treatment provides a much more sustainable route as it allows the efficient recycling of these nutrients [77].

Metal recovery from mining waste could be another lucrative secondary function of microalgae cultivation. A number of studies have explored the concept of algal remediation, and although most studies to date have employed microalgae only as a nonliving adsorbent, remarkable metal concentrations were observed in the recovered biomass, which could potentially be retrieved from the liquefaction residue. For example, one gram of biomass produced from *Oedogonium* sp. adsorbed up to 145 mg of Pb(II) when exposed to a lead solution containing a metal concentration of 200 mg/L for 90 min, corresponding to a lead recovery of 35% [80]. Chromium absorption by a strain of the filamentous algae *Spirogyra* was somewhat lower, at only 14.7 mg/g, however the initial concentration of chromium in the treated solution was also substantially depleted, with a reduction of 5 mg/L observed [81]. More recent studies have also looked at the direct cultivation of algae on industrial waste water, with acid mine drainage representing another large-scale source of water pollution in need of continuous remediation [82]. The diatom *Planothidium lanceolatum*, for example, could be grown at respective Cd, Zn, and Cu concentrations of 0.1, 0.2, and 0.4 mg/L, without significant effects on photosynthesis [83]. Metal toxicity is a clear challenge, however, and must be addressed before living microalgae can be used for mining waste treatment [84].

9.7.2 The Biorefinery Concept and Nutrient Recovery

Within the biorefinery paradigm it is necessary to consider upstream sustainability factors, as well as final product quality. Despite the advantages conferred by HTL, cultivation of algal biomass is still a relatively energy-intensive process and requires high inputs of water, nutrients and CO_2 [85]. The HTL paradigm goes some way to solving those issues, though the process must be optimized to render the process economically feasible. As well as optimizing bio-crude yields, it is also vital to maximize the carbon efficiency, retain efficient water and nutrient recycling, and ensure an inexpensive source of CO_2. These issues can be at least partially addressed by encompassing nutrient recycling from the aqueous phase, as well as augmenting algal growth using the CO_2 from gaseous products [13,32,86]. This nutrient-cycling concept has been discussed in recent literature, and has been described as 'Environment-Enhancing Energy' or E^2-Energy/E2E by Zhou et al. at the University of Illinois [87–89].

A reliable, low-cost water supply is critical to the success of biofuel production from microalgae to fulfil both economic and sustainability criteria [90]. Recirculating water is an obvious route to reducing overall water consumption. For nonthermochemical methods, this comes with a higher risk of infection and growth inhibition by bacteria, fungi, and viruses found in recycled water directly after biomass extraction. HTL confers the additional advantage of destroying biotic toxins (bacteria, viruses, and even prion proteins) in the bio-crude and aqueous phase [10], although non-living and inorganic growth inhibitors remain.

The HTL route also concentrates trace mineral matter and nutrients, such as nitrogen, phosphorus, and potassium, as well as Fe, Ca, Mg, and polar organics in the aqueous phase [12,38,86]. This could present a route for simultaneous nutrient recycling and remediation of waste water from a synergistic process, which could not be discharged without further treatment.

Recovering phosphorus is crucial for continuing global agriculture; as well as being a dwindling resource in itself, phosphorus is extracted using fossil fuels, therefore peak oil and peak phosphorus may be intrinsically connected [91]. Apart from the impact on mineral resources, the vast quantities of nutrients needed for cultivation can severely affect the energy balance owing to the energy-intensive production [92]. As a result, nutrient provision for algal cultivation is a key sustainability concern [93], and nutrient recovery is a crucial step in making third-generation biofuel production viable [94].

In hydrothermal processing, high protein levels in the feedstock can lead to accumulation of light organics in the aqueous phase [26] (up to 50% of biomass carbon has been found to accumulate in process water) [86], leading to reduced bio-crude yields and poor carbon efficiency. Although this may be detrimental in some ways (in addition to carbon losses from the bio-crude, it has been noted that phenols present in the aqueous process water may inhibit algal growth) [95], some studies have found that elevated levels of total organic carbon (TOC) in algal growth media may actually supplement algal mixotrophic growth, leading to higher biomass yields [96].

Consequently, reusing HTL process water for algal growth may also improve carbon efficiency, with nutrient recovery from the aqueous phase simultaneously being used to treat process waste water, before release into waterways. Biller et al. carried out a comprehensive study in 2012 examining the impact on nutrient recycling potential. Studies on HTL of *Chlorella vulgaris*, *Scenedesmus dimorphus*, and *Spirulina platensis* found significantly higher nutrient levels in HTL process water than in standard growth media, 3N-BBM + V (Table 9.4) [86].

TABLE 9.4 Nutrient Content of HTL Process Water (*Spirulina*, 300°C) Compared to Standard Growth Media 3N-BBM + V [86]

Nutrient	HTL Process Water Conc./ppm	3N-BBM + V Conc./ppm
TOC	15123	–
Total N	8136	124
NH_4^+	6295	–
PO_4^{3-}	2159	135
K	1506	63
Acetate	7131	–
NO_3^-	194	547
Phenols	98	–

The high levels of ammonium, phosphate, nitrates, and potassium ions are promising for potential nutrient recovery. However, elevated phenol levels could be problematic due to their inhibitory effects on algal growth [86], even at moderate concentrations (100–200 ppm) [97]. Additionally, high protein content in the starting biomass corresponds to elevated levels of nitrogen heterocycles (eg, pyrolidinones, piperidines, pyrroles, indoles) [15,23], thought to be produced via the Maillard reaction between amino acids and glucose/fructose [37]. This could, however, be beneficial if certain strains of algae are capable of utilizing organics in mixotrophic growth.

Several groups have already reported successful algal cultivation using HTL process water [32,84,86,89,98]. Jena et al. found that dilution was necessary to bring nutrient levels down to levels comparable to those in commercial growth media, but found that a 10-fold dilution was still too strong for algal growth to occur [32]. Consequently, Biller et al. performed tests using process water from the HTL of *Spirulina* at 300°C, diluted by a factor of 50, 100, and 400, for the cultivation of *Spirulina*, with algal growth measured using chlorophyll absorbance. The finding that no growth occurred at either 50× or 100× dilution (although at this point, nutrient levels are brought down significantly below typical concentrations encountered in growth media) suggests that the concentrations of phenols, fatty acids, and metals such as Ni were still above inhibitory levels. Finally, a 400× dilution was found to be effective, with *Spirulina* exhibiting very similar growth rates in diluted process water and 3N-BBM + V [86]. Notably, no growth was seen in pure distilled water, confirming the role of the process water as a nutrient source. For the algae *Chlorogloeopsis*, however, appreciable growth was seen at dilutions as low as 100×, while the cell counts at 400× dilution were actually increased by a third compared to those seen for standard growth media, suggesting that the organic carbon could supplement mixotrophic growth in this case. Growth does not seem to be inhibited by low concentrations of PO_4^{3-} (for *Chlorogloeopsis* at 200× dilution). The study demonstrates the potential of HTL process water to be used as a growth medium, although the optimum dilution was different for each strain. Additionally, Pham et al. demonstrated that recycling into algal cultivation systems, coupled with granular activated carbon treatment, was an effective means of removing organic toxins and significantly reducing HTL waste water cytotoxicity for environmental release [98]. Another recent investigation has demonstrated that HTL process water from liquefaction of *Spirulina* at 350°C is capable of supplementing growth of a strain of algae cultivated in acid mine drainage. The algae demonstrated appreciable growth over 15 days, and was able to utilize 75% of the phosphate provided by the HTL process water (1:100 dilution in synthetic acid mine drainage) [84].

These results suggest that there is significant potential to vastly reduce the amount of additional nutrients required for algal growth by using recycled HTL process water as a growth supplement. If the algal strain at the site is found to exhibit mixotrophic growth, water-soluble organics in the HTL aqueous phase could be incorporated (improving carbon efficiency and biomass yields). Alternatively, water-soluble organics could be extracted prior to process water recirculation and sold as high-value commodity chemicals.

9.8 CONCLUSIONS AND FUTURE DIRECTION

It is unlikely that the HTL of algae will give an aviation fuel alone, but it could produce a hydrocarbon-rich feedstock for upgrading into the standard fuel streams. This has the advantage of producing a bio-Jet A-1 drop-in replacement, rather than a fuel limited by blend level. Both seaweed and microalgae have potential as feedstocks to sufficiently supply the market, though currently both types of resources have specific issues that must be solved to produce an economic feedstock. There is huge potential for the commercialization of the HTL technology, where algae can be used as a feedstock. However, to advance the technical maturity of HTL, there are still a number of challenges that need to be addressed. Potential solutions could include reducing the risk of large-scale pumpability, reducing capital cost by moving away from a CSTR configuration to a scalable plug-flow reactor configuration, selection of more appropriate materials of construction for process design, upgrading bio-crude oil, and recycling of nutrients from the recovered by-products (P in the solids and N, K, and C in the aqueous phase) for process cost savings and improved sustainability.

References

[1] F. Alam, A. Date, R. Rasjidin, S. Mobin, H. Moria, A. Baqui, Biofuel from algae-is it a viable alternative?, Proc. Eng. 49 (2012) 221–227.

[2] J. Pirt, The thermodynamic efficiency (quantum demand) and dynamics of photosynthetic growth, New Phytol. 102 (1986) 3–37.

[3] L. Rodolfi, G.C. Zittelli, N. Bassi, G. Padovani, N. Biondi, G. Bonini, et al., Microalgae for oil: strain selection, induction of lipid synthesis and outdoor mass cultivation in a low-cost photobioreactor, Biotechnol. Bioeng. 102 (2009) 100–112.

[4] L. Gouveia, Microalgae as a Feedstock for Biodiesel, Springer, Heidelberg, 2011.

[5] A.A. Adenle, G.E. Haslam, L. Lee, Global assessment of research and development for algae biofuel production and its potential role for sustainable development in developing countries, Energy Policy 61 (2013) 182–195.
[6] D. López Barreiro, W. Prins, F. Ronsse, W. Brilman, Hydrothermal liquefaction (HTL) of microalgae for biofuel production: state of the art review and future prospects, Biomass Bioenergy 53 (2013) 113–127.
[7] L. Xu, D.W.F. Wim Brilman, J.A. Withag, G. Brem, S. Kersten, Assessment of a dry and a wet route for the production of biofuels from microalgae: energy balance analysis, Bioresour. Technol. 102 (2011) 5113–5122.
[8] P. Duan, X. Bai, Y. Xu, A. Zhang, F. Wang, L. Zhang, et al., Catalytic upgrading of crude algal oil using platinum/gamma alumina in supercritical water, Fuel 109 (2013) 225–233.
[9] X. Liu, B. Saydah, P. Eranki, L.M. Colosi, B. Greg Mitchell, J. Rhodes, et al., Pilot-scale data provide enhanced estimates of the life cycle energy and emissions profile of algae biofuels produced via hydrothermal liquefaction, Bioresour. Technol. 148 (2013) 163–171.
[10] A.A. Peterson, F. Vogel, R.P. Lachance, M. Fröling, M. J. Antal Jr., J.W. Tester, Thermochemical biofuel production in hydrothermal media: a review of sub-and supercritical water technologies, Energy Environ. Sci. 1 (2008) 32–65.
[11] J. Wagner, Production of transport fueld via the hydrothermal liquefaction of microalgae and subsequent upgrading reaction (PhD confirmation report), 2014.
[12] C. Tian, B. Li, Z. Liu, Y. Zhang, H. Lu, Hydrothermal liquefaction for algal biorefinery: a critical review, Renewable Sustainable Energy Rev. 38 (2014) 933–950.
[13] L. Garcia Alba, C. Torri, C. Samorì, J. Van Der Spek, D. Fabbri, S.R.A. Kersten, et al., Hydrothermal treatment (HTT) of microalgae: evaluation of the process as conversion method in an algae biorefinery concept, Energy Fuels 26 (2012) 642–657.
[14] S. Zou, Y. Wu, M. Yang, C. Li, J. Tong, Bio-oil production from sub- and supercritical water liquefaction of microalgae *Dunaliella tertiolecta* and related properties, Energy Environ. Sci. 3 (2010) 1073.
[15] P. Biller, A.B. Ross, Potential yields and properties of oil from the hydrothermal liquefaction of microalgae with different biochemical content, Bioresour. Technol. 102 (2011) 215–225.
[16] A. Demirbas, Mechanisms of liquefaction and pyrolysis reactions of biomass, Energy Convers. Manage. 41 (2000) 633–646.
[17] Y.F. Yang, C.P. Feng, Y. Inamori, T. Maekawa, Analysis of energy conversion characteristics in liquefaction of algae, Resour. Conserv. Recycl. 43 (2004) 21–33.
[18] S. Zou, Y. Wu, M. Yang, C. Li, J. Tong, Thermochemical catalytic liquefaction of the marine microalgae *Dunaliella tertiolecta* and characterization of bio-oils, Energy Fuels 23 (2009) 3753–3758.
[19] D. Zhou, L. Zhang, S. Zhang, H. Fu, J. Chen, Hydrothermal liquefaction of macroalgae *Enteromorpha prolifera* to bio-oil, Energy Fuels 24 (2010) 4054–4061.
[20] K. Anastasakis, A.B. Ross, Hydrothermal liquefaction of the brown macro-alga *Laminaria Saccharina*: effect of reaction conditions on product distribution and composition, Bioresour. Technol. 102 (2011) 4876–4883.
[21] G. Yu, Y. Zhang, L. Schideman, T. Funk, Z. Wang, Distributions of carbon and nitrogen in the products from hydrothermal liquefaction of low-lipid microalgae, Energy Environ. Sci. 4 (2011) 4587–4595.
[22] G. Yu, Y. Zhang, L. Schideman, T.L. Funk, Z. Wang, Hydrothermal liquefaction of low lipid content microalgae into bio-crude oil, Am. Soc. Agric. Biol. Eng 54 (2011) 239–246.
[23] P. Biller, R. Riley, A.B. Ross, Catalytic hydrothermal processing of microalgae: decomposition and upgrading of lipids, Bioresour. Technol. 102 (2011) 4841–4848.
[24] D.R. Vardon, B.K. Sharma, J. Scott, G. Yu, Z. Wang, L. Schideman, et al., Chemical properties of biocrude oil from the hydrothermal liquefaction of Spirulina algae, swine manure, and digested anaerobic sludge, Bioresour. Technol. 102 (2011) 8295–8303.
[25] S.S. Toor, H. Reddy, S. Deng, J. Hoffmann, D. Spangsmark, L.B. Madsen, et al., Hydrothermal liquefaction of *Spirulina* and *Nannochloropsis* salina under subcritical and supercritical water conditions, Bioresour. Technol. 131 (2013) 413–419.
[26] H. Li, Z. Liu, Y. Zhang, B. Li, H. Lu, N. Duan, et al., Conversion efficiency and oil quality of low-lipid high-protein and high-lipid low-protein microalgae via hydrothermal liquefaction, Bioresour. Technol. 154 (2014) 322–329.
[27] J.G. Speight, Handbook of Petroleum Analysis, first ed., Wiley Interscience, New York, NY, 2001.
[28] T.M. Brown, P. Duan, P.E. Savage, Hydrothermal Liquefaction and Gasification of Microalga *Nannochloropsis* sp., Energy Fuels 24 (2010) 3639–3646.
[29] G.W. Huber, S. Iborra, A. Corma, Synthesis of transportation fuels from biomass: chemistry, catalysts, and engineering, Chem. Rev. 106 (2006) 4044–4098.
[30] J.S. Ball, M.L. Whisman, W.J. Wenger, Nitrogen content of crude petroleums, Ind. Eng. Chem. 43 (1951) 2577–2581.
[31] C. Jazrawi, P. Biller, A.B. Ross, A. Montoya, T. Maschmeyer, B.S. Haynes, Pilot plant testing of continuous hydrothermal liquefaction of microalgae, Algal Res. 2 (2013) 268–277.

[32] U. Jena, N. Vaidyanathan, S. Chinnasamy, K.C. Das, Evaluation of microalgae cultivation using recovered aqueous co-product from thermochemical liquefaction of algal biomass, Bioresour. Technol 102 (2011) 3380–3387.

[33] A. Hirano, K. Hon-Nami, S. Kunito, M. Hada, Y. Ogushi, Temperature effect on continuous gasification of microalgal biomass: theoretical yield of methanol production and its energy balance, Catal. Today 45 (1998) 399–404.

[34] D.C. Elliott, T.R. Hart, A.J. Schmidt, G.G. Neuenschwander, L.J. Rotness, M.V. Olarte, et al., Process development for hydrothermal liquefaction of algae feedstocks in a continuous-flow reactor, Algal Res. 2 (2013) 445–454.

[35] Y. Dote, S. Sawayama, S. Inoue, T. Minowa, S. Yokoyama, Recovery of liquid fuel from hydrocarbon-rich microalgae by thermochemical liquefaction, Fuel 73 (1994) 1855–1857.

[36] T. Minowa, S. Yokoyama, M. Kishimoto, T. Okakura, Oil production from algal cells of *Dunaliella tertiolecta* by direct thermochemical liquefaction, Fuel 74 (1995) 1735–1738.

[37] C. Gai, Y. Zhang, W.-T. Chen, P. Zhang, Y. Dong, Energy and nutrient recovery efficiencies in biocrude oil produced via hydrothermal liquefaction of *Chlorella pyrenoidosa*, RSC Adv. 4 (2014) 16958.

[38] A.B. Ross, P. Biller, M.L. Kubacki, H. Li, A. Lea-Langton, J.M. Jones, Hydrothermal processing of microalgae using alkali and organic acids, Fuel 89 (2010) 2234–2243.

[39] M. Sasaki, T. Adschiri, K. Arai, Production of cellulose II from native cellulose by near- and supercritical water solubilization, J. Agric. Food Chem. 51 (2003) 5376–5381.

[40] S. Karagoz, T. Bhaskar, A. Muto, Y. Sakata, M.D. Azhar Uddin, Low temperature hydrothermal treatment of biomass: effect of reaction parameters on products and boiling point distributions, Energy Fuels 18 (2004) 234–241.

[41] G. Yu, Y. Zhang, L. Schideman, T.L. Funk, W. Wang, Hydrothermal liquefaction of low lipid content microalgae into bio-crude oil, Trans. ASABE 54 (2011) 239–246.

[42] J. Akhtar, N.A.S. Amin, A review on process conditions for optimum bio-oil yield in hydrothermal liquefaction of biomass, Renewable Sustainable Energy Rev. 15 (2011) 1615–1624.

[43] J.L. Faeth, P.J. Valdez, P.E. Savage, Fast hydrothermal liquefaction of *Nannochloropsis* sp. to produce biocrude, Energy Fuels 27 (2013) 1391–1398.

[44] T.M. Yeh, J.G. Dickinson, A. Franck, S. Linic, L.T. Thompson Jr, P.E. Savage, Hydrothermal catalytic production of fuels and chemicals from aquatic biomass, J. Chem. Technol. Biotechnol. 88 (2013) 13–24.

[45] J.X. Han, J.Z. Duan, P. Chen, H. Lou, X.M. Zheng, H. P. Hong, Nanostructured molybdenum carbides supported on carbon nanotubes as efficient catalysts for one-step hydrodeoxygenation and isomerization of vegetable oils, Green Chem. 13 (2011) 2561–2568.

[46] G. Wang, W. Li, B. Li, H. Chen, Direct liquefaction of sawdust under syngas, Fuel 86 (2007) 1587–1593.

[47] J.J. Milledge, B. Smith, P.W. Dyer, P. Harvey, Macroalgae-derived biofuel: a review of methods of energy extraction from seaweed biomass, Energies 7 (2014) 7194–7222.

[48] T. Lundquist, I. Woertz, N. Quinn, J. Benemann, A Realistic Technology and Engineering Assessment of Algae Biofuel Production, Energy Biosciences Institute, University of California, Berkley, CA, 2010.

[49] P. Biller, A.B. Ross, Hydrothermal processing of algal biomass for the production of biofuels and chemicals, Biofuels 3 (2012) 603–623.

[50] F. Murphy, G. Devlin, R. Deverell, K. McDonnell, Biofuel production in Ireland—an approach to 2020 targets with a focus on algal biomass, Energies 6 (2013) 6391–6412.

[51] K. Anastasakis, A.B. Ross, Hydrothermal liquefaction of four brown macro-algae commonly found on the UK coasts: an energetic analysis of the process and comparison with bio-chemical conversion methods, Fuel 139 (2015) 546–553.

[52] Q.V. Bach, M.V. Sillero, K.Q. Tran, J. Skjermo, Fast hydrothermal liquefaction of a norwegian macro-alga: screening tests, Algal Res. 6 (2014) 271–276.

[53] N. Neveux, A.K.L. Yuen, C. Jazrawi, M. Magnusson, B.S. Haynes, A.F. Masters, et al., Bioresour. Technol. 155 (2014) 334–341.

[54] B. Jin, P. Duan, Y. Xu, F. Wang, Biocrude yield and productivity from the hydrothermal liquefaction of marine and freshwater green macroalgae, Bioresour. Technol. 149 (2013) 103–110.

[55] G. Roesijadi, S.B. Jones, Y. Zhu, Macroalgae as a biomass feedstock: a preliminary analysis, pacific northwest national laboratory. Report no.: PNNL-19944. Sponsored by the US Department of Energy; September 2010, 2010.

[56] T. Bruton, H. Lyons, Y. Lerat, M. Stanley, M.B. Rasmussen, A Review of the Potential of Marine Algae as a Source of Biofuel in Ireland, Sustainable Energy Ireland, Dublin, 2009.

[57] W.J. Lenstra, J.H. Reith, J.W.V. Hal. Proceedings of the 4th International Algae Congress, Amsterdam, The Netherlands, 1–2 December 2010.

[58] D.J. McHugh, A guide to the seaweed industry, food and agricultural organization: 106, 2003.

[59] R. Morchio, C. Cáceres, Macroalgae current state in Latin America. International Workshop on Sustainable Bioenergy from Algae. Berlin, Germany, 2009.

[60] K.A. Jung, S.-R. Lim, Y. Kim, J.M. Park, Potentials of macroalgae as feedstocks for biorefinery, Bioresour. Technol. 135 (2013) 182–190.

[61] T.M. Lammens, M.C.R. Franssen, E.L. Scott, J.P.M. Sanders, Availability of protein-derived amino acids as feedstock for the production of bio-based chemicals, Biomass Bioenerg. 44 (2012) 168–181.

[62] Z. Hu, Y. Zheng, F. Yan, B. Xiao, S. Liu, Bio-oil production through pyrolysis of Blue-Green Algae Blooms (BGAB): product distribution and bio-oil characterization, Energy 52 (2013) 119–125.

[63] U. Jena, K.C. Das, J.R. Kastner, Effect of operating conditions of thermochemical liquefaction on biocrude production from *Spirulina platensis*, Bioresour. Technol. 102 (2011) 6221–6229.

[64] A.E. Harman-Ware, T. Morgan, M. Wilson, M. Crocker, J. Zhang, K. Liu, et al., Microalgae as a renewable fuel source: fast pyrolysis of *Scenedesmus* sp., Renewable Energy 60 (2013) 625–632.

[65] Z. Li, P.E. Savage, Feedstocks for fuels and chemicals from algae: treatment of crude bio-oil over HZSM-5, Algal Res. 2 (2013) 154–163.

[66] P. Duan, P.E. Savage, Catalytic treatment of crude algal bio-oil in supercritical water: optimization studies, Energy Environ. Sci. 4 (2011) 1447.

[67] P. Duan, P.E. Savage, Upgrading of crude algal bio-oil in supercritical water, Bioresour. Technol. 102 (2011) 1899–1906.

[68] C. Torri, D. Fabbri, L. Garcia-Alba, D.W.F. Brilman, Upgrading of oils derived from hydrothermal treatment of microalgae by catalytic cracking over H-ZSM-5: a comparative Py–GC–MS study, J. Anal. Appl. Pyrolysis 101 (2013) 28–34.

[69] P.-Q. Yuan, Z.-M. Cheng, X.-Y. Zhang, W.-K. Yuan, Catalytic denitrogenation of hydrocarbons through partial oxidation in supercritical water, Fuel 85 (2006) 367–373.

[70] P. Duan, P.E. Savage, Hydrothermal liquefaction of a microalga with heterogeneous catalysts, Ind. Eng. Chem. Res. 50 (2011) 52–61.

[71] P. Duan, P.E. Savage, Catalytic hydrotreatment of crude algal bio-oil in supercritical water, Appl. Catal. B Environ. 104 (2011) 136–143.

[72] S.G. Roussis, R. Cranford, N. Sytkovetskiy, Thermal treatment of crude algae oils prepared under hydrothermal extraction conditions, Energy Fuels 26 (2012) 5294–5299.

[73] X. Bai, P. Duan, Y. Xu, A. Zhang, P.E. Savage, Hydrothermal catalytic processing of pretreated algal oil: a catalyst screening study, Fuel 120 (2014) 141–149.

[74] K.S. Ocfemia, Y. Zhang, T. Funk, Hydrothermal processing of swine manure into oil using a continuous reactor system: development and testing, Am. Soc. Agric. Biol. Eng. 49 (2006) 533–541.

[75] United States International Trade Commission, Environmental and Related Services—Investigation No. 332-53, USITC Publication *4389*, 2013.

[76] W.T. Chen, Y. Zhang, J. Zhang, G. Yu, L.C. Schideman, P. Zhang, et al., Hydrothermal liquefaction of mixed-culture algal biomass from wastewater treatment system into bio-crude oil, Bioresour. Technol. 152 (2014) 130–139.

[77] J.K. Pittman, A.P. Dean, O. Osundeko, The potential of sustainable algal biofuel production using wastewater resources, Bioresour. Technol. 102 (2011) 17–25.

[78] Aquastat, Food and Agriculture Organization of the United Nations, 2014.

[79] L. Wang, M. Min, Y. Li, P. Chen, Y. Chen, Y. Liu, et al., Cultivation of green algae *Chlorella* sp. in different wastewaters from municipal wastewater treatment plant, Appl. Biochem. Biotechnol. 162 (2010) 1174–1186.

[80] V.K. Gupta, A. Rastogi, Biosorption of lead(II) from aqueous solutions by non-living algal biomass *Oedogonium* sp. and Nostoc sp.—a comparative study, Colloids Surf. B Biointerfaces 64 (2008) 170–178.

[81] V.K. Gupta, A.K. Shrivastava, N. Jain, Biosorption of chromium (VI) from aqueous solutions by green algae *Spirogyra* species, Water Res. 35 (2001) 4079–4085.

[82] D.B. Johnson, K.B. Hallberg, Acid mine drainage remediation options: a review, Sci. Total Environ. 338 (2005) 3–14.

[83] K. Sbihi, O. Cherifi, A. El gharmali, B. Oudra, F. Aziz, Accumulation and toxicological effects of cadmium, copper and zinc on the growth and photosynthesis of the freshwater diatom Planothidium lanceolatum (Brébisson) lange-bertalot: a laboratory study, J. Mater. Environ. Sci. 3 (2012) 497–506.

[84] S. Raikova, H. Smith-Baedorf, R. Bransgrove, O. Barlow, F. Santomauro, J.L. Wagner, et al., Assessing hydrothermal liquefaction for the production of bio-oil and enhanced metal recovery from microalgae cultivated on acid mine drainage, Fuel Process. Technol. 142 (2016) 219–227.

[85] D. Aitken, B. Antizar-Ladislao, Achieving a green solution: limitations and focus points for sustainable algal fuels, Energies 5 (2012) 1613–1647.

[86] P. Biller, A.B. Ross, S.C. Skill, A. Lea-Langton, B. Balasundaram, C. Hall, et al., Nutrient recycling of aqueous phase for microalgae cultivation from the hydrothermal liquefaction process, Algal Res. 1 (2012) 70–76.

[87] Y. Zhou, L. Schideman, G. Yu, Y. Zhang, A synergistic combination of algal wastewater treatment and hydrothermal biofuel production maximized by nutrient and carbon recycling, Energy Environ. Sci. 6 (2013) 3765.

[88] Y. Zhou, L. Schideman, Y. Zhang, G. Yu, In: Proceedings of the Water Environment Federation, 2011, pp. 7268–7282(15).

[89] Y. Zhou, L. Schideman, Y. Zhang, G. Yu, Z. Wang, M. Pham, Resolving bottlenecks in current algal wastewater treatment paradigms: a synergistic combination of low-lipid algal wastewater treatment and hydrothermal liquefaction for large-scale biofuel production, Proc. Water Environ. Fed. 6 (2011) 347–361.

[90] R. Slade, A. Bauen, Micro-algae cultivation for biofuels: cost, energy balance, environmental impacts and future prospects, Biomass Bioenergy 53 (2013) 29–38.

[91] C.J. Rhodes, Peak phosphorus—peak food? The need to close the phosphorus cycle, Sci. Prog. 96 (2013) 109–152 (44)

[92] M. Bagnoud-Velásquez, U. Schmid-Staiger, G. Peng, F. Vogel, C. Ludwig, First developments towards closing the nutrient cycle in a biofuel production process, Algal Res. 8 (2015) 76–82.

[93] National Research Council of the National Academies, Sustainable Development of Algal Biofuels in the United States, The National Academies Press, Washington, DC, 2012.

[94] Y. Chisti, Constraints to commercialization of algal fuels, J. Biotechnol. 167 (2013) 201–214.

[95] S. Nakai, Y. Inoue, M. Hosomi, Growth inhibition effects and inducement modes by plant-producing phenols, Water Res. 35 (2001) 1855–1859.

[96] A. Bhatnagar, S. Chinnasamy, M. Singh, K.C. Das, Renewable biomass production by mixotrophic algae in the presence of various carbon sources and wastewaters, Appl. Energy 88 (2011) 3425–3431.

[97] A.H. Scragg, The effect of phenol on the growth of Chlorella vulgaris and Chlorella VT-1, Enzyme Microb. Technol. 39 (4) (2006) 796–799.

[98] M. Pham, L. Schideman, J. Scott, N. Rajagopalan, M.J. Plewa, Chemical and biological characterization of wastewater generated from hydrothermal liquefaction of *Spirulinau*, Environ. Sci. Technol. 47 (4) (2013) 2131–2138.

CHAPTER

10

Aviation Turbine Fuels Through the Fischer–Tropsch Process

A. de Klerk

Department of Chemical and Materials Engineering, University of Alberta, Edmonton, AB, Canada

10.1 INTRODUCTION

Fischer–Tropsch (F-T) synthesis is a conversion technology that was developed to convert any carbon-based material into an oil product that can be refined to conventional transportation fuels and petrochemical products. In order to make use of biomass as feed material and produce aviation fuels as products, there are three process steps (Fig. 10.1): conversion of biomass to synthesis gas, conversion of synthesis gas to oil, and oil refining to aviation fuels.

In the first step the biomass is gasified to produce synthesis gas [1]. Pure synthesis gas is a mixture of hydrogen (H_2) and carbon monoxide (CO). The product from gasification is raw synthesis gas and it contains many other compounds that must be removed from the synthesis gas before it can be used as feed for the second step, namely, F-T synthesis. The F-T process is an indirect liquefaction process [2], and it is blind to the origin of the synthesis gas. The cleaned synthesis gas produced from biomass can consequently be used as feed material to an F-T process with the same confidence as the current industrial practise of using synthesis gas from coal and natural gas. The composition of the oil product from F-T synthesis depends on the nature and operation of the F-T technology, much like the composition of crude oil depends on its origin. The third step involves oil refining to produce the aviation fuels. The nature of the oil product from the F-T process is different to that of crude oil, as is the refining process to produce on-specification aviation fuels [3].

The development and qualification of aviation fuels from the F-T process employed products from coal-derived synthesis gas, but the aviation fuel qualification is not linked in any way to the raw material used to produce the synthesis gas. The requirements are generic and captured by the specifications for synthetic aviation fuels. These points are made to emphasize that the aviation fuels produced from biomass through the F-T process have no barrier to market entry. In fact, F-T-derived aviation fuels have been in commercial use for some time already.

The developments leading to the qualification of the semi- and fully synthetic jet

Biofuels for Aviation.
DOI: http://dx.doi.org/10.1016/B978-0-12-804568-8.00010-X

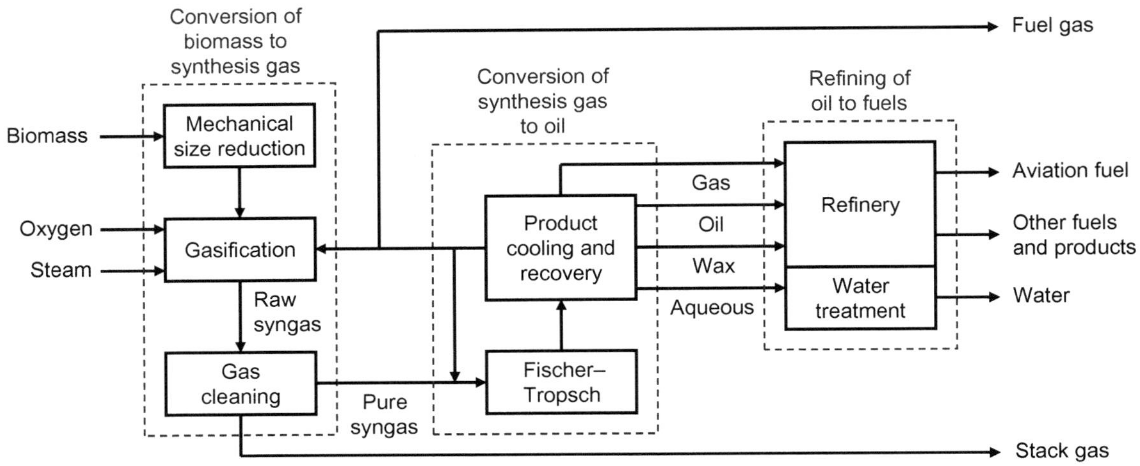

FIGURE 10.1 Block flow diagram of a generic biomass-to-liquids process based on Fischer–Tropsch synthesis to produce aviation fuels.

fuels-containing F-T material were explained by Morgan and Roets [4]. Semi-synthetic jet fuel, which is a mixture containing at most 50% F-T-derived kerosene in a petroleum-derived kerosene has been qualified and approved for use since 1999 in DEF STAN 91-91 Issue 3, and has been used commercially since then. Fully synthetic jet fuel, which contains kerosene obtained only from an F-T facility, was qualified and approved for use in 2008 in DEF STAN 91-91 Issue 6. Initially, individual fuels had to be approved for commercial use as aviation turbine fuels, which restricted market-development for such fuels. Since then a standardized approval process has been developed for aviation turbine fuels-containing material not derived from petroleum refining. The qualification process is outlined in ASTM D 4054 [5].

The objective of this chapter is to show how aviation fuels can be produced from the F-T process. The approach that will be taken is to identify the molecular properties that are required to meet the main performance characteristics of aviation fuels, and what conversion processes are needed to convert the products from an F-T process into aviation fuels.

10.2 MOLECULAR PROPERTIES REQUIRED BY AVIATION TURBINE FUELS

Current specification requirements for aviation turbine fuels-containing material that was not derived from petroleum refining are DEF STAN 91-91 Issue 7 [6] and ASTM D 7566 [7]. The specification requirements are discussed in detail in the chapter 'Certification and Performance: What Is Needed From an Aviation Fuel?'. The objective of this section is to point out the molecular properties that are required to meet the main jet fuel specifications.

10.2.1 Volatility

The kerosene fraction of oil is refined to aviation turbine fuel, or jet fuel. The kerosene distillation range varies somewhat between refineries, but for the purpose of jet fuel there are specification constraints. These are illustrated in Fig. 10.2.

a. At least 10 vol.% of the kerosene must distill at a temperature of 205°C, that is, $T_{10} \leq 205°C$. This determines the minimum amount of lighter boiling material that must be present in the kerosene. The initial boiling point (IBP) and amount of lighter boiling material in the kerosene is indirectly determined by the flash point specification. The minimum flash point temperature for commercial Jet A-1 is 38°C, but higher values for the minimum flash point can be found in some military jet fuels, such as JP-5, used on aircraft carriers. The higher the flash point, the less volatile the kerosene becomes. In terms of molecular properties, these specifications regulate the amount of C_9-C_{11} material in the jet fuel.

b. The final boiling point (FBP) temperature may not exceed 300°C. This is a straightforward limit and means that little material heavier than C_{16} can be included in jet fuel.

c. There are two additional volatility specifications that are specific to jet fuel-containing material not derived from petroleum refining, such as kerosene derived from F-T synthesis. Both

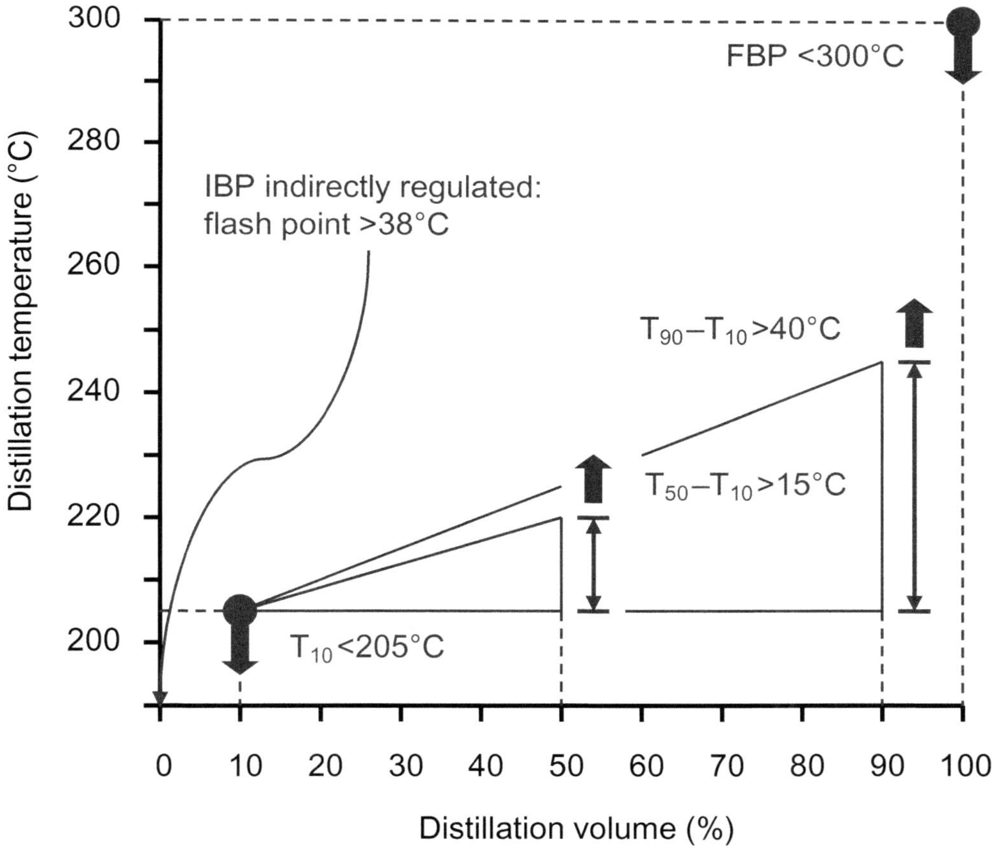

FIGURE 10.2 Volatility specifications for aviation turbine fuels-containing material not derived from petroleum refining.

specifications relate to the slope of the distillation curve: $T_{90}-T_{10} \geq 40°C$ and $T_{50}-T_{10} \geq 15°C$. These specifications are included to avoid cuts that are too narrow or the use of single component synthetic products that would not exhibit smooth vaporization during continuous combustion.

10.2.2 Density

The density of jet fuel when measured at 15°C must be between 775 and 840 kg/m^3 (Table 10.1). The density is affected by the distillation curve of the kerosene, with higher boiling hydrocarbons generally having a higher density. The density is also affected by the composition of kerosene. For molecules of the same boiling point the density increases in the order: alkanes <cycloalkanes <aromatics. It is impossible for kerosene containing only linear and branched alkanes to meet the minimum density requirement. In terms of molecular properties, the density specification requires some cycloalkanes and/or aromatics to be present in the kerosene.

TABLE 10.1 Fluidity- and Density-Related Specifications for Aviation Turbine Fuel

Specification	Specification Limit Minimum	Maximum
Density measured at 15°C (kg/m^3)	775	840
FREEZING POINT TEMPERATURE (°C)		
Jet A-1	–	−47
Jet A	–	−40
VISCOSITY (mm/s)		
Measured at −20°C	–	8
Measured at −40°C[a]	–	12

[a]*Specification only for fuel-containing material not derived from petroleum refining.*

10.2.3 Fluidity

The freezing point and viscosity specifications (Table 10.1) are critical and ensure adequate pumpability of the jet fuel in an aircraft during operation at high altitude. The freezing point specification is a maximum of −47°C, except for Jet-A (used on short distance domestic flights), which has a less stringent freezing point specification of −40°C. The maximum viscosity is 8 mm/s (8 cSt) measured at −20°C, with the additional requirement for jet fuel-containing material not derived from petroleum refining to have a viscosity of 12 mm/s (12 cSt) or lower when measured at −40°C.

The viscosity of most liquids in the kerosene range is low, but as a kerosene cut approaches its onset of freezing, the liquid viscosity normally increases significantly. Ensuring a freezing point that is well below the specification limit, rather than just meeting the specification, will usually ensure that the viscosity is also within specification. It is for this reason that jet fuel that is not derived from petroleum refining has an additional viscosity specification at a measurement temperature of −40°C.

Although both freezing point and liquid viscosity are easily measurable, prediction of freezing point and liquid viscosity based on structure is challenging [8,9]. This should be kept in mind when considering the relationships to molecular properties that are provided as guidelines. Generally speaking, most cycloalkanes and aromatics in the kerosene boiling range have low freezing points. Notable exceptions are durene (1,2,4,5-tetramethyl benzene) and naphthalene and derivatives of naphthalene. The *n*-alkanes have high freezing point temperatures. *n*-Decane and heavier *n*-alkanes all have freezing point temperatures well above the freezing point specification of jet fuel. Branched alkanes on the other hand have much lower freezing point temperatures and as a rule of thumb even a single methyl branch can decrease the freezing point temperature by 20°C or more. As long as a compound with high freezing point temperature remains in solution, the freezing points of individual compounds are not of concern. However, when compounds with high freezing point temperatures are present in high enough concentration, it is likely that they will freeze out of solution as the temperature is decreased beyond their freezing point temperature.

10.2.4 Composition

Composition-related requirements by definition are related to the molecular properties of the jet fuel. These specifications are summarized in Table 10.2.

The only additional specification requirement for jet fuel that contains material not derived from petroleum refining, is a mandatory minimum aromatics content. This specification is necessary for protection against differential elastomer swelling and shrinking that could cause leaks in the fuel system. Thus, jet fuel produced with material from F-T synthesis must contain at least 8 vol.% aromatics.

Semi-synthetic jet fuel has more specification requirements than fully synthetic jet fuel, possibly for historical reasons. Semi-synthetic jet fuel is limited to the inclusion of maximum 50 vol.% of components not derived from petroleum in the final jet fuel blend. The specification requirements in the DEF STAN 91-91 Issue 7 [6] are listed under the section dealing with 'Specific Manufacturer Approvals'. Two different F-T blending components were qualified, iso-paraffinic kerosene (IPK) and heavy naphtha #1 (HN1). IPK is the hydrotreated product from a solid phosphoric acid-catalysed alkene oligomerization process, such as the UOP CatPoly process. HN1 is the hydrotreated product of the straight run heavy naphtha cut from high-temperature Fe-based F-T synthesis. The main additional restrictions on such blends are that the final blend component should have 7 vol.% or less aromatics, not contain 'synthetic aromatics', and when it contains HN1, it should contain at least 25% IPK as well. The ASTM D 7566 [7] specification makes reference to synthetic paraffinic kerosene (SPK), which is more broadly defined. SPK is essentially any refining product derived from F-T synthesis, and it may be a blend of different streams in an F-T refinery. The additional composition requirements for SPK derived from an F-T process are given in Table 10.3. Note that these additional requirements are for the SPK blend component only, and it is not for the final blended semi-synthetic jet fuel.

The limitation on metals content is particularly strict (Table 10.3), and due to the ubiquitous presence of water in F-T processes, this is an important constraint for final quality control.

TABLE 10.2 Composition-Related Specifications for Aviation Turbine Fuel

	Specification Limit	
Specification	**Minimum**	**Maximum**
Aromatic content by ASTM D 1319 (vol.%)	8[a]	25
Sulphur content (mg/kg)	–	3000
Thiol content (mg/kg)	–	30
Acidity (mg KOH/g)	–	0.10/ 0.015[b]

[a] *Specification only for fuel-containing material not derived from petroleum refining.*
[b] *ASTM D 7566/DEF STAN 91-91 Issue 7.*

10.2.5 Other Properties

It is not the intent to provide an exhaustive list of the other specifications, but some that have important bearing on the aviation fuels produced from F-T products are listed in Table 10.4.

TABLE 10.3 Additional Specifications in ASTM D 7566 for Fischer–Tropsch-Derived Synthetic Paraffinic Kerosene (SPK) as Blending Component for Semi-Synthetic Jet Fuel

Specification	Specification Limit Minimum	Maximum
HYDROCARBONS		
Cycloalkanes by ASTM D 2425 (wt.%)	–	15
Aromatics by ASTM D 2425 (wt.%)	–	0.5
Total H and C mass (wt.%)	99.5	–
NONHYDROCARBONS		
Nitrogen content (mg/kg)	–	2
Sulphur content (mg/kg)	–	15
Water content (mg/kg)	–	75
Halogen content (mg/kg)	–	1
Metal content (mg/kg)[a]	–	0.1

[a]*Specification limit per metal (not total metal content) for each of Al, Ca, Co, Cr, Cu, Fe, K, Li, Mg, Mn, Mo, Na, Ni, P, Pb, Pd, Pt, Sn, Sr, Ti, V, and Zn.*

TABLE 10.4 Selected Other Specifications for Aviation Turbine Fuel

Specification	Specification Limit Minimum	Maximum
Net heat of combustion (MJ/kg)	42.8	–
Smoke point (mm), *or*	25	–
Smoke point (mm), *and*	18/19[a]	–
Naphthalene content (vol.%)	–	3.0
Wear scar, lubricity by ASTM D 5001 (mm)[b]	–	0.85
Thermal stability filter pressure drop (mmHg)[c]	–	25

[a]*ASTM D 7566/DEF STAN 91-91 Issue 7.*
[b]*Specification only for fuel-containing material not derived from petroleum refining and fuels according to DEF STAN 91-91; also severely refined petroleum-derived materials.*
[c]*ASTM D 3241 (JFTOT): 2.5 hours at 260°C (at 325°C for F-T-derived fuels).*

The gravimetric energy density indirectly controls the hydrogen-to-carbon ratio. More hydrogen-rich fuels have higher gravimetric energy densities. Smoke point is also indirectly a control to ensure an adequate hydrogen-to-carbon ratio. Practically, the smoke point is also affected by the content of multinuclear aromatics and other multicyclic compounds that can more readily form coke particles during combustion.

Surface-active compounds provide boundary layer lubricity, which reduces wear. Surface-active compounds usually contain heteroatoms, such as oxygen and sulphur. F-T and petroleum-derived kerosenes tend to lose their inherent lubricity during hydroprocessing because heteroatoms are eliminated during hydrotreating.

The thermal stability test conditions for F-T and other fuels not derived from petroleum are more stringent than for petroleum-derived fuels. This is used as a quality control measure to ensure batch-to-batch consistency. The need for this is likely related to the oxygenate content of the kerosene, which can make the fuel more susceptible to free radical initiation and free radical reactions when it comes into contact with hot surfaces.

10.3 COMPOSITION OF PRODUCTS FROM FISCHER–TROPSCH SYNTHESIS

F-T synthesis is not a singular technology with only one product slate. Since the invention of F-T synthesis, many different combinations of catalysts, operating conditions, and reactor designs were developed to reach demonstration scale or commercial implementation. Despite the diversity of technologies, when F-T synthesis is viewed from a product perspective, there are only two basic types of products that are industrially produced by F-T synthesis. The first type is the product from

TABLE 10.5 Main Product Types Produced Industrially by F-T Synthesis

Product Property	HTFT[a]	LTFT[b]
Carbon number range	C_1–C_{30}	C_1–C_{120}
COMPOUND CLASSES		
Alkanes (paraffins)	20–30%	Major product (>70%)
Cycloalkanes (naphthenes)	<1%	<1%
Alkenes (olefins)	Major product (>50%)	15 to 20%
Aromatics	1–5%	<1%
Oxygenates	10–15%	~5%

[a]*HTFT (High-Temperature Fischer–Tropsch) using Fe-based catalyst.*
[b]*LTFT (Low-Temperature Fischer–Tropsch) using Fe- or Co-based catalysts.*

TABLE 10.6 Mass Percentage of Straight Run Fischer–Tropsch Products Available for the Production of Jet Fuel Blending Components

Straight Run Product	HTFT Fe-catalyst	LTFT Fe-catalyst	Co-catalyst
Light alkenes (C_3–C_5)	20–30%	5–10%	3–5%
Heavy naphtha and kerosene (C_9–C_{14})	10–15%	10–15%	10–15%
Heavy distillate and wax (C_{23} +)	None	45–50%	40–50%

high-temperature Fischer–Tropsch (HTFT) synthesis employing Fe-based catalysts operated at 320°C and higher temperature. The second type is the product from low-temperature Fischer–Tropsch (LTFT) synthesis employing mainly Fe- and Co-based catalysts. The operating temperatures for LTFT depend on the catalyst and reactor design, and in practise range from around 170°C to 270°C.

The main composition characteristics of the two types of F-T products are given in Table 10.5 [10]. These are generic characteristics, and aspects such as the carbon number distribution and relative abundance of alkenes and oxygenates depend on the F-T technology and its operation. Generally speaking for LTFT synthesis, Fe-based catalysts that are operated at higher temperatures than Co-based catalysts produce products that contain a higher percentage of alkenes. Only HTFT synthesis produces meaningful amounts of aromatics.

It is important to point out that F-T synthesis can be manipulated to make the refining production of aviation turbine fuel easier. Unlike crude oil, where the composition is dictated by the natural resource, it is possible to manipulate the composition of the oil product from F-T synthesis.

The blending components used in the production of semi-synthetic and fully synthetic jet fuels from F-T synthesis are produced in three ways: hydrogenation of light alkene oligomerization products, hydrogenation of straight run heavy naphtha and kerosene, and the hydrocracking and hydroisomerization of distillate and wax. The relative contribution of each depends on the F-T technology. Table 10.6 [3] gives an indication of the relative fraction of the straight run organic products of each type that is available for the production of these blending components. Note that these amounts exclude additional material that can be made available by refining of the remainder of the product.

10.4 JET FUEL BLENDING COMPONENTS FROM FISCHER–TROPSCH SYNTHESIS

10.4.1 Hydrogenated Alkene Oligomers

The first blending component to be certified for the production of semi-synthetic jet fuel was IPK. A block flow diagram of the process

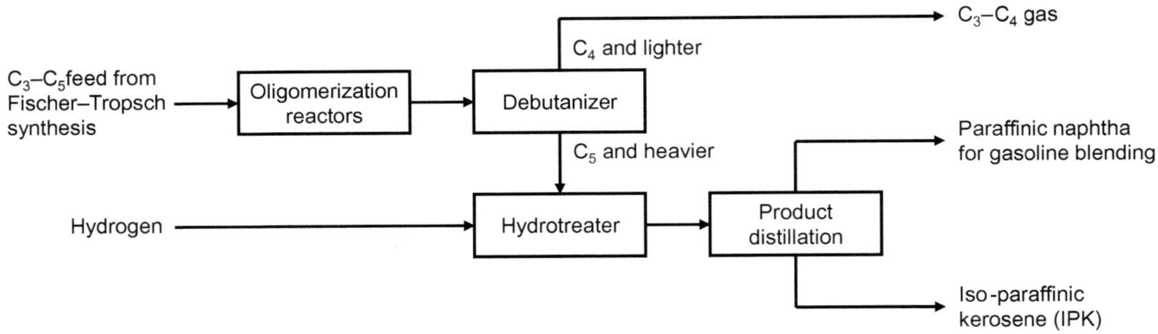

FIGURE 10.3 Block flow diagram of the process to produce iso-paraffinic kerosene from light Fischer–Tropsch alkenes.

to show how this material is produced is shown in Fig. 10.3.

The first step is the oligomerization of the light alkenes. The feed is comprised of mixed C_3 to C_5 alkenes and alkanes that were produced during F-T synthesis. The composition of this feed material depends on the F-T technology and prior separation steps, for example, recovery of propene as a petrochemical product. The oligomerization catalyst is solid phosphoric acid, which consists of a phosphoric acid phase on kieselguhr. The product from this type of oligomerization is predominantly branched alkenes in the C_7 to C_{13} carbon number range [11]. In principle, other acid catalysts can also be employed for the oligomerization, but solid phosphoric acid is particularly well-suited for this task, because the product is highly branched and the process has a high kerosene yield.

After the oligomerization reaction, the light C_3–C_4 alkanes are separated from the heavier product by distillation in a debutanizer column. The olefinic product is then hydrogenated in a hydrotreater. Any catalyst that is suitable for alkene-to-alkane hydrogenation can in principle be employed for this purpose. The hydrotreated product is fractionated by conventional distillation to produce kerosene that is comprised almost entirely of branched alkanes, hence the name 'IPK'.

Characterization of the IPK (Table 10.7) is available from the work performed to qualify it as a blend component for semi-synthetic jet fuel [12]. The product is a narrow distillation cut and more heavy naphtha could in principle be included to give the material better $T_{50}-T_{10}$ and $T_{90}-T_{10}$ characteristics. The exceptionally low freezing point reflects the highly branched nature of the product. The low density is typical of kerosene range alkanes. Although the IPK on its own does not meet jet fuel specifications, it meets specifications in 50:50 blends with conventional petroleum-derived kerosene.

10.4.2 Hydrogenated Straight Run Material

When straight run F-T material is hydrotreated, the hydrotreated product will reflect the carbon structure of the straight run oil. In this respect there is a considerable difference between HTFT- and LTFT-derived material (Table 10.5).

The aromatics content of straight run C_9–C_{14} HTFT material is close to 10 wt.%. Even when deep hydrodearomatization is performed, the hydrotreated HTFT material will have around 10% cycloalkanes. The hydrogenated straight run material from HTFT synthesis will therefore have a higher density and better cold flow properties than the same boiling range material prepared from LTFT synthesis. Irrespective, the inherently high *n*-alkane content of hydrotreated straight run F-T products will limit the amount of such kerosene that can be blended in semi-synthetic jet fuel.

Straight run HTFT material also has a higher oxygenate content than that of LTFT material. Hydrotreating is made more challenging by oxygenates in the straight run material, which necessitates the use of sulfided base metal catalysts to avoid deactivation [13,14]. Oxygenates can undermine storage stability, cause gum formation, and lead to poor thermal stability. As a result, deep hydrodeoxygenation is necessary. Hydrodeoxygenation is more difficult than alkene saturation. More severe hydrotreating conditions are needed to produce a jet fuel blending component from straight run HTFT material than from straight run LTFT material due to the higher oxygenate content in the HTFT material.

TABLE 10.7 Properties of Sasol Iso-Paraffinic Kerosene

Fuel Property	Sasol IPK[a]
DISTILLATION CURVE (°C)	
IBP	174
T_{10}	181
T_{50}	188
T_{90}	210
FBP	232
$T_{50}-T_{10}$ (°C)	7
$T_{90}-T_{10}$ (°C)	29
Density measured at 15°C (kg/m^3)	768.8
Freezing point temperature (°C)	< -65
Sulphur content (mg/kg)	<10
Acidity (mg KOH/g)	0.001
Lubricity, ASTM D 5001 wear scar (mm)	0.72[b]

[a]*Industrial product from Sasol Synfuels (Fe-based HTFT).*
[b]*Neat, no lubricity additives added.*

10.4.3 Hydrocracked and Hydroisomerized Distillate and Wax

Several LTFT-based gas-to-liquid facilities have started production in the past decade. Wax is the major product from LTFT synthesis. Hydrocracking of F-T wax produces a kerosene fraction that consists of linear and branched alkanes. Despite the similarity to the kerosene material obtained from hydrotreating alkene oligomers, it is not referred to as IPK. In the ASTM D 7566 [7] it is referred to under the umbrella term of 'SPK'. A block flow diagram of the process is shown in Fig. 10.4.

The heavy hydrocarbon material is co-fed with hydrogen to the hydrocracker. The hydrocracker performs both hydroisomerization and hydrocracking simultaneously using a bifunctional catalyst. Two classes of catalysts can be employed, either a noble metal on an acidic support, or a base metal (usually a sulfided base metal) on an acidic support. Insufficiently converted material is recycled to the hydrocracker, and the remainder of the product is fractionated by distillation. One of the product fractions obtained by distillation is SPK.

The degree of isomerization of the product depends on the catalyst and reactor operation, and it is often expressed in terms of the branched-to-linear alkane ratio (ie, the iso-to-normal ratio). The freezing point of the kerosene is affected both by the distillation range and the degree of isomerization. In fact, the properties of the SPK are determined mainly by process design and operation, and in particular by the hydrocracker technology that is employed to produce the kerosene.

Various products prepared from the hydrocracking of F-T wax were evaluated as

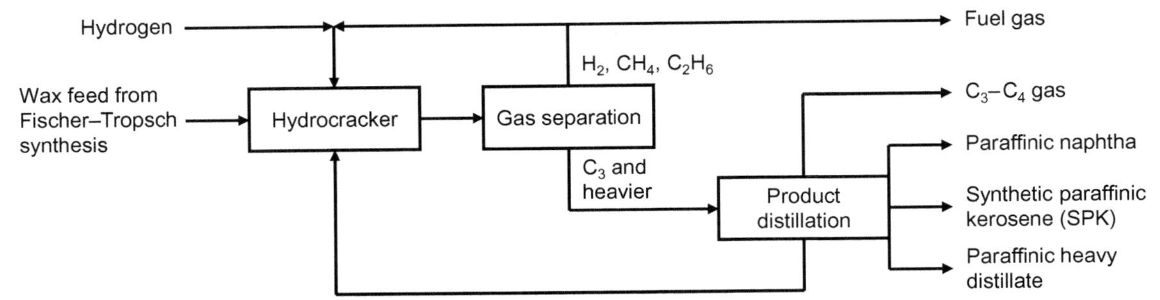

FIGURE 10.4 Block flow diagram of the process to produce synthetic paraffinic kerosene from Fischer–Tropsch heavy hydrocarbons.

semi-synthetic jet fuel blending components. Characterization of these SPKs is shown in Table 10.8 [12]. The influence of hydrocracking technology and distillation are evident. In the case of the Shell GTL and Sasol GTL-1 blending components, their densities were too low for a 50:50 blend with petroleum-derived jet fuel to meet the minimum density specification of 775 kg/m^3 for jet fuel. Nevertheless, some blends of semi-synthetic jet fuel meet all of the specification requirements in blends up to 50 vol.% with petroleum-derived kerosene [12,15].

When the hydrocracking technology is very isomerizing, a wider boiling range can be employed for blending into semi-synthetic jet fuel. For example, the Syntroleum S-5 blending component that was evaluated for blending with JP-5 jet fuel for aircraft carrier use had on average 1.7 methyl groups per molecule, low *n*-alkane content, and wide carbon number range, C_9 to C_{18} [16,17]. The importance of having a more isomerizing hydrocracking technology for the production of SPK was also highlighted by the yield constraints imposed by the flash point specification in combination with the freezing point specification [18].

TABLE 10.8 Properties of Different Synthetic Paraffinic Kerosenes Produced by Hydrocracking of Heavy Hydrocarbons from Co-based LTFT Synthesis

Fuel Property	Syntroleum S-8	Shell GTL	Sasol GTL-1	Sasol GTL-2
DISTILLATION CURVE (°C)				
IBP	144	156	144	179
T_{10}	167	162	153	196
T_{50}	206	169	170	226
T_{90}	256	184	193	256
FBP	276	198	208	266
$T_{50}-T_{10}$ (°C)	39	7	17	30
$T_{90}-T_{10}$ (°C)	89	22	40	60
Density measured at 15°C (kg/m^3)	755	736.1	733.3	761.6
Freezing point temperature (°C)	−51	−54	−49	−60
Sulphur content (mg/kg)	<100	<3	<1	1.3
Acidity (mg KOH/g)	0.004	0.001	0.021	0.003
Lubricity, ASTM D 5001 wear scar (mm)	–	0.92[a]	–	–

[a] *Neat, no lubricity additives added.*

10.5 ON-SPECIFICATION JET FUEL FROM FISCHER–TROPSCH SYNTHESIS

10.5.1 Sasol Fully Synthetic Jet Fuel

Fully synthetic jet fuel contains no petroleum-derived blending components. Thus far the only fully synthetic jet fuel-containing F-T material that was prepared in large quantities is the Sasol fully synthetic jet fuel from the Sasol Synfuels refinery. Unlike semi-synthetic jet fuel, fully synthetic jet fuel is not commonly used commercially. The first passenger-carrying flight employing only fully synthetic jet fuel took place in Sept. 2010 [4].

Various fully synthetic jet fuel blends have been tested [19,20], all of which were blended from five different refinery streams at the Sasol Synfuels facility (Fig. 10.5). The five streams are IPK, heavy F-T naphtha, light F-T distillate, heavy coal tar naphtha, and light creosote distillate. All of these streams are hydroprocessed products.

Three of the blending components are derived from F-T synthesis. The production of IPK (Section 10.4.1) and heavy F-T naphtha (Section 10.4.2) have already been discussed.

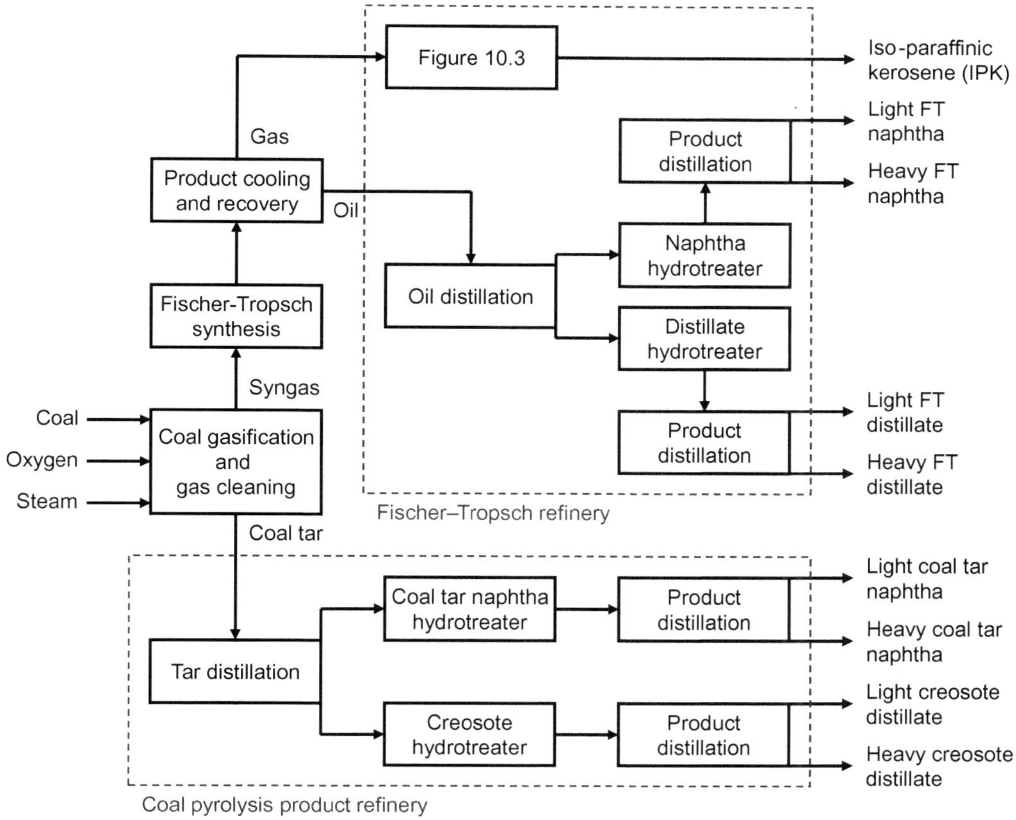

FIGURE 10.5 Block flow diagram of the Sasol Synfuels facility showing the streams used to blend Sasol fully synthetic jet fuel.

The light F-T distillate is produced in the same way as the naphtha, but it is hydrotreated in a different hydrotreater, which is dedicated to the hydrogenation of the F-T distillate fraction.

Two of the blending components are produced by the hydrogenation of coal pyrolysis products from moving bed coal gasifiers. The coal pyrolysis products are recovered by condensation from the raw synthesis gas and are not produced by F-T synthesis. The coal tar naphtha and the creosote distillate that are employed as blending components are hydrotreated products. The refining steps for the heavy coal tar naphtha and the light creosote distillate are similar to those of the F-T fractions. However, the hydrotreating is more severe. For example, the creosote hydrotreater is operated at 18.5 MPa hydrogen pressure [3]. Unlike the F-T material, the coal pyrolysis products are rich in aromatics, and in addition to oxygenates, there are also sulphur- and nitrogen-containing compounds. Hydrotreating removes most of the heteroatom-containing compounds, and the hydrotreated products are rich in cycloalkanes and aromatics.

The characterization of different fully synthetic jet fuel blends that were tested is shown in Table 10.9 [19,20]. Not all of the blends met all of the specifications for Jet A-1, and they reflect the developmental nature of the fuels at the time of their preparation. The importance of the hydrotreated fractions derived from coal tar is evident in both the density and aromatic content of the fuels. A more detailed chromatographic analysis was reported on blend FSJF #5 in Table 10.9 [21]. The compound classes identified and quantified were *n*-alkanes (2.7 wt.%), iso-alkanes (42.8 wt.%), cycloalkanes (13.8 wt.%), bicycloalkanes (20.4 wt.%), polycyclic alkanes (8.7 wt.%), alkylbenzenes (2.7 wt.%), naphtheno-monoaromatics (8.5 wt.%), and the remainder, which was heavier aromatics.

10.5.2 Refining Requirements for Fully Synthetic Jet Fuel

The extensive fuel testing that preceded the acceptance of aviation turbine fuels that contained refined F-T material highlighted some requirements that have to be met by a fully synthetic jet fuel. These requirements relate to both distillation profile as well as composition. At least some of these requirements can also be inferred from the physical and molecular requirements in the jet fuel specifications. From a refining perspective it is the bulk fuel properties that cannot be fixed by an additive that are important to address in the refinery.

Considering the composition of F-T material (Table 10.5) in relation to the jet fuel specifications (Fig. 10.2 and Tables 10.1–10.4), the following important refining targets can be outlined:

a. The acyclic aliphatic material must contain a high fraction of branched alkanes and as little as practicable *n*-alkanes in order to meet the freezing point specification. F-T synthesis produces predominantly linear hydrocarbons, which places a restriction on the amount of hydrotreated straight run material that can be included. In this respect, straight run HTFT products are somewhat more forgiving than LTFT products because the HTFT products contain some cyclic compounds too. Kerosene range material produced by alkene oligomerization is inherently branched, as is the kerosene range material produced by hydrocracking. However, it should be kept in mind that different oligomerization technologies result in different degrees of branching, just like different hydrocracking technologies result in different degrees of branching. These are mainly consequences of the catalysts employed [22]. It is therefore important to select the appropriate catalyst, particularly

TABLE 10.9 Properties of Different Fully Synthetic Jet Fuel Blends Prepared From the Refinery Streams Indicated in Fig. 10.5

	Sasol Fully Synthetic Jet Fuel Blend				
Fuel Property	**FSJF #1**	**FSJF #2**	**FSJF #3**	**FSJF #4**	**FSJF #5**
BLEND COMPOSITION, FIG. 10.5 (VOL.%)					
Iso-paraffinic kerosene	67	55	57	5	70
Heavy F-T naphtha	15	15	13	59	0
Light F-T distillate	18	0	10	0	0
Heavy coal tar naphtha	0	0	0	0	30
Light creosote distillate	0	30	20	36	0
DISTILLATION CURVE (°C)					
IBP	183	174	179	166	170
T_{10}	189	185	188	175	176
T_{50}	196	198	200	188	187
T_{90}	216	220	223	222	218
FBP	246	245	250	256	272
$T_{50}-T_{10}$ (°C)	7	13	12	13	11
$T_{90}-T_{10}$ (°C)	20	22	23	34	31
Flash point temperature (°C)	69	63	62	53	51
Density measured at 20°C (kg/m^3)	772.6	797.2	789.8	799.3	781.2
Freezing point temperature (°C)	−57	<−60	−59	−61	−55
Viscosity measured at −20°C (mm/s)	4.68	4.53	4.69	3.48	4.09
Net heat of combustion (MJ/kg)	43.8	43.4	43.5	43.1	43.7
Aromatic content (vol.%)	7.2	13.6	12.5	16.9	10.9
Lubricity, ASTM D 5001 wear scar (mm)	0.69	0.73	0.68	0.72	–

when heavier material is included in the kerosene.

b. A minimum amount of 8 vol.% aromatics is required in jet fuel to ensure compatibility with elastomer seals in aircraft fuel systems. In a mixed fuel usage environment the jet fuel can be changed from conventional petroleum-derived jet fuel to synthetic jet fuel. If the synthetic jet fuel contains too few aromatics, the fuels will decrease swelling in elastomer seals, which in turn can cause leaks in the fuel system. Various studies on elastomer swelling due to exposure to F-T-derived jet fuels and blend components can be found in the literature (eg, Refs [23–27]). In addition to the need for aromatics for elastomer swelling compatibility, aromatics also increase the density of the jet fuel. This is an important benefit because F-T materials inherently have little or no cycloalkanes, and the cycloalkanes present in hydrotreated heavy HTFT naphtha come from the hydrogenation of aromatics. As pointed out before, it is impossible for purely paraffinic kerosene to meet the minimum density requirement of jet fuel without blending of cycloalkanes or aromatics. The nature of the aromatics matter because mono-nuclear aromatics are less likely to cause deposits during thermal stressing than binuclear aromatics [28]. In the kerosene range the free radical addition propensity, irrespective of whether it is oxidatively or thermally induced, is high in naphtheno-aromatics, and is particularly high in homologues of indan [29]. The refining origin of the aromatics will therefore affect the thermal stability of the jet fuel.

c. The jet fuel should preferably have a wide distillation range and the blending of single components, or narrow distillation cuts should be avoided. This may seem a simple enough refining objective to achieve, but any inherent deficiencies in the molecular properties of the streams used as blending components may necessitate the use of narrower distillation cuts to meet the other specifications. The characterization data in Table 10.9 shows that meeting the $T_{90}-T_{10}$ and $T_{50}-T_{10}$ specifications can be a challenge if the refinery was not originally designed to produce fully synthetic jet fuel.

d. The fuel must be severely hydroprocessed to limit alkenes and oxygenates in the final jet fuel. Alkenes in combination with oxygenates have poor high-temperature thermal stability, because such materials are susceptible to form free radical addition products.

A useful approach to determine a viable composition range of material for fully synthetic jet fuel is the graphical method of Cookson, et al. [30] Although it is restricted to three compound classes, it can be adapted as a numeric approach, which can employ all four relevant compound classes for fully synthetic jet fuel blending: *n*-alkanes, iso-alkanes, cycloalkanes, and aromatics.

10.5.3 Fischer–Tropsch Jet Fuel Refining

The demonstration of fully synthetic jet fuel production by Sasol [4] relied on access to coal pyrolysis products as blending components to meet the Jet A-1 minimum aromatic and density specifications. These materials are not available in F-T facilities that employ biomass as raw material feed. The question should therefore be posed: Is it possible to produce a fully synthetic jet fuel in an F-T-based biomass-to-liquids facility? The short answer is 'yes'.

When planning an F-T facility to maximize production of on-specification jet fuel, the refinery can be designed to meet this objective without the aid of material that is not derived from F-T synthesis. If it is additionally required that all of the transportation fuels from the refinery must meet fuel specifications, the best F-T technology platform is Fe-based LTFT synthesis [3]. Fe-based LTFT synthesis strikes the most efficient balance between alkene content in the lighter products and wax production on the heavy end.

Various refinery designs that maximize jet fuel production from F-T synthesis without external blending components have been described in the literature [3,31]. In practise, a refinery design will seldom be geared to such an extreme, but the aforementioned work demonstrated that it is possible to refine more than 50 wt.% of HTFT material to fully synthetic Jet A-1 and meet fuel specifications for all other transportation fuels. Likewise, it is possible to refine around 60 wt.% of LTFT material to fully synthetic Jet A-1 and meet fuel specifications for all other transportation fuels.

The refining technologies that are required to produce fully synthetic jet fuel follow from the molecular requirements of jet fuel. A minimum of four of the following five chemical conversions are required to produce Jet A-1 from F-T material (Fig. 10.6):

a. Alkene oligomerization. Alkene oligomerization produces a highly branched product. There are different oligomerization technologies to choose from Refs [3,32]. For the purpose of jet fuel production from light F-T alkenes, solid phosphoric acid-catalysed oligomerization is ideal because it produces a carbon number-constrained product with good kerosene yield. This process forms the basis of IPK production as a semi-synthetic jet fuel-blending component, which was discussed in Section 10.4.1. It is also an important blending component in fully synthetic jet fuel.

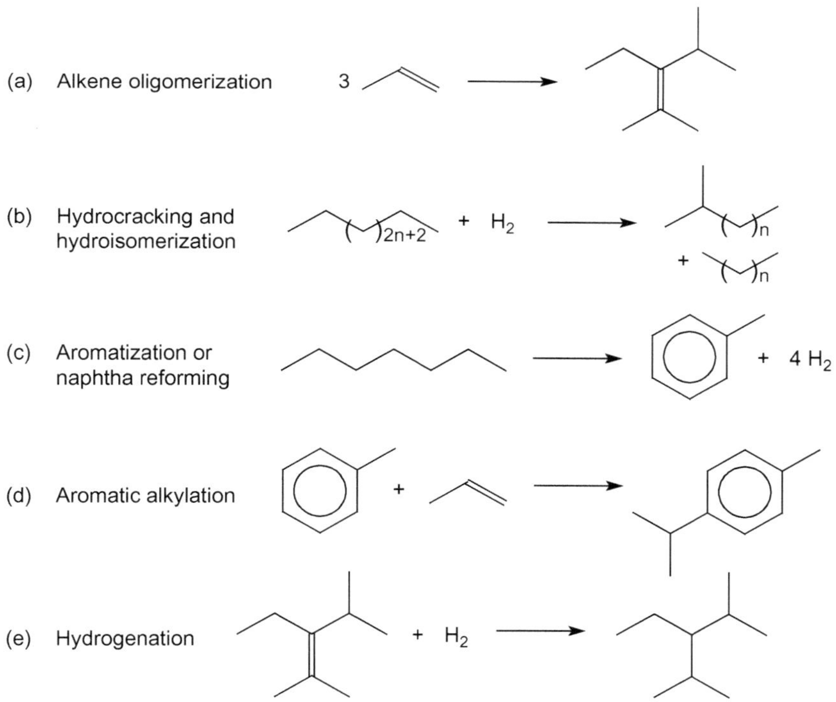

FIGURE 10.6 Chemical conversions needed in the production of fully synthetic jet fuel from Fischer–Tropsch material. (A) Alkene oligomerization; (B) Hydrocracking and hydroisomerization; (C) Aromatization or naphtha reforming; (D) Aromatic alkylation; (E) Hydrogenation.

b. Hydrocracking/hydroisomerization. The semi-synthetic jet fuel blending component derived from heavy hydrocarbon hydrocracking was discussed in Section 10.4.3. A hydrocracker is particularly important when refining LTFT material, since most of the organic product is wax (Table 10.5). Without a hydrocracker, very little LTFT material would be available for jet fuel. For HTFT synthesis, which does not have a meaningful heavy boiling fraction, a hydrocracker or hydroisomerization unit may still be necessary to ensure that the heavier material in the kerosene boiling range is sufficiently branched. Straight run F-T kerosene is too rich in linear hydrocarbons to meet the freezing point specification without hydroisomerization. The jet fuel distillation range requirements ($T_{90}-T_{10} \geq 40°C$ and $T_{50}-T_{10} \geq 15°C$) necessitate the inclusion of some heavier material. Unless such material can be produced from a source other than straight run F-T kerosene, some hydroisomerization would be required.

c. Aromatization/naphtha reforming. The minimum aromatics content for jet fuel is 8 vol.%. There must consequently be a source of aromatics. There are different technologies that can be considered, and the type of technology used to produce aromatics will depend on refinery streams that are earmarked for aromatics production, as well as the overall need for aromatics in the refinery [3,32]. In this discussion we are concerned only with jet fuel, but in a practical refinery design all transportation fuels will be considered. Aromatics are also needed to produce on-specification motor-gasoline (petrol).

d. Aromatic alkylation. Refining technologies that produce aromatics from alkanes produce aromatics mainly in the C_6-C_{10} range. In order to obtain aromatics that span the kerosene boiling range, as well as to make more aromatics available for use in jet fuel, the aromatics can be alkylated with light alkenes. This can be performed in a dedicated unit, but it can also be achieved by co-feeding aromatics with light alkenes to an alkene oligomerization unit [33].

e. Hydrogenation. Hydrotreating is necessary to hydrogenate alkenes from the alkene oligomerization unit. Hydrotreating is likewise needed to include straight run F-T material for the production of semi-synthetic jet fuel (Section 10.4.2). Depending on the technology selected to produce aromatics, hydrotreating might also be necessary as feed pretreatment for catalytic naphtha reforming.

An uncomplicated refinery design focused on the production of fully synthetic jet fuel is shown in Fig. 10.7 to illustrate the use of the different conversion types. This refinery design is for refining LTFT material, but an analogous design can be employed for HTFT material. This uncomplicated refinery design demonstrates the production of on-specification fully synthetic jet fuel, but the naphtha range material does not meet motor-gasoline specifications. In order to meet these specifications, the octane number of the naphtha range alkanes must be further improved.

The hydrocracker unit is responsible for converting the heavy hydrocarbons into branched kerosene range products. The heavy hydrocarbon fraction is recycled to extinction. Naphtha and lighter boiling alkanes are co-produced with the kerosene. The lighter material can be further refined (as shown in Fig. 10.7), but may also be recovered as naphtha and fuel gas products.

Aromatics can be produced in different ways. The most common refining technology is conventional catalytic naphtha reforming, but this technology is not very efficient in combination with F-T materials [3,32]. An alternative is

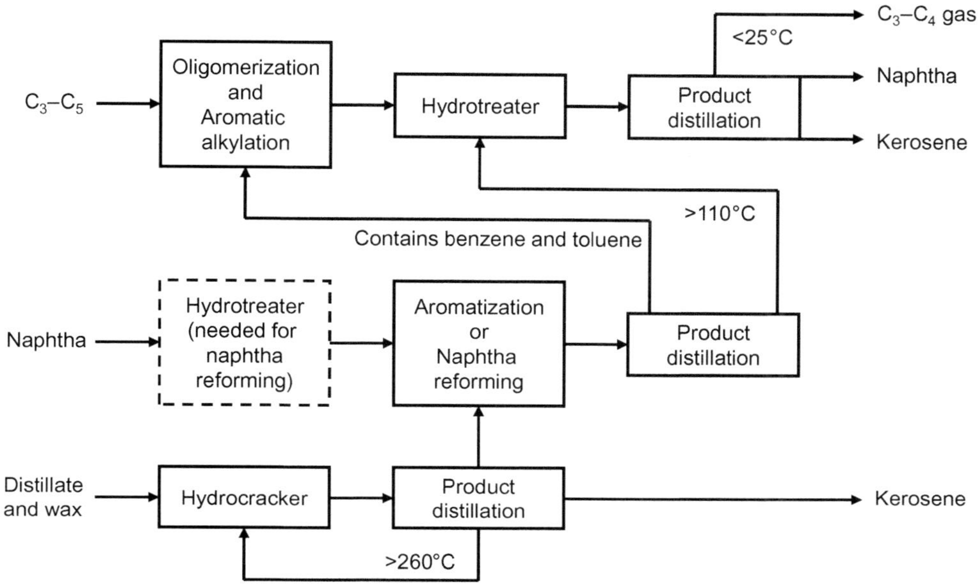

FIGURE 10.7 Block flow diagram of an uncomplicated Fischer–Tropsch refinery design focused on the production of fully synthetic jet fuel.

catalytic naphtha reforming employing a Pt on nonacidic L-zeolite catalyst, which is efficient for F-T naphtha. In either case, the feed is naphtha and the catalytic naphtha reformer feed must be hydrotreated first. Another possibility is to employ an aromatization technology based on a metal-promoted ZSM-5 catalyst, which can also convert material in the C_3–C_5 range. Irrespective of the technology selected, an aromatic-rich product is produced.

A portion of the light aromatic-rich product can be co-fed with the C_3–C_5 material to the alkene oligomerization unit, which will also perform aromatic alkylation. This type of operation was demonstrated with a solid phosphoric acid catalyst. Not all oligomerization technologies are capable of performing both oligomerization and aromatic alkylation in tandem. The product is a mixture of highly branched alkenes, alkanes, and alkyl aromatics. This material must be hydrotreated before it is fractionated by distillation to produce a kerosene product for jet fuel blending.

In all of the product distillation steps, the cut points can be adjusted to ensure that the final blended kerosene meets the fully synthetic jet fuel specifications. The IBP of the kerosene from the hydrocracker product distillation controls the flash point and influences the density of the paraffinic kerosene. The final boiling point of the kerosene from the hydrocracker product distillation can be adjusted to meet the freezing point specification, with a broader cut being possible if the hydrocracking technology is more isomerizing. The initial and final boiling point of the kerosene from the hydrotreater after alkene oligomerization and aromatic alkylation controls the aromatic content and density.

10.6 CONCLUSIONS

The production, refining, and qualification of aviation turbine fuel (jet fuel) from products derived from F-T synthesis was discussed. The following conclusions can be drawn for aviation turbine fuel production in a biomass-to-liquids facility that employs F-T synthesis:

a. Biomass gasification to produce synthesis gas must be designed with specific biomass feed in mind, but once synthesis gas is produced, the rest of the process becomes independent of the nature of the raw material feed.

b. Current DEF STAN 91-91 and ASTM D 7566 aviation turbine fuel specifications make provisions for the inclusion of F-T-derived material in jet fuel. Kerosene from the refining of F-T products can either be included up to 50 vol.% in blends with kerosene from petroleum refining to produce semi-synthetic jet fuel, or blended without any petroleum products to produce a fully synthetic jet fuel.

c. Irrespective of the concentration of F-T-derived material that is included in jet fuel, there are some additional jet fuel specifications that must be met beyond that required for petroleum-derived jet fuel. The additional specification requirements in DEF STAN 91-91 Issue 7 [6] and ASTM D 7566-15b [7] are minimum 8 vol.% aromatics, boiling range $T_{90}-T_{10} \geq 40°C$, boiling range $T_{50}-T_{10} \geq 15°C$, viscosity measured at $-40°C \leq 12$ mm/s, and more stringent requirements for acidity and metals content.

d. The main molecular transformations needed to produce jet fuel from F-T products are isomerization of *n*-alkanes to iso-alkanes, synthesis of aromatics, and hydrotreatment of alkenes and oxygenates to produce alkanes.

e. Refining of F-T products is required to produce jet fuel or jet fuel blend components. The following conversion types will typically be required in a biomass-to-liquids F-T refinery to produce fully synthetic jet fuel: (1) alkene oligomerization, (2) heavy hydrocarbon hydrocracking/hydroisomerization, (3) aromatization or catalytic naphtha reforming, (4) aromatic alkylation, and (5) hydrotreating.

References

[1] P. Basu, Biomass gasification and pyrolysis, Practical Design and Theory, Elsevier, Amsterdam, 2010.

[2] P.M. Maitlis, A. De Klerk (Eds.), Greener Fischer–Tropsch Processes for Fuels and Feedstocks, Wiley-VCH, Weinheim, 2013.

[3] A. De Klerk, Fischer–Tropsch Refining, Wiley-VCH, Weinheim, 2011.

[4] P. Morgan, P.N.J. Roets, The synthetic jet fuel journey, Proc. World Petrol. Congr. 20 (2011), WPC-20-2265.

[5] Standard Practice for Qualification and Approval of New Aviation Turbine Fuels and Fuel Additives. American Society for Testing and Materials, West Conshohocken, PA, 2014.

[6] United Kingdom Ministry of Defence, Defense Standard 91-91, Issue 7. Turbine Fuel, Kerosine Type, Jet A-1, NATO Code: F-35, Joint Service Designation: AVTUR; Ministry of Defence, London, 18 February 2011.

[7] ASTM D 7566–15b, Standard Specification for Aviation Turbine Fuel Containing Synthesized Hydrocarbons, American Society for Testing and Materials, West Conshohocken, PA, 2015.

[8] A. Bondi, Physical Properties of Molecular Crystals, Liquids, and Glasses, John Wiley & Sons, New York, NY, 1968.

[9] R.C. Reid, J.M. Prausnitz, B.E. Poling, The Properties of Gases and Liquids, fourth ed., McGraw-Hill, New York, NY, 1987.

[10] A. De Klerk, Fischer–Tropsch process, in: A. Seidel (Ed.), Kirk-Othmer Encyclopedia of Chemical Technology, John Wiley and Sons, New York, NY, 2013.

[11] R. Van der Westhuizen, H. Potgieter, N.M. Prinsloo, A. De Villiers, P. Sandra, Chromatography–mass spectrometry for analysis of cyclics in oligomerisation products of Fischer–Tropsch derived light alkenes, J. Chromatogr. A 1218 (2011) 3173–3179.

[12] C.A. Moses, Comparative evaluation of semi-synthetic jet fuels; Final report, Coordinating Research Council Project No. AV-2-04a, September 2008.

[13] D. Lamprecht, Hydrogenation of Fischer-Tropsch synthetic crude, Energy Fuels 21 (2007) 2509–2513.

[14] A. De Klerk, Hydroprocessing peculiarities of Fischer-Tropsch syncrude, Catal. Today 130 (2008) 439–445.

[15] D. Lamprecht, P.N.J. Roets, Sasol slurry phase distillate™ semi-synthetic aviation turbine fuel, Prepr. Pap. Am. Chem. Soc. Div. Petrol. Chem. 49 (4) (2004) 426–430.

[16] R.L. Freerks, P.A. Muzzell, Production and characterization of synthetic jet fuel produced from Fischer-Tropsch hydrocarbons, Prepr. Pap. Am. Chem. Soc. Div. Petrol. Chem. 49 (4) (2004) 407–410.

[17] P.A. Muzzell, R.L. Freerks, J.P. Baltrus, D.D. Link, Composition of syntroleum S-5 and conformance to JP-5 specification, Prepr. Pap. Am. Chem. Soc. Div. Petrol. Chem. 49 (4) (2004) 411–413.

[18] D. Lamprecht, Fischer-Tropsch fuel for use by the US military as battlefield-use fuel of the future, Energy Fuels 21 (2007) 1448–1453.

[19] P.N.J. Roets, C.A. Moses, J. Van Heerden, Physical and chemical properties of fully synthetic aviation turbine fuel from coal, Prepr. Pap. Am. Chem. Soc. Div. Petrol. Chem. 49 (4) (2004) 403–406.

[20] C.A. Moses, P.N.J. Roets, Properties, characteristics, and combustion performance of Sasol fully synthetic jet fuel, J. Eng. Gas Turbines Power 131 (2009) 041502.

[21] R. Van der Westhuizen, M. Ajama, P. De Coning, J. Beens, A. De Villiers, P. Sandra, Comprehensive two-dimensional gas chromatography for the analysis of synthetic and crude-derived jet fuels, J. Chromatogr. A 1218 (2011) 4478–4486.

[22] A. De Klerk, E. Furimsky, Catalysis in the Refining of Fischer-Tropsch Syncrude, Royal Society of Chemistry, Cambridge, UK, 2010.

[23] J.L. Graham, R.C. Striebich, D.K. Minus, W.E. Harrison III, The swelling of selected O-ring materials in jet propulsion and Fischer-Tropsch fuels, Prepr. Pap. Am. Chem. Soc. Div. Petrol. Chem. 49 (4) (2004) 435–439.

[24] J.L. Graham, R.C. Striebich, K.J. Myers, D.K. Minus, W.E. Harrison III, Swelling of nitrile rubber by selected aromatics blended in a synthetic jet fuel, Energy Fuels 20 (2006) 759–765.

[25] J.L. Graham, D.K. Minus, W.E. Harrison III, An investigation of aviation components material compatibility with blends of a Fischer-Tropsch-derived jet fuels with petroleum derived jet fuels, Prepr. Pap. Am. Chem. Soc. Div. Fuel Chem. 52 (2) (2007) 454–456.

[26] M.J. DeWitt, E. Corporan, J.L. Graham, D. Minus, Effects of aromatic type and concentration in Fischer-Tropsch fuel on emissions production and material compatibility, Energy Fuels 22 (2008) 2411–2418.

[27] E. Corporan, T. Edwards, L. Shafer, M.J. DeWitt, C. Klingshirn, S. Zabarnick, et al., Chemical, thermal stability, seal swell, and emissions studies of alternative jet fuels, Energy Fuels 25 (2011) 955–966.

[28] M.J. DeWitt, Z. West, S. Zabarnick, L. Shafer, R. Striebich, A. Higgins, et al., Effect of aromatics on the thermal-oxidative stability of synthetic paraffinic kerosene, Energy Fuels 28 (2014) 3696–3703.

[29] M.N. Siddiquee, A. De Klerk, Hydrocarbon addition reactions during low-temperature autoxidation of oilsands bitumen, Energy Fuels 28 (2014) 6848–6859.

[30] D.J. Cookson, C.P. Lloyd, B.E. Smith, Investigation of the chemical basis of kerosene (jet fuel) specification properties, Energy Fuels 1 (1987) 438–447.

[31] A. De Klerk, Fischer-Tropsch fuels refinery design, Energy Environ. Sci. 4 (2011) 1177–1205.

[32] A. De Klerk, Fischer-Tropsch refining: technology selection to match molecules, Green Chem. 10 (2008) 1249–1279.

[33] T.M. Sakuneka, A. De Klerk, R.J.J. Nel, A.D. Pienaar, Synthetic jet fuel production by combined propene oligomerization and aromatic alkylation over solid phosphoric acid, Ind. Eng. Chem. Res. 47 (2008) 1828–1834.

CHAPTER

11

Current and Potential Aviation Additives for Higher Biofuel Blends in Jet A-1

S.L. Repetto[1], *J.F. Costello*[2] *and D. Parmenter*[3]

[1]Department of Chemical Engineering, University of Bath, Bath, United Kingdom, [2]Faculty of Health and Applied Sciences, University of the West of England, Bristol, United Kingdom, [3]Airbus Operations Ltd, Bristol, United Kingdom

11.1 INTRODUCTION

The chemical composition of aviation fuel is expected to vary as a result of the diverse range of alternative sources and processes used for their generation. Stringent standards governing bulk physico-chemical properties are necessary in order to ensure such fuels are fit for purpose and do not impact aircraft and engine operation. This chapter considers the range of additives sometimes deployed in order to improve the ancillary properties of jet fuel – that is, thermal and storage stability, lubricity, conductivity, inhibition of bacterial growth, and ice formation. Potential future additives to improve the properties of biofuels, and more generally alternative fuels, are also considered.

Typically, additives are largely fuel-soluble chemicals derived from petroleum-based raw materials; their function and chemistry are highly specialized. They produce the desired effect in the parts per million (ppm) concentration range. These additives fall into general categories such as thermal stability improvers, antioxidants, metal deactivators, and lubricity improvers [1]. Additives are used to varying degrees in all petroleum-derived fuels, but the situation with aviation fuels is unique in that only those additives specifically approved by industry committees may be added to jet fuel. Most jet fuel specifications list the approved additives along with allowed concentration ranges [2]. Some approved additives are mandatory while others are optional, only to be used by agreement between customer and vendor. There are variations depending on the specification. In this chapter we will mainly consider the Def Stan 91-91 of the UK Ministry of Defence. Some classes of additives are

Biofuels for Aviation.
DOI: http://dx.doi.org/10.1016/B978-0-12-804568-8.00011-1

approved by the chemical compositions, so are therefore available from more than one supplier, whereas others are proprietary products approved as a commercial package on an individual basis [2]. The type and mandatory use of additives is also the principal difference between commercial and military jet fuels. Military jet fuels will normally contain three or more additives, such as antioxidants, fuel system icing inhibitors (FSII), or thermal stability improvers because of the longer-term storage and performance requirements [1].

In the evaluation of a potential additive, several key properties need to be met in order for it to be appropriate for use in jet fuel, including being effective below 1% (v/v), being chemically stable, not having any negative effects on the engine and fuel system components, having the required solubility in fuel, possessing a low environmental impact, and being cost effective [3,4]. The approval process is clearly outlined in ASTM D4054 [5], and it is a regimented, lengthy, and costly process.

11.2 APPROVED ADDITIVES FOR ALTERNATIVE FUELS

All jet fuel specifications list approved additives along with allowed concentration ranges; approved additives vary depending on the specification. In this chapter we consider the Def Stan 91-91 of the UK Ministry of Defence [2]. These additives fall into the following general categories: antioxidants, metal deactivators, lubricity improvers, static dissipators, biocides, FSII, and thermal stabilizers [1]. The use of additives is the principal difference between commercial and military jet fuels. The static dissipator *Stadis 450* is mandatory for Jet A-1, with antioxidant addition required for severely hydroprocessed fuels and synthetic blends, whereas military jet fuels will contain three or more additives – that is, antioxidant, metal deactivator, and FSII – because of the high performance requirements [1]. A list of Def Stan 91-91 approved additives for Jet A-1, with corresponding concentrations, is presented in Table 11.1.

11.2.1 Antioxidants

Antioxidants are added to improve fuel storage stability and inhibit formation of peroxides, soluble gums, and insoluble particulates resulting from oxidation reactions. Peroxides can chemically attack sealants in the aircraft fuel systems, gums can lead to engine deposits, and particulates can plug fuel filters [7]. Hydroprocessing of fuels removes naturally occurring antioxidants that provide protection from peroxidation. Therefore, antioxidants are mandatory for aviation fuels derived from the Fischer–Tropsch (F–T) process, hydroprocessed esters and fatty acids (HEFA), and other hydroprocessed feedstock [2].

Oxidation, which occurs in the presence of molecular oxygen, forming peroxides and hydroperoxides, is known as autoxidation. This process has been known to contribute to the oxidative degradation of hydrocarbon-based fuels during storage. There are two main stages in the oxidation process: a free-radical chain reaction forming the hydroperoxide species, and a subsequent decomposition of the hydroperoxides [8]. The chain reaction is initiated by abstraction of a hydrogen atom to form a radical species, which can then react with molecular oxygen to form hydroperoxides. Hydrogen abstraction can be instigated by light or heat, the presence of metal ions, or free radicals. In general, tertiary C–H bonds are more susceptible to oxidative attack than secondary, which are in turn more so than primary C–H bonds. This is because of the increased stability of the carbon radical with increasing substitution. Olefins are also susceptible to hydride abstraction at the carbon atom adjacent to the double bond, creating

TABLE 11.1 Approved Additives for Jet A-1 in the Def Stan 91-91[a]

Additive	Category	Approved Concentrations
2,6-Ditertiary-butyl-phenol	Antioxidants	17–24 mg/L
2,6-Ditertiary-butyl-4-methyl-phenol	Antioxidants	17–24 mg/L
2,4-Dimethyl-6-tertiary-butyl-phenol	Antioxidants	17–24 mg/L
2,6-Ditertiary-butyl-phenol (≥75%), tertiary and tritertiary-butyl-phenols (≤25%)	Antioxidants	17–24 mg/L
2,4-Dimethyl-6-tertiary-butyl-phenol (≥55%), 4 methyl-2,6-ditertiary-butyl-phenol (≥15%), mixture of monomethyl and dimethyl-tertiary-butyl-phenols (≤30%)	Antioxidants	17–24 mg/L
2,4-Dimethyl-6-tertiary-butyl-phenol (≥72%), mixture of tertiary-butyl-methyl-phenols and tertiary-butyl dimethyl phenols (≤28%)	Antioxidants	17–24 mg/L
Diethylene glycol monomethyl ether	Fuel system icing inhibitors	0.10–0.15% (v/v)
Hitec 580	Lubricity improvers	15–23 mg/L
Octel DCI-4A	Lubricity improvers	9–23 mg/L
Octel DCI-6A	Lubricity improvers	9–15 mg/L
Nalco 5403	Lubricity improvers	12–23 mg/L
Nalco 5405	Lubricity improvers	11–23 mg/L
Tolad 4410	Lubricity improvers	9–23 mg/L
Tolad 351	Lubricity improvers	9–23 mg/L
Unicor J	Lubricity improvers	9–23 mg/L
Spec Aid 8Q22	Lubricity improvers	9–23 mg/L
N,N′-Disalicylidene 1,2-propanediamine	Metal deactivators	<5.7 mg/L
Stadis 450	Static dissipators	<3 mg/L
Tracer A	Leak detection	<1 mg/kg

[a]*Information about other international specifications can be found in World Jet Fuel Specifications [6].*

stability of the radicals via hyperconjugation. Stability of fuels with respect to hydrocarbon composition decreases from paraffins > naphthenes > aromatics > olefins > di-olefins [8], therefore, unsaturated alternative fuels will possess lower stability.

An induction period is commonly encountered prior to the initiation of autoxidation. In pure hydrocarbons, this is a result of the time required to build up a sufficient concentration of hydroperoxide in solution to allow propagation of the chain reaction to proceed. In fuels, the induction period may also be the result of resonance stabilization of hydroperoxide radicals by impurities. The second phase of the process involves decomposition of the hydroperoxide species formed during the radical chain reaction. There are several mechanisms by which hydroperoxide decomposition can occur, depending upon the conditions of the reaction: free-radical, acid- or base-catalysed, and metal-catalysed. The products formed from the decomposition of the hydroperoxides can then produce a range of oxygenated species, such as alcohols, ketones, esters, ethers, and carboxylic acids, as well as insoluble oligomeric species [9]. Insoluble products can agglomerate to form gums, sediments, or deposits detrimental to fuel/engine performance [10]. Further, the physical and chemical properties of the fuel are altered, potentially leading to a decrease in the energy density of the fuel [11] and a susceptibility to microbial contamination [10]. Production of carboxylic acids and peroxides provides an acidic environment, which can corrode metals, degrade elastomer materials, and further catalyse the oxidative degradation, creating a positive feedback effect.

Butyl-phenols are the only antioxidants approved for jet fuel [2]. The use of antioxidants effectively stabilizes the radicals through resonance effects and therefore retards the oxidation process. It has been postulated that while antioxidants may increase the induction period of a hydrocarbon by preventing the build-up of hydroperoxide species, they may not have a noticeable effect once the concentration of these radical species reaches steady-state levels. Antioxidants can also influence the termination of the oxidation reaction by forming inactive radical species that will no longer propagate. Def Stan 91-91 states that a minimum of 17 mg/L of an approved antioxidant must be added to the proportion of the fuel blend that has been hydroprocessed, with the total concentration for the final fuel batch not exceeding 24 mg/L. Researchers are studying substituted phenols obtained from extracted and depolymerized lignin as a green alternative to current antioxidants [12]. This could potentially improve the economic viability of alternative fuel production from biomass resources [13].

11.2.2 Metal Deactivators

Metal ions in fuel can be effective catalysts for oxidation reactions, contributing to poor thermal stability (see Section 11.2.1). Metal deactivators are chelating agents that form stable complexes with metal ions that would otherwise engage in oxidation reactions [2]. Although metal contamination is not a common problem in alternative fuels, transition metals such as iron, nickel, manganese, cobalt, and copper are commonly found in the metallurgy of storage tanks, barrels, ocean tankers, and rail tank cars. Metals known to be deleterious to thermal stability include copper, zinc, cadmium, iron, and cobalt. Though metals themselves can act as catalysts for hydrocarbon oxidation, organic substrates commonly encountered in fuel, such as naphthenates and stearates, for example, form fuel-soluble organometallic complexes with cobalt, manganese, copper, nickel, and iron that are effective homogeneous oxidation catalysts [8]. El Naga et al. determined the extent of the oxidation of various metals and a Ni−Cr alloy by measuring the resulting viscosity, acidity, and the amount of insoluble materials produced, and found that copper caused the largest changes in these properties as well as the fastest rate of change. Iron, Ni−Cr, and aluminium affected the oxidation, but to a lesser degree than copper. This difference is most likely due to the solubility of the metals in the fuels themselves as copper is much more soluble in the fuel than iron, Ni−Cr, and aluminium. Solubility, viscosity, and acidity increases, suggesting a positive feedback of the oxidative degradation as the reaction progresses [14].

A metal deactivator additive (MDA) may be added to a fuel to counteract the effects of metal ions, provided that the nature of the contamination is specified beforehand; however, the use of MDAs to remedy off-specification fuel is prohibited. Only *N,N′*-disalicylidene-2-propanediamine has been approved for aviation fuels (Fig. 11.1). MDA is a chelating agent that binds metal ions and prevents fuel degradation. The concentration of the active material during initial doping of the fuel should not exceed 2.0 mg/L. Cumulative addition of MDA when re-doping the fuel should not exceed 5.7 mg/L. It has also been observed that MDA improves thermal stability in the thermal stability test in the absence of metal ions. That it is why MDA can be used to improve thermal stability provided that the thermal stability test is determined before and after MDA addition and reported accordingly [2].

11.2.3 Lubricity Improvers/Corrosion Inhibitors

Lubricity improvers were originally designed as corrosion inhibitors to protect uncoated steel tanks and pipelines from rusting and from corrosion caused by water and oxygen [7]. It was then discovered that they also improve the lubricity of fuels, and for this reason they are mandatory in all military and some civil turbine fuels [1]. Many aircraft fuel system components, especially pumps, rely on the fuel to lubricate moving parts [1]. Hydroprocessing of fuels removes trace components that provide the fuel with natural lubricating properties. That is why lubricity improvers are used to improve the lubricity of severely hydroprocessed jet fuels. Aviation fuels derived from the Fischer–Tropsch

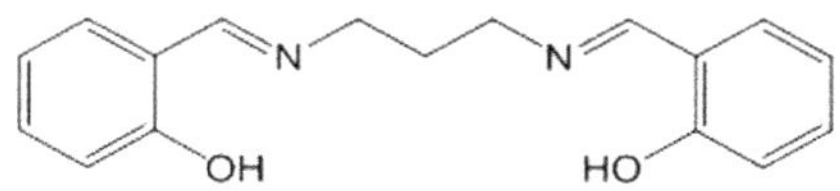

FIGURE 11.1 Structure of *N,N'*-disalicylidene-2-propanediamine.

(F–T) process, HEFA, and other hydroprocessed feedstock might require lubricity improvers. That is why Def Stan 91-91 demands, as part of the lubricity requirement, that the volume percentage of severely hydroprocessed components be reported. The term hydroprocessed includes hydrotreated, hydrofined, and hydrocracked fuels. Lubricity improvers are usually long-chain alkylphenol ethoxylates, carboxylic acids, or esters, such as linoleic acid derivatives (ie, dilinoleic acid; Fig. 11.2) [15].

In general, the greatest lubricity effects are afforded by compounds containing a carboxyl group, which possesses the highest adsorption activity in the series of functional groups including esters, amines, and amides [16]. In all cases, the polar group adheres to the metal surface creating a thin layer of the additive on the surfaces [7]. Because of their polar nature, these additives can have adverse effects on ground-based filtration systems and on fuel/water separation characteristics. Furthermore, most of the additives are so effectively adsorbed onto metal surfaces en route to the aircraft, that it creates issues regarding the correct additive loading in the fuel itself. This problem can only be rectified by either equilibration of the supply system downstream, or by additive injection at the point of entry to the aircraft [2]. A list of the approved lubricity improvers can be found in Table 11.1; Def Stan 91-91 allows the use of approved formulations without prior agreement between purchaser and supplier. The acidity of alternative fuels can also increase corrosion. Over time and the inevitable exposure to atmospheric oxygen, alternative fuel undergoes oxidation to produce weak carboxylic acids, which lower the fuel's pH [17].

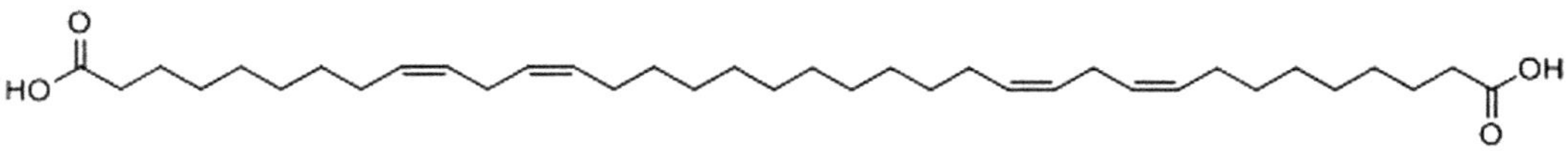

FIGURE 11.2 Structure of dilinoleic acid.

11.2.4 Static Dissipators

Certain fuel-handling procedures, in particular high-speed pumping and micro-filtration, can create charge separation at interfaces. This phenomenon is capable of producing high voltage gradients inside fuel storage areas and increases the risk of fire and explosion [18]. To eliminate this risk, a static dissipator additive is widely used in jet fuels. Static dissipators are added to improve the naturally poor conductivity of fuel, which is why they are also known as conductivity improvers. Jet fuel has very low natural electrical conductivity, usually less than 5 pS/m. The minimum and maximum fuel conductivity requirements for Def Stan 91-91 are 50 to 600 pS/m. Although the addition of a static dissipator is not mandatory under ASTM D1655, it is commonly injected into Jet A-1 at refineries [19]. Currently, *Stadis 450* is the only additive approved for use in aviation turbine fuels. Produced by Innospec LLC, the concentration on first doping of fuel is 3.0 mg/L, with a maximum concentration of 5.0 mg/L permitted. Although the composition is well protected by the manufacturer, the safety data sheet reports toluene, solvent naphtha, dinonyl-naphthalenesulphonic acid, propan-2-ol, naphthalene, dec-1-ene,and quaternary ammonium compounds, dicoco alkyl-dimethyl nitrites, and chlorides [20]. Previous safety [21] and 2011 product data sheets also report a polymer containing nitrogen and sulphur (ie, a polysulphone and a polyamine [22]). Polysulphone is a co-polymer of an alkene with sulphur dioxide, with a repeating unit as depicted in Fig. 11.3. Computational modelling suggests that when R_1 and R_2 are alkyl, the polymer may adopt local helical arrangements stabilized via non-bonded short-range interactions between nearby SO_2 groups [23]. A longitudinal dipole moment accompanies these helical arrangements, thereby imparting the unusual dielectric properties observed for these molecules. Importantly, when dissolved in apolar solvents such as benzene, toluene, or cyclohexane, the dipole moment of the polymeric material increases, presumably because intramolecular electrostatic attractive interactions lead to the formation of relatively stable helical domains within the polymer.

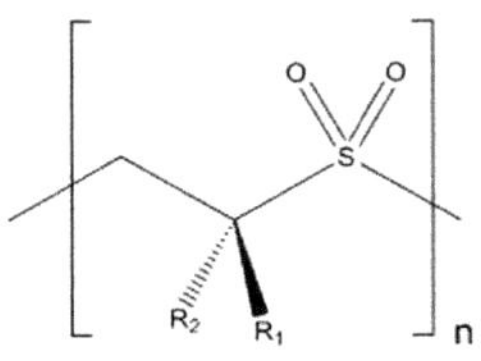

FIGURE 11.3 Structure of the polysulphone component.

The complex conductivity response of *Stadis 450* towards sodium salts and organic acids has received some attention; however, the precise dynamic molecular mechanisms responsible for these observations require further work [24].

11.2.5 Biocides

Biocides are added to prevent the growth of microorganisms that can produce bacterial films, clog fuel filters, and also generate corrosive by-products [25]. Fuel is sterilized by the refinery process because of the high temperatures, yet microorganisms such as bacteria and fungi can enter the fuel as soon as it is in contact with air and water.

Microbial contamination has been recognized as a problem in aviation fuel since the 1950s. A list of microbial contaminants found in jet fuel has been compiled by Denaro et al. [26]; bacteria are the predominantly recovered populations in JP-8 tank sumps [27]. All aviation fuels contain hydrocarbons, nitrogen, sulphur, phosphorus, and oxygenated organic compounds, organometallic species, and other metal salts. These molecules provide essential nutrients for bacterial and fungal growth [28]. Water is also an essential factor in microbial contamination, therefore, controlling the water content in fuel is the most recommended method for controlling microbial activities [29,30].

Oxygenated fuels are unlikely to be approved for use in aviation given the thermal stability of some, and the water solubility of others; they are also less energy dense. There is also the problem that oxygenated alternative fuels are more hydrophilic than conventional aviation fuels because of the presence of hydroxy and ester functional groups; they therefore dissolve more water and are at greater risk of microbial contamination. Furthermore, fatty acid esters, such as FAME, produce a favourable environment in which microbes can proliferate because the microbes can use them as an energy source [31]. Besides causing the degradation of fuel and the production of biosurfactants in the fuel system, the accumulation of biomass and biofilm can result in restriction of fuel flow, filter clogging, disruption of the operation of moving parts such as valves and pumps, and malfunction of automatic gauges [32]. Biofilms are also the root cause of microbiologically influenced corrosion. The microbes that comprise biofilms are generally embedded in a complex, heterogeneous, extracellular polymeric substance (EPS) matrix [33]. The current model for biofilm architecture suggests that biofilms contain channels and pores to increase the overall surface area and help with the transport of nutrients [28]. This causes biofilms to have greater population density and biochemical activity than the bulk fluid that they interface with [27]. Aerobic and facultative anaerobic microbes tend to grow at the interface of the EPS and the bulk fluid. These microbes scavenge oxygen, thereby creating an anoxic environment. This condition allows sulphate-reducing bacteria and other hydrogenase-positive anaerobes to thrive [28]. The metabolites produced by these microbes contain weak organic acids that can react with inorganic salts (eg, chlorides, nitrates, nitrites, and sulphates) in the fuel to form strong inorganic acids that can contribute to corrosion [34]. The inorganic acids produced include hydrochloric, sulphuric, nitrous, and nitric acids. The metabolites produced can negatively affect metals and metal alloys, consume corrosion inhibitors, degrade and dissolve protective coatings and films, and facilitate pitting corrosion [35]. Mitchell et al. [36] also observed that composite materials, regardless of polymer or fibre composition, and exposed to fungal growth, were subject to microbiologically influenced corrosion.

In the early 1960s, an unexpected solution to microbial contamination arose when ethylene glycol monomethyl ether (EGME) was introduced into JP-4 fuel because of its properties as a FSII; it was found to exhibit biostatic properties. However, the replacement of EGME with the less toxic diethylene glycol monomethyl ether (DiEGME) in the early 1980s resulted in a steady increase in microbial contamination of aviation fuel systems [26]. This led to the addition of biocides to jet fuel to help with the control of microbial contamination. Def Stan 91-91 makes no direct reference to biocides, however, there are currently two biocides approved for use in jet fuel [7]. A mixture of 5-chloro-2-methyl-4-isothiazolin-3-one and 2-methyl-4-isothiazolin-3-one is approved and commercialized as *Kathon FP1.5*. A mixture of 2,2′-oxybis(4,4,6-trimethyl-1,3,2-dioxaborinane) and 2,2′-(1-methyl-trimethylenedioxy)-bis-(4-methyl-1,3,2-dioxaborinane) is also approved outside the European Union, and commercialized as *Biobor JF*. The approved concentrations are 270 mg/kg for *Biobor JF*, and 100 ppm (v/v) for *Kathon FP1.5*. Biocides are permitted by engine and airframe manufacturers for intermittent use during maintenance turnaround. The aircraft are refilled and fully dosed, and as a general rule, will fly on the treated fuel until the additive has been depleted.

The biocidal mechanism of action for isothiazolinones – the active agents in *Kathon FP1.5* – is the oxidation of accessible cellular thiols and sulphydryl groups [37]. Because of this, the use of isothiazolinone-derived biocides has resulted in a large number of reported cases of human occupational exposure. Exposure of highly concentrated stock solutions (containing 1.5% of the active ingredients) during the dilution process can result in caustic burns, contact dermatitis, and allergic sensitization [38]. Although the neurotoxic consequences of both acute and chronic exposure to methylisothiazolone and related biocides in humans still requires investigation, studies demonstrate that they are potent neurotoxins [39]. Boron compounds are considered to have a nonspecific mode of action because they attack multiple targets in microbial cells. They can be linked to two possible major processes: inhibition of membrane proteins and inhibition of enzymes and co-enzymes within cells. Borate esters kill hydrocarbon-ingesting microorganisms with a lower concentration of boron than does boric acid, however, in aqueous solutions, organoborates are hydrolysed to boric acid and an organic alcohol. Boric acid can interact with or pass through the membrane of the microbial cell and interfere with many enzymatic processes in the cell by reacting with hydroxyl-rich compounds such as phospholipids, lipopolysaccharides, and glycoproteins [40]. Because of their toxicity, and preference for solubility in water phases, water collected from fuel tank drains containing biocides need to be disposed of appropriately [7].

Although biocides are highly effective in stopping microbial growth, they still have their limits. For example, if there is heavy accumulation of biofilm on the surface of the tank or other equipment, the biocide may not be able to reach the microbes living deep within the biofilm; in this case, the only remedy is to drain and manually clean the tank [7]. Furthermore, antimicrobial resistance continues to pose a serious problem not just to the aeronautical industry, but to society as a whole. The natural phenomenon of antimicrobial resistance is being accelerated by the misuse of agents [41]; therefore, the prophylactic, large-scale deployment of biocides for commercial purposes is difficult to justify.

11.2.6 Fuel System Icing Inhibitors

FSII are added to prevent ice formation. These additives dissolve in any free water that forms, lowering the freezing point of the mixture by preventing crystallization [7].

Dissolved water is ordinarily considered to be a constituent of jet fuel, and has no effect on the aircraft fuel system performance as it is vaporized during combustion [42]. Free water and water emulsions are, however, potentially hazardous and must be avoided [1]. Free water can enter aviation fuels from numerous sources, including during transportation or storage [1]. Even if jet fuel enters the aircraft tanks without free water, this does not prevent its formation. Indeed, dissolved water in fuel will precipitate out of solution in the form of micro droplets as the fuel temperature decreases during operation when the aircraft climbs to altitude [43]. Since the hygroscopic capacity of a fuel is determined by the water solubility of the individual chemical components, oxygenated alternative fuels such as alcohols, lactones, and furans will be able to dissolve more water compared to petroleum-based fuels.

On the contrary, the solubility of water in synthetic paraffinic kerosene blends is approximately 25% to 30% lower compared with Jet A-1 because they contain less heteroatoms, aromatics, and other polar compounds [44]. Jet A-1 containing a higher concentration of aromatics is also found to possess a higher affinity towards water, and tends to shed the greatest quantity of water as the temperature decreases [45]. At high altitudes, and therefore freezing temperatures, the separated water may form ice and the dispersed water may form a water-fuel slush or small particles of ice and snow suspended in the fuel [46]. This can create fuel flow problems. Microscopy experiments of water droplets in the micron range immersed in Jet A-1 demonstrate that fuel does not catalyse ice formation, and the droplets freeze homogeneously at temperatures between −36°C and −39°C [47]. However, ice nucleation can be encouraged by the presence of a foreign substrate, such as dust, rust, dirt, or other ice-forming nuclei [48]. This is called heterogeneous nucleation, and it is believed to be the most common process of ice formation in the fuel system. Factors such as turbulent flow, impurities, and cold metal surfaces in the aircraft feeding system can catalyse the nucleation process of supercooled water droplets at temperatures much higher than those required for homogenous nucleation [49].

FSII are hydrophilic substances that dissolve in free water. In doing so, the hydrogen-bonding networks responsible for molecular ordering are disrupted, and the freezing point of water is lowered by preventing crystallization [50]. Since EGME was banned because of its toxicity to humans and the environment, the only FSII currently approved for Jet A, Jet A-1, and military fuels is DiEGME (Fig. 11.4) [1]. The concentration of DiEGME must be in the range of 0.07–0.15% (v/v) [51].

Currently, the use of FSII is required in UK and US military jet kerosene, and although optional in many civilian specifications, it is very rarely used. This is because most commercial aircraft have heaters on their main fuel filters to melt any ice that is collected, whereas many military aircraft do not have these heaters and require FSII. As defined in Def Stan 91-91, concentrations of less than 0.02% by volume can be considered negligible and do not require agreement/notification; this facilitates the changeover from fuels containing FSII to those not containing FSII.

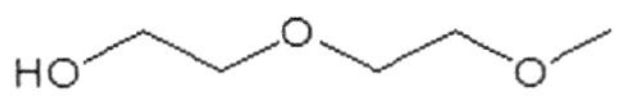

FIGURE 11.4 Structure of diethylene glycol monomethyl ether (DiEGME).

At the approved concentrations, Trohalaki et al. [52] found that water (0.007%, v/v) in jet fuel freezes below −33°C. DiEGME has also proved to be an effective deterrent to microbial growth [1]. However, the aviation industry has identified several problems related to the interaction of FSII and 'wet' fuel, which can actually undermine fuel protection:

- DiEGME is hygroscopic. This characteristic leads to an uptake of atmospheric water during blending operations, which affects its solubility in fuel.
- DiEGME has a very high water-fuel partitioning ratio, and it preferentially dissolves in water. This leads to denser water layers separating under gravity with up to 40–50% (m/m) of DiEGME. Unless replenished, the jet-fuel is essentially 'stripped' of its protection against icing.
- The presence of DiEGME in the water contained within the fuel alters interfacial properties, and can impair the performance of filters. Particularly, the separation efficiency of filter/coalescers can be compromised [53].
- A water phase that is sufficiently enriched with DiEGME will extract nitrogen and sulphur-containing compounds from the fuel and additives present, forming an 'Apple Jelly'. If the water/DiEGME-enriched phase is subsequently passed through water monitor filters, the mixture can extract the sodium polyacrylate water adsorption media, resulting in a higher viscosity non-Newtonian 'thick apple jelly' [54].

In addition, FSII are toxic at the concentrations required to be effective. Water drained from sumps of storage tanks and fuel systems inevitably contain higher concentrations of

DiEGME, creating concerns about the handling and disposal of these wastes [55]. FSII are only mildly irritating, but they are rapidly absorbed by the skin. Vapours can cause irritation to the eyes and the respiratory system. Long-term effects include damage to the central nervous system, blood, skin, eyes, and kidneys [56]. DiEGME is biodegradable in wastewater treatment systems; however, concentrations found in sump water could be high enough to disrupt microbiological decomposition. In the environment, the high oxygen demand required for decomposition results in there being less oxygen available for aquatic organisms [57]. FSII may also serve to inhibit fungal and bacterial growth in aircraft fuel systems, but are not approved as biocides because of their biostatic function. Over the years many other compounds have been studied as substitutes for DiEGME, however, none of them to date have gone through the lengthy approval process [52,55,58,59].

11.2.7 Thermal Stability Additives

Alternative fuels with high oxygen content will suffer from poor thermal stability. Currently, thermal stability additives are only permitted for use in certain military jet kerosene, and are not approved for use in Jet A or Jet A-1. Modern military jet engines require aviation fuel that has a higher thermal stability and heat sink capacity than is currently available. The US Military has, in conjunction with additive suppliers, developed an additive package that provides these benefits when added to jet fuel. The additive increases the thermal stability of JP8 jet fuel by 100°F, which is why it is referred to as '+100'. Unfortunately, it is incompatible with current commercially available filter/water separators, therefore it should be added after water coalescer systems, and as close as possible to the skin of the aircraft. The following additives are approved: *AeroShell Performance Additive 101, Spec Aid 8Q462, JFA-5, Turboline FS100 and FS100C,* and *BASF Kerojet 100*; the recommended dose is 256 mg/L [6]. Additives *BP/Lubrizol OS 169558F, Nalco VX-7603,* and *Infineum/ExxonMobil NB31011-33* are being evaluated as next-generation thermal stability additives [60]. Experience has demonstrated that JP-8 + 100 significantly reduces unscheduled maintenance, maintains cleaner engine components, and drastically reduces soot build-up on hot engine components.

There are various mechanisms to improve fuel thermal stability, including metal deactivation, suppression of oxidation, dispersancy, detergency, and precursor stabilization. The additives concentrated on the operation of the first four mechanisms listed above, through the use of a proprietary detergent/dispersant species, are formulated together with a commercial antioxidant (butylated hydroxytoluene) and MDA [61]. It is also believed that detergent-type additives may have long-term beneficial effects on emissions as fuel nozzles become cleaner and fuel atomization is improved [62].

11.3 LOOKING TO THE FUTURE: POTENTIAL ADDITIVES FOR ALTERNATIVE FUELS

11.3.1 Seal Swelling

The Def Stan 91-91 specification for Jet A-1 states a required minimum aromatic component of 8% (v/v) for fuels containing synthetic hydrocarbons [63]. One of the important properties of aromatic compounds in fuel is to permeate and swell the nitrile O-rings in the fuel delivery system, thus providing an airtight seal. As many potential drop-in alternative jet fuels are entirely aliphatic in nature, the lack of aromatics may cause some of these elastomers to shrink, leading to fuel leaks [64].

The current solution is to mandate the blending synthetic kerosenes with petroleum-derived jet fuel in order to ensure that a minimum aromatic content is maintained. However, the blending decreases many of the potential benefits of using a fully synthetic fuel, as adding aromatic species will also increase the amount of soot formed on combustion, water solubility, and the cost of blending.

Seal swelling additives could be an alternative option in order to fulfil this vital function, and at concentration levels much lower than those considered for blending. Since nitrile rubber O-rings used with fuel are typically made of polyacrylonitrile and polybutadiene, computational modelling has determined that species having the potential for dual interactions, one with the polar nitrile group and the other with the electron density in the butadiene system, will provide the highest degree of swelling [65]. Graham et al. conducted a study of the elastomer-swelling propensity of a range of aromatic species and found that it increased with increasing polarity and H-bond character, and decreased with increasing molecular weight [66]. Compounds possessing an aromatic ring with a polar substituent, such as phenols, naphthols, and benzyl alcohol, are all promising candidates, however, environmental and fuel performance concerns have determined benzyl alcohol to be the seal swell additive of choice. Synthetic fuels blended with 1.0, 0.75, and 0.5% (v/v) of benzyl alcohol have demonstrated a comparable capacity for seal swelling as petroleum jet fuel [65], however, aircraft fuel system and engine effects will need to be studied and quantified before such additives can be considered for use.

11.3.2 Soot Removers

Particulate matter can either be emitted directly or formed in the atmosphere by the transformation of gaseous emissions of compounds, including nitrogen oxides, volatile organic compounds, and sulphur oxides. Since these fine particles contain high amounts of carbon, they are usually referred to as soot. Studies have shown that these airborne particles pose both health and environmental risks. Most particulate matter from aircraft engines possess diameters less than 2.5 μm [67]. Diesel cetane improvers and commercial smoke abatement additives were tested in a helicopter engine and found to have minimal impact on particulate emissions, however, they are currently prohibited for aviation use. Positive reductions of particle number density were shown for some additives, but at relatively high concentrations (>1%, v/v) [62].

Commercial additives *Fuel Fix* and *Viscon* were also studied. They use high-molecular-weight polymers to prevent the rapid vaporization of the fuel's higher volatility components in the combustion chamber, and promote the vaporization and combustion of all fuel components uniformly. The manufacturers of *Fuel Fix* and *Viscon* have observed significant reductions in gaseous emissions (CO, HC, and NO_x) and fuel consumption with increased engine power with their additives in internal combustion engines.

Several commercial additives designed to reduce emissions by altering the combustion process were also studied. *Kleen Fuel* is a blend of aliphatic nitro compounds, which reportedly has active components that break down to release oxygen. The oxygen released from these nitroparaffin groups is used to burn the remaining hydrocarbon components in the combustion chamber to improve engine performance and reduce emissions. *RxP* is a hydrocarbon blend that reportedly contains a molecule that absorbs a large portion of the infrared radiation emitted during the combustion process. The unburned fuel is preheated (producing higher combustion temperatures), and the heat loss to the combustor chamber wall is reduced to produce more efficient engine operation and lower emissions. *Enviro Max* is a combination of a heterogeneous catalyst and a combustion promoter. The homogenous catalyst consists of micron-sized particles of solid zinc oxide/peroxide suspended in selected solvents, which are blended with the proprietary combustion promoter and *tert*-butyl hydroperoxide. It should be noted that in the pre-mixed EMP/JP-8 blend, a significant amount of the additive separated from the fuel and settled in the bottom of the fuel reservoir. This characteristic most likely prevented a homogenous fuel/additive mixture, and obviously makes the present formulation of EMP an undesirable additive for jet fuel. *Wynns Emissions Control* + Plus + additive is also reported to be an emissions reduction additive; however, specific details about its functionality are not defined [62].

There are many reasons why an additive that is effective in reducing emissions on a diesel engine may be ineffective in a jet-fuelled turbine engine. In general, turbine engines operate at higher combustion efficiencies than their internal combustion counterparts, and jet fuels have lower aromatics (proposed soot precursors) than both diesel and gasoline fuels. Furthermore, diesel engines operate at higher pressures and temperatures than jet engines, and the fuel distribution within their cylinders are non-uniform, which increases the fuel-rich zones and promotes the formation of soot particulate. Therefore, the challenge of reducing harmful PM2.5 emissions from aircraft turbine engines is significantly greater [62]. According to John Ambrust, publisher of *World Jet Fuel Report*, additives to cut down on carbon and sulphur emissions will soon become even more important in the effort to reduce contrails.

11.3.3 Water Scavenger Additives

Oxygenated alternative fuels have a higher water solubility compared to petroleum-based fuels, therefore water management can be a serious drawback. There are two novel approaches currently being examined to remove the potential free-water phase that can accumulate in the fuel system. *BASF Kerojet Aquarius* is a non-hygroscopic additive that encapsulates the water naturally present in the fuel by forming a molecular 'cage' or micelle around the water molecules. *Aquarius* does not chemically react with the fuel, nor with water, and the *Aquarius* micelle will only encapsulate the water molecules. Fuel treated with *Aquarius* remains as a single phase throughout the flight, and the water contained by *Aquarius* is removed and burnt with the fuel during flight. *Aquarius* is added to the fuel after the final filtration stage of the aircraft refuelling process [68].

Another approach is predicated upon the addition of lipophilic cyclic geminal ethers, which are readily hydrolysed by mild acidic conditions to produce hydrophilic ice inhibitors (Fuel Dehydrating Icing Inhibitors) [4]. Because oxygenated alternative fuel undergoes oxidation to produce weak carboxylic acids, the inherent acidity proves sufficient to catalyse the hydrolysis of cyclic geminal ethers. Although the development of this new class of additive is still at an early stage, it has created a certain amount of interest in the industry.

References

[1] CRC. Handbook of Aviation Fuel Properties (CRC Report No. 663, Amendment 1), 2014.

[2] Ministry of Defence, Defence Standard 91-91, Issue 7. Turbine Fuel, Kerosine Type, Jet A-1 NATO Code: F-35 Joint Service Designation: AVTUR. Technical Report. 2011.

[3] J.C. Toler, The use of biocides for improving storage conditions of fuels. Long term storage stabilities Liq. fuels, Tel Aviv, Israel, 1983.

[4] S.L. Repetto, J.F. Costello, B. De Lacy Costello, N.M. Ratcliffe, J.K.-W. Lam, The development of novel fuel dehydrating icing inhibitors, SAE Int. J. Fuels Lubr. 6 (2013) 553–563. Available from: http://dx.doi.org/10.4271/2013-01-2169.

[5] ASTM. ASTM D4054 – 14 Standard Practice for Qualification and Approval of New Aviation Turbine Fuels and Fuel Additives. 2014.
[6] ExxonMobil Aviation. World Jet Fuel Specifications with Avgas Supplement. 2008.
[7] G. Hemighaus, T. Boval, J. Bacha, F. Barnes, M. Franklin, L. Gibbs, et al., Aviation fuels technical review. Chevron Corporation. 2006.
[8] C.E. Frank, Hydrocarbon autoxidation, Chem. Rev. 46 (1950) 155–169. Available from: http://dx.doi.org/10.1021/cr60143a003.
[9] A. De Klerk, Fischer–Tropsch refining. Wiley Online Library, 2011.
[10] B.D. Batts, A.Z. Fathoni, A literature review on fuel stability studies with particular emphasis on diesel oil, Energy Fuels 5 (1991) 2–21. Available from: http://dx.doi.org/10.1021/ef00025a001.
[11] C.D. Bannister, C.J. Chuck, M. Bounds, J.G. Hawley, Oxidative stability of biodiesel fuel, Proc. Inst. Mech. Eng. Part D 225 (2011) 99–114. Available from: http://dx.doi.org/10.1243/09544070JAUTO1549.
[12] C.J. Chuck, H.J. Parker, R.W. Jenkins, J. Donnelly, Renewable biofuel additives from the ozonolysis of lignin, Bioresour. Technol. 143 (2013) 549–554. Available from: http://dx.doi.org/10.1016/j.biortech.2013.06.048.
[13] C. Wu, X. Dong, M. Dong, Y. Lu, A. Turley, T. Jin, Antimicrobial and antioxidant activities of lignin from residue of corn stover to ethanol production, Ind. Crops Prod. 34 (2011) 1629–1634. Available from: http://dx.doi.org/10.1016/j.indcrop.2011.06.002.
[14] H.H.A. El Naga, A.E.M. Salem, Effect of worn metals on the oxidation of lubricating oils, Wear 96 (1984) 267–283. Available from: http://dx.doi.org/10.1016/0043-1648(84)90041-3.
[15] B.H. Black, D.R. Hardy, The lubricity properties of jet fuel as a function of composition. Part 2. Application of analysis method. 1989.
[16] A.S. Kislenko, I.F. Krylov, G.I. Sokolova, V.P. Seregin, G.B. Skovorodin, Effects of monoesters of dicarboxylic acids on jet fuel lubricity, Chem. Technol. Fuels Oils 22 (1986) 286–290. Available from: http://dx.doi.org/10.1007/BF00719555.
[17] J.A. Dumesic, D.M. Alonso, J.Q. Bond, Catalytic conversion of biomass to biofuels, Green. Chem. 12 (2010) 1493. Available from: http://dx.doi.org/10.1039/c004654j.
[18] W.M. Bustin, W.G. Dukek, Electrostatic Hazards in the Petroleum Industry, Research Studies Press Ltd, England, 1983.
[19] Joint Inspection Group. Jet Fuel Conductivity. 2009.
[20] Innospec. Material Safety Data Sheet Stadis (R) 450. 2015.
[21] Octel. Material Safety Data Sheet Stadis (R) 450. 2004.
[22] J.C.P. Henry, Compositions of olefin-sulfur dioxide copolymers and polyamines as antistatic additives for hydrocarbon fuels, 1975.
[23] M.L. Mansfield, Conformational properties of poly (alkene sulfones), Macromolecules 15 (1982) 1587–1593. Available from: http://dx.doi.org/10.1021/ma00234a025.
[24] B. Dacre, J.I. Hetherington, The effects of contaminants on the behaviour of conductivity improvers in hydrocarbons, J. Electrost. 45 (1998) 53–68. Available from: http://dx.doi.org/10.1016/S0304-3886(98)00036-9.
[25] T. Hill, Microbial growth in aviation fuel, Aircr. Eng. Aerosp. Technol. 75 (2003) 497–502. Available from: http://dx.doi.org/10.1108/00022660310492582.
[26] T.R. Denaro, S.K. Chelgren, J.N. Lang, E.M. Strobel, L.M.T. Balster, M.D.Vangsness DNA Isolation of Microbial Contaminants in Aviation Turbine Fuel via Traditional Polymerase Chain Reaction (PCR) and Direct PCR. Preliminary Results. 2005.
[27] C.J. McNamara, T.D. Perry, R. Leard, K. Bearce, J. Dante, R. Mitchell, Corrosion of aluminum alloy 2024 by microorganisms isolated from aircraft fuel tanks, Biofouling 21 (2005) 257–265. Available from: http://dx.doi.org/10.1080/08927010500389921.
[28] F.J. Passman, Microbial contamination and its control in fuels and fuel systems since 1980 – a review, Int. Biodeterior. Biodegradation 81 (2013) 88–104. Available from: http://dx.doi.org/10.1016/j.ibiod.2012.08.002.
[29] S. Swift, Identification and control of microbial growth in fuel handling systems. in: H. Chesneau, M. Dorris, (Eds.), Distillate Fuel Contamination Storage and Handling, 100 Barr Harbor Drive, PO Box C700, West Conshohocken, PA 19428-2959: ASTM International, 1988. Available from: http://dx.doi.org/10.1520/STP1005-EB.
[30] D. Arnold, Water key to microbe problem. Mar Propuls (1991) 28–29.
[31] W.Siegert, Microbial contamination in diesel fuel – are new problems arising from biodiesel blends? IASH 2009, 11th International Conference Stability, Handling Use Liquid Fuels, Prague, Czech Republic, 2009, p. 20.
[32] IATA. Guidance Material on Microbiological Contamination in Aircraft Fuel Tanks. 2009.
[33] B. Lee, J.A.J. Haagensen, O. Ciofu, J.B. Andersen, N. Høiby, S. Molin, Heterogeneity of biofilms formed by nonmucoid *Pseudomonas aeruginosa* isolates from patients with cystic fibrosis, J. Clin. Microbiol. 43 (2005) 5247–5255. Available from: http://dx.doi.org/10.1128/JCM.43.10.5247-5255.2005.
[34] F.J. Passman, Manual 47 – Fuel and Fuel System Microbiology: Fundamentals, Diagnosis, and Contamination Control. ASTM International West Conshohocken, PA, 2003.

[35] H.A. Videla, Manual of Biocorrosion. 1996.

[36] R. Mitchell, J.-D. Gu, M. Roman, T. Esselman, The role of microbial biofilms in deterioration of space station candidate materials, Int. Biodeterior. Biodegradation 41 (1998) 25–33. Available from: http://dx.doi.org/10.1016/S0964-8305(98)80005-X.

[37] J.-Y. Maillard, Bacterial target sites for biocide action, J. Appl. Microbiol. 92 (2002) 16S–27S. Available from: http://dx.doi.org/10.1046/j.1365-2672.92.5s1.3.x.

[38] E. Primka, Three cases of contact allergy after chemical burns from methylchloroisothiazolinone/methylisothiazolinone: one with concomitant allergy to methyldibromoglutaronitrile/phenoxyethanol, Am. J. Contact Dermat. 8 (1997) 43–46. Available from: http://dx.doi.org/10.1016/S1046-199X(97)90036-1.

[39] S. Du, B. McLaughlin, S. Pal, E. Aizenman, In vitro neurotoxicity of methylisothiazolinone, a commonly used industrial and household biocide, proceeds via a zinc and extracellular signal-regulated kinase mitogen-activated protein kinase-dependent pathway, J. Neurosci. 22 (2002) 7408–7416.

[40] O. Borokhov, D. Schubert, New Biocides Development. vol. 967, American Chemical Society, Washington, DC, 2007. Available from: http://dx.doi.org/10.1021/bk-2007-0967.

[41] World Health Organization, Antimicrobial Resistance: Global Report on Surveillance. Geneva, Switzerland, 2014.

[42] A.V. Oreshenkov, Accumulation of water in jet fuels. Mathematical modeling of the process, Chem. Technol. Fuels Oils 40 (2004) 320–325. Available from: http://dx.doi.org/10.1023/B:CAFO.0000046266.83408.d7.

[43] L. Lao, C. Ramshaw, H. Yeung, M. Carpenter, J. Hetherington, J.K.-W.Lam, et al. Behaviour of water in jet fuel in a simulated fuel tank. SAE Tech Pap 2011-01-2794 2011. Available from: http://dx.doi.org/10.4271/2011-01-2794.

[44] C.A. Moses, Comparative evaluation of semi-synthetic jet fuels. Technical Report. 2008.

[45] J.K.-W. Lam, M.D. Carpenter, C.A. Williams, J.I. Hetherington, Water solubility characteristics of current aviation jet fuels, Fuel 133 (2014) 26–33. Available from: http://dx.doi.org/10.1016/j.fuel.2014.04.091.

[46] H.W. Schab, Problems associated with water contaminated jet fuels, J. Am. Soc. Nav. Eng. 72 (1960) 41–60. Available from: http://dx.doi.org/10.1111/j.1559-3584.1960.tb02358.x.

[47] B.J. Murray, S.L. Broadley, G.J. Morris, Supercooling of water droplets in jet aviation fuel, Fuel 90 (2011) 433–435. Available from: http://dx.doi.org/10.1016/j.fuel.2010.08.018.

[48] R.A. Shaw, A.J. Durant, Y. Mi, Heterogeneous surface crystallization observed in undercooled water, J. Phys. Chem. B 109 (2005) 9865–9868. Available from: http://dx.doi.org/10.1021/jp0506336.

[49] N.A. Ragozin, Jet Propulsion Fuels: Translated From the Russian, Pergamon Press, New York, 1961.

[50] O.P. Lykov, V.I. Slovetskii, N.V. Frolova, G.V. Kitashova, V.E. Emel'yanov, Mechanism of action of alcohol-ester anti-icing additives in automotive gasolines, Chem. Technol. Fuels Oils 25 (1989) 440–444. Available from: http://dx.doi.org/10.1007/BF00726105.

[51] USA Department of Defence. Detail specification: turbine fuel, aviation, kerosene type, JP-8 (NATO F-34), NATO F-35, and JP-8 + 100 (NATO F-37). MIL-DTL-83133H, Amendment 1, 2012.

[52] S. Trohalaki, R. Pachter, J.R. Cummings, Modeling of fuel-system icing inhibitors, Energy Fuels 13 (1999) 992–998. Available from: http://dx.doi.org/10.1021/ef9802577.

[53] S.E. Taylor, Component interactions in jet fuels: fuel system icing inhibitor additive †, Energy Fuels 22 (2008) 2396–2404. Available from: http://dx.doi.org/10.1021/ef800090p.

[54] J.A. Waynick, S.R. Westbrook, L. Hicks, P. Serino, Apple jelly contamination in aviation jet fuel storage and distribution systems. IASH 2003, 8th International Conference Stability Handling Liquid Fuels, 2003, p. 29.

[55] G.W. Mushrush, E.J. Beal, D.R. Hardy, J.M. Hughes, J. C. Cummings, Jet fuel system icing inhibitors: synthesis and characterization, Ind. Eng. Chem. Res. 38 (1999) 2497–2502. Available from: http://dx.doi.org/10.1021/ie9807755.

[56] R.P. Pohanish, Sittig's Handbook of Toxic and Hazardous Chemicals and Carcinogens (Google eBook), William Andrew, 2011.

[57] C.E. Meshako, C.A. Bleckmann, M.N. Goltz, Biodegradability and microbial toxicity of aircraft fuel system icing inhibitors, Environ. Toxicol. 14 (1999) 383–390. Available from: http://dx.doi.org/10.1002/(SICI)1522-7278(1999)14:4 < 383::AID-TOX2 > 3.0.CO;2-M.

[58] J.H. Chang, E.J. Beal, W.M. Stalick, G. Mushrush, Fuel system icing inhibitors: the synthesis of esters of oxoacids, Pet. Sci. Technol. 16 (1998) 979–1000. Available from: http://dx.doi.org/10.1080/10916469808949821.

[59] S. Zabarnick, M.J. DeWitt, R. Adams, Z.J. West, L.M. Shafer, T.F.Williams, et al., Evaluation of Triethylene Glycol Monomethyl Ether (TRIEGME) as an alternative fuel system icing inhibitor for JP-8 Fuel. 2010.

[60] J.Morris, W.Robert, Evaluation of next generation thermal stability-improving additives for JP-8, Phase 1, Thermal Stability Impact Characterization, 2012.

[61] P. Heneghan, S. Zabarnick, D.R. Ballal, W.E. Harrison, JP-8+100 – The development of high thermal stability jet fuel (AIAA). AIAA 34th Aerospace Science Meeting Exhibit, Reno, NV, 1996. Available from: http://dx.doi.org/10.2514/6.1996-403.

[62] E. Corporan, M. DeWitt, M. Wagner, Evaluation of soot particulate mitigation additives in a T63 engine, Fuel Process. Technol. 85 (2004) 727–742. Available from: http://dx.doi.org/10.1016/j.fuproc.2003.11.016.

[63] ASTM, ASTM D7566 – 13 Standard Specification for Aviation Turbine Fuel Containing Synthesized Hydrocarbons, 2013.

[64] G. Hemighaus, T. Boval, C. Bosley, R. Organ, J. Lind, R.Brouette, et al., Alternative Jet Fuels. Technical Report. Chevron Corporation, 2006.

[65] D.D. Link, R.J. Gormley, J.P. Baltrus, R.R. Anderson, P.H. Zandhuis, Potential additives to promote seal swell in synthetic fuels and their effect on thermal stability, Energy Fuels 22 (2008) 1115–1120. Available from: http://dx.doi.org/10.1021/ef700569k.

[66] J.L. Graham, R.C. Striebich, K.J. Myers, D.K. Minus, W.E. Harrison, Swelling of nitrile rubber by selected aromatics blended in a synthetic jet fuel, Energy Fuels 20 (2006) 759–765. Available from: http://dx.doi.org/10.1021/ef050191x.

[67] Agency UEP, Evaluation of air pollutant emissions from subsonic commercial jet aircraft. EPA420-R-99-013, 1999.

[68] BASF Corporation. Kerojet® Aquarius Water Scavenger Additive Product Data Sheet, 2013.

SECTION III

TESTING, ASSESSMENT AND THE FUTURE

CHAPTER

12

Life Cycle Assessment of Biojet Fuels

R.S. Capaz and J.E.A. Seabra

Universidade Estadual de Campinas (UNICAMP), Campinas, SP, Brazil

12.1 INTRODUCTION

In a reactive perspective, the consumer starts to worry about pollution only after using the product, without considering the embodied impacts in the product – that is, the amount of resources and the pollution generated along its production process and distribution. However, the search for an efficient consumption, under increasing environmental awareness, motivates the analysis throughout the life cycle of the products, known as Life Cycle Assessment (LCA). In recent years, LCA has been frequently used to provide an environmental basis for the decision making, while offering consumers a wider view of the product.

LCA is a structured, comprehensive, and internationally standardized method, yet there is no single method for conducting it. It quantifies all relevant emissions and resources consumed, and the related environmental and health impacts and resource depletion issues that are associated with any goods or services [1,2]. The comprehensive scope of LCA is useful in order to avoid problem-shifting, for example, from one phase of the life cycle to another, from one region to another, or from one environmental problem to another [3].

Several studies have presented this kind of analysis for the production and use of fuels, including biojet fuels, although only a few. But LCA will become increasingly important for international aviation as it will subsidize the assessment of the progress towards achieving the sector's aspirational goal on greenhouse gas (GHG) emissions [4]. This chapter presents the main methodological aspects of LCA and discusses the main published studies regarding biojet fuels.

12.2 HISTORICAL ASPECTS AND DEFINITIONS

According to ISO 14040:2009 [2], which provides the basic guidelines for conducting this analysis, LCA can be defined as *a compilation and evaluation of inputs, outputs, and the potential environmental impact of a product throughout its life cycle*. In turn, EPA [5] defines LCA as a 'cradle-to-grave' approach to industrial systems – that is, starting with the natural resources used as

Biofuels for Aviation.
DOI: http://dx.doi.org/10.1016/B978-0-12-804568-8.00012-3

raw materials in the processes, and ending with the residues generated by the consumption of the products that return to nature.

By this definition, it is noted that the focus of the LCA is directly on the product/service, and it results from two groups of parameters:

1. Inputs and outputs on or at the boundaries of the analysed system from which the respective product is obtained.
2. Environmental impacts associated with these inputs and outputs.

Among the advantages commonly observed with the use of LCA, we can cite the following:

- Allows a quantitative assessment of the environmental performance of the entire production process indicating critical points.
- Supports decision makers in the productive sector to select the best alternatives for the process considering the environmental aspects.
- Supports product labelling programs.
- Supports consumers in selecting the best product for the environment.

According to the work in Refs [5,6], the first LCA was conducted in 1969 by the Midwest Research Institute, when different drink containers for Coca Cola Company were analysed with the aim of selecting the option with the lowest effect on natural resources. This methodology was named *Resource and Environmental Profile Analysis* (REPA) in the United States and *Ecobalance* in Europe, consisting primarily of the inventory of inputs and outputs of the production system. Between 1970 and 1975, approximately 15 REPAS were developed, together with protocols for standardization of methodologies. With the oil crisis (1973 and 1975), the interest in the accounting of energy flows in production processes has intensified, and many *Energy Analysis* and *Energy Balances*, with the guidelines of REPA, were conducted. After all, for the industrial sector, it became strategic to quantify the characteristic energy flows of production processes, checking the feasibility of alternatives to reduce losses or mitigate the dependence on fossil fuels.

The *Energy Analysis* can be defined as the systematic study of energy flows through a production system in order to estimate the amount of energy inputs required to produce some goods, and to identify and locate the associated losses. With this analysis, the embodied energy in a product can be determined – that is, the energy required to produce it, and to compare it, in the case of fuels, with the amount of net energy delivered by the product (*net energy analysis*). In addition to the criterion for selecting processes with the lowest energy cost, this analysis provided consistency to programs aimed at the reduction of energy consumption or use of renewable sources.

In this context, as mentioned by Wilting [7], several researchers in the 1970s and 1980s began to account for and analyse the energy requirements in various industries. For instance, Berry and Fels [8] estimated the energy cost of cars, while Chapman [9] evaluated copper and aluminium, as well as some fuels. Employing economic input-output analysis, Wright [10] and Bullard and Herendeen [11] assessed the energy cost of goods and services. Hannon [12] compared the energy intensities of economic sectors as to the function of the intensity of work, and Leach [13] accounted for the energy flows in food production.

In the mid-1980s, the high production of domestic waste, due to the increasingly heavy use of packages and food waste, encouraged the resurgence of LCA in the United States with interest in analysing packages. From the 1990s, according to EPA [5], some American attorneys highlighted the concerns about the use of LCA methodologies for classifying products without standardization. After all, discordant results due to different databases and lack of a general methodology justified the need for standardization of the technique. At the same time, the increased interest in

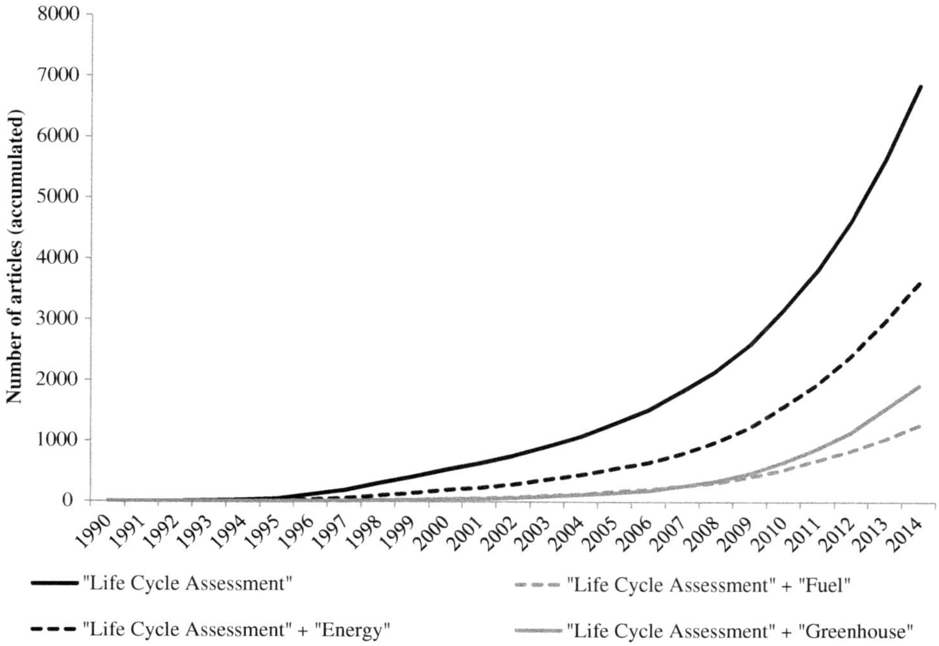

FIGURE 12.1 Accumulated publication of scientific articles cited in *Scopus Database* using specific terms (in between quotes, ' ') expressed in the title, abstract, or keywords.

environmental issues has stimulated researchers and experts to refine and expand the methodology beyond a simple inventory, seeking to define standardized guidelines.

During the 1990s, the Society of Environmental Toxicology and Chemistry played a leading and coordinating role in bringing LCA practitioners, users, and scientists together to collaborate on the continuous improvement and harmonization of the LCA framework, terminology, and methodology [14]. In 1994, ISO engaged in LCA focused on the formal task of standardization of methods and procedures. As a result, ISO produced the ISO 14040:1997 standard (which provided the principles and framework for LCA studies), among other related standards. These standards were later replaced by the two current standards: ISO 14040:2006 and ISO 14044:2006 [2,15]. It must be noted, however, that ISO never aimed to standardize LCA methods in detail [14], and points out that there is no single method for conducting LCA.

Today LCA is a much disseminated tool, being largely used in scientific publications, with a remarkable increase in usage since the 2000s (Fig. 12.1). From the *Scopus Database*,[1] up until 2014, around 7000 articles were tracked using 'Life Cycle Assessment' in the title, abstract, or keywords. Approximately 60% of these articles covered 'energy', while more than 40% addressed GHG emissions. Articles about jet fuel LCA, however, are more recent and are still limited in number.

LCA also became part of policy documents and legislation. For instance, for a biofuel to qualify as a renewable fuel under the US

[1]Available in http://www.scopus.com/.

Renewable Fuel Standard Program (RFS2),[2] among other requirements, it must achieve an established life cycle GHG threshold compared to a 2005 petroleum baseline. In California, the primary goal of the Low Carbon Fuel Standard regulation is to reduce the life cycle carbon intensity of transportation fuels used in the state by at least ten percent by 2020 from a 2010 baseline [16]. In Europe, the Renewable Energy Directive also established required life cycle GHG emissions savings as a sustainability criterion for biofuels and bioliquids [17].

Regarding the aviation sector, the 37th Assembly of the International Civil Aviation Organization (ICAO) [18] defined a global aspirational goal of keeping the global net carbon emissions associated to international aviation operations from 2020 at the same level. The use of biofuels is accounted among the alternatives. Assessing fuel life cycle emissions is a particular topic for which increased harmonization amongst aviation stakeholders is important to acquire a shared understanding of the potential benefits of alternative fuels. Therefore, the Alternative Fuels Task Force was created within the ICAO technical body on environment, and was tasked to develop a methodology to assess fuel life cycle emissions, which should be applied for the quantification of the emissions associated to a projected production of alternative jet fuels by 2050 [19].

Despite the traditional focus on environmental issues, other perspectives have emerged in LCA applications. The work in UNEP/SETAC [20] suggested a wider LCA, covering all pillars of sustainability, called Life Cycle Sustainability Assessment (LCSA). In this case, similar approaches to existing methodologies for assessing the environmental dimension would cover the economic (Life Cycle Costing, LCC) and social (Social Life Cycle Assessment, S-LCA) dimensions. But LCSA also broadens the scope from predominantly product-related questions (product level) to questions related to sector (sector level) or even economy-wide levels (economy level). As mentioned by Guinée et al. [14], unlike LCA, LCSA is a framework of models rather than a model in itself. The need to refine and standardize methodologies, as seen in the early stages of LCA, would also apply to these newer approaches. Thus, structuring, selecting, and making the plethora of models practically available in relation to different types of life cycle sustainability questions is then the main challenge [14].

12.3 LCA GUIDELINES

According to ISO [2], LCA consists of four phases, as shown in Fig. 12.2, which are detailed below.

12.3.1 Scope – Goal and Scope Definition

In this phase, the goals of the intended study are defined, together with some items that will feature the analysis, such as the product system, the functional unit, system boundaries, allocation procedures, and others.

The product system is composed by the unit processes that are within the system boundary. Fig. 12.3 presents a generic product system composed by two production stages (in addition to the end use), which are composed by four processes: A, B, C, and D. When establishing the system boundary, the unit processes within it are defined, as well as the input and output flows to be assessed. Note that in Fig. 12.3 general flows are listed; in a real analysis such flows can be disaggregated. Ideally, the product system should be modelled in such a manner that inputs and outputs at its boundaries are elementary flows. But some care must be taken to avoid expending

[2] Available in http://www2.epa.gov/renewable-fuel-standard-program.

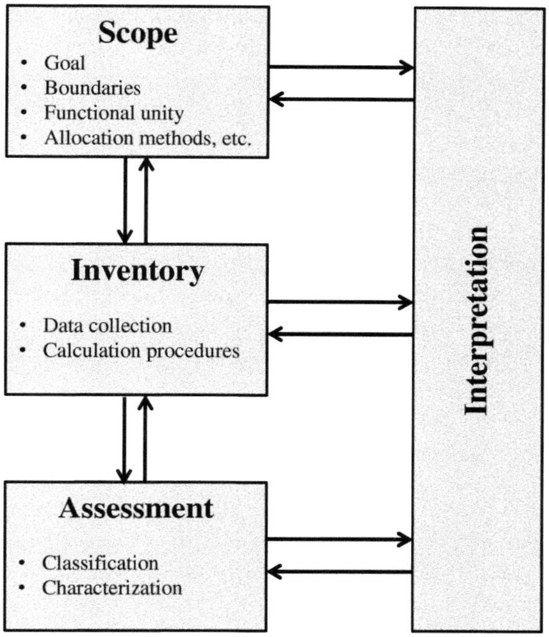

FIGURE 12.2 LCA phases according to ISO 14040:2006.

resources on the quantification of inputs and outputs that will not significantly change the overall conclusions of the study.

The use of different boundaries is one of the limitations when comparing different LCA studies. The assessment can be conducted from the 'cradle-to-grave' – or well-to-wake (WTWa) for aviation fuels, or well-to-well (WTW) for road transportation fuels – when the boundaries go from acquisition of raw materials up to end-use and disposal, if applicable. 'Cradle-to-gate' studies do not include the use of the product, and the 'gate-to-gate' approach considers only one step of the production process. In well-to-tank studies, the boundary goes up to the delivery of the fuel to market.

Another important point is: LCA is a relative approach that is structured around a functional unit. This functional unit defines the quantification of the identified functions (performance characteristics) of the product and provides a reference to which the inputs and outputs are

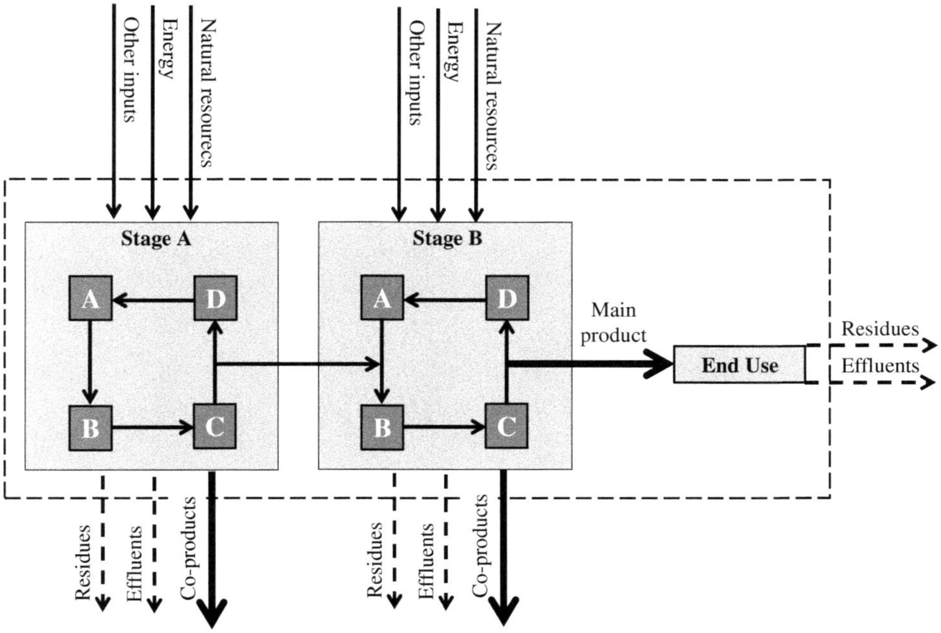

FIGURE 12.3 Example of system boundary.

related. All subsequent analyses are then relative to that functional unit. In the case of biofuels, reference flows commonly used refer to their mass, volume, heating value, or the service/function they provide (eg, kilometres travelled).

An emerging issue regarding the goal and scope definition is the differentiation between attributional LCA (ALCA) and consequential LCA (CLCA), with the former being more widely used for historical and practical reasons. They represent from their logic the two fundamentally different situations of modelling the analysed system [1]. ALCA is defined by its focus on describing the environmentally relevant physical flows to and from a life cycle and its subsystems, while CLCA is defined by its aim to describe how environmentally relevant flows will change in response to possible decisions [3]. In attributional modelling, the system is hence modelled as it is or was (or is forecasted to be), whereas CLCA models a hypothetical, generic supply-chain prognosticated along market-mechanisms, and potentially includes political interactions and consumer behaviour changes [1].

The different focuses of attributional and consequential LCA are reflected in several methodological choices in LCA, so the distinction is relevant when discussing system boundaries, data collection, and allocation [3]. In relation to this last aspect, the allocation procedures are common when the product system produces more than one product (multifunctional process): the main product or the interested product, and other co-products. In these cases, the question to be solved is: what is the effective environmental impact associated to the main product when other products are obtained together with it? Or how to allocate all inputs and outputs to the main product? In fact, the approach for solving multifunctional processes is one of the most critical issues in LCA. According to the ISO hierarchy [2], wherever possible, allocation should be avoided by process subdivision, or by expanding the product system to include the additional functions related to the co-products. This can mean to add another, not-provided function to make systems comparable (ie, system expansion in the stricter sense), or to subtract the not-required function(s) substituting them by the ones that are superseded/replaced (ie, substitution by system expansion). Strictly speaking, system expansion and substitution are the corresponding approaches under consequential modelling for solving multifunctionality [1].

Where allocation cannot be avoided, the ISO standards advise that the inputs and outputs of the system should be partitioned between its different products or functions in a way that reflects the underlying physical relationships between them (ie, they should reflect the way the inputs and outputs are changed by quantitative changes in the products or functions delivered by the system). Lastly, where physical relationship alone cannot be established or used as the basis for allocation, the inputs and outputs should be allocated between the products and functions in a way that reflects other relationships between them (eg, economic value of the products).

It is clear that the use of different allocation methodologies can result in different environmental performances, especially where significant quantities of co-products are obtained, like biofuel processes. Regardless of which method is adopted, it is important to make clear the considerations for future comparisons with other analyses [21].

12.3.2 Inventory – Life Cycle Inventory Analysis

In this phase, the flows, considered in the analysis according to the previously defined boundaries, are evaluated using qualitative and quantitative data. These flows must always refer to the same functional unit, paying attention to the data collection method, allocation procedures, reliability, and representativeness requirements. The result from the Life Cycle Inventory Analysis is therefore a compilation of the inputs (resources) and the outputs (eg, pollutants) through the boundaries of the product system in relation to the functional unit. If necessary, changes in the system boundaries must be made to ensure a higher quality of the study.

In the case of biofuels and bioproducts, an important element of the assessment concerns the impacts associated with Land Use Change (LUC), so distinguishing direct and indirect LUC is relevant. According to ISO/TS 14067 [22], the direct LUC (DLUC) is a change in the use or management of land within the product system being assessed, while the indirect LUC (ILUC) refers to a change in the use or management of land which is a consequence of direct LUC, but which occurs outside the product system being assessed.

Since the second half of the 2000s, LUC assessment through CLCA modelling has boomed, mainly stimulated by the debate on the environmental consequences of the expected expansion of biofuels. As pointed out by Macedo et al. [23], LUC has been the most contentious issue in evaluating GHG effects of biofuels. It can lead to a reduction or an increase of C stocks in biomass and soil, and the quantification is inherently uncertain, as can be seen in the widely different results reported in the literature. The inclusion of ILUC especially adds greatly to the uncertainty in quantifications of LUC effects, provided it can only be determined using global models, which are highly uncertain, once the available models are based on unobservable and unverifiable parameters and are dependent on assumed policy, economic contexts, and exogenous inputs.

12.3.3 Assessment – Life Cycle Impact Assessment

Life Cycle Impact Assessment is aimed at understanding and evaluating the magnitude and significance of the potential environmental impacts of the studied system. This phase involves accounting, assessing, and interpreting the potential environmental impacts generated by the product through *Categorization* and *Characterization* of the flows. A distinction must be made between midpoint and endpoint, where endpoint indicators are defined at the level of the areas of protection (natural environment, human health, and natural resources), and midpoint indicators indicate impacts somewhere between the emission and the endpoint [3].

For categorization, in function of the goals that are defined initially, the environmental impact categories are selected, such as depletion of abiotic resources, global warming, and ozone layer depletion, among others. The characterization aims to quantify the potential impacts concerning each selected category due to the flows identified in the product system. These potential impacts are estimated at the mid- or endpoint levels through characterization factors. Table 12.1 shows some categories

TABLE 12.1 Some Environmental Categories Assessed in LCAs

Category	Flows	Characterization Factor	Reference
Global warming	CO_2, CH_4, N_2O, CFC	Global warming potential	CO_2
Ozone depletion	CFC, HFC, Halons	Ozone depleting potential	CFC-11
Eutrophication	PO_4, NO, NO_2, NH_4	Eutrophication potential	PO_4^{-3}
Acidification	SO_x, NO_x, HCl, NH_4	Acidification potential	SO_2
Photochemical smog	NMHC (Non methane hydrocarbons)	Potential to create photochemical oxidation	C_2H_6

Adapted from: EPA, Environmental protection agency, in: Life Cycle Assessment: Principles and Practice. Ohio. EPA/600/R-06/060. 88p. Available from: http://nepis.epa.gov, 2006.

typically assessed in LCA studies, and their respective inventory flows and characterization factors.

12.3.4 Interpretation

In the interpretation phase, the results from the previous phases are evaluated in relation to the goal and scope in order to reach conclusions and recommendations. When producing a comparative study, the scope of the study must naturally be defined in such a way that the systems can be consistently compared, implying the use of equivalent methodological considerations. This is one of the critical aspects when producing broad comparisons from different studies, given the wide range of assumptions and methodological considerations.

12.4 LCA OF BIOJET FUELS

As detailed in the previous chapters, the main biojet fuel production technologies can be classified into three groups (Fig. 12.4) [24,25]. These are discussed below.

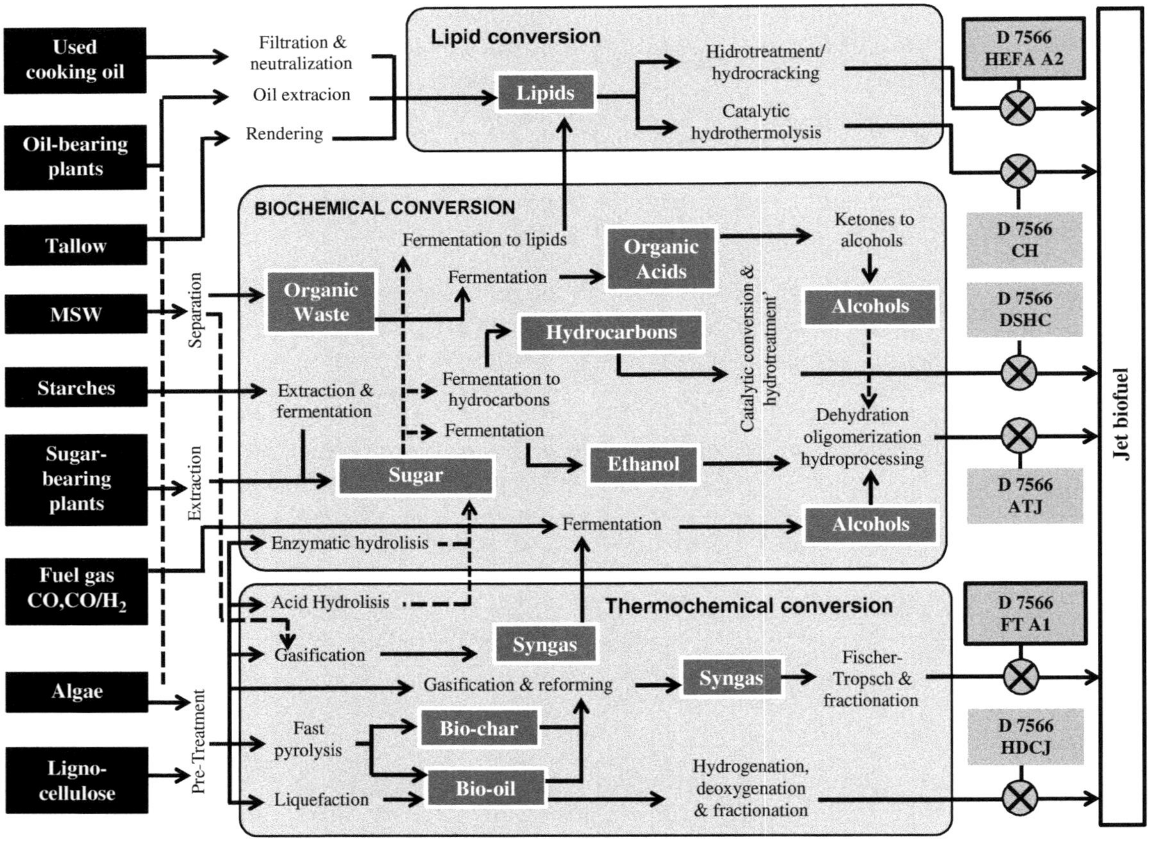

FIGURE 12.4 Main pathways to produce biojet fuels. *Source: Adapted from: FAPESP, Fundação de Amparo à Pesquisa do Estado de São Paulo. Plano de Vôo para biocombustíveis de aviação no Brasil: Plano de Ação. 2013. Available from: http://www.fapesp.br/publicacoes/plano-de-voo-biocombustiveis-brasil-pt.pdf. Acesso em setembro de 2014.*

12.4.1 Lipids Conversion

The hydrotreating/hydrocracking of vegetable oils – Hydroprocessed Esters and Fatty Acids or Hydroprocessed Renewable Jet (HRJ) or Hydrogenated Vegetable Oil – is currently the best-known process for aviation biofuel production, and is tested on a large-scale. It consists of decomposing vegetable oils in the presence of hydrogen, saturating carbon double-bonds, and removing oxygen as water. The oil is cracked and isomerized to produce jet fuel, as well as other products like naphtha, diesel, and gasoline.

12.4.2 Thermochemical Conversion

By this route, in the main studied process, the solid biomass is gasified at elevated temperatures so that it becomes a mixture of gases – mostly CO, H_2, CO_2, and water – which is called *syngas*. After purification, these gases can be synthesized into liquid hydrocarbons, usually by the Fischer–Tropsch process (F-T), using catalysts. Depending on the adopted design and processed streams, a similar hydrocarbon jet fuel can be obtained.

12.4.3 Biochemical Conversion

In this group, one possible pathway is the Alcohol To Jet (ATJ) process, where the alcohol molecules, which are obtained either from sugar, starch, or lignocellulosic materials, are submitted to dehydration and oligomerization to remove water and oxygen via the addition of hydrogen. Other pathways employ genetically modified microorganisms to convert sugar into hydrocarbons (Direct Sugar to Hydrocarbons, DSHC) or lipids. In one of these cases, yeasts, instead of producing ethanol, produce isoprenoids, such as farnesene, which is then hydrogenated into farnesane, which has good characteristics as a fuel.

In the last several years, LCA studies on biojet fuels have been elaborated upon, although in a still-small number. Through a survey conducted in *Scopus* and *Web of Science Database*, 16 articles were selected for comparison here, along with one academic Thesis [26] and two Reports [21,27]. Due to the challenges from climate change and the consequent aspirational goals from the aviation sector, almost all studies (17 in total) addressed GHG emissions, mostly for a farm-to-gate approach, while some studies also included the use phase. Exceptions are in Trivedi et al. [28], where pathways were analysed in terms of energy indicators (Energy Return on Energy Investment, EROI), and Staples et al. [29], where water consumption and land requirements were assessed.

In terms of conversion technology, 15 studies reported the use of lipid conversion routes, specifically HRJ, followed by six studies on thermochemical conversion, mainly F-T, and five studies on biochemical conversion (BC). Regarding feedstocks, algae were evaluated in eight studies, seven through the HRJ pathway, followed by camelina and jatropha in five studies. Soybean, palm, rapeseed, tallow, used cooking oil, sugarcane, corn grains, and corn stover were assessed as well. Table 12.2 presents the selected studies divided by pathway, feedstock, and environmental issue. Some studies also estimated the cost of the product, but this is not the focus of the present discussion. Fig. 12.5 summarizes the main GHG estimations from the analysed studies in comparison to the range for the conventional jet fuel. Each point in the figure refers to a published result, but there are studies that present more than one result for a particular pathway because of the adoption of different scenarios or allocation methods.

Microalgae were the most reported feedstock among the selected studies, following the number of analyses exploring their potential for fuel production. The GHG emissions for HRJ from microalgae have been estimated within a wide range (14.1–476 g CO_2/MJ), highly influenced

TABLE 12.2 Summary of the Selected Studies by Feedstock and Pathway

Technology	Feedstock	Study	Environmental Issue
HRJ	Camelina	[30]	GHG/energy
		[31]	GHG/cost
		[32]	GHG
		[33]	GHG/energy
		[34]	GHG
	Jatropha	[35]	GHG
		[21]	GHG
		[32]	GHG
		[29]	Water/land
		[34]	GHG
		[28]	Energy
	Microalgae	[21]	GHG
		[31]	GHG/cost
		[36]	GHG
		[27]	GHG/energy
		[37]	GHG
		[38]	GHG/energy/water/other
		[34]	GHG
	Palm	[26]	GHG/energy
		[21]	GHG
		[32]	GHG
		[28]	Energy
	Pongamia pinnata	[38]	GHG/energy/water/other
	Rapeseed	[21]	GHG
		[29]	Water/land
		[32]	GHG
		[28]	Energy
	Salicornia	[21]	GHG
	Soybean	[26]	GHG/energy
		[21]	GHG
		[27]	GHG/energy
		[29]	Water/land
		[32]	GHG
		[28]	Energy
	Thlaspi arvense L.	[39]	GHG/energy
	Tallow	[40]	GHG/cost
	Yelow grease	[40]	GHG/cost
FT	Corn stover	[26]	GHG/energy
		[21]	GHG
		[31]	GHG/cost
		[32]	GHG
	Switchgrass	[21]	GHG
		[31]	GHG/cost
		[28]	Energy
	Wood crops	[31]	GHG/cost
	Forest residue	[21]	GHG
	Biomass	[27]	GHG/energy
	Herbaceous biomass	[26]	GHG/energy
BC	Corn grains	[29]	Water/land
		[41]	GHG/cost
		[28]	Energy
	Sugarcane	[29]	Water/land
		[42]	GHG
		[41]	GHG/cost
		[38]	GHG/energy/water/other
		[28]	Energy
	Switchgrass	[29]	Water/land
		[28]	Energy
		[41]	GHG/cost
HLT	Microalgae	[43]	GHG

HRJ, Hydroprocessed Renewable Jet; *F-T*, Fischer–Tropsch; *BC*, Biochemical Conversion; *HTL*, Hydrothermal Liquefaction.

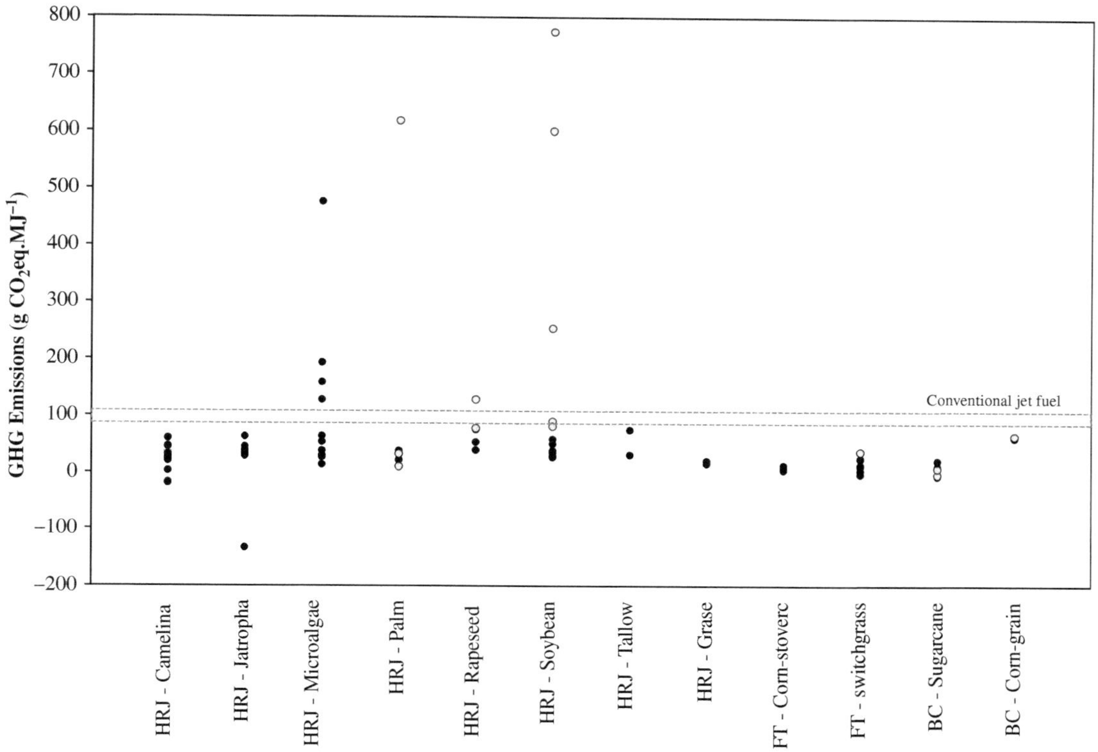

FIGURE 12.5 Range of main results from LCA studies for biojet fuels. Grey dots include LUC effects (*HRJ*, Hydroprocessed Renewable Jet; *F-T*, Fischer–Tropsch; and *BC*, Biochemical Conversion).

by the growth conditions and allocation methods. In Handler et al. [34], for instance, indicated WTW results from 29.0 g CO_2eq/MJ (mass allocation) to 40.1 g CO_2eq/MJ (energy allocation). Using energy allocation, the research in Ou et al. [37] found a much higher value (160 g CO_2eq/MJ) for the production of biojet fuel from microalgae ponds in China. Considering the system expansion method and including the CO_2 emitted in power plants as an input for algae growth, the work in Stratton et al. [21] found 14.1 g CO_2eq/MJ for a scenario of high yield and algal lipid content, in contrast to 193.2 g CO_2eq/MJ for an opposite scenario.

Hydrothermal Liquefaction (HTL) – the process that employs subcritical water to convert biomass to a carbon-rich biocrude – of microalgae cultivated in wastewater effluent was evaluated by Fortier et al. [43]. The GHG performance was a function of the HLT plant location, and was estimated as 35.2 g CO_2eq/MJ for a plant located at the wastewater treatment site, and 86.5 g CO_2eq/MJ for the plant located at a refinery. The logistics required to recycle aqueous products for algae cultivation (former case) or transportation of dewatered algae to refinery (later case) were determinant in the study.

Camelina pathways were reported in six studies, including Fan et al. [39], which assessed the biojet fuel from *Thlaspi arvense* L. (part of the camelina family). Stimulated by the non-direct competition with food due to the possible cultivation as a rotation crop, the research in Shonnard et al. [30] evaluated the life cycle performance of the camelina jet fuel between 20.2 g CO_2/MJ (mass allocation) and 22.4 g CO_2/MJ (energy allocation). The work in Lokesh et al. [34], in turn, found values from 24 g CO_2/MJ (mass allocation) to 31.4 g CO_2/MJ (energy allocation), while the research in Han et al. [32] reported 46 g CO_2/MJ, considering energy allocation. Lower emissions were verified with the system expansion approach, provided the displacement of more carbon-intensive products by the co-products obtained together with the biojet fuel. For example, estimations from Refs. [30,39] reached negative values, respectively, evaluated as −18.3 and −17 g CO_2eq/MJ.

The production of Synthetic Paraffinic Kerosene (SPK, ie, HRJ) in the United States from *jatropha curcas* cultivated in Brazil was assessed by researchers in Bailis et al. [35]. The results ranged from 33 g CO_2/MJ (mass allocation) to 40 g CO_2/MJ (energy allocation), while values based on the system expansion approach varied from −134 to 40 g CO_2/MJ. Considering DLUC effects and energy allocation, authors estimated that when jatropha was planted in agro-pastoral lands, the emissions would be 13 g CO_2/MJ, but they would jump to 141 g CO_2/MJ in the case where jatropha was cultivated in cerrado woodlands. Although the analysis resulted in interesting mitigation effects for most of the cases, this industry in Brazil is relatively young and 'best practices' can still be achieved.

Similar to jatropha, the performance of soybean and palm-derived fuels are largely affected by LUC. The work in Wong [26] estimated emissions from 29.9 to 50.8 g CO_2eq/MJ for HRJ from soybean, which would increase to 90.4 and 600.3 g CO_2eq/MJ, assuming the conversion of cerrado and tropical rainforest, respectively. For HRJ from palm, another typical feedstock for biofuel, emissions were evaluated in the 22–32 g CO_2eq/MJ range (without LUC), and from 10.2 to 617.3 g CO_2eq/MJ, assuming, respectively, the conversion of logged-over forest and peatland rainforest. In Stratton et al. [21], similar ranges for soybean and palm were found, but with more drastic effects from LUC in the case of conversion of cerrado and tropical rainforest. The cultivation on set-aside land would also lead to higher life cycle emissions in the case of rapeseed pathways [21].

For HRJ pathways involving the conversion of tallow, in Seber et al. [40] two system boundaries were considered. In the first one, tallow was treated as a co-product of the rendering industry, and in the second, as a co-product of the meat production industry, considering emissions from cattle husbandry and slaughtering. By the energy allocation method, the results for tallow were between 25.7 and 37.5 g CO_2/MJ for the first option and 67.1 to 83.9 g CO_2/MJ for the second. Emissions for HRJ from grease from waste cooking oil were estimated to be between 16.8 and 21.4 g CO_2/MJ [40].

Corn stover, representing agricultural residues, and switchgrass, representing non food energy crops, were the main feedstocks studied in the F-T pathways. In Wong [26] the life cycle emissions were estimated between 5 and 15 g CO_2/MJ for jet fuels from corn stover, forest residue, and switchgrass. In Stratton et al. [21] changes of soil carbon were considered in the analysis, due to the soil carbon sequestration is a potential strategy for offsetting CO_2 emissions, depending on the rate of carbon additions, the capacity of soil for carbon storage, and the stability of sequestered soil carbon over time. At the same time, the removal of agricultural residues from the field (as corn stover) can lead to an incremental use of fertilizer or, if left on the field, to N_2O emissions from the soil. Considering the effect of carbon sequestration, in Stratton et al. [21] was estimated that F-T fuel from switchgrass would move from 17.7 g CO_2/MJ to −2.0 g CO_2/MJ. For corn stover, on the other hand, they reported 9.0 g CO_2/MJ (without soil carbon changes) and 13.6 g CO_2/MJ (with the soil carbon change).

BC pathways using mainly sugarcane and corn grains were also reported in the literature. Moreira et al. [42] analysed the production of farnesane, which is obtained by the direct bioconversion of sugarcane sugars, with posterior hydrogenation. Effects from direct and indirect LUC were addressed, resulting in a life cycle emission of 8.5 g CO_2eq/MJ in the reference case. However, the results of the Monte Carlo analysis indicated life cycle emissions of 21 ± 11 g CO_2eq/MJ, with substantial influence from the LUC factor and the level of electricity exports.

The ATJ process for sugarcane, corn grain, and switchgrass were evaluated by Staples et al. [41]. Without LUC considerations, emissions for the sugarcane pathway varied from −4.9 (system expansion) to 12.7 g CO_2eq/MJ (market allocation), for corn grain they were from 62.6 (system expansion) to 65.6 g CO_2eq/MJ (market), and for switchgrass they were 37.4 g CO_2eq/MJ (market and system expansion). With LUC considerations and market allocation, the results were 48.1, 113.8, and 40.3 g CO_2eq/MJ for sugarcane, corn grain, and switchgrass, respectively.

Regarding the contribution from the different life cycle stages (Fig. 12.6), similar to other

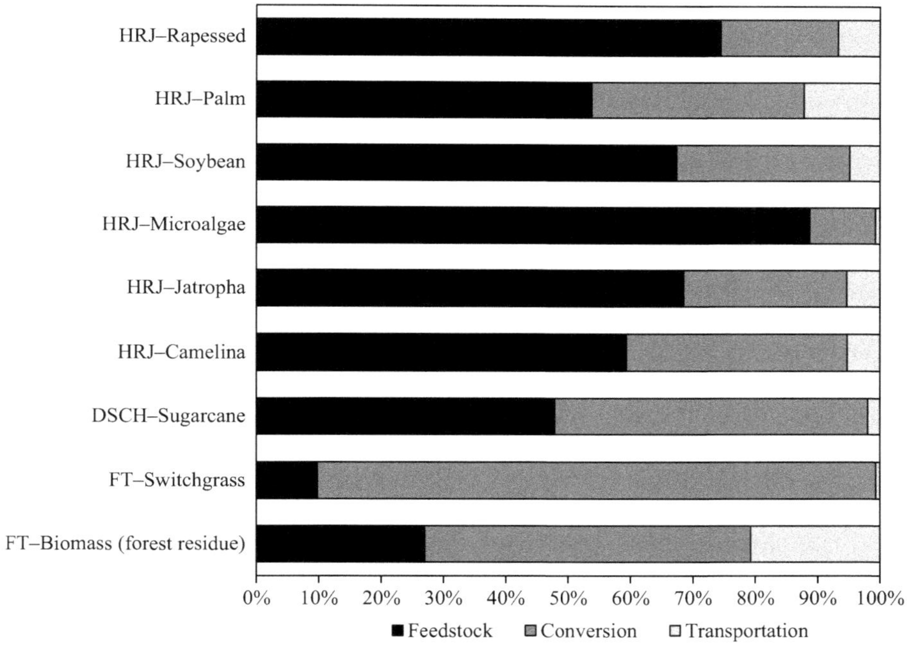

FIGURE 12.6 Contributions to GHG emissions by life cycle stage (excluding credits from co-products and LUC). *Source: Based on R.W. Stratton, H.M. Wong, J.I Hileman, Life Cycle Greenhouse Gas Emissions From Alternative Jet Fuels. Partnership for Air Transportation Noise and Emission Reduction. Massachusetts Institute of Technology: Cambridge, MA, 2010. Available from: http://web.mit.edu/aeroastro/partner/reports/proj28/partner-proj28-2010-001.pdf, H.M. Wong, Life-cycle assessment of greenhouse gas emissions from alternative jet fuels. Thesis submitted to the Engineering Systems Division in Partial Fulfillment of the Requirements for the Degree of Master of Science in Technology and Policy. Massachusetts Institute of Technology. September Available from: http://lae.mit.edu/uploads/LAE_report_series/2008/LAE-2008-005-T.pdf, 2008, Moreira, Marcelo, Angelo C. Gurgel, and Joaquim E. A. Seabra. 2014. "Life Cycle Greenhouse Gas Emissions of Sugar Cane Renewable Jet Fuel." Environmental Science & Technology 48: 14756–14763. K. Lokesh, V. Sethi, T. Nikolaidis, E. Goodger, D. Nalianda, Life cycle greenhouse gas analysis of biojet fuels with a technical investigation into their impact on jet engine performance. Biomass Bioenergy (2015) 77 26–44. Available from: http://dx.doi.org/10.1016/j.biombioe.2015.03.005, and R.M. Handler, C.E. Canter, T.N. Kalnes, F.S. Lupton, O. Kholiqov, D.R. Shonnard, et al., Evaluation of environmental impacts from microalgae cultivation in open-air raceway ponds: analysis of the prior literature and investigation of wide variance in predicted impacts. Algal Res. (2012) 1 (1). 83–92. Available from: http://dx.doi.org/10.1016/j.algal.2012.02.003.*

biofuels [23], the feedstock production was usually the main contributor to GHG emissions. The conversion steps become more relevant in the F-T pathways, as well as in the direct conversion of sugars for farnesane production, because of the requirement for a hydrogenation step. In turn, the contributions from the transportation of raw materials or products were normally small.

12.5 FINAL COMMENTS

In addition to the intrinsic performance of each biofuel pathway, the different methodological considerations among the studies also explain the differences in the life cycle emissions. Partly due to the maturity level of the technology, HRJ from vegetable oils has been the most studied pathway. The analyses showed that their environmental performance strongly depended on the agricultural conditions and where the crops were cultivated.

Almost all pathways presented better performances in terms of GHG emissions than the conventional jet fuel, which produced between 88 and 106 g CO_2eq/MJ. HRJ from palm was evaluated as an attractive route, as well as HRJ from yellow grease and *T. arvense* L. HRJ from jatropha and microalgae featured a wide range of variation among the five studies, in part because of the allocation method and the cultivation conditions. On the other hand, the ranges for HRJ from soybean and camelina (which were assessed by more studies), were much narrower, also because these pathways involved mature supply chains.

The LUC effect was considered in five studies, and in all cases represented relevant implications for the overall performance. For HRJ from palm and soybean, this effect alone can negate all the benefit of the biofuel in terms of GHG emissions mitigation. For the direct conversion of sugarcane in Brazil, however, despite the uncertainties, the mitigation effect would be verified even considering the emissions from the indirect LUC.

In summary, due to the relatively small number of studies, the low maturity level or scale of the pathways, and the fact that GHG emissions have been the almost exclusive environmental issue under analysis, at this moment it is difficult to draw conclusions about the real benefits of the different routes for the production of biojet fuels. Furthermore, the methodological differences between the studies pose significant barriers for comparisons. This makes the harmonization efforts particularly relevant in terms of painting a clear picture for decision makers and society when comparing the performances of different biojet fuel pathways.

References

[1] ILCD, International Reference Life Cycle Data System. General Guide for Life Cycle Assessment – Detailed Guidance, first ed., European Commission, Joint Research Centre, Institute for Environment and Sustainability, 2010, 394 p. Available from: <http://publications.jrc.ec.europa.eu/repository/bitstream/JRC48157/ilcd_handbook-general_guide_for_lca-detailed_guidance_12march2010_isbn_fin.pdf>.

[2] ISO, International Standardization Organization. ISO 14040 Environmental Management – Life Cycle Assessment – Principles and Framework. 2006a. Available from: <http://www.iso.org/iso/home.htm>.

[3] G. Finnveden, M.Z. Hauschild, T. Ekvall, J. Guinée, R. Heijungs, S. Hellweg, et al., Recent developments in life cycle assessment, J. Environ. Manage. 91 (2009) 1–21.

[4] ICAO, Assembly – 38th Session, 24 September to 4 October 2013. International civil aviation organization. Available from: <http://www.icao.int/Meetings/a38/Pages/WP_Agenda.aspx>, 2013.

[5] EPA, Environmental protection agency, in: life cycle assessment: principles and practice. Ohio. EPA/600/R-06/060. 88p. Available from: <http://nepis.epa.gov/Exe/ZyPDF.cgi/P1000L86.PDF?Dockey = P1000L86.PDF>, 2006.

[6] R.G. Hunt, J.D. Sellers, W.D. Franklin, Resource and environmental profile analysis: a life cycle environmental assessment for products and procedures, Environ. Impact Assess. Rev. 12 (1992) 245–269.

[7] H.C.Wilting, An energy perspective on economic activities. Available from: <http://dissertations.ub.rug.nl/faculties/science/1996/h.c.wilting/>, 1996.

[8] R.S. Berry, M.F. Fels, The energy cost of automobiles, Science and public affairs, Bull. At. Sci. (1973), pp. 11–17, 58–60.

[9] P.F. Chapman, The energy costs of producing copper and aluminium from primary sources, Metals Mater. Vol. 8 (1974) 107–111.

[10] D.J. Wright, Goods and services: an input-output analysis, Energy Policy Vol. 2 (1974) 307–315.

[11] C.W. Bullard, R.A. Herendeen, The energy cost of goods and services, Energy Policy Vol. 3 (1975) 484–493.

[12] B. Hannon, Options for energy conservation, Technol. Rev. (1974) 24–31.

[13] G. Leach, Energy and food production, I.P.C. Science and Technology Press Limited, Surrey, United Kingdom, 1976.

[14] J. Guinée, R. Heijungs, G. Huppes, A. Zamagni, P. Masoni, R. Buonamici, et al., Life cycle assessment: past, present, and future, Environ. Sci. Technol. 45 (2011) 90–96.

[15] ISO, International Standardization Organization. ISO 14044 Environmental Management – Life Cycle Assessment – Requirements and Guidelines. 2006b. Available from: <http://www.iso.org/iso/home.htm>.

[16] CARB, Staff report: initial statement of reasons for rulemaking proposed re-adoption of the Low Carbon Fuel Standard Regulation. California Air Resources Board, Industrial Strategies Division, Transportation Fuels Branch, Oil & Gas and GHG Mitigation Branch. 2 January 2015.

[17] EU, Renewable energy directive. Available from: <https://ec.europa.eu/energy/en/topics/renewable-energy/renewable-energy-directive>, 2015.

[18] ICAO, International civil aviation organization. Available from: <http://www.icao.int/Meetings/AMC/Assembly37/Documents/ProvisionalEdition/a37_res_prov_en.pdf>.

[19] IATA, IATA 2014 report on alternative fuels. Effective December 2014, ninth ed., International Air Transport Association, Montreal-Geneva, 2014.

[20] UNEP/SETAC, United nations environment programme/society of environmental toxicology and chemistry. Towards a Life Cycle Sustainability Assessment – Making informed choices on products. Available from: <http://www.unep.org/publications/contents/pub_details_search.asp?ID = 6236>, 2011.

[21] R.W. Stratton, H.M. Wong, J.I. Hileman, Life Cycle Greenhouse Gas Emissions From Alternative Jet Fuels, Partnership for Air Transportation Noise and Emission Reduction. Massachusetts Institute of Technology, Cambridge, MA, 2010. Available from: <http://web.mit.edu/aeroastro/partner/reports/proj28/partner-proj28-2010-001.pdf>.

[22] ISO, International Standardization Organization. ISO/TS 14067 Greenhouse gases – Carbon footprint of products — Requirements and guidelines for quantification and communication. 2013. Available from: < http://www.iso.org/iso/home.htm >.

[23] I.C. Macedo, A.M. Nassar, A.L. Cowie, J.E.A. Seabra, L. Marelli, M. Otto, et al., Greenhouse gas emissions from bioenergy, in: G.M. Souza, R. Victoria, C. Joly, L. Verdade (Eds.), Bioenergy & Sustainability: Bridging the gaps, Vol. 72, SCOPE, Paris, 2015, p. 779. Available from: < http://bioenfapesp.org/scopebioenergy/index.php >.

[24] FAPESP, Fundação de Amparo à Pesquisa do Estado de São Paulo. Plano de Vôo para biocombustíveis de aviação no Brasil: Plano de Ação. Available from: <http://www.fapesp.br/publicacoes/plano-de-voo-biocombustiveis-brasil-pt.pdf>, 2013 Acesso em setembro de 2014.

[25] L.A.B. Cortez, Roadmap for Sustainable Aviation Biofuels for Brazil: A Flighpath to Aviation in Brazil, Blucher, São Paulo, 2014.

[26] H.M.Wong, Life-cycle assessment of greenhouse gas emissions from alternative jet fuels. Thesis submitted to the Engineering Systems Division in Partial Fulfillment of the Requirements for the Degree of Master of Science in Technology and Policy. Massachusetts Institute of Technology. September Available from: <http://lae.mit.edu/uploads/LAE_report_series/2008/LAE-2008-005-T.pdf>, 2008.

[27] A. Elgowainy, J. Han, M. Wang, N. Carter, R. Stratton, J. Hileman, et al., Life-cycle analysis of alternative aviation fuels in GREET, ANL/ESD/12-8, 2012. Available from: <http://www.osti.gov/scitech/biblio/1046913 >.

[28] P. Trivedi, H. Olcay, M.D. Staples, M.R. Withers, R. Malina, S.R.H. Barrett, Energy return on investment for alternative jet fuels, Appl. Energy 141 (2015) 167–174. Available from: http://dx.doi.org/10.1016/j.apenergy.2014.12.016.

[29] M.D. Staples, H. Olcay, R. Malina, P. Trivedi, M.N. Pearlson, K. Strzepek, et al., Water consumption footprint and land requirements of large-scale alternative diesel and jet fuel production, Environ. Sci. Technol. 47 (21) (2013) 12557–12565. Available from: http://dx.doi.org/10.1021/es4030782.

[30] D.R. Shonnard, L. Williams, T.N. Kalnes, *Camelina*-derived jet fuel and diesel: sustainable advanced biofuels, Environ. Prog. 29 (3) (2010) 382–392. Available from: http://dx.doi.org/10.1002/ep.

[31] D.B. Agusdinata, F. Zhao, K. Ileleji, D. DeLaurentis, Life cycle assessment of potential biojet fuel

production in the United States, Environ. Sci. Technol. 45 (21) (2011) 9133–9143. Available from: http://dx.doi.org/10.1021/es202148g.

[32] J. Han, A. Elgowainy, H. Cai, M.Q. Wang, Life-cycle analysis of bio-based aviation fuels, Bioresour. Technol. 150 (2013) 447–456. Available from: http://dx.doi.org/10.1016/j.biortech.2013.07.153.

[33] X. Li, E. Mupondwa, Life cycle assessment of *Camelina* oil derived biodiesel and jet fuel in the Canadian prairies, Sci. Total Environ. 481 (1) (2014) 17–26. Available from: http://dx.doi.org/10.1016/j.scitotenv.2014.02.003.

[34] K. Lokesh, V. Sethi, T. Nikolaidis, E. Goodger, D. Nalianda, Life cycle greenhouse gas analysis of biojet fuels with a technical investigation into their impact on jet engine performance, Biomass Bioenergy 77 (2015) 26–44. Available from: http://dx.doi.org/10.1016/j.biombioe.2015.03.005.

[35] R.E. Bailis, J.E. Baka, Greenhouse gas emissions and land use change from *Jatropha curcas*-based jet fuel in Brazil, Environ. Sci. Technol. 44 (22) (2010) 8684–8691. Available from: http://dx.doi.org/10.1021/es1019178.

[36] R.M. Handler, C.E. Canter, T.N. Kalnes, F.S. Lupton, O. Kholiqov, D.R. Shonnard, et al., Evaluation of environmental impacts from microalgae cultivation in open-air raceway ponds: analysis of the prior literature and investigation of wide variance in predicted impacts, Algal Res. 1 (1) (2012) 83–92. Available from: http://dx.doi.org/10.1016/j.algal.2012.02.003.

[37] X. Ou, X. Yan, X. Zhang, X. Zhang, Life-cycle energy use and greenhouse gas emissions analysis for bioliquid jet fuel from open pond-based micro-algae under China conditions, Energies 6 (9) (2013) 4897–4923. Available from: http://dx.doi.org/10.3390/en6094897.

[38] K. Cox, M. Renouf, A. Dargan, C. Turner, D. Klein-Marcuschamer, Environmental Life Cycle Assessment (LCA) of aviation biofuel from microalgae, *Pongamia pinnata*, and sugarcane molasses, Biofuels, Bioprod. Biorefin. 8 (2014) 579–593.

[39] J. Fan, D.R. Shonnard, T.N. Kalnes, P.B. Johnsen, S. Rao, A life cycle assessment of pennycress (*Thlaspi arvense* L.) – derived jet fuel and diesel, Biomass Bioenergy 55 (2013) 87–100. Available from: http://dx.doi.org/10.1016/j.biombioe.2012.12.040.

[40] G. Seber, R. Malina, M.N. Pearlson, H. Olcay, J.I. Hileman, S.R.H. Barrett, Environmental and economic assessment of producing hydroprocessed jet and diesel fuel from waste oils and tallow, Biomass Bioenergy 67 (2014) 108–118. Available from: http://dx.doi.org/10.1016/j.biombioe.2014.04.024.

[41] M.D. Staples, R. Malina, H. Olcay, M.N. Pearlson, J.I. Hileman, A. Boies, et al., Lifecycle greenhouse gas footprint and minimum selling price of renewable diesel and jet fuel from fermentation and advanced fermentation production technologies, Energy Environ. Sci. 7 (5) (2014) 1545. Available from: http://dx.doi.org/10.1039/c3ee43655a.

[42] M. Moreira, A.C. Gurgel, J.E.A. Seabra, Life Cycle Greenhouse Gas Emissions of Sugar Cane Renewable Jet Fuel, Environ. Sci. Technol. 48 (2014) 14756–14763.

[43] M.-O.P. Fortier, G.W. Roberts, S.M. Stagg-Williams, B. S.M. Sturm, Life Cycle Assessment of Bio-Jet Fuel from Hydrothermal Liquefaction of Microalgae, Appl. Energy 122 (2014) 73–82.

CHAPTER

13

Government Policy on Delivering Biofuels for the Aviation Sector

D. Thrän[1,2,3] and J. Ponitka[2]

[1]Department Bioenergy, Helmholtz Center for Environmental Research – UFZ, Leipzig, Germany, [2]Deutsches Biomasseforschungszentrum gGmbH – DBFZ, Leipzig, Germany, [3]University of Leizpig, IIRM, Leipzig, Germany

13.1 BIOFUELS IN THE AVIATION SECTOR – CURRENT STATUS AND EXPECTATIONS

This section highlights the current state-of-the-art of biofuel utilization within the aviation sector, as well as expectations for the use of biojet fuels in the future – that is, development of the aviation sector, greenhouse gas (GHG) effects from aviation, feedstock availability for biojet fuels, cost of biojet fuels, scenarios for market implementation, and supply conditions for biojet fuels.

13.1.1 Development of the Aviation Sector

Air transport is a developing business sector with rapidly increasing rates in transport loads and fuel demand (Fig. 13.1). For example: (1) air passenger transport increased from 2.072 billion in 2006 to 3.213 billion passengers in 2014 [1]; (2) the related revenue passenger kilometres (RPK) performed grew about 6% per year on average between 2001 and 2013, reaching 5.8 trillion RPK, international and domestic services combined [2]; (3) world scheduled air freight traffic expressed in freight tonne-kilometres performed grew by 0.4% in 2013, and is expected to increase by about 4% per year in 2014, 2015, and 2016, respectively [2]; and (4) kerosene demand increased steadily to 6.3 mb/day in 2013, which is equivalent to about 10 EJ/a (see Fig. 13.1).

A strong increase in air transport is also expected for the future – that is, Airbus and Boeing project passenger transport growth of about 5% per year from 2010 to 2030 [5]. Air traffic is expected to increase from 3.8% to 5.1% [3] per year, which will lead to over 18 billion RPK in 2040 compared to around 5 billion RPK

Biofuels for Aviation.
DOI: http://dx.doi.org/10.1016/B978-0-12-804568-8.00013-5

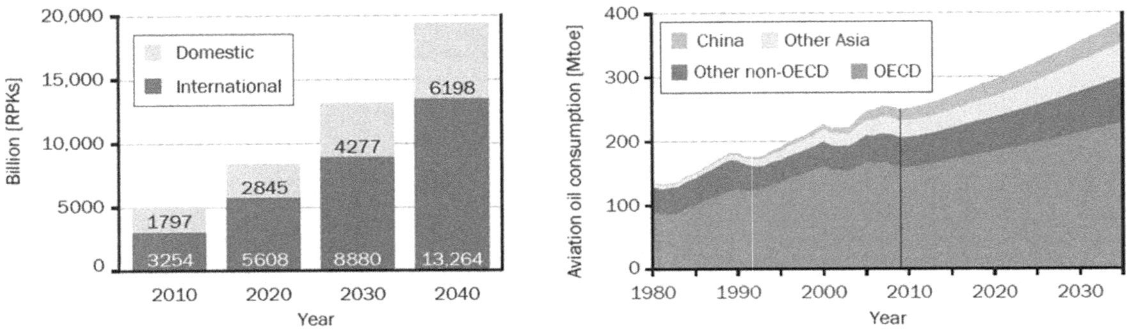

FIGURE 13.1 Expected growth of the aviation sector. Left: Passenger Traffic Forecast in billion revenue passenger kilometers (RPK) (most likely scenario/central forecast) [3]. Right: Aviation oil consumption [Mtoe] by region (New Policies Scenario) [4].

in 2010 (Fig. 13.1). The related increase in kerosene demand is expected with 'only' 1.5% per year between 2012 and 2035 [6] because of several improvements in the specific fuel demand. Due to the long life of aircraft products, particularly engines (typically 30–40 years), the speed at which these improvements are incorporated into the total fleet is a slow process. Besides the fuel efficiency of aircraft, another major area that offers considerable potential for significantly reducing fuel demand is the optimization of air traffic management [7].

13.1.2 Greenhouse Effects From Aviation

Aircraft emissions, in conjunction with other anthropogenic sources, are impacting GHGs and hence inducing climate change, though the extent of such impacts are difficult to predict and are heavily debated. Atmospheric changes from aircraft result from three types of processes: (1) direct emission of radioactive substances (eg, CO_2 or water vapour); (2) emission of chemical species that produce or destroy radioactive substances (eg, NO_x, which modifies O_3 concentration); and (3) emission of substances that trigger the generation of aerosol particles or lead to changes in natural clouds (eg, contrails). The largest areas of scientific uncertainty in aircraft-induced climate effects lie with persistent contrails, with tropospheric ozone increases and consequent changes in methane, with potential particle impacts on 'natural' clouds, and with water vapour and ozone perturbations in the lower stratosphere (especially for supersonic transport) [8].

Aviation contributed about 2% of anthropogenic CO_2 emissions in the year 2000, which is equivalent to 10% of the direct emissions of the transport sector [9]. In the absence of additional measures, an increase in emissions of 3–4% per year is projected. There is also considerable concern regarding aviation's total climate impact on the global climate, which has been estimated by the Intergovernmental Panel on Climate Change (IPCC) as being two to four times higher than the effect of carbon dioxide emissions alone due to release of nitrogen oxides, water vapour, and sulphate and soot particles (excluding cirrus cloud effects) [10]. Lee et al. [11] estimated radiative forcing from aviation in the year 2005 to be 3.5% (range 1.3–10%, 90% likelihood range) of current anthropogenic forcing, or, including cirrus cloud enhancement, 4.9% of current forcing (range 2–14%, 90% likelihood range).

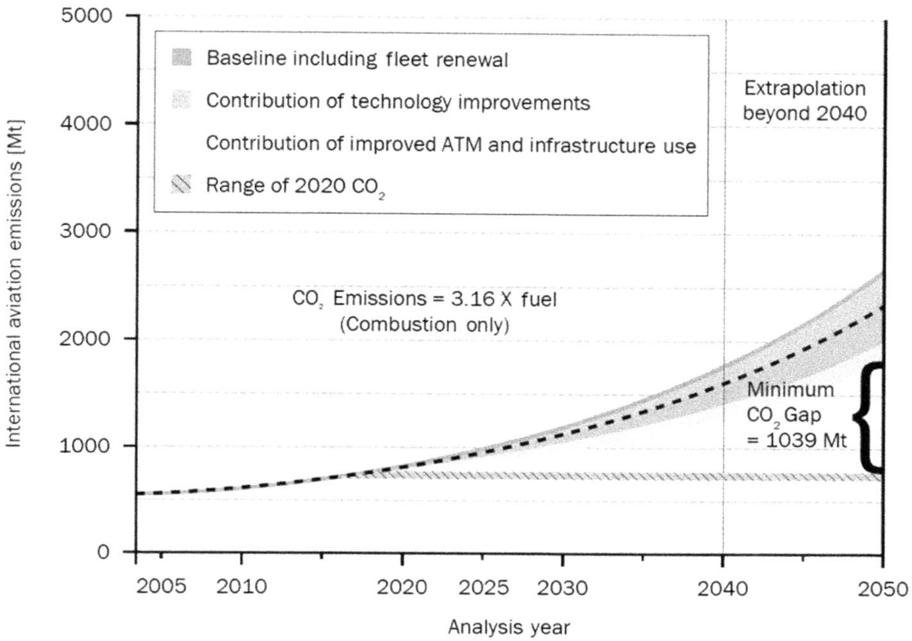

FIGURE 13.2 CO_2 emission trends from international aviation from 2005 to 2050 [3].

The public and political pressure on the sector to decrease its GHG emissions is increasing, particularly in Europe. The International Air Transport Association (IATA) has therefore already agreed on global targets for GHG emissions, as follow [3,12–14]:

- Improve fuel efficiency by 1.5% annually to 2020;
- Cap net emissions from 2020 onward through carbon-neutral growth;
- Cut emissions in half by 2050 as compared to 2005.

To support these targets, in 2050 technologies and procedures need to be available that allow a 75% reduction in CO_2 emissions per passenger kilometre [15].

Several studies have investigated the development of aviation-related global CO_2 emissions, but found that the suggested CO_2 reduction needs to incorporate additional efforts to change the currently expected trends [5]. Along with this increase, and despite improvements in efficiency, consumption of jet fuels, and also absolute emissions, will steadily increase [16]. This was also pointed out by the International Civil Aviation Organization (ICAO) in 2013 (see Fig. 13.2). Biojet fuels are seen as one approach to close the CO_2 gap.

This is not only discussed by the aviation sector, but also pointed out in different energy scenarios with ambitious GHG reduction targets (eg, Refs [17–19, 70]). Policy makers argue that due to persistent reliance on fossil fuels, it is often posited that transport, especially aviation relying entirely on kerosene, is more difficult to decarbonize than other sectors [20]. So climate policy is an important driver for the implementation of biojet fuels.

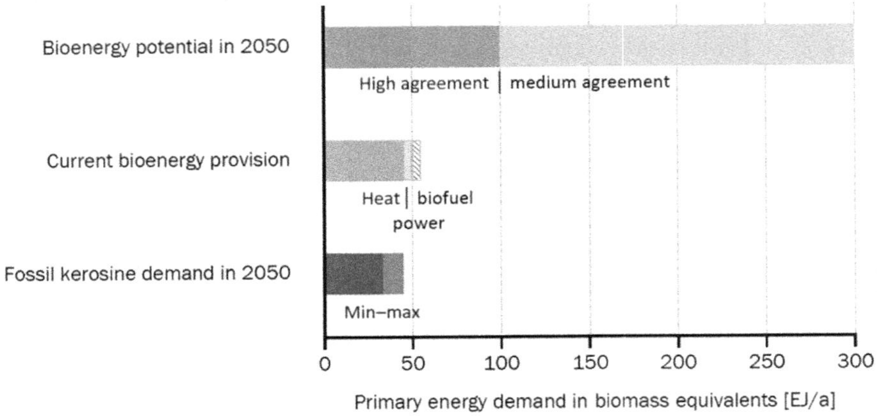

FIGURE 13.3 Current bioenergy provision and estimated future fossil kerosene demand and biomass resource potentials on a global scale. *Source: Data from F. Rosillo-Calle, S. Teelucksingh, D. Thrän, M. Seiffert, The Potential and Role of Biofuels in Commercial Air Transport – Biojetfuel, IEA Bioenergy Task 40 Sustainable International Bioenergy Trade, 2012; A. Faaij, M. van Dijk, White Paper on Sustainable Jet Fuel. SkyNRG, 2012; [22] F. Creutzig, E. McGlynn, J. Minx, O. Edenhofer, Climate policies for road transport revisited (I): evaluation of the current framework, in: Energy Policy, vol. 39 (no. 5), 2011, pp. 2396–2406, doi:10.1016/j.enpol.2011.01.062, ISSN 0301-4215; [24] D. Thrän (Ed.), Smart Bioenergy, Technologies and Concepts for a more Flexible Bioenergy Provision in Future Energy Systems, Springer-Verlag, Heidelberg, 2015, ISBN 978-3-319-16192-1.*

13.1.3 Feedstock Availability for Biojet Fuels

Going for a biojet fuel application, a reasonable question is, what is the overall biomass feedstock availability in the medium- and long-term? It proves difficult to provide any reliable forecast, but different scenarios suggest an overall aviation fuel demand of 16 to 25 exajoules (EJ) in 2050, which is equivalent to a biomass demand of about 33 to 45 EJ/a [5,21] if all the aviation fuel was be provided via biojet fuels (see Fig. 13.3).

This amount is in the range of about 65–80% of the overall biomass use of 55 EJ/a [23] for bioenergy today. The future biomass potential is estimated to be in the range of 100–300 EJ/a in 2050 [22]. So in a very first, theoretical approach, the biomass resource potential is seen as potentially sufficient to meet future kerosene demands if other bioenergy applications (heat, power, transport fuels) are not extended significantly in the future. Those theoretical numbers do not give any information about specific feedstock availabilities for different conversion routes.

13.1.4 Cost of Biojet Fuels

There is a wide range of potential costs for biojet fuels, from as low as 21 EUR/GJ up to 90 EUR/GJ (Table 13.1), usually greatly exceeding current kerosene prices. Feedstock costs not only vary substantially by region [24], but are usually an important component of total provision cost. As feedstock for the biojet fuel technologies is the major cost factor, the future costs of biojet fuels will be highly dependent on feedstock cost development.

As kerosene is one of the major cost factors in an airline's total operating costs, the fossil fuel price and price development are also relevant for the market development of biojet fuels. In the past several years, the price of jet kerosene has fluctuated between less than $1

TABLE 13.1 Costs of Biojet Fuels (Exemplarily)

	Today	Range (Approximate)	Projections
BTL (Fischer–Tropsch fuel)	Approximately 29 €/GJ [5] (but no commercial plant)	35–65€/GJ [25–27]	18–23€/GJ (based on Ref. [5])
HEFA (vegetable oil)	26–29 €/GJ [28]	21–50€/GJ [5,25–27]	20–26€/GJ (based on Ref. [5])
Bio-GTL	–	55–90€/GJ [25,26]	
Kerosene (Jet A-1)	10.8 €/GJ[a]	8–20 €/GJ[b] (10/2005–10/2015) (based on [29])	2050: 19.4€/GJ (based on 2% annual increase, following IEA WEO projections [6])

[a] *30 Oct. 2015: 1.543 US$ per gallon aviation jet fuel [approximately 460€/t or 10.8 €/GJ] [67, 68].*
[b] *1.26–3.89 $/gallon, exchange rates: http://www.finanzen.net/devisen/dollarkurs/historisch.*
Note: Due to changes in price and currency exchange rates, the price gap between fossil kerosene and biojet fuel provision costs may vary.

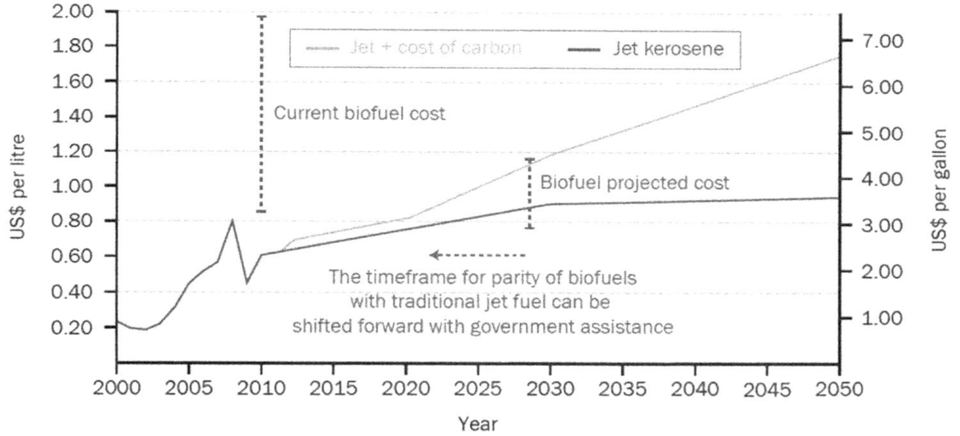

FIGURE 13.4 Jet fuel and carbon prices [14].

and up to $4 per gallon, with a recent declining trend.

Due to the recent decrease in oil prices, kerosene price projections (eg, Fig. 13.4), and the possible parity of renewable kerosene with fossil kerosene, proves difficult. In the longer term, price parity could be achieved with additional measures (eg, carbon price).

The production costs of renewable kerosene will, for the foreseeable future, be considerably higher than the corresponding fossil kerosene costs [30].

Another promising renewable, nonbio-based option in the future could be jet fuels based on power (PTL, Power-to-Liquid), for which the first demonstration projects (eg, Sunfire or SOLAR-Jet) have already reported the feasibility of various concepts [31–32]. Nevertheless, for those renewable biojet fuels, even higher costs are reported, such as PTL (via Fischer–Tropsch (F-T) or methanol) with a cost range of 25–150€/GJ [25, 26]. Consequently, policy action is needed in the short term to induce a sufficient demand.

13.1.5 Supply Conditions for Aviation Fuels

The main aviation fuels (Aviation Turbine Fuel) are Jet A-1 (standard commercial aviation fuel outside the United States) and Jet A (in the United States). Other jet fuels (eg, Jet B, JP-8) for military uses or cold climates do exist. Aviation fuels are based on kerosene fractions as a complex mixture of hydrocarbons that varies depending on crude source and manufacturing process, typically derived from crude oil [33]. The requirements are specified in the fuel standards DEF STAN 91-91 [33] and American Society for Testing Materials (ASTM) D1655[1]. The standard ASTM D7566 specifies Aviation Turbine Fuel Containing Synthesized Hydrocarbons, and is relevant for renewable jet fuels. Important specifications of the main fuel JET A-1 are the following:

- Flash Point (38°C minimum);
- Freeze Point (−47°C);
- Net Heat of Combustion (42.8 MJ/kg minimum);
- Density (775–840 kg/m^3).

From a variety of feedstocks and their pretreatment, extraction, or separation steps, the main conversion (physico-chemical, thermochemical, bio-chemical) and subsequent processing or upgrading provide the final product (examples in Table 13.2). Three of these

TABLE 13.2 Conversion Pathways for Biojet Fuels

Pathway	Feedstock/Technology	ASTM Certification
HEFA	Hydro-processed esters and fatty acids (vegetable oils); future: algae oils	X
F-T	Fischer–Tropsch kerosene (BTL) from lignocellulosics	X
DSCH/SIP	Direct sugar to hydrocarbon/Synthesized Iso-Paraffinic fuel (sugar to farnesene from, eg, sugar cane)	X
Bio-GTL	Biogas to liquid via biomethane and a wide range of biomass feedstock	–
ATJ	Alcohol to jet	–
HDCJ	Hydrotreated depolymerized cellulosic jet	–
PTL	Power-to-Liquid; electrolysis and subsequent synthesis either via F-T or methanol	–

Data from U. Lambrecht, F. Müller-Langer, Themenbereich 3 "Alternative Energieträger im Verkehr und ihre Infrastruktur". Presentation Hold: Mobilitäts-und Kraftstoffstrategie (MKS) Jahreskonferenz 2015. On 09-06-2015 in Berlin, 2015; F. Müller-Langer, K. Zech, K. Naumann, Alternative Fuels for Aviation – Status quo and Perspectives of Different Options. Presentation Hold: CORE JetFuel Workshop I Sustainable Alternative Aviation Fuels – Innovative Conversion Technologies and Deployment. On 01-06-2015 in Vienna, 2015; [36] AIREG (Ed.), Klimafreundlicher Fliegen: Zehn Prozent Alternative Flugkraftstoffe bis 2025. Aireg – Aviation Initiative for Renewable Energy in Germany e.V., 2012; [37] AIREG e.V., http://www.aireg.de/en/production.html. Biofuel Production, Aviation Initiative for Renewable Energy in Germany e.V. Retrieved from: <http://www.aireg.de/en/production.html>, 2015 (26.11.15); [38] L.A.B. Cortez, G.M. Souza, C.H.de B. Cruz, R. Maciel, An Assessment of Brazilian Government Initiatives and Policies for the promotion of biofuels through research, commercialization, and private investment support, in: S.S. da Silva, A.K. Chandel (Eds.), Biofuels in Brazil, Springer International Publishing, pp. 31–60. ISBN 978-3-319-05019-5, 2014.

[1]A revision to ASTM International's Aviation Turbine Fuel Standard (ASTM D1655) adapts to the growing presence of biodiesel in the petroleum industry and distribution systems. The level of allowable cross-contamination of fatty acid methyl ester in jet fuel was recently increased from 5 parts per million to 50 parts per million without compromising safety or adversely affecting aircraft operation (http://www.astmnewsroom.org/default.aspx?pageid = 3656).

renewable fuels (hydroprocessed esters and fatty acids (HEFA), F-T, synthesised iso-paraffins (SIP)) are already ASTM (ASTM D7566) certified, which allows a blending of up to 50% (SIP up to 10%) with conventional commercial and military jet fuels. In different projects (burnFAIR; High Biofuel Blends in Aviation (HBBA)), the test of blended jet fuels showed no negative effects on noise emissions, no effects or positive effects on CO_2 and NO_x, and positive effects on soot and particle emissions [34]. The commercial availability of the fuels is still limited, and to date, there has been little routine production of alternative jet fuel. Commercial flights thus far have been operated with specially produced batches of fuel [35].

The supply chain of Jet A-1 and biojet fuels includes extraction, conversion, and distribution [39]. The aim of the aviation industry is that the drop-in characteristics of biojet fuels will allow the fuels to be fully integrated into the conventional jet fuel storage and distribution systems, and therefore be used by all aircraft refuelling from those systems with no barriers [40]. As biojet fuels meet the technical standards (Jet A-1, ASTM), no modifications within the current distribution system are necessary. The certified biojet fuel (and conventional Jet A-1 from different sources) can be blended at any point (preferably at the refinery) of the supply chain with Jet A-1.

In most cases the biojet fuel will be delivered to and blended at a small number of international consumer points. International agreements are therefore a precondition to delivering the biojet fuels to the big international airports and selling them to international airlines. As the future sees increasing quantities of biofuels available at the main aviation hubs, logistics remain a challenge, and so does transport from the producer to the airport or the location of fuel blending. About 50% of current cargo and passenger transport are organized in less than 50 airports worldwide (see Fig. 13.5). Adjusted frame condition is also demanded for international trade for the majority of the biojet fuels [5].

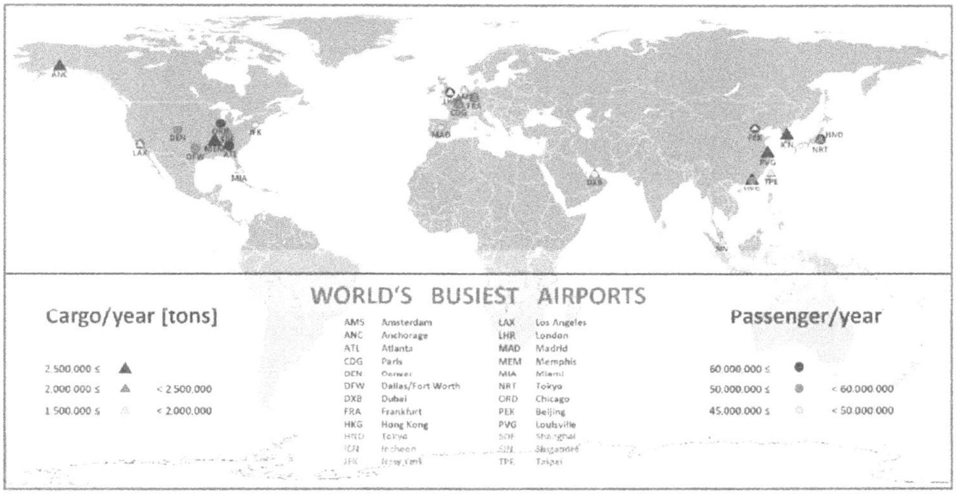

FIGURE 13.5 Map of the world's busiest airports [5].

13.2 POLICIES INFLUENCING THE MARKET INTRODUCTION OF BIOJET FUELS

This section describes the current situation in government policies for delivering biojet fuels, including the target definition and the different instruments in general, and the more detailed description regarding government policies in different countries/regions in particular.

13.2.1 Agreements on Targets for Biojet Fuel Implementation

Renewable energy policy in general, and biojet fuel implementation interests in particular, follow different overall targets in different nations and regions. Those include climate policies, policies on energy security, agriculture policies, and economic policies. So it is not surprising that there is no globally agreed-upon target regarding the amount or share of biojet fuels.

There are some national targets and usage goals to encourage the production of biojet fuels: in the United States, '1 billion gallons of alternative jet fuels annually by 2018' [41]; in Indonesia, 2% aviation biofuels by 2018; in Norway, 10% biojet fuel by 2020; and in Mexico (2011), 1% by 2015 and 45% by 2020 [42].

Different goals were formulated within the European Union (EU) regarding biojet fuels. According to the Biofuel FlightPath Initiative [43], the targeting of two million tonnes annual production of fuel derived from renewable sources by 2020 would facilitate the deployment of sustainably produced advanced biofuels for the EU aviation sector [43]. One white paper [9] formulated the goal of 40% low-carbon sustainable fuels in aviation by 2050.

Stakeholders also propose national goals, such as the use of 10% biojet fuels in 2025 in Germany [36]. Several airlines have announced that they have already concluded long-term (up to 10 years) off-take agreements with biofuel suppliers within the range of 10,000 to 100,000 tonnes per year from different feedstocks (waste, forest residues, HEFA) [12].

The International Energy Agency (IEA) has come up with expectations for the deployment of biofuels and biojet fuels as well (Fig. 13.6). The IEA sees advanced aviation fuels as a middle-term solution, with the demand expected to increase significantly [44]. It must be mentioned that there are many other projections of future supply of biojet fuels, and the expected share of renewables in the aviation sector differs wildly from 10% of the overall fuel demand up to 100% [5].

13.2.2 Instruments for Biojet Fuel Implementation

A wide variety of policy instruments and measures is available to achieve the desired goal of limiting emissions of GHG and hence influence the market introduction of biojet fuels. This includes, for example, raising communication or awareness, research and development support (innovation), guidelines, governance and coordination, voluntary agreements, market-based instruments, private and nonprivate financing schemes, tradable permits, taxes and charges, subsidies and incentives, and new regulations and standards (see Refs [9,10,45]). The main stimulus for current biofuels is blending mandates. Their outlook is highly sensitive to possible changes in government subsidies [6].

The market implementation for renewable energies is supported by a wide range of instruments. In general, they can be classified into five major groups, which are described in Table 13.3. Those systems are prototypes, while in practice the instruments are often combined.

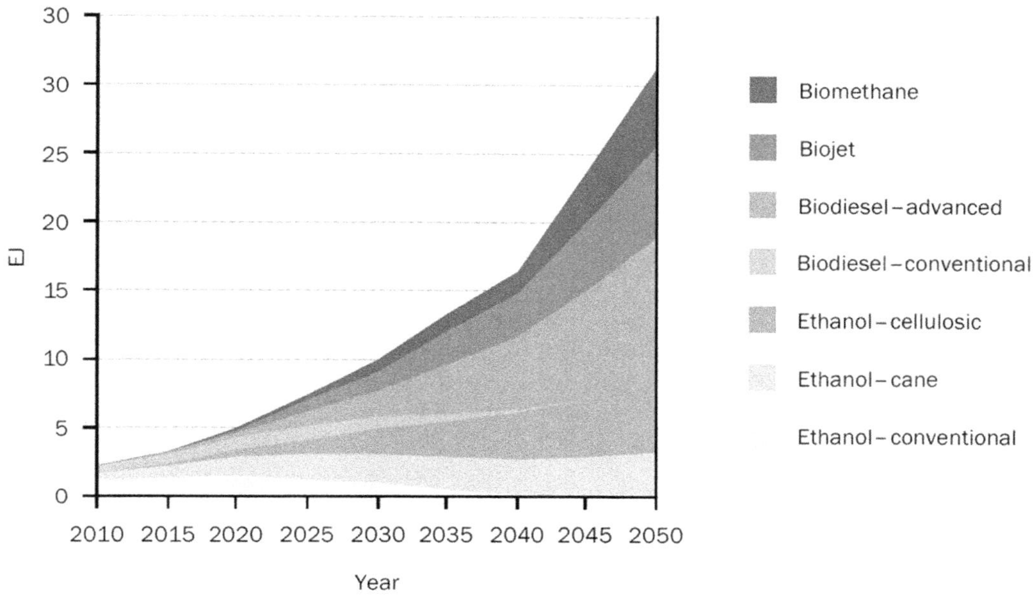

FIGURE 13.6 Demand for biofuels in the IEA biofuel technology roadmap [44]; biojet and partly advanced biodiesel (via the F-T process, a fuel mix including kerosene is usually produced) as alternative kerosene.

TABLE 13.3 Typical Support Schemes for Renewable Energies

Instrument	Principle	Cost Burden for...
GENERAL INSTRUMENTS FOR CLIMATE POLICY		
European ETS	For the emission of GHG, certificates need to be posessed/bought.	Supplier/end user
Taxation	For products with higher GHG-related emissions, higher tax rates have to be paid.	Supplier/end user
DEDICATED INSTRUMENTS FOR RENEWABLE ENERGY IMPLEMENTATION		
Feed-in tariffs	Producer gets a guaranteed price for the provided renewable energy.	End user
Investment support programs	Producer gets investment support for certain parts of the conversion plant and/or infrastructure.	Government
Taxation	For renewable products, lower tax rates have to be paid (tax credit, tax exemption).	Government
Quotas/blending mandates	Suppliers have to provide a certain share of renewables in their product portfolio.	Supplier/end user

Data from IEA, World Energy Outlook 2013, International Energy Agency. Paris, ISBN 978-92-64-20130-9, 2013; [46] D. Thrän, M. Edel, T. Seidenberger, S. Gesemann, M. Rohde, Identifizierung strategischer Hemmnisse und Entwicklung von Lösungsansätzen zur Reduzierung der Nutzungskonkurrenzen beim weiteren Ausbau der energetischen Biomassenutzung. 1. Zwischenbericht für das deutsche Minisiterium für Umwelt, Naturschutz und Reaktorsicherheit, Deutsches Biomasseforschungszentrum, Leipzig, 2009.

Within the scope of the market implementation of alternative aviation fuels, the main policies discussed or implemented are the following:

- Quotas/blending mandates;
- Taxation (eg, VAT or ticket tax, carbon tax, tax exemptions/credits, kerosene tax);
- Emission charges (emissions cap-and-trade system; Emissions Trading System (ETS)).

Worldwide quotas are often proposed, as EU-wide, national, or regional measures may lead to certain market distortions, competitive disadvantages, and adverse effects, such as the shift of passengers or airlines to unaffected airport hubs.

In addition, research and commercialization funding and investment (promotion actions) are considered to be a required action to support a range of potentially successful biofuel technologies, from which a winner, or winners, may emerge [14]. This, for example, includes the planning and construction of biojet fuel refineries to achieve economies of scale and to gain experience.

13.2.3 European Union

The EU is currently responsible for 35% of global aviation emissions [47]. With acknowledgement for the Kyoto Protocol's project mechanisms in 2004, the EU established a scheme for GHG emission allowance trading within the Community (Directive 2004/101/EC). The EU ETS remains the world's largest scheme, covering all 28 member states of the EU, plus Norway, Iceland, and Liechtenstein [6].

To attempt to mitigate emissions in the aviation sector, although a Europe-wide emissions charge 'has proved difficult' [48], the emissions from international aviation were included as a binding policy from the 1 Jan. 2012 into the carbon market in the EU ETS [48]. Emissions from all flights arriving and departing from airports in the EU were to be incorporated into the scheme, covering around a third of global aviation emissions. The aviation sector cap has been defined as 210 $MtCO_2e$ per year for 2013–2020 (and not decreasing). Also in 2012, 85% of the allowances were allocated for free based on benchmarks [49]. For the time period 2012–2020, 15% of allowances are to be auctioned and 82% allocated for free based on benchmarks. The remaining 3% constitutes a special reserve for new entrants and fast-growing airlines [49]. In response to a backlash from the aviation sector, in Apr. 2013 the EU temporarily suspended enforcement of the EU ETS requirements for flights operating from or to non-European countries, while continuing to apply the legislation to flights within and between countries in the European Economic Area, regardless of the carriers' origin [49, 50]. The EU institutions will decide on how to further regulate aviation emissions within the EU ETS after 2016 based on progress within the ICAO on developing a global market-based mechanism to address international aviation emissions from 2020 [49].

The EU Directive 2009/30/EC sets targets for fuel suppliers to gradually reduce life cycle GHG emissions by up to 10% per unit of energy from fuel and energy supplied by 2020. This reduction should be obtainable through the use of biofuels and alternative fuels. It is recognized that production of biofuels in certain developing countries might not meet minimum environmental or social requirements, and these countries are encouraged to develop agreements and voluntary international or national schemes that cover key environmental and social considerations in order to promote the production of biofuels worldwide in a sustainable manner.

In order to ensure the sustainability of alternative fuels, a range of legal policies were implemented in Europe, in particular, the EU Renewable Energy Directive (EU RED) 2009/28/EC [51] for the promotion of the use of energy from renewable sources, including sustainability criteria for biofuels and bioliquids. The RED sets a 10% target for the use of renewable energy in transport by 2020, and is currently realizing a new public consultation related to the 'preparation of a new Renewable Energy Directive for the period after 2020'. For aviation biofuels to be counted, this requires implementation at the member state level. At present this has only been implemented by the Netherlands [52].

Alongside the RED, the EU Fuel Quality Directive 2009/30/EC (FQD) sets a 6% target of GHG emission reduction from all energy used in transport for 2020 as compared to 2010. The FQD target does not apply to aviation fuel, but is expected to be a driver for increased road biofuels, alongside the RED [52]. The RED and the FQD have harmonized requirements regarding biofuel sustainability [40]. National implementation of the EU RED is necessary and leads to differences between the policies in the EU member states, especially with regard to priorities in market implementation of renewables in different energy sectors.

Further progress of the EU RED with respect to aviation will depend on the commission's report (scheduled for 2021), which reviews the application of this directive. That report shall in particular address 'technological developments in energy from renewable sources, including the development of the use of biofuels in commercial aviation' [51] on the best cost-benefit basis. Also, the results of the COP21 negotiations in Paris might influence the further development of European policies on biofuel delivery.

13.2.4 United States

In the United States, several initiatives and broad national strategies of the federal government promote the development of a variety of alternative fuels – partly including alternative jet fuel – to help achieve national goals, such as securing energy independence, fostering economic development, and reducing GHG emissions. In addition, the renewable fuel program – established by law in 2005 to encourage greater use of renewable fuels – requires that US transportation fuels contain certain amounts of renewable fuels annually. Other federal agencies – that is, Department of Transportation's (DOT) Federal Aviation Administration (FAA), US Department of Agriculture (USDA), Department of Energy (DOE), and Department of Defense (DOD) – directly support alternative jet fuels through targeted goals, initiatives, and interagency and industry coordination [41].

For example, the FAA set a goal for the US aviation industry to use 1 billion gallons of alternative jet fuels annually by 2018. The FAA and DOD support research to determine the technical feasibility of using new alternative jet fuels on aircraft and in the existing infrastructure. Also, the USDA, DOE, and DOD have coordinated their activities to support the future construction or retrofit of multiple domestic commercial- or precommercial-scale production facilities to produce alternative fuels, including alternative jet fuels. One example is the Farm to Fly initiative that has brought together the US aviation community, government stakeholders, the USDA, the DOE, the DOT, and the DOD, and has already been extended to Farm to Fly 2.0 [53, 54]. Specifically, in May and Jun. 2013, four private fuel producers received awards totalling $20.5 million in federal funds, with private industry paying at least 50% of the cost [41].

The USDA recently announced a loan guarantee, provided through the Biorefinery Assistance Program, to help innovative companies turn waste into renewable jet fuel [55]. Congress established the Biorefinery Assistance Program within in the 2008 Farm Bill. It reauthorized and extended the program in the 2014 Farm Bill, and expanded the program to include bio-based renewable chemicals and bio-based product manufacturing through regulations to set forth upcoming application terms for additional loan guarantees under the program.

One of the more prominent forms of federal policy support is the Renewable Fuel Standard (RFS), where a minimum volume of biofuels is to be used in the national transportation fuel supply each year. Congress first established the RFS with the enactment of the Energy Policy Act of 2005. The Energy Independence and Security Act of 2007 greatly expanded the biofuel mandate volumes and extended the date through 2022. The expanded RFS (RFS2) required the annual use of 9 billion gallons of biofuels in 2008, rising to 36 billion gallons in 2022, with at least 16 billion gallons from cellulosic biofuels, and a cap of 15 billion gallons for corn-starch ethanol. The RFS is a market-based compliance system in which involved parties have to submit credits (Renewable Identification Numbers (RINs)) to cover their obligations. Nevertheless, jet fuel, among other things, is excluded from RFS2's national transportation fuel supply. Renewable fuels used for these purposes may, however, count towards the RFS2 mandates [56].

Therefore, a combination of incentives according to the RFS, and incentives for agriculture, under the right conditions, may open the possibility for price-competitive biojet fuel being available [12].

13.2.5 Brazil

Brazil has long experience with biofuels in the transport sector. Currently, Brazil does not have a specific policy for alternative aviation fuels, however, initiatives exist on the state level. As a member of ICAO, Brazil supports its goal of carbon-neutral growth from 2020 [57].

In recent years, several initiatives have promoted alternative aviation fuels in Brazil. These include (among others) ABRABA, Sustainable Aviation Biofuels for Brazil, and the Brazilian Biojetfuel Platform [12,49,58]. In addition to these initiatives, in 2011 the US and Brazil signed a Memorandum of Understanding to cooperate on the development of renewable aviation biofuels [52]. Also, an action plan was formulated by industry and research stakeholders in 2013 [59].

Brazil's first commercial biojet fuel flight (HEFA) was performed on 24 Oct. 2013. In 2014, the Brazilian airline GOL made its first flight on a 10% blend of ASTM-certified SIP jet fuel [60]. Over 200 flights were made in 2014 using alternative jet fuel in Brazil.

A global aviation fuel supplier and an industrial partner providing conversion technology currently collaborate to produce an alternative jet fuel blend containing 10% farnesane fermented from Brazilian sugar cane [57]. The farnesene molecule is already produced at an initial commercial scale in Brotas in Brazil; it can deliver 40 kt of fuel per year. The plant mainly uses sugar cane, but the process can be applied to other feedstocks, in particular, cellulosic sugars from woody biomass and agricultural or forest residues [35].

For policies concerning sustainability of Brazilian biojet fuels, only the decree on agro-ecological zoning for palm oil and sugar cane, as well as the decree on working conditions in the sugar cane industry, concern biofuel production. No laws have been implemented on GHG targets or chain of custody requirements [52].

The main policies and actions required for Brazil, formulated by [61], are as follows: '(1) Establish the "drop-in" sites as far as possible downstream in the distribution chain without compromising fuel quality and technical certification requirements of aviation; (2) Establish legal mechanisms to ensure that incentives for aviation biofuels are only available where demonstrated to fully implement national laws and regulations, especially environmental and social safeguards, natural forest and other habitat protections, land use zoning and worker protections; (3) Observe closely and anticipate regulatory actions by ICAO in such a way to take advantage of international regulations to promote a jet biofuel industry in Brazil; (4) Establish a governmental long-term program for integrated use of biofuels in all transportation modes in the country, to neutralize the cost difference of producing a "drop-in" fuel versus a product for biofuel-adapted engines as is the case for road transportation' [61].

13.2.6 Other Countries

A number of other countries have announced initiatives to stimulate the development of biojet fuel, or have in place sustainability requirements. These include Australia (Australian Initiative for Sustainable Aviation Fuels, (AISAF)), New Zealand, China, and Korea (emission trading schemes) [12,49,58]. Indonesia has introduced a biojet fuel mandate of 2% commencing in 2016, rising to 5% by 2025 [12].

Besides funding of research and development, several initiatives and targets formulated by the industry have been examined. However, the inclusion of aviation and biojet fuels into ETS, and crediting for biofuel mandates, remain the only direct policy measures.

13.3 SUSTAINABILITY AND CERTIFICATION ISSUES

Sustainability certification is an additional relevant field for governmental policies. The current ongoing debate on adopted sustainability criteria for bioenergy in general, and biojet fuels specifically, with first hand experience with available certification schemes, is briefly described in the following section.

13.3.1 Sustainability Criteria for Biojet Fuels

Sustainable biojet fuel provision is one of the major prerequisites of bio-based jet fuels as land and biomass are limited resources. Much uncertainty has recently developed with regards to how biofuel policies in several key markets will evolve. The discussions are driven by sustainability issues, including concern that feedstock production for biofuels contributes to deforestation or uses land that could be used to grow food. This could impact production, trade, and use of biofuels [6,62]. Sustainability criteria and standards have to be met along the whole complex provision chain of biomass to energy, from cultivation and harvest/collection to conversion and use [69].

There is currently no dedicated indicator for biojet fuels, but a wide range of sustainability criteria and indicators for bioenergy is available, and could be applied to biojet fuels. The following are two approaches for the application of sustainability criteria in the bioenergy field: (1) The EU's RED, and (2) The Global Bioenergy Partnership (GBEP).

When introducing the biofuels quota, the EU defined a set of sustainability indicators that needs to be fulfilled to get biofuels counted for the biofuels quota. So, biofuels and bioliquids produced in installations have to fulfill the following three indicators [51]:

- Conversion of natural habitats: Energy crops must not be produced on biodiversity-rich areas, carbon-rich areas, and peat lands;
- Sustainable agricultural production: Energy crop cultivation within the EU has to follow cross compliance requirements and maintain a good agricultural and environmental condition;
- GHG mitigation potential: The biofuels have to achieve GHG emission savings of at least 60% relative to fossil fuels, starting from 1 Jan. 2018.

Those indicators have to be verified by certification of the whole value chain. Different certification schemes have to be introduced to trace the value chains back to the biomass production and to provide the demanded information for each unit of biofuel (see next section). An issue that is often discussed in relation to biofuels in general is indirect land use change (ILUC). Within the EU in 2009 [51], a factor was proposed for indirect land use changes in the calculation of GHG emissions that uses concrete methodology to minimize those GHG emissions. Scientific work indicates that emissions from indirect land use change—emissions estimates are calculated through modelling, and remain sensitive to (and may vary according to) the modelling framework and assumptions made—can vary substantially between feedstocks, and can negate some or all of the GHG savings of individual biofuels relative to the fossil fuels they replace. The EU Commission was therefore of the 'view that in the period after 2020 biofuels which do not lead to substantial GHG savings (when emissions from indirect land use change are included) and are produced from crops used for food and feed should not be subsidized' [63]. In 2015, members of the European Parliament adopted an EU Council text that sets a new target regime to limit the amount of crop-generated biofuels used in the transport sector [64]. This includes, for example, a cap for first-generation biofuels, national targets for advanced biofuels, and reporting on ILUC.

Another approach was more holistic, taking not only environmental, but also economic and social factors, into consideration; it was developed under the GBEP in 2011 [65]. Those indicators refer more to national bodies and aim to monitor the development of different effects from bioenergy use over time. In the first round, GBEP proposed 24 indicators distributed over environmental, economic, and social principals (see Table 13.4). Those indicators are currently being tested in case studies.

TABLE 13.4 GBEP Indicators to Describe Sustainable Bioenergy Use

Environmental	Social	Economic
Lifecycle GHG emissions (1)	Allocation and tenure of land for new bioenergy production (9)	Productivity (17)
Soil quality (2)	Price and supply of a national food basket (10)	Net energy balance (18)
Harvest levels of wood resources (3)	Change in income (11)	Gross added value (19)
Emissions of non-GHG air pollutants, including air toxics (4)	Jobs in the bioenergy sector (12)	Change in consumption of fossil fuels and traditional biomass (20)
Water use and efficiency (5)	Change in time spent by women and children for biomass collection (13)	Training and re-qualification of the workforce (21)
Water quality (6)	Bioenergy used to expand access to modern energy services (14)	Energy density (22)
Biological diversity in the landscape (7)	Change in mortality and burden disease attributed to indoor smoke (15)	Infrastructure and logistics for distribution of bioenergy (23)
Land use and land use change related to bioenergy feedstock production (8)	Incidence of occupational injury, illness, and fatalities (16)	Capacity and flexibility of use of bioenergy (24)

Data from GBEP, The Global Bioenergy Partnership Sustainability Indicators for Bioenergy. Global Bioenergy Partnership (GBEP), Food and Agriculture Organization of the United Nations (FAO), Rome, 2011.

13.3.2 Certification Schemes and First Experiences from Practice

Besides a great variety of voluntary sustainability initiatives and standards (eg, ISO, CEN, and forest or biomass certification schemes), mandatory standards do apply on national/regional (eg, EU ETS, EU RED, US RFS, Brazil Agroecological zoning) and international levels (UN CBD convention). The implementation of the standards is ensured through third-party certification systems.

Three important voluntary standards and schemes are the following:

- Roundtable on Sustainable Biomaterials (RSB);
- International Sustainability and Carbon Certification (ISCC);
- US Renewable Fuel Standard (RFS).

The ISCC and the RSB EU RED are the main voluntary certification schemes recognized under the European RED. The US RFS is a mandatory approach that also provides an effective incentive system based on tradable certificates (RINs) that are generated for each batch of biofuel [61].

Further voluntary schemes commonly used to certify agricultural, forestry, and waste feedstock are Biomass Biofuel Sustainability Voluntary Scheme (2BSvs), Bonsucro EU, and Roundtable on Sustainable Palm Oil (RSPO-RED) [52].

There are considerable differences between the standards in terms of the sustainability criteria. The applied methods for lifecycle analysis of GHG emissions are also not directly comparable to each other. Therefore, at the ICAO level (ie, 191 states), it will be challenging according to IATA [66] to 'define a preferred set of sustainability criteria that have a chance to be accepted' [35].

The certification itself is applicable. There is still a need to have international agreements or mutual recognition of different standards (EU RED and RFS2), and to develop a metastandard for aviation biofuels that was formulated by Refs [52,40].

When it comes to practical experiences, the RSB has already certified a biojet fuel supplier, and several aviation biofuel initiatives have recommended using RSB standards for their biofuel supplies. Also, feedstock supply for biojet fuel has already been RSB certified [61].

13.4 CONCLUSION AND OUTLOOK: WHAT IS NECESSARY TO BRING BIOJET FUELS INTO THE MARKETS

This chapter has assessed the main determining factors that shape the use of biofuels in commercial air transport. As stated, aviation is a global industry with global problems and challenges, and it demands global actions and solutions. Additionally, it is an industry for which an increase in fuel demand is expected. In answering those expectations, the IATA is committed to achieving carbon-neutral growth by 2020 and 50% reduction by 2050. To achieve those targets, biojet fuels are seen as one necessary element.

Some relevant preconditions of the market implementation of biojet fuels have been achieved during the last decade, especially the following:

- Demonstration of a wide range of biojet fuels that have been tested successfully;
- Definition of technical standards for drop-in biojet fuels, which are already fulfilled by three different biojet fuels (HEFA, F-T, and SIP);
- Establishment of sustainability assessment approaches and certification schemes for the biofuel sector, which can be implemented in the biojet fuel sector in principle;

- Multistakeholder groups around the world are already working together on initiatives for the deployment of biojet fuels. Some examples include CAAFI (US); Ubrabio (Brazil); Initiative Towards sustAinable Kerosene for Aviation, (mainly EU); aireg (Germany); Bioqueroseno (Spain); Bioport Holland (The Netherlands); Plan de Vuelo (Mexico); AISAF (Australia); NISA (Nordic countries); and BioFuelNet Canada [12].

On the other hand, there are many challenges to solve, and uncertainties to consider, before biojet fuels can be used in aviation on any significant scale, which can be summarized as follows:

- So far no mandatory international targets, international standards, or international certification schemes exist [40, 52];
- Production costs remain the main stumbling block, and, even if they are expected to fall when more capacity is installed, additional governmental policy is necessary to make biojet fuels price-competitive by 2020–2030. Having the sustainability certification process being adapted to incorporate the use of biojet fuels is not expected to be an obstacle for its deployment in the medium term, though this will add additional costs;
- Mainly GHG savings, land conversion restrictions, and biodiversity and carbon stock protection are included in national legislation and voluntary schemes, whereas economic and social criteria and soil, air, and water protection are not (or only partially) included [35, 52];
- Uncertainties on the achievable environmental effects of biojet fuel substitution are reported, concerning the no-carbon greenhouse effect of the air traffic and the GHG emissions of land use change caused by biojet fuel feedstock provision. Therefore, even in an advanced biojet fuel scenario, the aviation sector will continue to cause an increasing portion of greenhouse effects until 2050.

Government policies that bring biojet fuels to the market need to address different issues.

One key issue for delivering biofuels for the aviation sector is the development of an appropriate support mechanism. To achieve economies of scale and cost reduction for biojet fuel provision, a stable, long-term policy framework to build investor confidence and induce demand is necessary. This could be achieved, for example, by blending mandates, higher carbon taxes or penalties, taxation of fossil jet fuel, or by integration into international ETSs. In the past, the implementation of the EU ETS for biojet fuels did not succeed due to international competition issues. Those experiences clearly show that international agreements are a key factor. The best way might be to negotiate the frame condition for biojet fuel implementation within the international negotiation on GHG emission reduction.

Another key issue is the provision of sustainable biojet fuels. With the existing certification schemes in general, a sustainable production of biojet fuels under today's definition of suitability can be certified. The next necessary step is to harmonize those schemes with regard to their applicability in the aviation sector. Additionally, it is necessary to also include the major concern of biomass availability. Therefore, biojet fuel policies need to be linked with other bioenergy application fields (biofuels for land transport, heat production from biomass, etc.) in order to make sure that restricted land is not promised to different applications in parallel. Such an overall bioenergy strategy is often missing on a country level, and also in the international debate. Additionally, GHG emission reduction from the use of biojet fuels is still discussed on a scientific basis and might change the picture in the future. Monitoring systems for bioenergy (with biojet fuels as part of them) need to be implemented internationally to make sure that the intended effects on GHG reduction occur, and leakage effects (such as direct and ILUC) are detected at early stages.

The third key issue is to provide long-term stable support for research and development in the different fields of biojet fuels, including: (1) improvements in costs and efficiency of biojet fuel production, logistics, and use; (2) further development of sustainability criteria and certification approaches; and (3) integration of biojet fuels in overall concepts of an efficient (and mainly based on) renewable energy supply in the future.

From an industry perspective, IATA [67] sees the role of government as especially important in helping to:

- Adopt globally recognized sustainability standards and work proactively to harmonize global standards;
- Allow biojet fuel to compete on an equal basis with land transport-based fuels through equivalent public incentives (ie, 'level the playing field');
- Encourage user-friendly biofuel accounting methods, and work to proactively harmonize global standards;
- Support biojet R&D and demonstration plants;
- Design effective policy to de-risk investments into biojet production plants;
- Engage in public-private partnerships for biojet fuel production and supply;
- Pursue a harmonized transport and energy policy;
- Commit to policy certainty, or at a minimum, policy timeframes that match investment timeframes.

Ref. [12].

References

[1] WORLD BANK, Air Transport, Passengers Carried, The World Bank Group, Washington, 2015.

[2] ICAO, Forecasts of Scheduled Passenger and Freight Traffic. International Civil Aviation Organization. Retrieved from: <http://www.icao.int/sustainability/pages/eap_fp_forecastmed.aspx>, n.d. (17.11.15).

[3] ICAO, ICAO Environmental Report 2013. AVIATION AND CLIMATE CHANGE, International Civil Aviation Organization (ICAO), Montreal, 2013.

[4] IEA, World Energy Outlook 2010, International Energy Agency, Paris, 2010, ISBN 978-92-64-08624-1.

[5] F. Rosillo-Calle, S. Teelucksingh, D. Thrän, M. Seiffert, The Potential and Role of Biofuels in Commercial Air Transport - Biojetfuel, IEA Bioenergy Task 40 Sustainable International Bioenergy Trade, London, 2012.

[6] IEA, World Energy Outlook 2013, International Energy Agency, Paris, 2013, ISBN 978-92-64-20130-9.

[7] S. Blakey, L. Rye, C.W. Wilson, Aviation gas turbine alternative fuels: a review, in: Proceedings of the Combustion Institute, vol. 33 (no. 2), 2011, pp. 2863–2885, doi:10.1016/j.proci.2010.09.011, ISSN 1540–7489.

[8] IPCC, (1999) Aviation and the Global Atmosphere. (Online) <http://www.ipcc.ch/ipccreports/sres/aviation/index.htm>.

[9] IPCC, Climate Change 2007: Impacts, Adaptation and Vulnerability: Contribution of Working Group II to the Fourth Assessment Report of the Intergovernmental Panel on Climate Change, Cambridge University Press, Cambridge, UK; New York, NY, 2007, ISBN 978-0-521-88010-7.

[10] EC (Ed.), COMMISSION STAFF WORKING PAPER. IMPACT ASSESSMENT – Accompanying Document to the WHITE PAPER: Roadmap to a Single European Transport Area – Towards a Competitive and Resource Efficient Transport System. SEC(2011) 358 Final, European Commission, Brussels, 2011.

[11] D.S. Lee, D.W. Fahey, P.M. Forster, P.J. Newton, R.C. N. Wit, L.L. Lim, et al., Aviation and global climate change in the 21st century, Atmos. Environ. 43 (2009) 3520–3537.

[12] IATA, IATA Sustainable Aviation Fuel Roadmap, International Air Transport Association, Montreal, Geneva, 2015, ISBN 978-92-9252-704-4.

[13] ATAG, The Right Flightpath to Reduce Aviation Emissions. A Position Paper Presented by the Global Aviation Industry. UNFCCC CLIMATE TALKS, 2010.

[14] ATAG, Powering the Future of Flight. The Six Easy Steps to Growing a Viable Aviation Biofuels Industry, Air Transport Action Group, Geneva, 2012.

[15] EC (Ed.), WHITE PAPER – Roadmap to a Single European Transport Area – Towards a Competitive and Resource Efficient Transport System. COM (2011) 144, European Commission, Brussels, 2011.

[16] EC, EU Energy, Transport and GHG Emissions Trends to 2050, 2013, ISBN 978-92-79-33728-4.

[17] S. Teske, et al., Energy [r]evolution a Sustainable Global Energy Outlook, European Renewable Energy Council, Greenpeace International (publisher), Amsterdam, Brussels, 2010.

[18] World Energy Council, Global Transport Scenarios 2050, 2011. ISBN 978-0-946121-14-4.

[19] WWF, ECOFYS, OMA, The Energy Report 100% Renewable Energy by 2050, WWF International, Switzerland, 2011, ISBN 978-2-940443-26-0.

[20] R.C. Pietzcker, T. Longden, W. Chen, S. Fu, E. Kriegler, P. Kyle,, et al., Long-term transport energy

demand and climate policy: alternative visions on transport decarbonization in energy-economy models, in: Energy. vol. 64, pp. 95–108, doi:10.1016/j.energy.2013.08.059, ISSN 0360-5442.

[21] A. Faaij, M. van Dijk, White Paper on Sustainable Jet Fuel. SkyNRG, 2012.

[22] F. Creutzig, E. McGlynn, J. Minx, O. Edenhofer, Climate policies for road transport revisited (I): evaluation of the current framework, in: Energy Policy, vol. 39 (no. 5), 2011, pp. 2396–2406, doi:10.1016/j.enpol.2011.01.062, ISSN 0301-4215.

[23] D. Thrän (Ed.), Smart Bioenergy. Technologies and Concepts for a more Flexible Bioenergy Provision in Future Energy Systems, Springer-Verlag, Heidelberg, 2015, ISBN 978-3-319-16192-1.

[24] H. Chum, A. Faaij, J. Moreira, G. Berndes, P. Dhamija, H. Dong, et al., Bioenergy, in: O. Edenhofer, R. Pichs-Madruga, Y. Sokona, K. Seyboth, P. Matschoss, S. Kadner, T. Zwickel, P. Eickemeier, G. Hansen, S. Schlömer, C. von Stechow (Eds.), IPCC: IPCC Special Report on Renewable Energy Sources and Climate Change Mitigation. Prepared by Working Group III of the Intergovernmental Panel on Climate Change, Cambridge University Press, Cambridge, United Kingdom and New York, NY, USA, 2011, p. 1075.

[25] U. Lambrecht, F. Müller-Langer, Themenbereich 3 "Alternative Energieträger im Verkehr und ihre Infrastruktur". Presentation Hold: Mobilitäts-und Kraftstoffstrategie (MKS) Jahreskonferenz 2015. On 09-06-2015 in Berlin, 2015.

[26] F. Müller-Langer, K. Zech, K. Naumann, Alternative Fuels for Aviation – Status quo and Perspectives of Different Options. Presentation Hold: CORE JetFuel Workshop | Sustainable Alternative Aviation Fuels – Innovative Conversion Technologies and Deployment. On 01-06-2015 in Vienna, 2015.

[27] P. Ponitka, O. Arendt, V. Lenz, J. Daniel-Gromke, W. Stinner, A. Ortwein, et al., Konversionspfade – zur energetischen Biomassenutzung im 21. Jahrhundert, in: D. Thrän, J. Ponitka, O. Arendt (Eds.), Focus on: Bioenergie-Technologien, 2015. Leipzig 2015 – ISSN 2192-1156 (2015) ISSN: 2192-1156. <https://www.energetische-biomasse-nutzung.de/fileadmin/user_upload/Downloads/Ver%C3%B6ffentlichungen/fh_technologie_web_small.pdf>.

[28] K. Naumann, K. Oehmichen, M. Zeymer, K. Meisel Monitoring Biokraftstoffsektor. DBFZ Report No. 11 (2. Auflage). DBFZ Deutsches Biomasseforschungszentrum gemeinnützige GmbH, Leipzig, 2014. Nelles, M. (publisher). ISSN 2197-4632.

[29] IndexMundi, Jet Fuel Daily Price. Retrieved from: <http://www.indexmundi.com/commodities/?commodity = jet-fuel>, 2015 (26.11.15).

[30] BMVI (Ed.), Untersuchungen der wissenschaftlichen Begleitung Jahreskonferenz zur Mobilitäts- und Kraftstoffstrategie der Bundesregierung "Nachhaltige Konzepte für die Energiewende im Verkehr", 2015.

[31] Aireg, Flugkraftstoffe aus Sonnenenergie – Staatsminister Dr. Söder verleiht 1. aireg-Award an SOLAR-JET. Retrieved from: <http://www.aireg.de/de/pressemeldungen/410-flugkraftstoffe-aus-sonnenenergie-1-aireg-award.html>, 2015 (10.12.15).

[32] Forschungszentrum Jülich, Sunfire – BMBF Forschungsprojekt. Retrieved from: <http://www.fz-juelich.de/iek/iek-1/DE/Projekte/Proj_Beschreibungen/Sunfire.html>, 2015 (10.12.15).

[33] Ministry of Defence, DEF STAN 91-91 Defence Standard 91-91, Issue 7, Turbine Fuel, Kerosine Type, Jet A-1. Ministry of Defence, 2011.

[34] A. Zschocke, Biokerosin im Luftverkehr –ein Überblick. Presentation hold: Gemeinschaft ehemaliger Lufthanseaten. On 13-11-2015 in Frankfurt, 2015. < http://gelhfrankfurt.de/PDF%20DATEIEN/2015-11-13%20%20%C3%9Cberblick%20Biokerosin%20im%20Luftverkehr.pdf >.

[35] IATA, Overview of Alternative Jet Fuels in 2014 – Update of the Text that was Originally Published in the IATA 2014 Report on Alternative Fuels as a Contribution from ICAO Secretariat, 2014.

[36] AIREG (Ed.), Klimafreundlicher Fliegen: Zehn Prozent Alternative Flugkraftstoffe bis 2025. Aireg – Aviation Initiative for Renewable Energy in Germany e.V., 2012.

[37] AIREG e.V., http://www.aireg.de/en/production.html. Biofuel Production, Aviation Initiative for Renewable Energy in Germany e.V. Retrieved from: <http://www.aireg.de/en/production.html>, 2015 (26.11.15).

[38] L.A.B. Cortez, G.M. Souza, C.Hde B. Cruz, R. Maciel, An Assessment of Brazilian Government Initiatives and Policies for the promotion of biofuels through research, commercialization, and private investment support, in: S.S. da Silva, A.K. Chandel (Eds.), Biofuels in Brazil, Springer International Publishing, Switzerland, 2014, pp. 31–60. ISBN 978-3-319-05019-5.

[39] H. An, W.E. Wilhelm, S.W. Searcy, Biofuel and petroleum-based fuel supply chain research: a literature review, in: Biomass and Bioenergy, vol. 35 (no. 9), pp. 3763–3774, 2011, doi:10.1016/j.biombioe.2011.06.021, ISSN 0961-9534.

[40] G. Toop, M. Cuijpers, B. Borkent, M. Spöttle, Accounting Methods for Biojet Fuel Final Report - Final Report, ECOFYS Germany GmbH, Berlin, 2014.

[41] GAO, ALTERNATIVE JET FUELS – Federal Activities Support Development and Usage, but Long-term Commercial Viability Hinges on Market Factors. No. GAO-14-407, United States Government Accountability Office, 2014.

[42] A. Lang, H. Farouk, The Worldwide Production of Bio-Jet Fuels – the Current Developments Regarding

Technologies, Feedstocks, and Innovative New R&D Developments, 2014.

[43] K. Maniatis, M. Weitz, A. Zschocke, 2 Million tons per year: A Performing Biofuels Supply Chain for EU Aviation, The Biofuels FlightPath Initiative (2013 Update: Revision of the version initially published 2011), 2013.

[44] OECD/IEA, Technology Roadmap - Biofuels for Transport, International Energy Agency, Paris, 2011.

[45] S. Siddiqui, A. Christensen, Determining energy and climate market policy using multiobjective programs with equilibrium constraints, in: Energy. vol. 94, 2016, pp. 316–325. Available from: http://dx.doi.org/10.1016/j.energy.2015.11.002, ISSN 0360-5442.

[46] D. Thrän, M. Edel, T. Seidenberger, S. Gesemann, M. Rohde, Identifizierung strategischer Hemmnisse und Entwicklung von Lösungsansätzen zur Reduzierung der Nutzungskonkurrenzen beim weiteren Ausbau der energetischen Biomassenutzung. 1. Zwischenbericht für das deutsche Minisiterium für Umwelt, Naturschutz und Reaktorsicherheit, Deutsches Biomasseforschungszentrum, Leipzig, 2009.

[47] IPCC, Climate Change 2014: Mitigation of Climate Change. Contribution of Working Group III to the Fifth Assessment Report of the Intergovernmental Panel on Climate Change, 2014.

[48] A. Seely, Taxing aviation fuel, Research briefings – UK Parliament, Standard Note: SN00523, 2012.

[49] ICAP, Emissions Trading Worldwide – International Carbon Action Partnership (ICAP) Status Report 2015, 2015.

[50] R. Elsworth, P. MacDonald, L. Watson, Aviation and the EU ETS – What Happened in 2012 During "Stop the Clock"? Sandbag, London, 2013.

[51] EU RED, Directive 2009/28/EC of the European Parliament and of the Council of 23 April 2009 on the Promotion of the use of Energy from Renewable Sources and Amending and Subsequently Repealing Directives 2001/77/EC and 2003/30/EC, 2009.

[52] S. Alberici, M. Spöttle, G. Toop, Assessment of Sustainability Standards for Biojet Fuel – Final Report, ECOFYS Germany GmbH, Berlin, 2014.

[53] USDA, Agriculture and Aviation: Partners in Prosperity, U.S. Department of Agriculture (USDA), Washington D.C., 2012.

[54] USDOE, Farm to Fly 2.0: Energy Department Joins Initiative to Bring Biofuels to the Skies | Department of Energy. Retrieved from: <http://energy.gov/eere/bioenergy/articles/farm-fly-20-energy-department-joins-initiative-bring-biofuels-skies>, 2014 (07.12.15).

[55] USDA, News Release, United States Department of Agriculture. Retrieved from: <http://www.usda.gov/wps/portal/usda/usdahome?contentidonly = true&contentid = 2014/09/0195.xml>, 2014 (04.12.15).

[56] R. Schnepf, B. Yacobucci, Renewable Fuel Standard (RFS): Overview and Issues. Congressional Research Service (R40155), 2013.

[57] R. Janssen, D. Rutz, J. Michel, Sustainable Aviation Fuels Forum- Summary Core-JetFuel Day, 2014.

[58] IATA, IATA Fact Sheet: Alternative Fuels, 2015.

[59] Boeing, Embraer, FAPEPS, UNICAMP, Flightpath to Aviation - Biofuels in Brazil: Action Plan. An Initiative of Boeing/Embraer/ FAPESP and UNICAMP, 2013.

[60] T. Radich, The Flight Paths for Biojet Fuel. Working Paper Series. U.S. Energy Information Administration, 2015.

[61] IATA, IATA 2013 report on alternative fuels, eighth ed., International Air Transport Association. Montreal—Geneva, 2013, ISBN 978-92-9252-259-9.

[62] D. Thrän, U.R. Fritsche, Standards for Biobased Fuels and Resources – Status and Need. Presentation Hold: IEA Bioenergy Conference. On 27-10-2015 in Berlin, 2015.

[63] EC, Proposal for a Directive of the European Parliament and of the Council Amending Directive 98/70/EC Relating to the Quality of Petrol and Diesel Fuels and Amending Directive 2009/28/EC on the Promotion of the use of Energy from Renewable Sources. COM(2012) 595 Final, 2012.

[64] European Council, Preparing the Transition to Advanced Biofuels – Consilium. Retrieved from: <http://www.consilium.europa.eu/en/press/press-releases/2015/04/28-preparing-transition-advanced-biofuels/> 2015 (8.12.15).

[65] GBEP, The Global Bioenergy Partnership Sustainability Indicators for Bioenergy, Global Bioenergy Partnership (GBEP), Food and Agriculture Organization of the United Nations (FAO), Rome, 2011.

[66] IATA, IATA 2014 Report on Alternative Fuels, ninth ed., 2014, ISBN 978-92-9252-508-8.

[67] IATA, IATA Fuel Price Analysis. Retrieved from: <http://www.iata.org/publications/economics/fuel-monitor/Pages/price-analysis.aspx>, 2015 (26.11.15).

[68] EIAU.S. Kerosene Prices. U.S. Energy Information Administration. Retrieved from: <http://www.eia.gov/dnav/pet/hist/LeafHandler.ashx?n = pet&s = ema_eppk_pwg_nus_dpg&f = m>, 2015 (24.11.15).

[69] L. Iriarte, U.R. Fritsche, Benchmark and gap analysis of criteria and indicators (C&I) for legislation, regulations and voluntary schemes at international level and in selected EU Member States. S2Biom Deliverable 5.2, 2015.

[70] W. Reh, H. Breidenbach, M. Jäcker-Cüppers, M. Lege, A. Mahler, M. Müller-Görnert, et al., NGO-Luftverkehrskonzept. Schritte zu einem zukunftsfähigen und umweltverträglichen Luftverkehr in Deutschland, 2015.

CHAPTER

14

Sustainable Aviation: A UK Roadmap for Sustainable Aviation Fuels

L. Hudson[1], A. Jefferson[1], A. Bauen[2] and L. Nattrass[2]

[1]Sustainable Aviation Group, London, United Kingdom, [2]E4tech Management Consultancy, London, United Kingdom

14.1 INTRODUCTION

The UK's Sustainable Aviation (SA) group was launched in 2005 and brought together the main players from UK airlines, airports, manufacturers, and air navigation service providers. SA has established a long-term strategy that sets out the collective approach of UK aviation to tackling the challenge of ensuring a sustainable future for our industry. The aerospace and aviation sectors make a vital contribution to the UK economy, generating a turnover of more than £60 billion and contributing £52 billion (3.4%) to the UK's GDP [1]. This chapter is an overview of a detailed UK roadmap developed in 2014 by the SA group [2].

14.1.1 The UK Sustainable Aviation Carbon Dioxide Roadmap

In 2012, SA published its Carbon Dioxide (CO_2) Roadmap [3], setting out the UK's aerospace industry's combined view that UK aviation is able to accommodate significant growth up to 2050 without a substantial increase in absolute CO_2 emissions. We also support the reduction of net CO_2 emissions to 50% of 2005 levels through internationally agreed-upon carbon trading. This is shown in the chart below (Fig. 14.1).

SA estimated that by 2050 sustainable fuels will offer between 15% and 24% reduction in CO_2 emissions, attributable to UK aviation. This assumption was based on a 25–40% penetration of sustainable fuels into the global aviation fuel market, coupled with a 60% life-cycle CO_2 savings per litre of fossil fuel displaced. For the purposes of our Roadmap, SA assumed an 18% reduction in CO_2 emissions from UK aviation through the use of sustainable fuels.

In recent years, significant technical progress has been made towards the commercialization of sustainable alternatives to fossil fuel. Several demonstration flights have been undertaken, using 'drop-in' solutions that replicate the characteristics of fossil fuel. Many different fuel production and conversion processes are being developed. These have used an array of different feedstocks, including

Biofuels for Aviation.
DOI: http://dx.doi.org/10.1016/B978-0-12-804568-8.00014-7

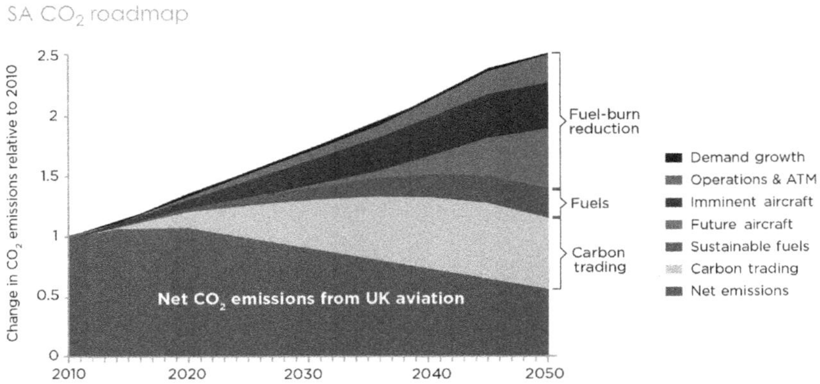

FIGURE 14.1 The sustainable aviation CO_2 roadmap (2012).

municipal waste and waste gases from heavy industry, algae, and sustainable crops (ie, those grown on nonagricultural land and crop residues). The ASTM International Fuels Standards Committee has now approved three production processes that can generate sustainable fuels, and airlines have flown revenue flights (ie, carrying fare-paying passengers) using sustainable fuels blended with fossil fuel. Thus technically feasible solutions already exist. However, at the moment, the market is at a nascent stage of development, and these fuels are produced in small volumes at relatively high cost, and are therefore are not yet commercially competitive. The ultimate goal is to develop commercial, sustainable, 'drop-in' fuel solutions to supply an increasingly significant proportion of the fuel at airports in the United Kingdom, and at airports with flights to the United Kingdom.

Since publishing its CO_2 Roadmap in 2012, SA has continued to review its earlier assumptions about the role of sustainable fuels in achieving an 18% reduction in total CO_2 emissions from UK departing flights by 2050. In order to develop a more detailed assessment of the contribution from sustainable fuels and their pathway towards commercialization, SA commissioned the sustainable fuels consultancy E4tech to conduct a more detailed study. The output of this work has been used to outline the economic and policy conditions that would maximize this potential.

Developing this roadmap presented some challenges as it is inherently difficult to forecast the size and shape of a new supply market. In the case of sustainable fuels, the industry is very much in its infancy. The International Civil Aviation Organization (ICAO) is presently working at a global level to model scenarios setting out the global potential of sustainable fuels. Inevitably, the technical potential for this new industry will be realized only if governments support the development of sustainable fuel supply chains.

The UK's SA group is committed to working collaboratively to accelerate the development and commercialization of advanced sustainable fuels. To create this new industrial sector, significant government support for innovation and investment in a new bioeconomy will be required. The UK government has identified this potential within their report 'Building a High Value Bioeconomy (2014)' [4].

14.1.2 Priorities for the Roadmap

The roadmap identifies the opportunities that exist to move sustainable fuels forward, and offers proposals on the roles that industry and government can play in realizing this opportunity. The aerospace industry already has the experience of successful collaborative initiatives with the government to grow and promote high-value technologies and jobs – SA firmly believes that this model would form a good foundation to develop a similar approach for sustainable fuels.

The specific objectives of this roadmap were the following:

1. To identify and forecast global and UK potential sustainable fuel production volumes out to 2050 and to relate these to the assumptions made in the 2012 SA CO_2 Roadmap;
2. To show the extent to which UK aviation alternative fuels can contribute to the decarbonization of the UK economy and enhance UK fuel security;
3. To demonstrate a viable market potential for advanced sustainable fuels to producers, refiners, investors, and other stakeholders;
4. To highlight the potential for this new industry in terms of job creation, growth, and improved fuel security in the United Kingdom;
5. To secure government engagement and support for sustainable fuels for aviation.

14.2 SUSTAINABILITY

14.2.1 Definition of Sustainable Fuels

Over the last eight years, concerns over the sustainability of certain biofuel production pathways have grown, with an increasing body of evidence that some biofuels can lead to higher greenhouse gas (GHG) emissions than the fossil fuels they are replacing. In addition, impact on biodiversity and land use – as well as other socioeconomic impacts – have led to concerns about increasing the deployment of first-generation fuels. One of the key priorities for the roadmap was to ensure that the sustainable fuels used in the modelling exercise would not lead to unintended negative consequences for environmental systems.

SA members are committed to the development of sustainable fuels that offer significantly reduced life cycle GHG emissions over fossil fuels. These fuels ideally possess the following characteristics:

- Meet stringent sustainability standards with respect to land, water, and energy use;
- Avoid Direct and Indirect Land Use Change (ILUC) impacts (eg, tropical deforestation);
- Do not displace or compete with food crops;
- Provide a positive socioeconomic impact;
- Exhibit minimal impact on biodiversity.

14.2.2 Indirect Impacts

Direct impacts can generally be measured and attributed to the fuel's production process within a GHG life cycle analysis (LCA). However, there is also concern that fuel production can have negative, unintended consequences, with the most cited examples being ILUC, in which the production of newly demanded products on existing croplands displaces other agricultural activities to previously nonproductive land. ILUC displaces other agricultural activity to other high carbon stock land (eg, tropical forests), and is therefore a significant concern.

Members of the aviation industry have been heavily engaged in discussions with nongovernmental organizations (NGOs) and policymakers to ensure that the sustainable fuels they are considering using do not generate risks of negative ILUC impacts. One major way of mitigating the risk of ILUC is through the use of feedstocks based on waste materials and residues. Other ways in which the risk of ILUC impacts can be mitigated are the following:

- Production of feedstock on unused land, that is, land that is not currently used to provide provisioning services;

- Increasing feedstock availability for sustainable fuels without increasing the pressure on land use through increased yield or land productivity;
- Feedstock production on underused land – that is, land that falls between the above two categories;
- Increased feedstock availability through reduction in post-harvest waste;
- Integrating food and fuel production in ways that lead to higher overall land productivity;
- Using feedstocks that require minimal land, such as algae.

Sometimes fuels are referred to in terms of the generation of the technology with which they are associated (1st, 2nd, 3rd, etc.), or as 'Advanced' compared to early types of feedstock and processes. This is a shorthand way of distinguishing sustainable from nonsustainable fuels. Such classifications can sometimes be overly simplistic, and do not provide a consistent indication of sustainability.

There is a considerable body of research, including the 2008 Gallagher Review [5], which has highlighted the potential indirect effects of directing biomass resources to energy and fuel applications. Some of these impacts are not captured within existing LCA methods, and this has been a concern for European Union (EU) policymakers. The widening use of food for fuel has become controversial as biofuels have been linked to higher food prices and consequently incentivizing the increased conversion of land to agricultural use. For these reasons, SA decided on a conservative approach to the types of feedstocks that would be used in future fuel production.

For the purpose of this roadmap, SA has chosen to focus on those fuels that are produced from wastes, residues, and nonfood crops grown on degraded lands. This is referred to more fully in Section 14.4.1.

14.2.3 Sustainability Standards and Certification

Sustainability standards have been developed to provide reassurances that supply chains are robust and provide genuine environmental and climate benefits. Increasingly, they also address socioeconomic aspects of fuel production. Some of these standards are national or regional standards that are embodied in legislation. Others are voluntary standards developed by trade-based bodies or NGOs.

SA members are actively supportive of the Roundtable on Sustainable Biomaterials (RSB), widely recognized as the most robust global sustainability standard. The RSB is an international, multistakeholder standard organization that has developed a feedstock-and-technology-neutral global standard for sustainability [6].

The possible extent of negative ILUC and the resulting GHG emissions that may occur as a result of additional demands for a variety of crops in different regions is not fully understood. Considerable work has been done to identify policies and practices that mitigate the risk of causing negative ILUC, notably by WWF International, Ecofys, and EPFL [7]. One way to mitigate the risk of ILUC is through the use of feedstock based on waste materials and residues for fuel production. The RSB has also introduced a new sustainability standard based on these principles [8].

SA members have been focused on understanding best practices in this area, and are working with ICAO, the International Air Transport Association, Ecofys, and NGOs to develop globally harmonized sustainability criteria. Ecofys [9] SA believes that careful feedstock selection that leads to minimal land use and ILUC impacts, coupled with the use of a robust independent standard which is independently audited, provides the highest level of assurance.

The UK government was at the forefront of development of sustainability standards within in the EU, and the aviation sector and governments should work together to ensure that sustainability of fuel production globally is prioritized as a complement to the development of a global climate change policy for aviation.

14.3 DEVELOPING SUSTAINABLE FUELS

Developing sustainable fuels requires a number of processes to come together. This is depicted in the diagram below (Fig. 14.2).

14.3.1 The Need for 'Drop-In' Fuels?

There has been great progress in the development of sustainable alternatives to fossil fuel. These new fuels are often referred to as 'drop-in' fuels as they have a similar chemical composition to fossil fuels and can be completely blended with existing fuels. New fuels that meet the performance specifications (referred to as certified fuels) can be delivered into bulk storage and delivery systems at airports, and can be used alongside fossil fuel.

The fact that the fuels are 'drop-in' means that no modifications have to be made to existing aircraft and engines, or to airport infrastructure. This innovation has led to the

FIGURE 14.2 Elements required to deliver sustainable fuels.

creation of a new generation of fuels that are able to deliver emissions benefits alongside sustainability benefits.

Both aircraft and engine manufacturers are working on new, innovative propulsion systems (eg, Boeing and Airbus) [10,11]. These include electric hybrid systems, full electric propulsion systems, and the use of solar power. All these have the long-term potential to reduce and possibly eliminate the need for liquid hydrocarbon fuels, however, these are not likely to reach commercial production within the next thirty years. The timeframe of the roadmap (ie, up to 2050) will focus exclusively on liquid hydrocarbon fuels that can be manufactured from renewable feedstock.

14.3.2 Feedstock Availability and Sustainability

Aviation fuels are produced by taking a feedstock and converting it via industrial processes into a fuel. A range of different potential feedstocks have been identified. Some are based on oil crops such as algae and other nonfood feedstocks, while others are based on waste sources such as municipal solid waste, used cooking oil, and waste industrial gases. The sophisticated processing technologies necessary for the manufacturing of aviation fuels will widen the number of available feedstock types; many of these are low-grade, low-value materials.

The European Climate Framework and the International Council on Clean Transportation (2014) [12] published a report on the availability of cellulosic wastes and residues in the EU that identified a potential of 900 million tonnes of available waste material in the EU. Of this, it is estimated that 220–230 million tonnes can sustainably be recovered for energy production. The conversion of this into sustainable fuels could generate 36.7 million tonnes of biofuel for all transport modes.

There have been a number of reports over recent years outlining both feedstock availability and the significant potential for the development of conversion technologies in the United Kingdom. Low Carbon Innovation Coordination Group [13], House of Commons, Science and Technology Select Committee (2014) [14].

14.3.3 EU Legislation

The EU has a target to obtain 10% of its transport fuels from renewable sources by 2020. To achieve this goal, the EU Renewable Energy Directive (RED) established renewable energy mandates and financial incentives for eligible fuels. The RED specified a number of sustainability criteria that fuels must meet in order to be deemed 'sustainable' within the EU:

- Minimum level of GHG emissions savings of 35% compared to fossil fuels, rising to 50% from 1 Jan. 2017, and 60% from 1 Jan. 2018 for new plants that begin production beginning in 2017;
- Areas of high carbon stock (eg, wetland, forest, and peatland) should not be used for fuel production;
- Land with high biodiversity should not be used for biofuel production.

The EU (2015) [15] also implemented changes to existing legislation to address concerns about ILUC impact:

- Limit the volume of crop-based biofuels counting towards the RED target to 7% by volume;
- Introduce an introductory target for advanced biofuels for introduction by member states into national targets in 2017;
- Harmonize the feedstock types that qualify for double counting within the RED;
- Introduce a GHG savings of 60% minimum for new installations.

These changes could provide greater potential for the production of aviation fuels as the restriction on crop-based fuels and a sub-target for advanced fuels are likely to lead to a modest boost in production in the period up to 2020.

14.3.4 The UK's Legislative Framework

The EU RED is then transposed into national legislation by member states. In the United Kingdom the RED is implemented through the Renewable Transport Fuel Obligation (RTFO). Under the RTFO, those supplying biofuels must meet specified sustainability criteria in order for their fuels to be recognized as being entitled to the benefit of Renewable Transport Fuel Certificates (RTFCs). Obliged fuel suppliers are required to redeem a number of RTFCs in proportion to the volume of unsustainable fuel (eg, fossil fuels) they supply. RTFCs may be earned by any company supplying sustainable fuels. They may be bought or sold on an open market. Obligated suppliers also have the option to 'buy out' their obligation, paying a fixed fee per litre of sustainable fuel that would otherwise have to have been supplied to earn RTFCs. Fuels from wastes and residues (and lignocellulosic and nonfood cellulosic feedstocks) earn double credits. These fuels receive twice as many RTFCs per litre. For all fuels, sustainability data supplied must be independently verified by a qualified third party.

A number of different voluntary standards exist that relate to biofuels. The most robust of these have been developed through multistakeholder frameworks and follow the ISEAL Principles. The ISEAL Alliance is an NGO whose mission is to promote best practices in standards organizations and certification systems. Membership is open to all multistakeholder sustainability standards and accreditation bodies that demonstrate their ability to meet the ISEAL Codes of Good Practice and accompanying requirements, and who commit to learning and improving.

The EU has formally recognized a number of voluntary standards approved to demonstrate compliance with the EU RED sustainability requirements. These include the RSB standard.

14.4 SUSTAINABLE FUEL SUPPLY POTENTIAL

Rapid progress in the technological development of sustainable fuels for aviation has been made since the first sustainable fuel gained certification for commercial aviation in 2009. SA appointed E4tech to assess the potential role of sustainable fuels in achieving carbon reductions from UK departing flights[1] in the period up to 2030, and to estimate the role played by these fuels from 2030 to 2050.

14.4.1 Scope

In this study, E4tech provided an estimate of how much sustainable fuel could be produced globally and in the United Kingdom up to 2030, and the CO_2 emissions savings that could be achieved through the use of sustainable fuels. From this, E4tech also estimated the volume of sustainable fuel required to meet the SA projections for 2050, and discussed the viability of these aims.

SA defines sustainable fuels as those fuels that are produced from wastes (including the biogenic and nonbiogenic fractions of postconsumer and industrial wastes), residues, and nonfood crops from degraded land (avoiding food versus fuel conflicts and the impact of ILUC). This report therefore considered fuels fitting this definition, and also additional sustainability criteria.

[1]This is consistent with UK government accounting principles and is the approach taken in previous SA roadmaps.

Due to the fuel supply infrastructure and fuel quality requirements, sustainable fuels are blended with fossil fuel. As a result, this report includes only sustainable fuels that have been, or are expected to be, approved as fossil fuel blending components by the relevant standardization organizations.

The scope of the SA CO_2 Roadmap includes all commercial flights departing from UK airports, and therefore the potential UK supply of sustainable fuels is presented relative to the projected fuel demand for all those flights. At present, the majority of fuel used in flights departing from UK airports is dispensed within the United Kingdom, and fossil fuel is imported to meet current UK demands. But in the case of sustainable fuels, the industry believes that in the near term they will be used near the point of production, at least while volumes remain relatively low and prices high. In the longer term, sustainable fuels may become commoditized and traded internationally as supply and demand dictates.

The approach focused on estimating global and UK production of sustainable fuels up to 2030, and the potential CO_2 emissions savings that may be realized through their use. If the scale of ambition of the UK aviation industry to reduce CO_2 emissions remains high, and appropriate support mechanisms are in place, it is possible that UK aviation could access greater volumes of sustainable fuels to those produced in the United Kingdom, uplifting sustainable fuels at other airports, or importing sustainable fuels into the United Kingdom.

14.4.2 Approach

The approach taken assumed that the main constraint to the supply of sustainable fuels up to 2030 would be limited production capacity, due to the low levels of existing capacity and the early stage of development of the majority of technologies.

2020: Sustainable fuel potential is equal to estimated UK and global production capacity, based on an inventory of operational and planned plants (see Table 14.1). This table reflects the state of play in the latter half of 2014.

2030: Assumed sustainable fuels are produced by processes that have been proven at pilot scale as a minimum. Production capacity is estimated based on scenario analysis, which defines plant build rates and production capacity growth between 2020 and 2030. The scenarios are defined based on the economic capacity for increased production capacity across the sustainable fuels sector and the scale of ambition for sustainable fuels in the aviation sector. We used the following three scenarios for sustainable fuel availability up to 2030:

- *Business as Usual (BAU):* This scenario describes a situation in which continued policy uncertainty leads to low growth in sustainable fuel production capacity for all transport modes.
- *Low:* This scenario describes a future in which strong growth in the production of sustainable fuels for road transport results in availability of fuels for aviation, although aviation continues to consume only a small proportion of fuel production. This scenario may emerge from a policy framework that continues to promote the use of sustainable fuels in road transport, but where the implementation of policy to support the use of sustainable fuels in aviation is largely delayed beyond 2030.[2]
- *High:* This scenario describes an ambitious trajectory where low financial constraints

[2]This analysis found constrained growth in the production of sustainable fuels and strong demand from the aviation sector lead to a similar result.

TABLE 14.1 Summary of Operational and Planned Plants

Process	Producer	Location	Status	TPA
HEFA	Neste	Finland	Operational	38,000
				38,000
		Singapore	Operational	160,000
		Netherlands	Operational	160,000
	UOP	United States	Operational	2000
	ENI	Italy	Operational	62,000
	Dynamic fuels	United States	Operational	47,000
	Darling Int. and valero	United States	Under construction	90,000
	Alt air	United States	Under construction	56,000
	Emerald biofuels	United States		53,000
				94,000
F-T	Haldor topsoe	United States	Operational	350
	Syntroleum	China	Operational	1400
	TRI	Canada	Operational	7000
		United States		7000
				13,000
	UPM	Finland	Under construction	20,000
		France		37,000
	Red rock biofuels	United States		17,000
	Solena	United Kingdom		50,000
		Italy		50,000
		Australia		50,000
		Ireland		50,000
		India		150,000
		Turkey		50,000
		Germany		50,000
		Sweden		50,000
	Forest BTL	Finland		29,000
ATJ	Terrabon	United States	Operational	80
				800
	Fulcrum biofuels		Operational	Demonstration

(Continued)

TABLE 14.1 (Continued)

Process	Producer	Location	Status	TPA
		United States		5000
	Swedish biofuels	Europe		5000
	LanzaTech	China		40,000
SIP	Virent	United States	Operational	<10
	Solazymes	Galva, Iowa, United States	Operational	50,000
	Amyris	Brazil	Operational	40,000
				40,000
	LS9	United States	Operational	700
				16,000
		Brazil		120,000
HDCJ	Allenotech	United States	Operational	Pilot
	KiOR	United States	Operational	7800
				23,000
	UOP	Hawaii	Under construction	30
	BTG	Netherlands		3500
Other	Licella	Australia	Operational	Pilot
				8500
	Blue sun	United States		2000
	Biochemtex	Europe		2000

result in strong growth in the production capacity for sustainable fuels and strong market pull from the aviation sector due to, for example, strong support policies. It results in a high proportion of new capacity producing sustainable fuels.

2050: Estimated the volume of sustainable fuel required to achieve GHG emissions savings of between 18% and 24%[3] in the United Kingdom, and quantified the annual growth rates required to achieve this aim.

UK aviation is not alone in its ambition to achieve carbon-neutral growth, and its interest in using sustainable fuels in order to achieve these ambitions. We have therefore estimated global demand for sustainable fuels based on equal international GHG emissions savings targets.

Under this approach, the UK's use of sustainable fuel is assumed to be proportionate to its share of the commercial aviation market. UK aviation could, however, play a greater role in stimulating the demand for sustainable fuels.

[3]Corresponding to the central and high projections of the SA CO_2 Roadmap.

14.4.3 Assumptions

The sustainable fuel potential is based on an inventory of operational and planned plants that could produce 'drop-in' fuels suitable for aviation, based on publically available information on pilot, demonstration, and commercial plants. The technology pathways are presented in other chapters of this book, and are therefore not replicated here. Projected capacity is given in Table 14.1.

The inventory provides an estimate of the pipeline up to 2020. However, there is uncertainty over how many plants will be successfully established due to a wide range of technical and market barriers. Individual projects have therefore been assigned a realization potential factor based on the stage of development.[4]

Total production capacity in 2020 is calculated as a function of reported plant size, potential aviation fuel fraction, and realization potential. Where the volume of fuel is not stated, we have assumed the fossil fuel fraction is equal to the low boundary of the ranges presented in Table 14.2. For the United Kingdom, 2020 production capacity is based on planned plants, and a higher estimate is based on the UK share of the commercial aviation market. In 2020, the UK share of commercial aviation is estimated by SA on the basis of Revenue Passenger Kilometres at 4.8%.

The global production capacity increase between 2020 and 2030 is calculated based on the lower and upper build rates, average plant size, and aviation fuel fraction presented in Table 14.2. Under the BAU scenario, production capacity increase is calculated as the product of the lower build rate, average plant size, and lower aviation fuel fraction, summed for all technology groups. Under the low scenario, production capacity increase is calculated as the product of upper build rate, average plant size, and lower aviation fuel fraction for technologies; and under the high scenario, production capacity increase is calculated as the product of the upper build rate, average plant size, and upper aviation fuel fraction for all technologies. The plant build rate assumptions are based on the current technology status, historical build rates, and the number of technology providers [16].

Between 2020 and 2030, UK production capacity is estimated to grow in line with global production capacity, at average annual growth rates of between 8% and 26%. Due to there being a small number of plants, we have added an integer analysis to ensure total volumes represent full-scale plants.

The GHG emissions for each fuel type have been estimated based on reported literature values for a representative range of feedstocks. We have assumed annual improvements in life cycle GHG emissions of 1% per year for established technologies (hydroprocessed esters and fatty acids, HEFA; and Fischer–Tropsch, F-T), and 2% per annum for technologies at an earlier stage of development which may achieve greater efficiency improvements (SIP; alcohol to jet, ATJ; and hydrotreated depolymerized cellulosic jet, HDCJ). Emission savings are calculated based on a GHG emissions factor for fossil fuel of 3.15 tCO_2eq/t.

14.4.4 Results

The potential global supply of sustainable fuels for aviation is estimated to be 1.3 million

[4]Operational plants and those under construction allocated a realization factor of 1, and 0.75 is allocated to projects at the advanced stages of development, which may be indicated by site identification, planning and permits, and/or off-take agreements. Projects at earlier stages of development are at 0.5, and announcements that indicate an aim or interest in developing plants are at 0.25.

TABLE 14.2 Assumptions for Global Production Capacity Growth for Each Technology Group (2030)

	Build Rate		Average Plant Size	Sustainable Fuel Fraction	
	Lower	Upper	TPA	Lower (%)	Upper (%)
HEFA	8	16	280,000	20	60
F-T	12	28	124,000	20	50
SIP	36	67	40,000–115,000[a]	20	60
ATJ	12	52	55,000–155,000[b]	50	50
HDCJ	6	14	120,000	20	60

[a]*Synthesized isoparaffin (SIP) represents a group of technologies expected to operate at different scales. We have therefore applied different assumptions on plant size under the BAU and low/high scenarios.*
[b]*For ethanol plant sizes, the range represents different plant locations.*

tonnes in 2020. This includes between 50,000 and 60,000 tonnes of UK production, and represents approximately 0.45% of total aviation fuel demand.

The prospects for the United Kingdom are largely dependent on the success of the planned British Airways (BA) waste to fuels plant in London. In addition, some other technologies, such as the Virgin Atlantic/LanzaTech plans to process waste gases from steel production, could also be implemented in the United Kingdom before 2020.

The main focus of this study is the estimation of the potential supply of sustainable fuels up to 2030. This is estimated based on scenarios for growth that consider these two factors:

- *Economic capacity:* This is the capacity of the sustainable fuels industry to invest in new production capacity and to commercialize technologies at early stages of development. It determines the potential growth in production of sustainable fuels.
- *Scale of the ambition for sustainable aviation:* This is the market pull for sustainable fuel, which includes the scale of ambition of the industry, political focus, and regulatory support for the use of sustainable fuels in aviation. This determines the volume of sustainable fuel allocated to aviation.

The purpose of the scenario analysis is to determine the range of possible futures in 2030 rather than to focus on the potential for specific technologies or products. For this reason, growth has been modelled based on five categories of sustainable fuels that may be suitable for aviation, which represent the wide range of technologies currently under development.

It is estimated that in 2030, between 3 and 13 million tonnes of sustainable fuels may be produced globally (3.6–16.3 billion litres). This is equivalent to 0.7–3.3% of global aviation fuel use, and GHG emissions savings of between 8 and 35 million tonnes of CO_2 eq. In comparison, the International Energy Agency (IEA) [17] estimates that global biofuel production increased from 16 billion litres in 2000 to more than 100 billion litres in 2011, based on commercially mature technologies, to provide 3% of total road transport fuel globally (on an energy basis).

In the United Kingdom, we estimate that sustainable fuel production could reach between 100,000 and 640,000 tonnes per annum (TPA) in 2030. This translates to between 5 and 12 sustainable fuel plants in the United Kingdom, producing between 20% and 60% of aviation fuel in combination with road transport fuels (Fig. 14.3).

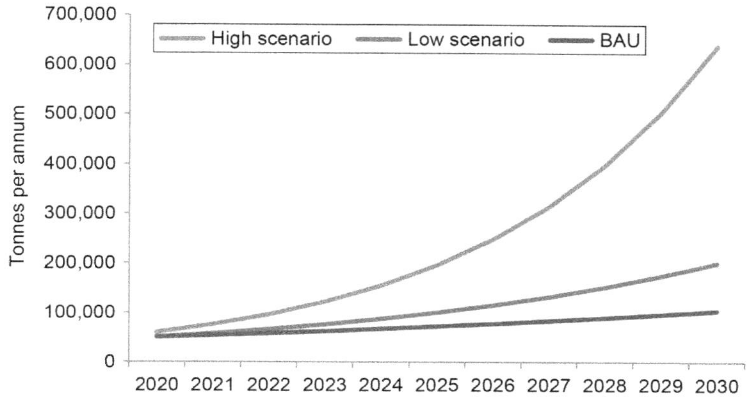

FIGURE 14.3 Estimated UK sustainable fuels production up to 2030 (tonnes per annum).

14.4.5 Trajectory to 2050 – Sustained Growth

In 2012, SA estimated that by 2050 sustainable fuels should offer between 15% and 24% reduction in the CO_2 emissions attributable to UK aviation, with a central scenario of 18% CO_2 emissions reduction. This was based on the assumption that sustainable fuels contribute between 25% and 40% of the aviation fuel market, and achieve life cycle CO_2 emissions savings of 60% compared to fossil fuel.

Most sustainable fuels may actually achieve GHG emissions savings of between 80% and 95%, and we estimate that *the volume of sustainable fuel required to meet the SA GHG aims in the United Kingdom will be between 3.3 and 4.5 million tonnes per year in 2050. International commitment to these targets would require between 140 and 190 million tonnes of sustainable fuel in 2050 (equivalent to 17–28% of total fuel consumption).*

To achieve the high 24% GHG emissions savings target (based on the high scenario for production up to 2030) would require a sustained annual growth rate of around 14% per year between 2030 and 2050. To achieve the central aim of 18% GHG emissions savings from the low scenario for production up to 2030 would require a sustained annual growth rate of 18% between 2030 and 2050.

These annual growth rates are not dissimilar to historic growth rates in global biofuel production. For example, between 2001 and 2011, biofuel production grew from 16 billion litres to 100 billion litres, corresponding to an average annual growth rate of 20%. However, to achieve this level of growth, the production and use of sustainable fuel must be mainstream by 2030, having overcome both technical and market entry barriers. For comparison, the IEA BLUE Map Scenario (2011) [17] estimates that global demand for sustainable fuels will reach 760 million tonnes per annum by 2050, including almost 200 million tonnes for aviation (Fig. 14.4).

14.5 THE UK VALUE IN DEVELOPING SUSTAINABLE FUELS

As the sustainable fuels industry grows, there is potential for UK deployment to deliver revenues and support jobs in the United Kingdom. Additional economic value could be realized by UK-based businesses competing successfully in non-UK markets. This section provides an estimate of the value to the UK of keeping pace with international deployment of sustainable fuels to 2030.

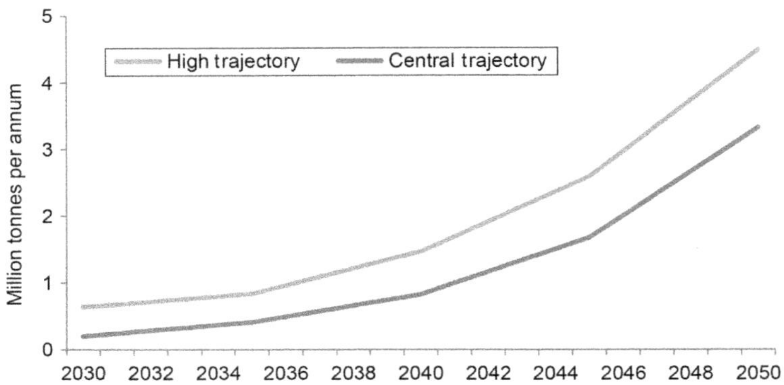

FIGURE 14.4 Trajectory for UK sustainable fuel use up to 2050 (million tonnes per year).

Scenario analyses estimate that in 2030 there may be 90–160 operational sustainable fuel plants globally, producing aviation fuels in combination with fuels for road transport and other modes. Global revenue for these sustainable fuel plants is estimated to be £8–17 billion in 2030.

There may be 5–12 sustainable fuel production plants producing 20–60% sustainable fuels in combination with road transport fuels in the United Kingdom by 2030. The potential value of the sustainable fuels industry to the United Kingdom is based on a methodology developed in the context of the UK Department of Energy and Climate Change's Technology Innovation Needs Assessment (TINA) [13]. This assessment uses global deployment figures to estimate the potential gross value added (GVA) contribution to the UK economy for technology areas.

Turnover figures are calculated from global deployment scenarios and expected technology costs. These are then converted to GVA figures based on known turnover-to-GVA ratios for similar industries (eg, basic manufacturing, agriculture, high tech services, etc.), which range from 10% to 65%. Then the proportion of the global market that might be accessible (or 'tradable') to UK-based companies is estimated. Finally, the proportion of the globally accessible market that the United Kingdom can capture is estimated based on its competitive advantage. The UK's competitive advantage is graded from low to high, which is used to estimate a percentage of the available market that the United Kingdom could potentially be expected to take.

The figures have been adjusted to take into account an additional displacement effect because part of the value created in the export market will be due to a shift of resources and thus partly cancelled out by loss of value in other sectors. Expert opinion has roughly assessed this effect to be between 25% and 75%, so we have applied a flat 50%, including this displacement.

Development of a domestic industry for the production of sustainable fuels could generate a GVA of £70–265 million in 2030. To realize these scenarios, the capital investment in production plant capacity required between 2020 and 2030 is estimated to be £0.5–1.3 billion (equivalent to £50–130 million per year).

The successful development of a domestic industry is estimated to support 900–3400 jobs in plant construction and operation, feedstock supply and fuel distribution, and the design and development of conversion technology components and processing plants.

The successful capture of global sustainable fuel opportunities could generate additional value for the United Kingdom. The value of global exports is estimated at £100–220 million in 2030 with the support of 500–1000 jobs. The majority of the value and jobs are in the

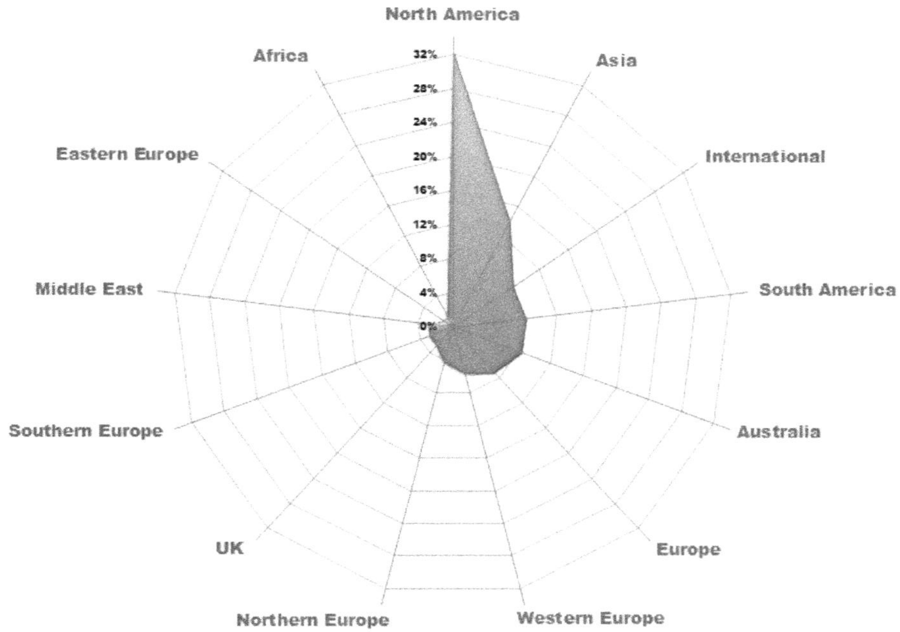

FIGURE 14.5 Global sustainable fuel project announcements between 2006 and 2013.

supply of technology components and engineering services related to the design and development of conversion technology components and plants – these are exportable, protectable through IP, and well aligned with the UK's commercial strengths (Fig. 14.5).

14.6 OVERCOMING DEVELOPMENT BARRIERS

This analysis has estimated the potential contribution that sustainable fuels may make to the aviation industry. The E4tech study outlines the importance of the period up to 2030 in establishing an industry for sustainable transport fuels. Realization of the potential outlined is dependent on overcoming barriers that currently constrain the supply of sustainable fuels and the market demand or pull.

Both the central and high scenarios presented in Section 14.5 require a combination of conversion technologies, which utilize a range of feedstocks. These conversion technologies are at different stages of development, and therefore the roadmap must recognize the need to overcome demand-side barriers – which currently limit the volumes of sustainable fuel used in aviation – and supply side barriers that will continue to influence the production capacity and availability. This chapter outlines the most significant barriers to the increased production and use of sustainable fuels, and identifies the most pressing actions for the UK government and industrial stakeholders to take to enable increased uptake of sustainable fuels.

The emissions savings and economic benefits of sustainable fuels can only be realized if steps are taken to actively promote and support their development. This includes coherent policy, a level playing field, incentives for investment, prioritized research and development (R&D), and support for commercialization of projects.

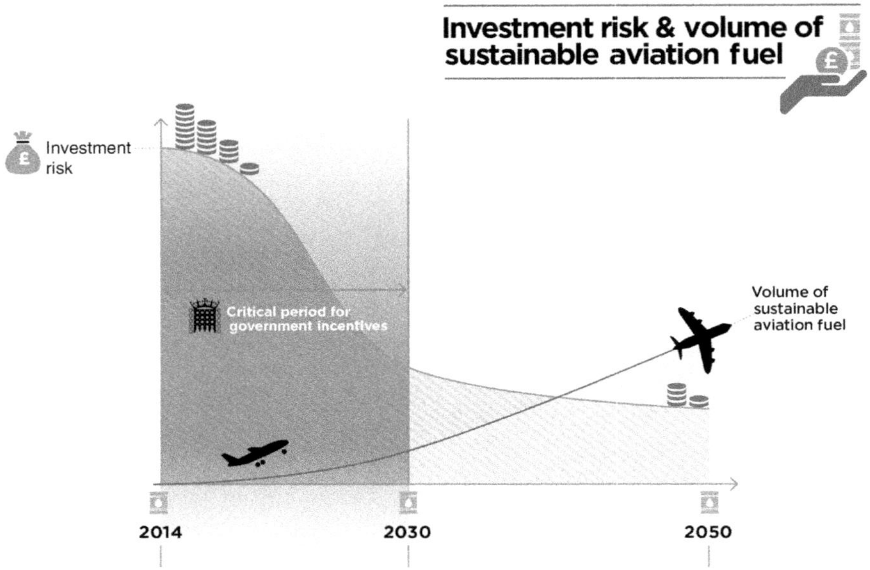

FIGURE 14.6 Enabling sustainable fuels production by de-risking investment.

14.6.1 Economic Barriers

As with all emerging technologies, there are significant initial barriers to commercialization. In the near term, sustainable fuel pathways are not economical in their own right, and carry with them material first-of-a-kind technological risk for investors and developers. Support is needed to reduce investor risk in bringing the technology to commercial scale, to stimulate market demand in the medium-term while production costs are high, and to level the playing field with fuel incentives in other sectors. Over the long-term, support can be scaled back as the fuels must become competitive without intervention.

The cost of currently available sustainable fuels are reported to be considerably higher than fossil fuel; for example, HEFA, the only fuel currently produced on a commercial scale today is reportedly 2.5 times the cost of fossil fuel. There is potential for products at an earlier stage of development to reduce production costs as production is scaled up and optimized (especially those derived from wastes/residues). In the immediate-term, appropriate support mechanisms are required to enable process scale-up and optimization, and to reduce production costs (Fig. 14.6).

Fuel costs account for 30–50% of the operating costs for the aviation industry, and the ability of airlines to absorb an increase in fuel price is limited. Therefore, to remain an effective option, sustainable fuels must become economically viable and cost-competitive over the long-term compared to fossil fuel plus carbon costs in the short-term. However before scale-up occurs, a price disparity may exist, and additional short-term support is required to ensure that continued development of these critical technologies is pursued.

14.6.2 Policy Barriers

There are a number of noneconomic barriers relating to factors that either prevent deployment altogether or lead to higher costs than necessary (or distorted prices). These barriers may be related to a lack of, or distortions in, policy support. They can also relate to a lack of expertise in developing new technologies or a lack of public acceptance. The main barriers affecting the deployment of sustainable fuels are described here.

The strength of the market pull is of immediate importance as a number of conversion technologies are looking to progress from demonstration- to commercial-scale production. In the United Kingdom and across Europe, policy mechanisms generally drive use of sustainable fuels in the road transport sector, as mandates and incentive mechanisms create the market for investment. Immediate action is needed to create a level playing field between aviation and other markets for sustainable fuels.

In part as a result of this inequality in policy support, aviation appears to be a low priority for UK policy makers, despite recognition at a strategic level of the long-term need for sustainable fuels in aviation. Currently there is no specific focus in the Department for Transport (DfT) on the role of sustainable fuels for the aviation sector. This situation creates a perception that may undermine the commitments made by the aviation industry, and lead to missed opportunities to reduce GHG emissions in transport. The Bioenergy Review published by the Climate Change Committee [18] pointed to the need to prioritize the use of bioenergy for aviation in the period up to 2050.

Present policy uncertainty throughout the EU and the United Kingdom is damaging investor confidence. Without adequate targets for advanced transport fuels up to 2030, investments in domestic production capacity will cease. Long-term policy certainty and a focus on instruments that de-risk investment are needed. Long-term goals for sustainable fuels, together with an incentive framework to stimulate investment, send an important signal to the investment community and to project developers. The United Kingdom has an opportunity to address this in 2016/2017, as the amendments required by the recent ILUC directive changes need to be transposed into UK legislation.

14.6.3 Financing Barriers

The supply of sustainable fuels is limited by delays in the progression of conversion technologies from demonstration to commercial production. The main barriers to development relate to project financing and risk.

Global factors have reduced the availability of various forms of project financing over the last 5 years. The UK banking sector is currently averse to providing debt financing to high-risk projects as the criteria and terms have typically shortened, and as a result, payback hurdle rates cannot be achieved by the majority of sustainable fuel projects. Venture capital investment has also declined in Europe, as the United States is currently seen as a more attractive market for investors.

The risks associated with sustainable fuel projects inherently affect the availability of project financing and the associated terms. One of the main issues impacting project risk is the price and competing uses of feedstocks, which can represent 50–90% of production costs; project returns are therefore highly sensitive to feedstock price variations. Competition for feedstocks with energy and nonenergy sectors is likely to grow, and while high demand will stimulate increased production of sustainable feedstocks, this variability adds to project risk.

14.6.4 Technical Barriers

A technology push will also be important if the UK economy is to take advantage of the demand for new technologies and intellectual property. For new technologies, the scale-up of production volumes and reference plant data (obtained by demonstration at scale) is necessary to reduce the risk of commercial-scale development. However, securing the investment required to support these stages of development is a major barrier. The EU[5] has previously supported the development of first-of-a-kind biofuel plants through FP7, NER300, and EIBI, and will continue to do so through Horizon 2020. Experience from the NER300 call has demonstrated that successful funding applicants are still undergoing significant challenges, including a lack of fuel off-take agreements. In addition, the terms associated with Horizon 2020 may limit the suitability of the scheme to leverage the level of investment needed.

Fuel qualification and certification is necessary to ensure continued high levels of safety in the aviation industry. This demands high levels of investment from both airframe and engine manufacturers. To date, engine and airframe manufacturers have responded strongly to the requirement to test and approve new fuels. As a result, three synthetic fuel pathways were approved under ASTM D7566, and others are in the process of approval.

Previous technical approvals of new fuels have been facilitated by significant investment and testing done by the US Department of Defense (DOD). The budget for this work has been reduced, and the engine and airframe Original Equipment Manufacturers (OEMs) do not have sufficient resources (funding) to do the level of testing previously supported by the DOD. This will inevitably reduce the amount of testing carried out, and the availability of future fuels. This may act as an on-going barrier to the development of new fuels and conversion routes.

14.7 ENABLING SUSTAINABLE FUELS IN THE UNITED KINGDOM

14.7.1 Policy Certainty

The present RTFO structure officially extends to 2020, providing project developers no policy certainty beyond that date. The policy for renewables in the EU beyond 2020 also fails to provide policy certainty. Should the UK government set out long-term goals for sustainable fuels and establish an incentive framework to stimulate investment, this would send an important signal to the investment community and to project developers. Without adequate targets for advanced transport fuels up to 2030, investments in domestic production capacity will cease. Long-term certainty and bankability of policy incentives are needed.

We recommend the establishment of a public–private sector initiative to enable sustainable fuels. The taskforce should be low-cost and agile, building on the experience of existing initiatives such as the Commercial Aviation Alternative Fuels Initiative (CAAFI) in the United States. This partnership could also enable opportunities for making direct strategic investment. We already have experience as an industry of successful collaborative programmes, such as the Aerospace Growth Partnership, which has been a springboard for new technology development, skills development, and training. As bio-energy impacts a number of government departments, a government lead department needs to be assigned to ensure that a consistent approach to emerging policy support is maintained.

14.7.2 Market Incentives

The UK RTFO is a blending mandate (obligation) to fuel suppliers of road transport fuels to meet renewables commitments specified in the EU RED. The RTFO provides a market incentive to the producers of road transport fuels in the form of RTFCs, worth \$200–300 (USD) per tonne of fuel. This currently has the effect of dis-incentivizing investment in, and

[5] The EU has a number of funding instruments, including Framework Programme funding, the New Entrant Reserve (NER) fund, and the European Industrial Bioenergy Initiative (EIBI).

production of, sustainable aviation fuels because they are not eligible under the RTFO.

Aviation fuel suppliers should be eligible for market incentives in the form of RTFCs that are currently awarded to qualifying road transport fuels under the UK RTFO, but without directly obligating them in the system. The effect of this incentive is to reduce effective production costs, providing investors and suppliers with confidence to invest and to produce the fuels when costs would otherwise be uncompetitive.

By extending eligibility to aviation fuel suppliers, the incentives playing field would be levelled with road fuels, without the need to extend the obligation to those suppliers. This approach is already being successfully pursued in the Netherlands and the United States. In the United Kingdom, this option could be implemented by a straightforward extension of the RTFO, with no implications for additional costs to road fuel suppliers, road users, or the government.

If a mandate was applied directly to aviation fuel without consideration of market distortions, then competitive imbalance and carbon leakage[6] would result. This would happen if, for example, the United Kingdom were to impose an obligation on all aviation fuel uplifted to flights from the United Kingdom. The result would be an increase in the cost of fuel for UK-based airlines above the cost faced by competitor airlines in the same markets. Where an obligation can be applied without causing market distortions, it might be considered as a future policy option to support the development of sustainable fuel. While a zero rating for eligible fuels in the EU Emissions Trading System (EU ETS) is welcome, it only provides a very modest market incentive that does not make a material difference to investment decisions.

In the US, the DOD has been playing a significant role in stimulating the development of sustainable fuels by creating market demand. The UK government should consider how the military procurement policy for sustainable fuels can help to de-risk the investment case, as in the case of the US 'green fleet' programme, for example.

14.7.3 Investment Incentives

The DfT's £25 million grant for demonstration-scale advanced fuel production, awarded in 2015, was a positive first step. However, it failed to provide support to any aviation fuel initiatives. If the United Kingdom is to lead innovation and development of advanced aviation fuel technology and become an exporter rather than an importer, a more focused approach to sustainable fuels will be required to build on the UK's competitive advantage in relevant areas of science and technology.

Development support through the provision of grants and equity co-investment are needed to reduce investor risk in funding demonstration plants. The Green Investment Bank (GIB) can play a key role in helping initial commercial-scale production plants become established through the provision of financial support that reduces investor risk through, for example, loans and loan guarantees. At present, the GIB has a limited role in this area. Also, there is no obvious source of development funding within the EU to provide development support. Even if it becomes available, in the absence of long-term policy support, it is of limited use. A number of EU projects have been cancelled or delayed over concerns that there is insufficient policy support to provide a long-term market for advanced fuels.

[6]'Carbon leakage' is the term often used to describe the situation that may occur if, for reasons of costs related to climate policies, businesses were to transfer production to other countries which have more lax constraints on GHG emissions (see http://ec.europa.eu/clima/policies/ets/cap/leakage/index_en.htm for details).

14.7.4 Research and Development

The priority given to innovation into sustainable fuels should be reviewed. R&D is being pursued around the world to demonstrate new feedstock sources and processing technologies with the aim of reducing fuel costs and broadening the range of supply-chain opportunities. The pace of development has been rapid, with US-based companies playing a notable role. There are significant market failures during innovation, and action by the public sector has been critical to advancing the sustainable fuels industry (Table 14.3).

In the UK, the government has been funding some selected cross-department work through several bodies, including the Research Councils, the Carbon Trust, and the Centre for Process Innovation; their research involves feedstocks and development of conversion technologies. Bioenergy research support has been limited, but we understand that the Biotechnology and Biological Sciences Research Council is seeking to provide more funding in the future. If the United Kingdom is to be a developer of technology (rather than just a user of technology developed elsewhere), a more focused approach to sustainable fuels will be required, building on the UK's scientific and technological strengths in areas such as plant science, biochemistry, chemical engineering, and mechanical engineering.

Pyrolysis oils and gas capture of waste carbon gases from industrial processes also have potential in the United Kingdom. In order to result in commercial solutions, funding programmes should prioritize those technology pathways that have the best chance of reaching a commercial scale by using sustainable feedstock and processes that produce scalable solutions at cost parity with fossil fuel (Table 14.4).

A stronger innovation focus could contribute to re-balancing the UK economy through the nurturing of new businesses and the stimulation of

TABLE 14.3 GHG Emissions Savings Estimates for Each Technology Group (tCO_2eq/t)

	Emissions Savings Factor 2020	Emissions Savings Factor 2030	Emissions Savings Factor 2050
HEFA[a]	2.9	2.9	2.8
F-T	3.0	3.0	3.0
SIP	2.5	2.6	2.7
ATJ	2.3	2.5	2.6
HDCJ	2.2	2.4	2.5

[a] *Emissions savings for HEFA reduce up to 2050 as a result of feedstock diversification. As the demand for HEFA increases beyond the supply of waste oils, growth continues based on low ILUC oil crops and microalgae.*

TABLE 14.4 Sustainable Fuel Production in United Kingdom

Timeframe	Technology Pathway	Feedstock	UK Production Potential
To 2020	Biomass to liquid	Mixed MSW	High
To 2020	HEFA	Waste oils	Low
To 2020	ATJ	Waste gases	High
To 2020	Green diesel	Waste oils	Low
2020–30	ATJ	Lignocellulosic	Medium
2020–30	Pyrolysis oils	Mixed MSW	High
2020–30	Farnesene	Sugar cane/ LC residue	Low
2020–30	Co-processing	Wastes oils/ pyrolysis	Medium
2020–30	SIP	Sugars/LC materials	Med
2030–40	Novel hydro routes	Waste oils	Low/ Medium
2030–50	HEFA	Algae	Unknown
2030–50	Biotech conversion	Waste gases	Unknown

employment and technology export opportunities. Establishing a government–industry initiative could be a key first step to developing a more focused approach.

14.7.5 The Role of Industry

SA signatories and other industrial stakeholders have key roles to play in overcoming development barriers, in particular those relating to project risk, and also project financing. SA members have demonstrated the importance of effective partnerships with technology companies, and a continuation and expansion of these activities is required.

Opportunities to reduce project risk include ensuring that contracts spread risk in an appropriate way across the supply chain, including Engineering Procurement Construction services, feedstock supply contracts, and fuel off-take agreements. Airlines and airports have a role to play in establishing fuel off-take agreements. Joint purchasing agreements may have an additional benefit to fuel producers by decreasing project risk, and increasing the availability of off-take contracts (on a fuel volume basis), as this enables small airlines (who may otherwise find this challenging) to participate. The development of purchasing groups may provide an opportunity for fuel users to share the associated cost and risk, and to hedge the risk of higher prices.

The sustainable aviation fuels industry is at an early stage of development and therefore strategic investors have an important role to play in providing project financing. Strategic investors may include public sector organizations, and also industry stakeholders with a vested interest in sector growth. The scale of investment required for early commercial projects is large, and there will be a limited number of actors able to make such contributions to project financing. However, in the short-term, these strategic investments are being demonstrated as vital to technology development.

The development of sustainable fuels requires the concerted efforts of actors across the supply chain, as demonstrated by the ASTM fuel-testing programmes. SA brings together the main players from UK airlines, airports, manufacturers, and air navigation service providers to set a collective approach towards ensuring the sustainable future of the industry. However, additional action is required to bring together a wider collection of stakeholders, including sustainable fuel producers, researchers, and representatives from the UK government to specifically focus on the issues affecting sustainable fuels and to ensure the delivery of this roadmap.

The Department of Business, Innovation & Skills (BIS), the DfT, and the Civil Aviation Authority should play a central role in the interaction between this sector and the government. Existing successful collaborative programmes involving BIS and the aerospace sector cater to the need for a public-private approach.

14.8 THE SUSTAINABLE AVIATION FUELS ROADMAP

This SA Roadmap uses a different approach from its predecessors: CO_2 and noise. This is necessary as fuel/carbon efficiency and noise are already part of an established international regulatory regime, and there is greater certainty about technology development over the period up to 2050. In the case of sustainable fuels, much depends on how emerging technology deployment is supported. There is great potential for an advanced fuels industry able to meet the demands of modern aviation and to serve those other sectors dependent on liquid hydrocarbons for the medium-term (eg, freight). ICAO has established a global Advanced Fuels Task Force to estimate the technical potential of these fuels. They are also working on the harmonization of international sustainability standards and criteria.

14.8.1 The Roadmap

This SA Fuels Roadmap has explored the opportunities for reducing UK aviation carbon emissions through the use of new and rapidly evolving, sustainable sources of fuel. There is a role for aviation in supporting the transition to more sustainable pathways using wastes, residues, and low-value feedstock in the short-to-medium term. In the longer-term, as more novel feedstocks (eg, waste gases, algae) are developed and move to larger-scale production, we expect this opportunity to expand.

Future fuels offer the opportunity to achieve high GHG savings (60–95%) over the emissions associated with fossil fuel equivalents. One aspect highlighted by the work shows the critically important role that sustainable fuels development will play in the period up to 2030. Without this early policy support, the technologies needed post-2030 will not be technically proven, and scale-up will be slow (Fig. 14.7).

SA's early estimates assumed a very gradual and steady increase in the use of sustainable fuels up to 2050. However, the outcome of this study illustrated that early volumes will be fairly modest up to 2030, with a more rapid ramp-up of production beyond this date. It may be necessary to redraw the sections of the CO_2 roadmap to properly reflect this. The roadmap will need to be reviewed, as new sustainable fuel pathways are constantly evolving.

We predict a range of technologies will be deployed globally, but with respect to the United Kingdom, we see waste-derived fuels as the primary opportunity in the short-to-medium term. SA members are working on oil feedstock production models that prevent ILUC – for example Camelina production on contaminated land. If these methods are found to be successful, we expect greater volumes of sustainable oil feedstocks in the future. These projections are consistent with recent estimates from the IEA, who predicted a sustainable fuel production of 27% of transport fuels in 2050. (2011) [17].

14.8.2 Conclusions

As a result of reviewing the current, new, and emerging sustainable fuels market, this roadmap has determined that, with the right policy and investment framework, UK aviation can reduce its CO_2 emissions by up to 24% by

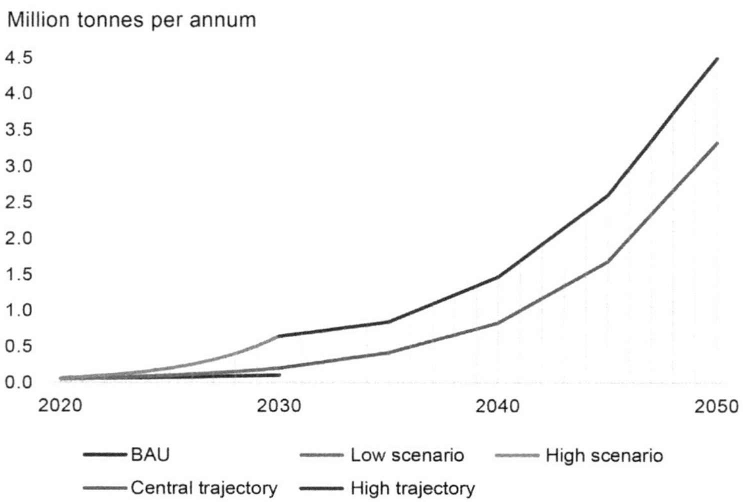

FIGURE 14.7 Sustainable fuel production in United Kingdom.

2050. Achieving this result will require a step change in the current policy and investment framework for sustainable aviation fuels, and SA welcomes the opportunity to collaborate with the government to establish the United Kingdom as a leader in this exciting new industry.

References

[1] Sustainable Aviation, UK Aviation Industry Socio-Economic Report, Sustainable Aviation, London, 2016.

[2] Sustainable Aviation, Sustainable Fuels UK Road-Map 2015, Sustainable Aviation, London, 2015. Available from: http://www.sustainableaviation.co.uk/wp-content/uploads/2015/09/SA-SAF-Roadmap-FINAL-24-Nov-2.pdf.

[3] Sustainable Aviation, *Sustainable Aviation* CO_2 *Road-Map 2012*, Sustainable Aviation, London, 2012. Available from: http://www.sustainableaviation.co.uk/wp-content/uploads/2015/09/SA-Carbon-Roadmap-full-report.pdf.

[4] Great Britain, Parliament, House of Commons, *Building a high value bioeconomy: opportunities from waste*. Available from: <https://www.gov.uk/government/uploads/system/uploads/attachment_data/file/408940/BIS-15-146_Bioeconomy_report_-_opportunities_from_waste.pdf>, 2014.

[5] Renewable Fuels Agency, The Gallagher review of the indirect effects of biofuels production. UK Renewable Fuels Agency. 2009.

[6] Roundtable on Sustainable Biomaterials, RSB Principles & Criteria for Sustainable Biofuel Production, RSB, Geneva, 2011. Available from: http://rsb.org/pdfs/standards/11-03-08%20RSB%20PCs%20Version%202.1.pdf.

[7] WWF International, Ecofys and EPFL, *The low indirect impact biofuels (LIIB) Methodology*. Available from: <http://www.ecofys.com/files/files/12-09-03-liib-methodology-version-0-july-2012.pdf>, 2012.

[8] Roundtable on Sustainable Biomaterials, RSB Low iLUC Risk Biomass Criteria and Compliance Indicators, RSB, Geneva, 2015. Available from: http://rsb.org/pdfs/standards/RSB-STD-04-001-ver0.3RSBLowiLUCCriteria Indicators.pdf.

[9] Ecofys, Assessment of Sustainability Standards for Biojet Fuel, Ecofys, Berlin, 2014.

[10] Boeing, *Here's watts up: we've got the power*. Available from: <http://www.boeing.com/features/2015/01/corp-hybrid-01-05-15.page>, 2015.

[11] Airbus Group, *The buzz around E-Fan. Latest electric aircraft news*. Available from: <http://www.airbusgroup.com/int/en/innovation-citizenship/airbus-e-fan-the-future-of-electric-aircraft/News.html>, 2015.

[12] European Climate Foundation, *Wasted: Europe's untapped resource*. Available from: <http://europeanclimate.org/wp-content/uploads/2014/02/WASTED-final.pdf>, 2014.

[13] Low Carbon Innovation Coordination Group, *Technology innovation needs assessment (TINA). Bioenergy Summary Report*. Available from: <www.lowcarboninnovation.co.uk/document.php?o = 9>, 2012.

[14] Great Britain, Parliament, House of Commons, Waste or Resource? Stimulating A Bioeconomy, Science and Technology Select Committee, London, 2014, HL 141 (2013-14). Available from: http://www.publications.parliament.uk/pa/ld201314/ldselect/ldsctech/141/141.pdf.

[15] European Commission, Proposal for a Directive of the European Parliament and of the Council Amending Directive 98/70/EC Relating to the Quality of Petrol and Diesel Fuels and Amending Directive 2009/28/EC on the Promotion of the Use of Energy From Renewable Sources COM(2012) 595 Final, EU, Brussels, 2015.

[16] E4tech, *A harmonised auto-fuel biofuel roadmap for the EU to 2030*. Available from: <http://www.E4tech.com/PDF/EU_Auto-Fuel-report.pdf>, 2013.

[17] IEA, *Technology Roadmap: Biofuels for Transport*. Available from: http://www.iea.org/publications/freepublications/publication/Biofuels_Roadmap_WEB.pdf>, 2011.

[18] Climate Change Committee, *Bioenergy review*. Available from: <http://archive.theccc.org.uk/aws2/Bioenergy/1463%20CCC_Bioenergy%20review_bookmarked_1.pdf>, 2011.

CHAPTER

15

Roadmap to a Sustainable Aviation Biofuel: A Brazilian Case Study

L.A.B. Cortez[1,2], *H. Cantarella*[3], *M.A.F.D. Moraes*[4], *L.A.H. Nogueira*[2], *U. Schuchardt*[5], *T.T. Franco*[2,6], *F.E.B. Nigro*[7], *A.M. Nassar*[8], *R.L.V. Leal*[2] *and R. Baldassin, Jr.*[2]

[1]School of Agricultural Engineering (FEAGRI), State University of Campinas (UNICAMP), Campinas, SP, Brazil, [2]Interdisciplinary Center for Energy Planning (NIPE), State University of Campinas (UNICAMP), Campinas, SP, Brazil, [3]Agronomic Institute of Campinas (IAC), Campinas, SP, Brazil, [4]"Luiz de Queiroz" College of Agriculture (ESALQ), University of São Paulo (USP), Piracicaba, SP, Brazil, [5]Institute of Chemistry (IQ), State University of Campinas (UNICAMP), Campinas, SP, Brazil, [6]School of Chemical Engineering (FEQ), State University of Campinas (UNICAMP), Campinas, SP, Brazil, [7]Polytechnic School (EPUSP), University of São Paulo (USP), São Paulo, SP, Brazil, [8]AGROICONE, São Paulo, SP, Brazil

15.1 INTRODUCTION

Aviation biofuels have a particularity that sets them apart from other biofuels, which is the requirement that they must have properties that characterize them as 'drop in'. This is for safety reasons and because aircraft fly over different continents where they have to be refuelled [1]. Therefore, the conversion technologies have a prominent role, and some feedstocks may have more complex transformation pathways in order to generate the highly regulated jet fuel standards. Notwithstanding, a variety of feedstocks, including sugars, oils, and cellulosic crops, and wastes, may be used to produce jet fuels [2].

Restriction to food crop feedstock may apply in certain countries or regions because of food versus fuel competition and other political reasons [3,4], although this is not a relevant issue under Brazilian conditions because of the land availability and the evidence that biofuel production has not been jeopardizing food production [2,5–7]. In fact, aiming at sustainability is crucial to producing bioenergy using efficiently the natural resources and other required inputs, such as land, water,

Biofuels for Aviation.
DOI: http://dx.doi.org/10.1016/B978-0-12-804568-8.00015-9

manpower, capital, etc., independently of the feedstock processed. The real competition between food and other agricultural products (feed, fibre, bioenergy) occurs at the production level, not between products.

Another relevant aspect to be taken into account in the definition of more feasible production routes is the trade-off between the price/cost of feedstock and the processing costs (capital and operational costs). Among the several alternative processes available for biofuel production, to use cheap raw material, such as cellulosic wastes and residues, expensive and complex processes should be adopted, while the more simple and straightforward processes generally require more valued feedstock, such as vegetable oils. This point is crucial and will be explored in the Brazilian context.

15.2 AGRICULTURAL FEEDSTOCKS IN BRAZIL: EXISTING AND POTENTIALLY AVAILABLE

Recent studies have examined the feedstock options in Brazil for jet biofuel with respect to availability, production feasibility, logistics, costs, and energy and greenhouse gas (GHG) balances, as compared to fossil fuels [2,6,7]. Because of extensive land availability, different climate conditions, and agriculture traditions, practically all sorts of feedstock can be produced at prices that are competitive with most other countries. This includes sugar and starch crops (eg, sugarcane, cassava, sorghum, corn). A variety of oil crops are also potential feedstock in Brazil, such as soybeans, palm oil, and an assortment of other oil crops with small cultivated areas – cotton, peanut, sunflower. Nonedible oil crops such as castor bean, or plants that are not commercially cultivated – jatropha, camelina, macauba palm – can also be cited [8], but it is unlikely that they will be options in the short- or medium-term.

Cellulosic feedstocks of great potential include wood from planted forests, especially eucalyptus, tropical grasses such as Elephant grass, energy cane, for which there is an intense search for varieties, and plant residues, including those of sugarcane.

The best options to start a jet biofuel industry in Brazil, considering the main feedstock groups, are sugarcane (sugar), eucalyptus (cellulose), and soybean (oil crops) [2,6,7]. Among the reasons for those choices are that they are widely cultivated – with well-known agricultural technology behind them – and high-yielding, cost-competitive crops [6].

The main constraints that should be considered are cost of feedstock and transformation, logistics, and also nontechnical aspects such as political restrictions on food crops used for bioenergy.

Brazil is the largest producer of sugarcane and exporter of sugar in the world, with more than 9 Mha dedicated to this crop. Therefore, it is highly competitive. In addition, there are 60 Mha of land suitable for sugarcane cultivation, even without opening forest areas or environmentally sensitive regions [9].

With almost 6 Mha of planted eucalyptus forest, Brazil is a leading exporter of cellulose and paper, and has the world's highest wood yields [10] due to favourable climate conditions and investment in research.

Brazil is about to become the leading exporter of soybeans, planted in about 30 Mha. In 2010, 2.2 Mm^3 of soy oil (out of a total of 6.9 Mm^3) was directed to biodiesel production [5]. Although the transformation routes of vegetable oils to jet biofuel are the simplest alternatives among all classes of feedstock [11], and most of the test flights with jet biofuel so far have been done with fuel produced with oil crops, soybean may not be widely acceptable internationally as a feedstock for aviation biofuel because of its food character and because of competing prices. About 70% of the biofuel cost is feedstock [5]. Cantarella et al. [6] compared the opportunity cost of the incorporated energy of several feedstocks to that of fossil jet biofuel. They found that eucalyptus, sugarcane, palm, and corn had competitive costs if just agricultural product prices were considered; soybean grains were not competitive at the 2012 prices. However, when the opportunity costs of final products (ethanol, soy oil, palm oil, sugar, and others) were considered, only sugarcane bagasse and wood were competitive, indicating that cost of feedstock will always be an important issue, even in Brazil. As the price of petroleum has declined in recent years, these price comparisons may not hold, and feedstock cost may be even more restrictive.

Other issues are logistics and storability. About 20% to 35% of the cost of sugarcane is represented by harvesting and transportation to the mill, usually not more than 50 km from the field [12,13]. In addition, sugarcane stalks (for sugar and ethanol production) must be processed within 3 days. However, storability is not a problem with sugarcane bagasse and trash, or wood and oil grains [6].

Other studies have pointed out the problem of competitiveness of jet biofuel made with different biomass feedstocks [14]. Even in a country with low comparative costs for agricultural feedstock, appropriate public policies and tax regimes, in addition to premium prices for jet biofuel and carbon taxes schemes may be necessary to stimulate an emerging biofuel industry until technological progress and scale make it economically sustainable [6].

15.3 AVAILABILITY OF LAND IN BRAZIL FOR BIOFUELS

Brazil has a unique combination of positive factors for deployment of the production of sustainable feedstocks for aviation biofuels: significant availability of land already cleared for agriculture, a dynamic and mature agriculture sector presenting strong productivity growth, large areas of legally protected native vegetation and strong conservation laws, and human health and safety regulations for rural activities equivalent to urban activities. However, it is clear that to develop an aviation biofuel programme in compliance with responsible principles and sustainable requirements, several public policies will have to be implemented.

Land use in Brazil is strongly protected by the National Forest Code (a federal legislation approved in 2012 that set the land use and native vegetation protection in rural private properties) and also by the delimitation of Legal Reserves, Conservations Units, and Indigenous Reserves. This effort results in a protection of approximately 483 Mha (56.9% of the Brazilian territory), as shown in Fig. 15.1. Despite the high production of Brazilian agriculture, only 60.5 Mha have been used for crops and planted forests, whereas livestock (mainly beef cattle) takes three times more land (approximately 198 Mha) with a low stocking rate (1.0 head/ha) [15].

Bioenergy production in Brazil is a successful reality, mainly considering sugarcane, wood (planted forest), and soybean. For instance, sugarcane (molasses, juice, and bagasse), the most important bioenergy crop in Brazil, has used only 8.5 Mha (1% of the Brazilian land), and was responsible for 15.7% of Brazil's domestic energy supply in 2014 [16]. Firewood and charcoal, mostly from planted forests (eucalyptus and pinus), and which use approximately 2 Mha (0.2% of Brazilian land and approximately 31% of the wood planted area) [10], were responsible for 8.1% of the Brazilian domestic energy supply in 2014 [16]. Soybean biodiesel, responsible for 80% of Brazilian biodiesel production, uses approximately 7.6 Mha (32% of the soybean production) [17] and was responsible for 4.2% of diesel consumption in 2014 [16].

In order to supply high volumes of feedstocks for aviation biofuels, it is clear that Brazil will need to provide new areas for agriculture expansion. So, the key question is whether Brazil can support the agriculture expansion for aviation biofuel production with no impact on food production and protected areas.

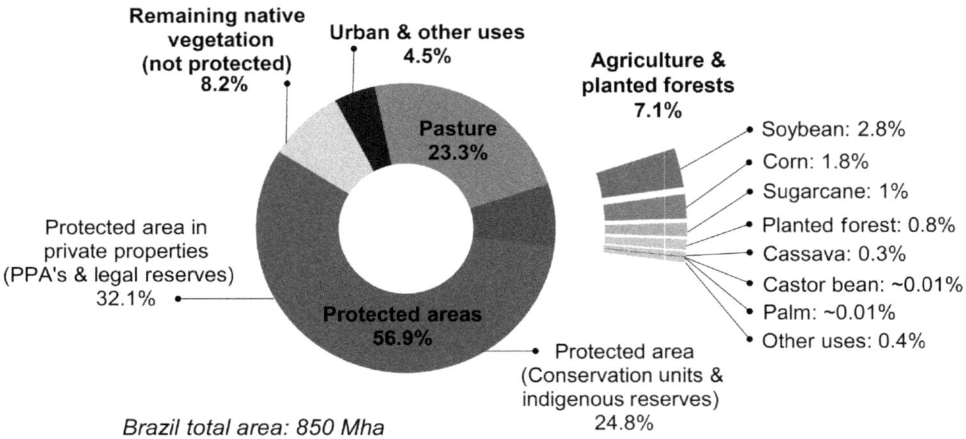

FIGURE 15.1 Land use in Brazil. *Notes:*
(1) Conservation Units exclude the areas called Environmental Protection Areas (APAs).
(2) The Permanent Protection Areas (PPAs) include buffer strips along rivers, high slope areas, public forests, and other remaining natural vegetation areas.
(3) The data for other natural vegetation and protected areas include former slave settlement areas ('quilombos'), public forests, and other such areas.
(4) The Protected Areas inside private properties were estimated based on the National Forest Code.
Source: Adapted from: L.A.B. Cortez, F.E.B. Nigro, A.M. Nassar, H. Cantarella, L.A.H. Nogueira, M.A.F.D. Moraes, et al., Roadmap for Sustainable Aviation Biofuels for Brazil, Blucher, São Paulo, 2014, 272 p. Available from: http://openaccess.blucher.com.br/article-list/roadmap-aviation-272/list#articles.

In the last two decades, Brazil has been experiencing a substantial increase in food production based essentially on productivity gains and requiring a relatively small expansion of cultivated areas (Fig. 15.2). For instance, the production of grain increased by 175% from 1992 to 2012, whereas the cultivated area expanded only by 49%. Moreover, there still is a lot of space for intensifying cattle production, taking into account that technical yields such as slaughter age, birth rate, and meat produced per hectare are still low in Brazil [15]. In addition, the double-cropping system has been developed successfully, allowing the integration of soybean and corn in the same year, and responsible for half of the total corn production; since 2000, most corn production expansion has taken place as a second crop.

The expansion of sugarcane for ethanol, although very strong, has not undermined the expansion of other annual and perennial crops. Moreover, the cultivation of oilseeds in rotation with sugarcane is also generating food and fuel in the same systems. Deforestation has been reduced since 2004, and the decreased deforestation levels in 2011 and early 2012 have been encouraging, around 75% lower than 2004 levels. Therefore, agricultural production is not dependent on the clearing of native forests in Brazil.

Facing this recent context, the data show that rather than 'food VERSUS fuel', in Brazil biofuel production is taking place as 'food AND fuel'. In many cases, these favourable results have been achieved thanks to the incorporation of modern agriculture technologies, such as the increasing use of fertilizers, better agriculture management practices, plant breeding and genetic gains, planting and harvesting improvements, etc.

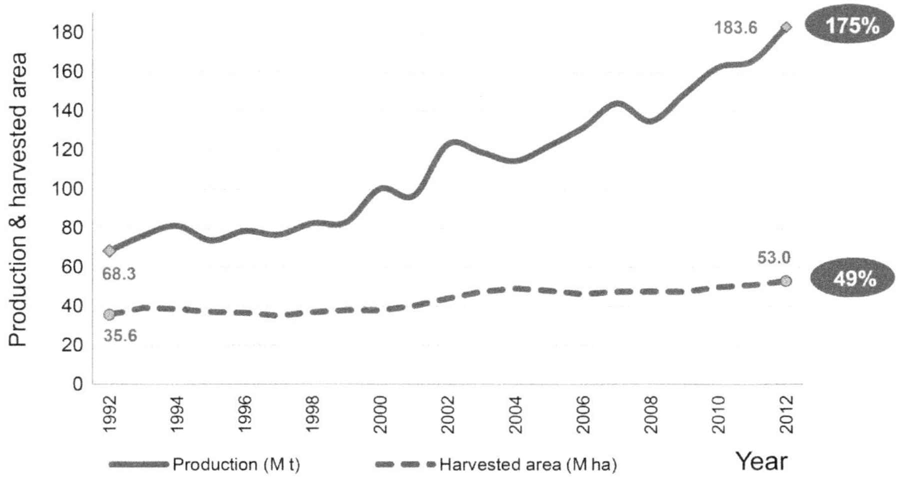

FIGURE 15.2 Production and harvested area in Brazil: grains (1999–2012). *Source: Adapted from: L.A.B. Cortez, F.E.B. Nigro, A.M. Nassar, H. Cantarella, L.A.H. Nogueira, M.A.F.D. Moraes, et al., Roadmap for Sustainable Aviation Biofuels for Brazil, Blucher, São Paulo, 2014, 272 p. Available from: http://openaccess.blucher.com.br/article-list/roadmap-aviation-272/list#articles.*

In energy terms, to replace 100% of Brazilian jet fuel consumption (around 7 Mm^3), and with sugarcane ethanol as feedstock, the land required is estimated at 1.9 Mha (ie, approximately 1% of Brazil's pasture land).

15.4 SUSTAINABILITY OF AVIATION BIOFUEL PRODUCTION IN BRAZIL

Sustainability assessment of biofuel production and use has been growing in importance in recent years, especially in a scenario including the need for reduction of GHG emissions, the food versus fuel debate, and the increasing need to respect environmental and social standards. So far, the most promising potential feedstocks for jet biofuel are plants that contain sugars, starch, and oil, as well as residues such as lignocellulose, municipal solid wastes, and industrial waste residues.

The assessment on the sustainability criteria for the production of feedstocks in Brazil (sugars and starch, oil, lignocelluloses, and wastes) was carried out according to the principles stated in RSB [18], as well as the criteria of the currently available and most well-known sustainability standards (Bonsucro, Roundtable on Sustainable Biomaterials (RSB), and the International Sustainability and Carbon Certification System (ISCC)).

The analysis included the evaluation of several parameters (when data was available): Energy Balance; GHG Emissions; Agricultural Productivity; Agricultural and Environmental Best Practices (Fertilizer Consumption, Pesticide Consumption, Soil Loss, Water Use); Biodiversity; Energy Self-Sufficiency; and Social, Economic, Institutional and Environmental concerns (eg, legislation and regulations) [19].

Although there are important differences among the four groups of feedstock analysed, some general considerations can be made regarding biofuel production, and as to the gaps of complying with sustainability requirements.

Concerning the social area, the main positive impacts are the high potential of job creation and income generation, and the positive impacts on regional development. As for the sustainability requirements gaps, the factors that make compliance more difficult are considered the great number of laws and rules, sometimes stricter than sustainability standards; different interpretations and the lack of knowledge on how to apply some laws; and labour laws not adapted to the rural context. It was also noted that there is a need for qualification and capacitation of workers.

As for the environmental aspects, the positive impact regarding compliance with requirements is related to reduction in GHG emissions compared to fossil fuels, especially in the sucrose and cellulosic groups.

The debate over food versus fuel (as presented in the previous topic) is not as important an issue in Brazil as elsewhere because there is enough available land for the production of biofuels from agricultural feedstock. Recent studies have shown that even with the need to increase food production to meet the demand of a growing population, land is not a constraint at the global level [4,20–24].

The great number of environmental laws and rules (sometimes stricter than sustainability standards) and legal uncertainty (changes in the Forest Code, law regulating the conditions for Foreign Investors) are considered barriers to compliance with legal requirements. It is possible to comply with national and international rules and norms, but this requires investment. The need for a clear price policy for fuels is important.

Environmental questions, especially concerning GHG emissions associated with biofuel production, were key in studies [21]. GHG savings from potential feedstock for bioenergy in Brazil ranged from 40% to 82%, depending on the source of biomass [6,7]. Calculations took into consideration the energy content of different feedstock, including sugars, starch, oil crops, wood, and other cellulosic materials. The best returns in terms of GHG emissions were obtained with sugarcane, eucalyptus, and Elephant grass [6].

15.5 POSSIBLE PATHWAYS FOR BIOFUELS FOR AVIATION IN BRAZIL

As already shown, Brazil has large quantities of a great variety of biomasses. Transformation of this biomass by different technologies into chemicals and fuels has been well described in the literature [25]. Depending on the feedstock, different refining technologies were considered to produce drop-in biofuels for aviation [7]. Jet biofuels can be obtained by hydrotreatment (hydroprocessed esters and fatty acids, HEFA) or catalytic hydrothermolysis (CH) of vegetable oils. They can also be obtained by alcohol dehydration/oligomerization (Alcohol To Jet, ATJ) or by the direct conversion of sugars to hydrocarbons (DSHC). Proven technologies for lignocellulosics are fast pyrolysis, gasification/Fischer–Tropsch Synthesis (FT A1), or solvent liquefaction and posterior upgrading of the bio-oil (hydrotreated depolymerized cellulosic jet, HDCJ). Acid or enzymatic hydrolysis of carbohydrates to sugars, which can be fermented to ethanol or butanol, are another possibility. These alcohols can then be converted to jet biofuels by ATJ technology. The least developed technologies are jet biofuels from wastes; however, fermentation of organic wastes to acids, ketones, and alcohols, fermentation of flue gases to alcohols, and hydrotreatment of industrial fatty residues are possibilities that should be further explored.

The different possible feedstocks, the identified pathways for their transformation, and the final products are shown in Fig. 15.3. Jet biofuels obtained by the HEFA and FT A1 processes (green gates (grey in print versions)) have already been approved by ASTM for many years and successfully used in

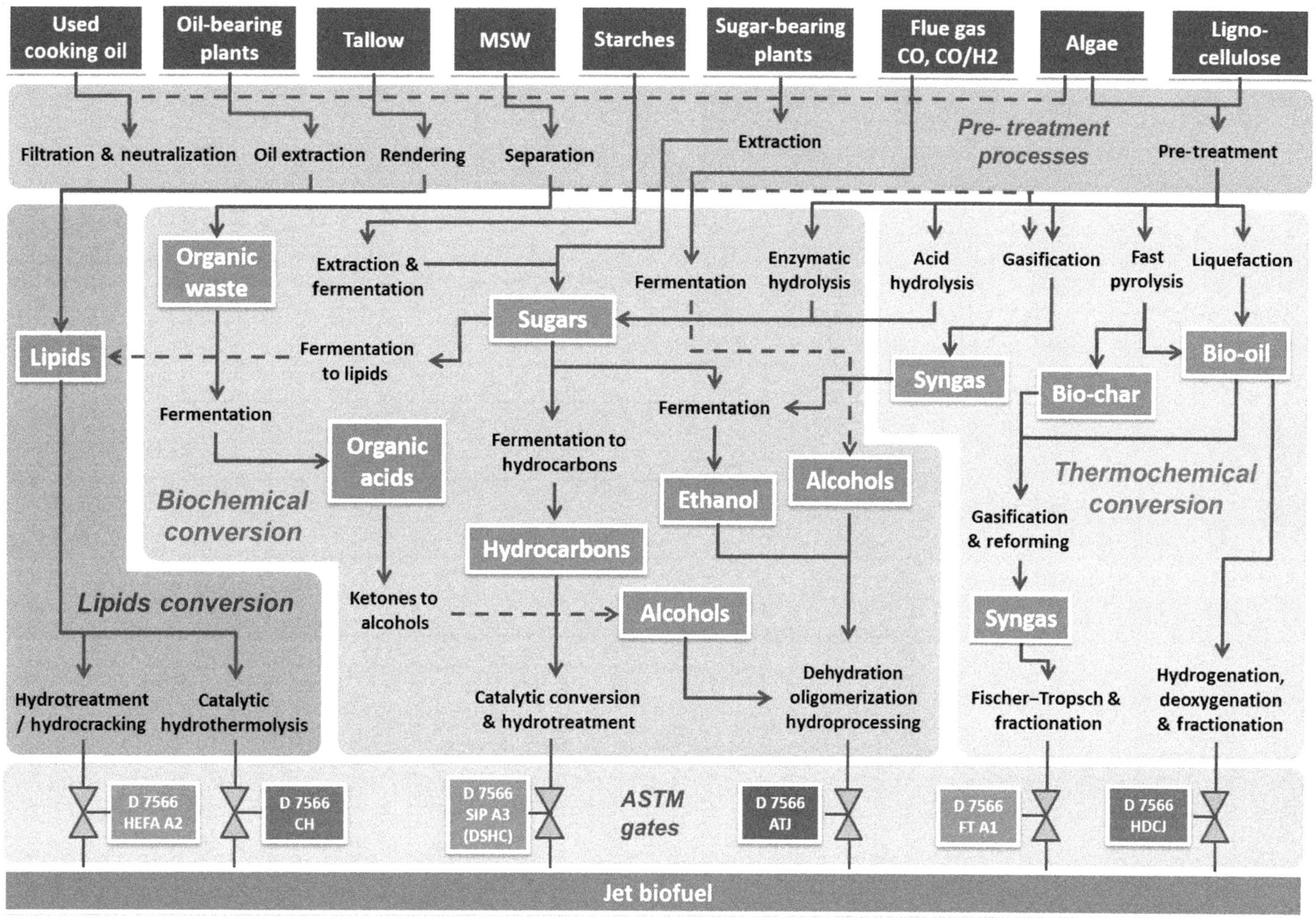

FIGURE 15.3 Jet biofuel production in Brazil: identified pathways. *Source: Adapted from: L.A.B. Cortez, F.E.B. Nigro, A.M. Nassar, H. Cantarella, L.A. H. Nogueira, M.A.F.D. Moraes, et al., Roadmap for Sustainable Aviation Biofuels for Brazil, Blucher, São Paulo, 2014, 272 p. Available from: http://openaccess.blucher.com.br/article-list/roadmap-aviation-272/list#articles.*

commercial aviation. Recently, DSHC (synthetic isoparaffins, SIP) (green gate (grey in print versions)) has also been approved, and ATJ (blue gate (dark grey in print versions)) is currently under ASTM's standard evaluation. No preference should be given to any technological alternative. All of them are in principle valid for jet biofuel production [7]. However, at present conditions, none of these technologies can produce jet biofuels at market price.

The main commercial gap to using sugar, ethanol, or vegetable oils to produce jet biofuel is that their market prices are high due to their potential uses as food or automotive biofuel [11]. However, their transformation into jet biofuel is well developed and does not require high investment. Cellulosic residues and wastes are cheap, and their transformations by thermochemical or biochemical processes into jet biofuel are potentially attractive. However, the costs of the equipment needed, or the costs of efficient and robust enzymes, are high, showing that these technologies have to be further developed.

Municipal solid wastes and flue gases are certainly prime candidates for conversion into jet biofuel. However, the technologies are not yet well developed, and nothing can yet be said about the effective costs. The relationship between feedstock costs and technical efforts needed for their transformation into jet biofuel is shown in Fig. 15.4. The prices of the feedstocks closer to the right are higher, but they are easier and less costly to convert into jet biofuel.

A perfect scenario would be a combination of a waste feedstock and a well-developed, cheaper conversion process. However, it seems that in the short-term only the sugar fermentation to hydrocarbons or to ethanol and subsequent ethanol conversion, as well as the hydrotreatment of esters and fatty acids, will be suitable for jet biofuel production [11]. For medium- or long-term second-generation technologies, processing cellulosic biomass or wastes may become economically more feasible. However, costs of these technologies have

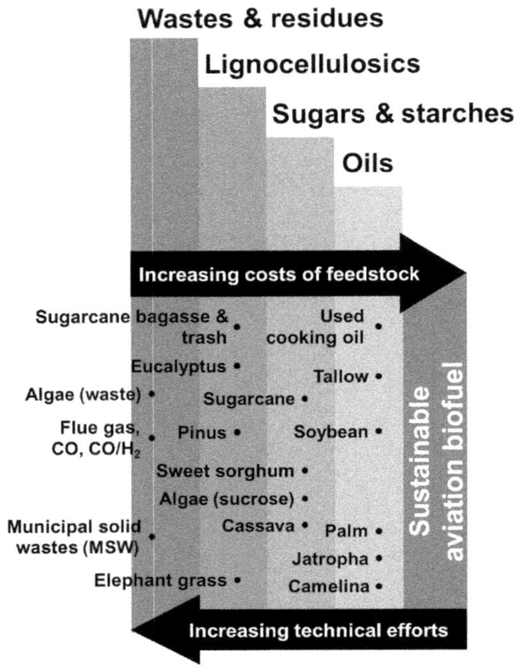

FIGURE 15.4 Feedstocks & aviation biofuel: a general overview of costs and technical efforts. *Note:* For illustrative purposes only. The feedstock prices and technical efforts may vary significantly. *Source: Adapted from: L.A.B. Cortez, F.E.B. Nigro, A.M. Nassar, H. Cantarella, L.A.H. Nogueira, M.A.F.D. Moraes, et al., Roadmap for Sustainable Aviation Biofuels for Brazil, Blucher, São Paulo, 2014, 272 p. Available from: http://openaccess.blucher.com.br/article-list/roadmap-aviation-272/list#articles.*

to be significantly reduced. At the present stage it is not possible to select a single solution.

More Research & Development efforts are needed to define the most promising and sustainable alternatives. Furthermore, little data on Life Cycle Analysis is available [26], and the production costs for the different pathways under Brazilian conditions are not available. Despite the previous experiences with the production and use of biofuels in Brazil, some important and relevant institutional issues have to be resolved in order to construct a new biofuels-for-aviation industry.

TABLE 15.1 Biofuels for Aviation in Brazil: Summary of Limiting Factors and Policy Recommendations

	Limiting Factors			Policy Recommendations		
Issue	Current	Future	Relevance/ Priority	Immediate/Short-Term	Medium-Term (2020)	Long-Term (2050)
Feedstock	• Limited information about species with potential for bioenergy. • Limited information on land zoning for bioenergy. • High costs for producers to comply with environmental and social regulations.	• Risk of constraints in natural resources supply (water, chemicals, etc.) for efficient biomass production. • Risks of expanding biofuel production with high impact on land use change.	Medium High	• Promote the development of human resources. • Promote life cycle analysis studies on crops with bioenergy potential. • To evaluate gaps and mechanisms to allow producers to legalize land tenure situation.	• Promote advanced agronomic studies on bioenergy cultures. • Develop assessment on residue availability and collection. • Promote agricultural yield improvements (productivity, energy and water use, fertilizer and pesticide use, land use changes, etc.).	• Promote studies on innovative sources of biomass for bioenergy.
Refining Technology	• Lack of information about process feasibility, high technology risk.	• Technology risk associated with development of innovative process.	High	• Promote development of human resources. • Support (financing) for pilot and demonstration plants.	• Support (financing/ regulatory) aviation biofuel demonstration programmes and commercial use.	
Logistics	• Infrastructure constraints.		High	• Evaluate needs of regions with potential for biofuel production.	• Promote logistics improvements. • Assess new productive schemes, reducing transport of bulky biomass.	

(*Continued*)

TABLE 15.1 (Continued)

	Limiting Factors			Policy Recommendations		
Issue	Current	Future	Relevance/ Priority	Immediate/Short-Term	Medium-Term (2020)	Long-Term (2050)
Sustainability	• Need for successful enforcement of social and environmental laws.	• Protect workers, and avoid potential loss of Brazil's major natural resources.	High	• Establish legal mechanisms to ensure that incentives for aviation biofuels are only available where national laws and regulations, especially natural forest and other habitat protections, land use zoning and worker protections, are demonstrated to be fully implemented.		
	• Need for monitoring of performance of aviation biofuels to international social and environment standards (assessing real benefits and impacts).	• Avoid difficulties to Brazil's aviation biofuels production.	High	• Consolidate the sustainability certification process. • Research and incentives only to feedstocks systems that increase overall productivity of energy and food/feed/fibre on same land.		
General	• Lack of co-ordination among governmental agencies and stakeholders in fostering aviation biofuels. • Lack of information about aviation biofuels among decision makers and public.	• Heterogeneity and lack of clarity in the sustainability evaluation of biofuels.	High	• Launch an aviation biofuel program, with a clear agenda of strategic actions. • Promote information campaign on potential, benefits, and implications of aviation biofuels.	• Assess and issue regularly indicators of the aviation biofuels program. • Consolidate the sustainability certification process.	

15.6 ROADMAP SUMMARY FOR BRAZIL: LIMITING FACTORS AND POLICY RECOMMENDATIONS

Table 15.1 summarizes the main outcomes of this study and presents a list of limiting factors and policy recommendations to take into account. They are grouped according to the value chain of the production route, different time frames for implementation, and level of priority.

15.7 CONCLUSIONS

Brazil has excellent conditions to harbour a jet biofuel industry, mainly because of land availability and competitive costs of biomass feedstocks, a strong component of the final price of fuel. The feedstocks with the greatest chance of kickstarting a jet biofuel industry are sugarcane, eucalyptus, and soybeans, or sugar, cellulose, and oil crops. They depend on the technical and economic feasibility of transformation technologies, all of which are still in development. Sugarcane and eucalyptus would give the highest GHG savings for a biofuel replacing fossil fuels. Notwithstanding, even with competitive Brazilian feedstocks, some type of subsidy or special tax scheme seems necessary for support of this emerging industry until technological progress can make it economically sustainable.

Acknowledgements

The Roadmap for Sustainable Aviation Biofuels for Brazil was funded by Boeing, Embraer, FAPESP, and several other sponsors. Any opinions, findings, and conclusions or recommendations expressed in this paper are those of the authors and do not necessarily reflect the views of the funding parties, including The Boeing Company and Embraer S.A. The authors also gratefully acknowledge the important contributions of the project members (ie, researchers, administrative team, and consultants).

References

[1] ASTM, ASTM D4054-09: Standard Practice for Qualification and Approval of New Aviation Turbine Fuels and Fuel Additives, ASTM International, West Conshohocken, PA, 2009.

[2] L.A.B. Cortez, F.E.B. Nigro, L.A.H. Nogueira, A.M. Nassar, H. Cantarella, M.A.F.D. Moraes, et al., Perspectives for sustainable aviation biofuels in Brazil, Int. J. Aerosp. Eng. 2015 (2015) 12.

[3] EU, European Parliament Backs Switchover to Advance Biofuels, European Union Press, Brussels, 2013. Service Available from: <www.europarl.europa.eu/pdfs/news/expert/infopress/201309061PR18831/2013090618831_en.pdf/download>. (accessed 08.05.15)

[4] P. Osseweijer, H.K. Watson, F.X. Johnson, M. Batistella, L.A.B. Cortez, L.R. Lynd, et al., Bioenergy and food security, in: G.M. Souza, R.L. Victoria, C.A. Joly, L.M. Verdade (Eds.), Bioenergy & Sustainability: Bridging the Gaps, SCOPE and FAPESP, São Paulo, 2015, pp. 90–137.

[5] L.A.H. Nogueira, R.S. Capaz, Biofuels in Brazil: evolution, achievements and perspectives on food security, Global Food Secur. 2 (2013) 117–125.

[6] H. Cantarella, A.M. Nassar, L.A.B. Cortez, R. Baldassin Jr., Potential feedstock for renewable aviation fuel in Brazil, Environ. Dev. 15 (2015) 52–63.

[7] L.A.B. Cortez, F.E.B. Nigro, A.M. Nassar, H. Cantarella, L.A.H. Nogueira, M.A.F.D. Moraes, et al., Roadmap for Sustainable Aviation Biofuels for Brazil, Blucher, São Paulo, 2014, 272 p. Available from: <http://openaccess.blucher.com.br/article-list/roadmap-aviation-272/list#articles>

[8] J. Poetsch, D. Haupenthal, I. Lewandowski, D. Oberländer, T. Hilger, *Acrocomia aculeata* – a sustainable oil crop, Scientific World Rural 21 (2012) 41–44.

[9] EMBRAPA, in: C.V. Manzatto, E.D. Assad, J.F.M. Bacca, M.J. Zaroni, S.E.M. Pereira (Eds.), Zoneamento Agroecológico da Cana-de-açúcar, Empresa Brasileira de Pesquisa Agropecuária – Embrapa, Rio de Janeiro, 2009, 58 p.

[10] ABRAF Associação Brasileira de Produtores de Florestas Plantadas, ABRAF Anuário Estatístico da ABRAF 2013 ano base 2012, ABRAF, Brasilia, 2013, p. 146.

[11] U. Schuchardt, T.T. Franco, J.C.P. Melo, L.A.B. Cortez, Sustainable aviation fuels for Brazil, Biofuels Bioprod. Biorefin. 8 (2014) 151–154.

[12] R.L.V. Leal, M.V. Galdos, F.V. Scarpare, J.E.A. Seabra, A. Walter, C.O.F. Oliveira, Sugarcane straw availability, quality, recovery and energy use: a literature review, Biomass Bioenergy 53 (2013) 11–19.

[13] A. Walter, M.V. Galdos, F.V. Scarpare, M.R.L.V. Leal, J.E.A. Seabra, M.P. da Cunha, et al., Brazilian

sugarcane ethanol: developments so far and challenges for the future, Wiley Interdiscip. Rev.: Energy Environ. 3 (2014) 70–92.

[14] SAFN, Sustainable Aviation Fuels Northwest: Powering Next Generation of Flight, Sustainable Aviation Fuels Northwest – SAFN, 2011. Available from: <www.wedc.wa.gov/Download%20files/SAFN_2011Report.pdf>

[15] G.B. Martha Jr., E. Alves, E. Contini, Land-saving approaches and beef production growth in Brazil, Agric. Syst. 110 (2012) 173–177.

[16] EPE, Brazilian Energy Balance: Year 2013, EPE – Empresa de Pesquisa Energética, Rio de Janeiro, 2014, 292 p.

[17] UBRABIO, Impactos do PNBP: Aspectos mais relevantes das externalidades positivas do PNPB (Programa Nacional de Produção e Uso do Biodiesel) e Impactos na Progressão da Mistura. União Brasileira do Biodiesel e Bioquerosene – UBRABIO. Available from: <http://www.ubrabio.com.br/sites/1800/1891/Temporada/IMPACTOSDOPNPB.pdf>, 2014.

[18] RSB, RSB Principles and Criteria for Sustainable Biofuel Production, Roundtable on Sustainable Biomaterials – RSB, Geneva, 2010. Available from: <http://rsb.org/pdfs/standards/11-03-08%20RSB%20PCs%20Version%202.1.pdf>

[19] M.A.F.D. Moraes, A.M. Nassar, P. Moura, R.L.V. Leal, L.A.B. Cortez, Jet biofuels in Brazil: sustainability challenges, Renewable Sustainable Energy Rev. 40 (2014) 716–726.

[20] L.A.H. Nogueira, J.R. Moreira, U. Schuchardt, J. Goldemberg, The rationality of biofuels, Energy Policy 61 (2013) 595–598.

[21] A. Karp, P.E. Artaxo Netto, G. Berndes, H. Cantarella, H. El-Lakany, T.E.M. Duque Estrada, et al., Environmental and climate change, in: G.M. Souza, R. L. Victoria, C.A. Joly, L.M. Verdade (Eds.), Bioenergy & Sustainability: Bridging the Gaps, SCOPE and FAPESP, São Paulo, 2015, pp. 138–182.

[22] G.M. Souza, R.L. Victoria, L.M. Verdade, C.A. Joly, P.E. Artaxo Netto, H. Cantarella, et al., Bioenergy & sustainability. bioenergy numbers, in: G.M. Souza, R.L. Victoria, C.A. Joly, L.M. Verdade (Eds.), Bioenergy & Sustainability: Bridging the Gaps, SCOPE and FAPESP, São Paulo, 2015, pp. 28–57.

[23] J. Woods, L.R. Lynd, M. Laser, M. Batistella, D.C. Victoria, K. Kline, et al., Bioenergy and food security, in: G.M. Souza, R.L. Victoria, C.A. Joly, L.M. Verdade (Eds.), Bioenergy & Sustainability: Bridging the Gaps, SCOPE and FAPESP, São Paulo, 2015, pp. 258–300.

[24] G. Berndes, H. Youngs, M.V.R. Ballester, H. Cantarella, A.L. Cowie, G. Jewitt, et al., Soils and water, in: G.M. Souza, R.L. Victoria, C.A. Joly, L.M. Verdade (Eds.), Bioenergy & Sustainability: Bridging the Gaps, SCOPE and FAPESP, São Paulo, 2015, pp. 618–659.

[25] M. Aresta, A. Dibenedetto, F. Dumeignil (Eds.), Biorefinery: From Biomass to Chemicals and Fuels, Walter de Gruyter, Berlin/Boston, 2012.

[26] M. Balakrishnan, E.R. Sacia, S. Sreekumar, G. Gunbas, A.A. Gokhale, C.D. Scown, et al., Novel pathways for fuels and lubricants from biomass optimized using life-cycle greenhouse gas assessment, Proc. Natl. Acad. Sci. U.S.A. 112 (2015) 7645–7649.

Appendix A: Standard and Alternative Fuel Specifications

The American Society for Testing Materials (ASTM) and the United Kingdom Ministry of Defence have leading roles for setting and maintaining standard specifications for civilian aviation [1]. ASTM Standard Specification D1655 describes the quality of conventional aviation turbine fuels from the refinery to the aircraft. ASTM Specification D7566, *Standard Specification for Aviation Turbine Fuels Containing Synthesized Hydrocarbons*, was developed to provide quality control for fuels of novel compositions that include synthesized hydrocarbons from new sources. It allows blending of conventional blending components, or Jet A or Jet A-1 fuel certified to ASTM Specification D1655, with up to 50 vol.% of FT-SPK or with HEFA-SPK. Military standard specification MIL-83133H provides quality control for the three kerosene aviation turbine fuels, JP-8 (NATO F-34), NATO F-35, and JP-8 + 100 (NATO F-37). ASTM Designation D4054-09 has been established as the standard practice for qualification and approval of new aviation turbine fuels and fuel additives to provide a more efficient way to evaluate and approve new fuels and fuel additives. It provides a framework for qualification and approval that is intended to streamline the approval process and to permit the cost-effective and timely introduction of new fuels and additives into the field [2] (Table A.1).

A.1 FUEL ADDITIVES

Jet fuel *energy content* and *combustion quality* are primary fuel performance properties, but *stability, lubricity, fluidity, volatility, non-corrosivity,* and *cleanliness* are also important performance properties [1]. In addition to its primary function as the energy source for the engine, jet fuel also functions as hydraulic fluid and coolant for aircraft systems. Fuel additives are used to maintain desired properties in the base fuel or to provide specific performance properties that are lacking in the base fuel. These include performance enhancing, fuel handling, and maintenance additives (ASTM D1655-13 2013). The types and amounts of additives are specified in the recognized standard specifications. For cases where the use of optional additives is desired, the supplier and purchaser must come to agreement on the additives to use, and any additives used must be approved by the aircraft certifying authority for the aircraft and engine they will be used in (ASTM D1655-13 2013).

A.2 JET A, JET A-1, AND JP-8 SPECIFICATIONS

Jet A became the ASTM baseline fuel specification for commercial jet aircraft in the early

TABLE A.1 Standard Specifications Used to Control Turbine Jet Fuel Quality

Turbine Jet Fuel Type	Standard Specification	Specification Title
Conventional Jet A and Jet A-1 for civil use	ASTM D1655-13	Standard Specification for Aviation Turbine Fuels
Jet A or Jet A-1 containing synthesized hydrocarbons from nonconventional sources	ASTM D7566-13	Standard Specification for Aviation Turbine Fuel Containing Synthesized Hydrocarbons
JP-8 (NATO F-34), NATO F-35, and JP-8 + 100 (NATO F-37)	MIL-DTL-83133H w/Amendment 1	Turbine Fuel, Aviation, Kerosene Type, JP-8 (NATO) F-34, NATO F-35, and JP-8 + 100 (NATO F-37)
ASTM D4054-09	ASTM D4054-09	Standard Practice for Qualification and Approval of New Aviation Turbine Fuels and Fuel Additives

1960s, and the standard jet fuel for all US and many international airlines [2]. Jet A is produced and supplied according to the standard specification requirements of International ASTM D1655. It is specified to a minimum flash point of 38°C and a freezing point of no greater than −40°C. Additives are allowed in the amount and composition specified by ASTM D1655.

Primary standard specifications for Jet A-1 grade fuel are provided by ASTM Standard Specification D1655, UK specification DEF STAN 91-91 (Jet A-1), and NATO Code F-35 [3], but Jet A-1 is also produced to Joint Inspection Group (JIG) Check List requirements. The Check List is recognized by eight major aviation fuel suppliers: BP, Chevron, Eni, ExxonMobil, Kuwait Petroleum, Shell, Statoil, and Total [4]. It defines the fuel quality requirements for supply into Jointly Operated Fueling Systems. The Aviation Fuel Quality Requirements for Jointly Operated Systems (AFQRJOS) for Jet A-1 combines the most stringent requirements of the British Ministry of Defence Standard DEF STAN 91-91/Issue 7 Amendment 2, 1 December 2012, for Turbine Fuel, Kerosene Type, Jet A-1, NATO Code F-35, Joint Service Designation: AVTUR, and ASTM Standard Specification D 1655-12 for Aviation Turbine Fuels 'Jet A-1' [3]. This approach simplifies supply arrangements for fuels that normally would have to be stored and distributed as separate fuels against two separate sets of standard specifications. Under this arrangement, with the agreement of the parties involved, a single fuel meets the requirements of multiple standard specifications and the needs of multiple fuel type users, which can reduce handling and storage costs [4]. Jet A-1 might require one or more additives not usually required for Jet A (eg, a static dissipater and/or an icing inhibitor).

JP-8 is produced to the requirements of the US Military Specification MIL-T-83188D, British military jet fuel specification DEF STAN 91-87 AVTUR/FSII, and NATO Code F-34 [5]. It is essentially Jet A-1, but with the addition of a static dissipater additive, corrosion inhibitor/lubricity improver, and fuel systems icing inhibitor [6]. It may also contain antioxidant and metal deactivators [6]. Like Jet A-1, it is specified to a maximum freezing point of −47°C and a flash point of 38°C.

A.3 APPROVED NONCONVENTIONAL FUELS

Two nonconventional fuels (containing non-conventional hydrocarbons) specified in Defence Standard 91-91 have been recognized as meeting the requirements of ASTM Standard Specification D1655 (ASTM D1655-13), and therefore are considered to be acceptable fuels from nonconventional sources. The first is SASOL semisynthetic fuel, which is a blend of conventionally produced kerosene and a synthetic iso-paraffinic kerosene by itself, or as combined with SASOL heavy naphtha #1 and specified in Defence Standard 91-91 (ASTM D1655-13 2013). The second is SASOL fully synthetic fuel that is a blend of up to five synthetic streams specified in Def Stan 91-91 and noted in ASTM Specification D1655, Annex A1. These semisynthetic jet fuels are said to have compositions, properties, and performance that are essentially identical to conventional jet fuels [7]. Selected specifications are shown in Tables A.2 and A.3 for conventional and alternative fuels.

TABLE A.2 Select Standard Specifications for Conventional Jet A and Jet A-1 With and Without FT-SPK and HEFA-SPK Blend Components

	Standard Specifications		
Fuel Property	**Conventional Jet A or Jet A-1 (ASTM D1655-13)**	**Jet A or Jet A-1 Blended with 50% FT-SPK or HEFA-SPK (ASTM D7566-13)**	**100% FT-SPK or HEFA-SPK Blending Component for Jet A or Jet A-1 (ASTM D7566-13)**
COMPOSITION			
Acidity, total mg KOH/g	≤0.10[a]	≤0.10	≤0.015[c]
Total Aromatics, volume %	≤25	≥8 ≤25[b]	≥8 ≤25
Aromatics, mass %	report	report	≤0.5
Cycloparaffins, mass %	report	report	≤15
Paraffins, mass %	report	report	report
Carbon and Hydrogen, mass %	report	report	≥99.5
Sulphur, total mass %	≤0.3	≤0.3	≤0.3
Sulphur, mg/kg	report	report	≤15
Sulphur, mercaptan, mass %	≤0.003	≤0.003	≤0.003
or Doctor test	report	report	report
Nitrogen, mg/kg	report	report	≤2
Water, mg/kg	report	report	≤75
Phosphorus, mg/kg	report	report	report
Metals (Al,Ca,Co,Cr,Cu,Fe,K, Mg,Mo,Na,Ni, Pb,Pd,Pt,Sn,Sr,Ti, V,Zn), mg/kg	report	report	≤0.1 per metal
Fatty Acid Methyl Ester, mg/kg			≤5
Halogens, mg/kg	report	report	≤1

(Continued)

TABLE A.2 (Continued)

Fuel Property	Standard Specifications Conventional Jet A or Jet A-1 (ASTM D1655-13)	 Jet A or Jet A-1 Blended with 50% FT-SPK or HEFA-SPK (ASTM D7566-13)	 100% FT-SPK or HEFA-SPK Blending Component for Jet A or Jet A-1 (ASTM D7566-13)
VOLATILITY			
Initial boiling point, °C report	report	report	report
10% recovered	≤205	≤205	≤205
20% recovered, report			
50% recovered, report	report	report	report
90% recovered, report	report	report	report
T50-T10, gradient recovery, °C	N/A	≥15	≥15
T90-T10, gradient recovery, °C	N/A	≥40	≥22
Final boiling point, °C	≤300	≤300	report
Distillation residue, %	≤1.5	≤1.5	≤1.5
Distillation loss, %	≤1.5	≤1.5	≤1.5
Flash point, °C	≥38	≥38	≥38
Density, 15°C, kg/m^3	775–840	775–840	730–770
Freezing point, °C	Jet A ≤ −40; A-1 ≤ −47	≤ −47	≤ −40
Viscosity at −20°C, mm^2/sec	≤8	≤8	≤8
COMBUSTION			
Net heat of combustion, MJ/kg	≥42.8	≥42.8	≥42.8
Hydrogen content, mass %			
One of the following requirements:			
1. Smoke point, mm or	≥25	≥25	≥25
2. Smoke point, mm and	≥18	≥19	≥19
naphthalenes, % volume	≤3	≤3	≤3
Calculated cetane index			
Copper strip corrosion, 2 hr, 100°C	No. 1	No. 1	No. 1
THERMAL STABILITY			
2.5 h at control temp of 260°C	260	260	≥325
Filter pressure drop, mm Hg	≤25	≤25	≤25
Heater Tube deposits	<3	<3	<3
CONTAMINANTS			
Existent gum, mg/100 mL	≤7	≤7	≤7

[a] *White cells indicate spec same as conventional fuel.*
[b] *Light & dark grey cells indicate spec different from conventional fuel.*
[c] *Dark grey cells indicate specification is different than for conventional and blended fuels.*

TABLE A.3 Select Standard Specifications for Conventional JP-8 Aviation Turbine Fuels With and Without FT-SPK and HEFA-SPK Blend Components

	Standard Specifications		
Fuel Property	**Conventional JP-8 (MIL-DTL-83133H w/Amendment 1)**	**Conventional JP-8 Blended with 50% FT-SPK or HEFA-SPK (MIL-DTL-83133H w/Amendment 1)**	**100% FT-SPK or HEFA-SPK Blending Component for JP-8 (MIL-DTL-83133H w/Amendment 1)**
COMPOSITION			
Acidity, total mg KOH/g	≤0.015[a]	≤0.015	≤0.015
Total aromatics, volume %	≤25	≥8 ≤25[b]	≥8 ≤25
Aromatics, mass %	report	report	≤0.5[c]
Cycloparaffins, mass %	report	report	≤15
Paraffins, mass %	report	report	report
Carbon and Hydrogen, mass %	report	report	≥99.5
Sulphur, total mass %	≤0.3	≤0.3	≤0.0015
Sulphur, mercaptan, mass %	≤0.002	≤0.002	≤0.002
or Doctor test, max	negative	negative	negative
Nitrogen, mg/kg	report	report	≤2
Water, mg/kg	report	report	≤75
Phosphorus, mg/kg	report	report	≤0.1
Metals (Al,Ca,Co Cr,Cu,Fe, K,Mg Mo Na,Ni,Pb,Pd,Pt, Sn,Sr,Ti,V,Zn), mg/kg	report	report	≤0.1 per metal
Halogens, mg/kg	report	report	≤1
Fatty Acid Methyl Esters, mg/kg	report	report	For HEFA-SPK ≤5
VOLATILITY			
Initial boiling point, °C report	report	report	report
10% recovered	≤205	≤205	≤205
20% recovered, report	report	report	report
50% recovered, report	report	report	report
90% recovered, report	report	report	≥22
T50-T10, gradient recovery, °C	N/A	≥15	N/A
T90-T10, gradient recovery, °C	N/A	≥40	N/A
Final boiling point, °C	≤300	≤300	≤300
Distillation residue, %	≤1.5	≤1.5	≤1.5
Distillation loss, %	≤1.5	≤1.5	≤1.5

(*Continued*)

TABLE A.3 (Continued)

Fuel Property	Standard Specifications Conventional JP-8 (MIL-DTL-83133H w/Amendment 1)	Conventional JP-8 Blended with 50% FT-SPK or HEFA-SPK (MIL-DTL-83133H w/Amendment 1)	100% FT-SPK or HEFA-SPK Blending Component for JP-8 (MIL-DTL-83133H w/Amendment 1)
Flash point, °C	≥38	≥38	≥38
Density, 15°C, kg/L	0.775–0.840	0.775–0.840	0.751–0.770
Gravity API at 60°C	≥37.0≤51.0	≥37.0≤51.0	≥52≤57
Freezing point, °C	≤−47	≤−47	≤−47
Viscosity at −20°C, mm^2/sec	≤8	≤8	≤8
COMBUSTION			
Net heat of combustion, MJ/kg	≥42.8	≥42.8	≥42.8
Hydrogen content, mass %	≥13.4	≥13.4	≥13.4
One of the following:			
1. Smoke point, mm **or**	≥25	≥25	≥25
2. Smoke point, mm **and**	≥19	≥19	≥19
Naphthalenes, % volume	≤3	≤3	≤3
Calculated cetane index		≥40	
Copper strip corrosion, 2 h, 100°C	No. 1	No. 1	No. 1
Filter pressure drop, mm Hg	≤25	≤25	≤25
Heater Tube deposits	<3	<3	<3
CONTAMINANTS			
Existent gum, mg/100 mL	≤7	≤7	≤7
Particulate matter, mg/L	≤1.0	≤1.0	≤1.0
Filtration time, minutes	≤15	≤15	≤15

[a] *White cells indicate spec same as conventional fuel.*
[b] *Light & dark cells indicate spec different from conventional fuel.*
[c] *Dark grey cells indicate specification is different from conventional and blended fuels.*

TABLE A.4 Select Specifications for JP-8, JP-9, and JP-10 Fuels

Property	JP-9	JP-10	JP-8
Composition	Methylcyclohexane (10–12%), 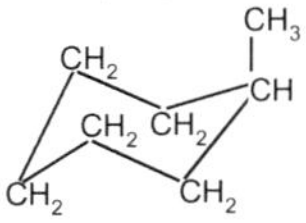 Perhydronorbornadiene dimer (20–25% wt), and 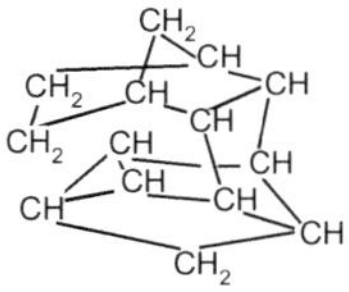 Exotetrahydrodicyclopentadiene (see JP-10) (65–70% wt.)	Exotetrahydrodicyclopentadiene 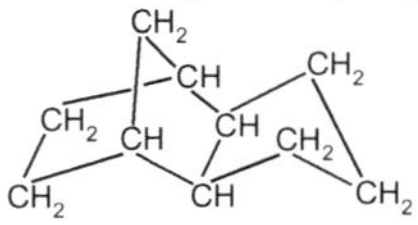	Paraffins (70–85% vol.) Aromatics (≤25% vol.) Olefins (≤5% vol.)
Minimum heat content	39573 MJ/m^3 (142,000 Btu/gal)	39434 MJ/m^3 (141,500 Btu/gal)	34835 MJ/m^3 (125,000 Btu/gal)
Density at 15°C	0.935–0.955 kg/L	0.935–0.943 kg/L	0.775–0.840 kg/L
Freeze point	≤ − 54°C (−65.2°F)	≤ − 79°C (−110.2°F)	≤ − 47°C (−52.6°F)
Flash point	≥ 21°C (70°F)	≥ 55°C (131°F)	≥ 38°C (100°F)

A.4 JET-P 9 AND 10 SELECT SPECIFICATIONS

Table A.4 describes the composition and select properties for JP-9 and JP-10 as compared to JP-8.

References

[1] Chevron, Aviation Fuels Technical Review, pp. 24–29. Accessed at: <https://www.cgabusinessdesk.com/document/aviation_tech_review.pdf>, 2006.

[2] U. Yildirim, S. Abanteriba, Manufacture, qualification and approval of new aviation turbine fuels and additives, Procedia Eng. 49 (2012) 310–315.

[3] Shell, Military Jet Fuel Grades and Specifications (NATO). Accessed at: <http://www.shell.com/global/products-services/solutions-for-businesses/aviation/shell-aviation-fuels/fuels/types/military-jet-fuel-grades.html>, 2013.

[4] JIG, Joint Inspection Group, (February 2013) Bulletin No. 60, "Aviation Fuel Quality Requirements for Jointly Operated Systems, (AFQRJOS)," p. 1–6. Joint Inspection Group. Accessed at: <www.jigonline.com>, 2013.

[5] CSG, Aviation Turbine Fuel (Jet Fuel). CSG Network. Accessed at: <www.csgnetwork.com/jetfuel.html>, 2013.

[6] DOD, "Detail Specification: Turbine Fuel, Aviation, Kerosene Type, JP-8 (NATO F-34), NATO F-35, and JP-8+100 (NATO F-37)," MIL-DTL-83133H w/Amendment 1, 2012, p. 2–18.

[7] FAA, Federal Aviation Safety, September 2011, Bulletin NE-11-56, "Special Airworthiness Information Bulletin," p. 1–2. Federal Aviation Administration, Washington, DC, 2011.

Appendix B: Commercial Biofuel Flight Demonstrations (Data Collected by Innovate Washington 2013)

TABLE B.1 Flights with Biofuel %, Distance (Miles), Gallons Biojet Consumed, Number of Flights

Flights	Biofuel (%)	Distance (Miles)	Gallons Biojet Consumed	Number of Flights
RECURRING OPERATIONS				
Finnair, A319, 20 Jul. 2011 (3 flights)	50	945	6000	3
Interjet, A320, 21 Jul. 2011 (2 flights)	27	866	333	2
DLH, A321, Jul. 2011–Dec. 2011 (1187 flights)	50	255	466,800	1187
KLM, 737, Sep. 2011–Feb. 2012 (200 flights)	50	247	17,310	200
Alaska, 737, Weekly flights Nov./Dec. 2011 (11 flights)	20	2320	7271	11
Alaska, 737, Weekly flights Nov./Dec. 2011 (64 flights)	20	129	2355	64
Aeromexico, 737, Weekly flights, Nov. 2011–2012 (~52 flights)	25	1190	17,628	52
KLM, 777-200, Weekly flights, Mar.–Sep. 2013 (26 flights)	25	3630		26
INDIVIDUAL EVENTS				
Virgin Atlantic, 747-400, 24 Feb. 2008	20	247	84	1
Cargolux, 747-8, 22 Jun. 2011	15	5000	6651	1
KLM, 737, 30 Jun. 2011	50	247	176	1
Iberia A320, 5 Oct. 2011	25	301	107	1
Aeromexico, 777, 1 Aug. 2011	30	5630	8912	1
Thomson Fly, 757, 6 Oct. 2011	50	1730	1736	1
Air France, A321, 21 Oct. 2011	50	376	252	1
United, 737, 7 Nov. 2011	40	926	528	1

(Continued)

TABLE B.1 (Continued)

Flights	**Biofuel (%)**	**Distance (Miles)**	**Gallons Biojet Consumed**	**Number of Flights**
Thai Airways, 777, 22 Dec. 2011	50	353	2400	1
Lufthansa, 747-400, 12 Jan. 2012	50	4070	12000	1
Etihad, 777-300ER, 19 Jan. 2012	50	7460	13,500	1
LAN, A320, 7 Mar. 2012	50	261	186	1
All-Nippon Airways	50	4790	10,435	1
Qantas, A330-300, 13 Apr. 2012	50	735	738	1
Porter Airlines, Q400, 17 Apr. 2012	49	218	85	1
Jetstar, A320, 19 Apr. 2012	50	378	189	1
Porter, Q400, ICAO Rio + 20 'Chain of Flights', 19 Jun. 2012	50	300	120	1
Air Canada, A319, ICAO Rio + 20 'Chain of Flights', 19 Jun. 2012	50	2020	1216	1
AeroMexico, 777-200 ER, ICAO Rio + 20 'Chain of Flights', 19 Jun., 2012	50	4620	13,500	1
GOL, 737-800, ICAO Rio + 20 'Chain of Flights', 19 Jun. 2012	50	213	152	1
KLM, 777-200ER, 19 Jun. 2012	20	6070	5400	1
Azul, EMB-195, 19 Jun. 2012	–	–	–	1
LAN Columbia, A320, 21 Aug. 2013	33	250	93	1
Total Biojet Consumed			596,155	1568

Index

Note: Page numbers followed by "*f*" and "*t*" refer to figures and tables, respectively.

B

N

O

P

R

S

T

CPI Antony Rowe
Chippenham, UK
2016-12-07 19:12